Natural Habitat Management

Natural Habitat Management

Editor: Jacob Wright

Callisto Reference
www.callistoreference.com

Callisto Reference,
118-35 Queens Blvd., Suite 400,
Forest Hills, NY 11375, USA

Visit us on the World Wide Web at:
www.callistoreference.com

ISBN: 978-1-63239-818-5 (Hardback)

Printed in China.

Cataloging-in-publication Data

Natural habitat management / edited by Jacob Wright.
p. cm.
Includes bibliographical references and index.
ISBN 978-1-63239-818-5
1. Habitat (Ecology). 2. Habitat conservation. 3. Ecosystem management. 4. Conservation of natural resources.
I. Wright, Jacob.
QH75 .N38 2017
639.92--dc23

Table of Contents

Preface

Natural habitat management is defined as the conservation and preservation of natural habitat of plant and animal species that rely on conservation programs to survive and repopulate in a given area. Rapid urbanization has fueled the study of habitat management and the migration of wildlife in the face of loss of land has become a cause for concern. This book is aimed at readers who are interested in the study of forest ecology, land management and habitat development. This book elucidates the concepts and innovative models around prospective developments with respect from natural resources management. It will also provide interesting topics for research which interested readers can take up. This book is appropriate for students seeking detailed information in this area as well as experts.

The purpose of the book is to provide a glimpse into the dynamics and to present opinions and studies of some of the scientists engaged in the development of new ideas in the field from very different standpoints. This book will prove useful to students and researchers owing to its high content quality.

At the end, I would like to appreciate all the efforts made by the authors in completing their chapters professionally. I express my deepest gratitude to all of them for contributing to this book by sharing their valuable works. A special thanks to my family and friends for their constant support in this journey.

Editor

1

Using Satellite and Airborne LiDAR to Model Woodpecker Habitat Occupancy at the Landscape Scale

Lee A. Vierling[1]*, Kerri T. Vierling[2], Patrick Adam[3], Andrew T. Hudak[4]

1 Department of Forest, Rangeland, and Fire Sciences, McCall Outdoor Science School, University of Idaho, Moscow, Idaho, United States of America, **2** Department of Fish and Wildlife Sciences, University of Idaho, Moscow, Idaho, United States of America, **3** Environmental Science, University of Idaho, Moscow, Idaho, United States of America, **4** Rocky Mountain Research Station, US Forest Service, Moscow, Idaho, United States of America

Abstract

Incorporating vertical vegetation structure into models of animal distributions can improve understanding of the patterns and processes governing habitat selection. LiDAR can provide such structural information, but these data are typically collected via aircraft and thus are limited in spatial extent. Our objective was to explore the utility of satellite-based LiDAR data from the Geoscience Laser Altimeter System (GLAS) relative to airborne-based LiDAR to model the north Idaho breeding distribution of a forest-dependent ecosystem engineer, the Red-naped sapsucker (*Sphyrapicus nuchalis*). GLAS data occurred within ca. 64 m diameter ellipses spaced a minimum of 172 m apart, and all occupancy analyses were confined to this grain scale. Using a hierarchical approach, we modeled Red-naped sapsucker occupancy as a function of LiDAR metrics derived from both platforms. Occupancy models based on satellite data were weak, possibly because the data within the GLAS ellipse did not fully represent habitat characteristics important for this species. The most important structural variables influencing Red-naped Sapsucker breeding site selection based on airborne LiDAR data included foliage height diversity, the distance between major strata in the canopy vertical profile, and the vegetation density near the ground. These characteristics are consistent with the diversity of foraging activities exhibited by this species. To our knowledge, this study represents the first to examine the utility of satellite-based LiDAR to model animal distributions. The large area of each GLAS ellipse and the non-contiguous nature of GLAS data may pose significant challenges for wildlife distribution modeling; nevertheless these data can provide useful information on ecosystem vertical structure, particularly in areas of gentle terrain. Additional work is thus warranted to utilize LiDAR datasets collected from both airborne and past and future satellite platforms (e.g. GLAS, and the planned IceSAT2 mission) with the goal of improving wildlife modeling for more locations across the globe.

Editor: Adina Maya Merenlender, University of California, Berkeley, United States of America

Funding: The authors would like to thank the USGS Gap Analysis Program (agreement Grant number 08HQAG0123; K. Vierling, PI) and the NASA-sponsored Idaho Space Grant Consortium (http://www.id.spacegrant.org; graduate fellowship awarded to P. Adam) for funding. The funders had no role in study design, data collection and analysis, decision to publish, or preparation of the manuscript.

Competing Interests: The authors have declared that no competing interests exist.

* E-mail: leev@uidaho.edu

Introduction

Remote sensing data have become fundamental for mapping actual or potential species distributions [1,2] to aid conservation efforts at local to global scales [3,4,5]. Current species distribution maps draw upon a variety of information sources, including land cover, to model habitat. However, land cover classified as forest is often too broadly defined and often does not account for structural variability that can significantly alter habitat suitability for avian species [6]. While passive spectral remote sensing data products are essential for delineating horizontal habitat heterogeneity (e.g. forest fragmentation) and vegetation productivity at spatial scales relevant to many species, they are of limited use for quantifying details of vertical habitat structure crucial to animal habitat (e.g. [7]). As a result, improving maps of animal distributions is likely to require remote quantification of vertical habitat structure across wide areas [7–9].

Light detection and ranging (LiDAR) provides unique data for characterizing vertical habitat structure, and has been utilized widely to address multiple animal-habitat relationships. For instance, airborne LiDAR has been used alone or in combination with passive remote sensing data to assist habitat modeling for terrestrial birds [10–17], mammals [18,19], and invertebrates [20,21] at spatial scales relevant to numerous organisms. The basis for these advances in habitat modeling stem from the fact that the energy returned in the LiDAR signal provides information to characterize forest structural characteristics such as vegetation height, density, and volume in discrete vertical slices above ground level [22,23]. The resultant data can provide information ranging from structural characteristics of individual trees [24] to area-wide maps of forest successional status [25,26] and biomass distribution [27,28]. As a result, these vertical forest characteristics afforded from airborne LiDAR can improve our understanding and mapping of animal-habitat relationships from the stand to the landscape scale [29].

Although LiDAR has provided significant leaps forward in modeling habitat, data are typically collected via aircraft, thereby imposing limits on the extent and frequency of spatial and temporal sampling relative to orbiting sensors mounted on satellites. A satellite based LiDAR remote sensing platform may provide information about 3-D structure for animal habitat modeling critical for broad-scale application [30]. However to our knowledge no empirical study has yet tested the utility of satellite LiDAR for habitat modeling.

The Geoscience Laser Altimeter System (GLAS) LiDAR instrument aboard the ICESat satellite, in operation between 2003 and 2009, provides a novel opportunity for delineating habitat. While the primary intent of the GLAS instrument was to measure polar ice-sheet thickness [31], GLAS data has been used to characterize vertical vegetation structure [32–34] and forest biomass [35]. Similar to airborne LiDAR in the underlying physics, a photo-detector within the GLAS receiving telescope records the time-of-flight (t) between emission of the transmit laser pulse and its return to the instrument. Using the speed of light (c), distance (D) from the instrument is then computed as $D = 0.5\ tc$. Vertical vegetation structure is derived from the varying intensity at discrete heights of the returned transmit signal [34]. The major attraction of GLAS data for informing ecological questions is that (1) the data samples forest environments across the globe and (2) the data (along with computer code to manipulate it) is freely available from the U.S. National Snow and Ice Data Center (NSIDC). However, LiDAR data from GLAS differs in both spatial extent and resolution from airborne LiDAR data. While airborne LiDAR data are typically scanned at high horizontal resolution (i.e. with each laser pulse sampling a 'small footprint' $<1\ m^{-2}$ in size and typically collected at a density of >1 pulse m^{-2}) continuously across landscapes, GLAS is a 'large footprint' instrument collecting data samples with a diameter in our study area of ~64 m each and separated by a minimum of 172 m. The utility of GLAS LiDAR to describe wildlife habitat at spatial scales relevant to animals is as yet unexplored.

Advances in the global description of vertical vegetation structure are increasingly important for understanding issues associated with species distribution, diversity, and habitat connectivity (e.g. [9]). The major objective of this study was to model breeding site occupancy of a forest dependent bird species, the Red-naped Sapsucker (*Sphyrapicus nuchalis*), using both GLAS and airborne LiDAR data. We chose the Red-naped Sapsucker because of its status as an ecosystem engineer, capable of creating cavities in trees that can be used by multiple other forest avian and mammalian species (e.g. [36,37]). To our knowledge, this study represents the first to utilize spaceborne LiDAR data in assessing the habitat associations of a forest dependent species, and the first to compare GLAS and airborne-LiDAR models in this ecological context.

Methods

Study Site

Field survey sites were located in temperate mixed-conifer forests dominated by Ponderosa pine (*Pinus ponderosa*), White pine (*Pinus monticola*), Grand fir (*Abies grandis*), Western larch (*Larix occidentalis*), Douglas fir (*Pseudotsuga menziesii*), and Western red cedar (*Thuja plicata*) located at the western extent of the Clearwater Mountain Range and within the Idaho Panhandle National Forest in northern Idaho, USA (Fig. 1). Site ground elevation (hereafter, elevation) ranged from 794–1357 m. Land ownership and management across the study area included private timber concessions, private land holders, and state and federal management agencies. Permits are not explicitly required for observational bird studies on U.S. Forest Service land. We obtained verbal permission and access to state owned property from the Idaho Department of Lands. Private land holdings included Potlatch Corporation and Bennett Lumber Products, both of whom granted permission to perform research on their respective properties. The diverse ownership and management matrix has created a mosaic of forest stands that exhibit a range of successional stages that can influence the habitat selection behavior of woodpeckers [38–41].

GLAS LiDAR

The GLAS instrument is a waveform LiDAR system with 15 cm vertical resolution that uses a near infrared (1064 nm) laser. The laser records data covering a ground footprint (i.e. sample area) of ca. 64 m diameter that can vary significantly in its ellipticity over time [42]. The semimajor axis and eccentricity are recorded for each laser shot and were used to define an elliptical mask for extraction of terrain features from the airborne and GLAS data using geographic information system software. The laser footprints are spaced 172 m between centers along the satellite transect, with a between-transect spacing of about 540 m with different acquisition years in our study area (Fig. 1). We acquired GLAS data from the NSIDC, which distributes data products from 18 laser sampling campaigns. We used the GLA01 product (L1A Global Altimetry) which provides the transmitted and received waveform, and the GLA14 product (L2 Land Surface Altimetry) which provides the laser footprint geolocation and geodetic, instrument, and atmospheric corrections. The two datasets are linked by record ID and shot number. We used 'release 531' data, which specifies a level of data pre-processing, from laser campaigns 2A (Sep–Nov 2003), 2B (Feb–Mar 2004), 3A (Oct–Nov 2004), 3B (Feb–Mar 2005), 3D (Oct–Nov 2005), 3H (Mar–Apr 2007), and 3I (Oct–Nov 2007). GLAS waveforms can be influenced by interaction with clouds and signals that are saturated due to high gain settings. We therefore removed waveforms that were affected by clouds and saturation on the laser signal following Chen [43].

We modified Interactive Data Language (IDL) code provided by NASA and available on the NSIDC GLAS site [44] to extract complete laser waveform return signals (from GLA01) and waveform summary statistics (from GLA14) (Fig. 2). To compute vertical vegetation statistics from the waveform signal, we first established the ground return provided by the summary statistics in the GLA14 product. The summary statistics include up to 6 primary energy peaks generated from a Gaussian decomposition algorithm [45] to identify the major components of the signal which include both vegetation and ground returns (Fig. 2). We estimated the ground return location following Chen [46] who investigated the correlation between GLAS and airborne lidar data in forested mountainous areas and found that the largest in magnitude of the two lowest peaks corresponded most closely with ground elevations.

Before deriving canopy height statistics (Fig. 3), we filtered the waveforms by retaining signals that were greater than 4 times the standard deviation of the noise (following [47]). Maximum waveform vertical resolution at 1 ns pulse duration is 15 cm. We selected a Gaussian filter of 60 cm (4 ns) in height because we were interested in primary structural features of the canopy. We did not evaluate other filter widths, but narrower widths may allow noise to pass or may produce results that are difficult to interpret due to their vertical length scale being finer than individual branches. Canopy height was estimated from an algorithm developed by Chen [43] for a coniferous montane location in the Pacific Northwest that accounts for the effect of terrain slope on waveform vegetation indices by incorporating the maximum terrain relief distance within the laser footprint (obtained from the 10 m National Elevation Dataset [48]) and the difference between the start and end of the filtered and smoothed waveform signal. The

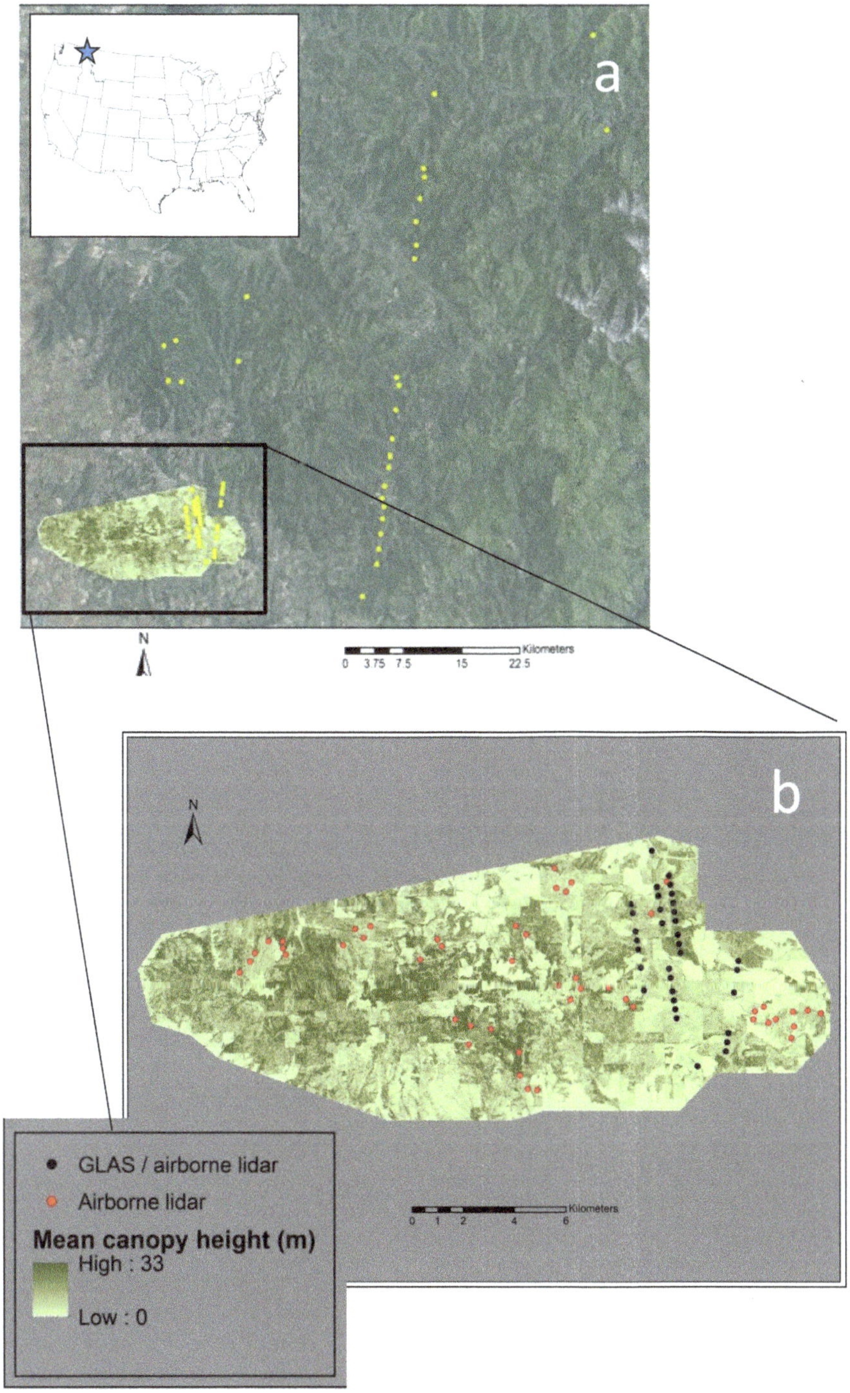

Figure 1. Map of woodpecker survey sites within GLAS LiDAR footprints (yellow dots) (a) in northern Idaho, and (b) survey sites within both GLAS and airborne LiDAR (black dots) and airborne only (red dots).

complete suite of canopy metrics derived from GLAS data is listed in Table 1.

Airborne LiDAR

LiDAR data for Moscow Mountain (Fig. 1) was acquired in July, 2009 using a discrete return airborne Leica ALS50 Phase II laser system scanning with a pulse rate of ~150 kHz at ±14° from nadir. Adjacent survey transects were flown in opposing directions with ≥50% overlap to ensure 100% terrain overlap for all transects, resulting in a mean data density of ~12 points m^{-2}. A maximum of 4 range measurements, or returns, were recorded for each laser pulse transmission. To geolocate laser point positions, aircraft attitude (pitch, roll, yaw) was recorded at 200 Hz by an onboard inertial measurement unit (IMU), while sensor position (x, y, z) was recorded at 2 Hz by an onboard differential GPS. Aircraft attitude and sensor position were indexed by GPS time to enable post-processing correction and calibration. Data were projected in UTM Zone 11 North, NAD 83 using NAVD88

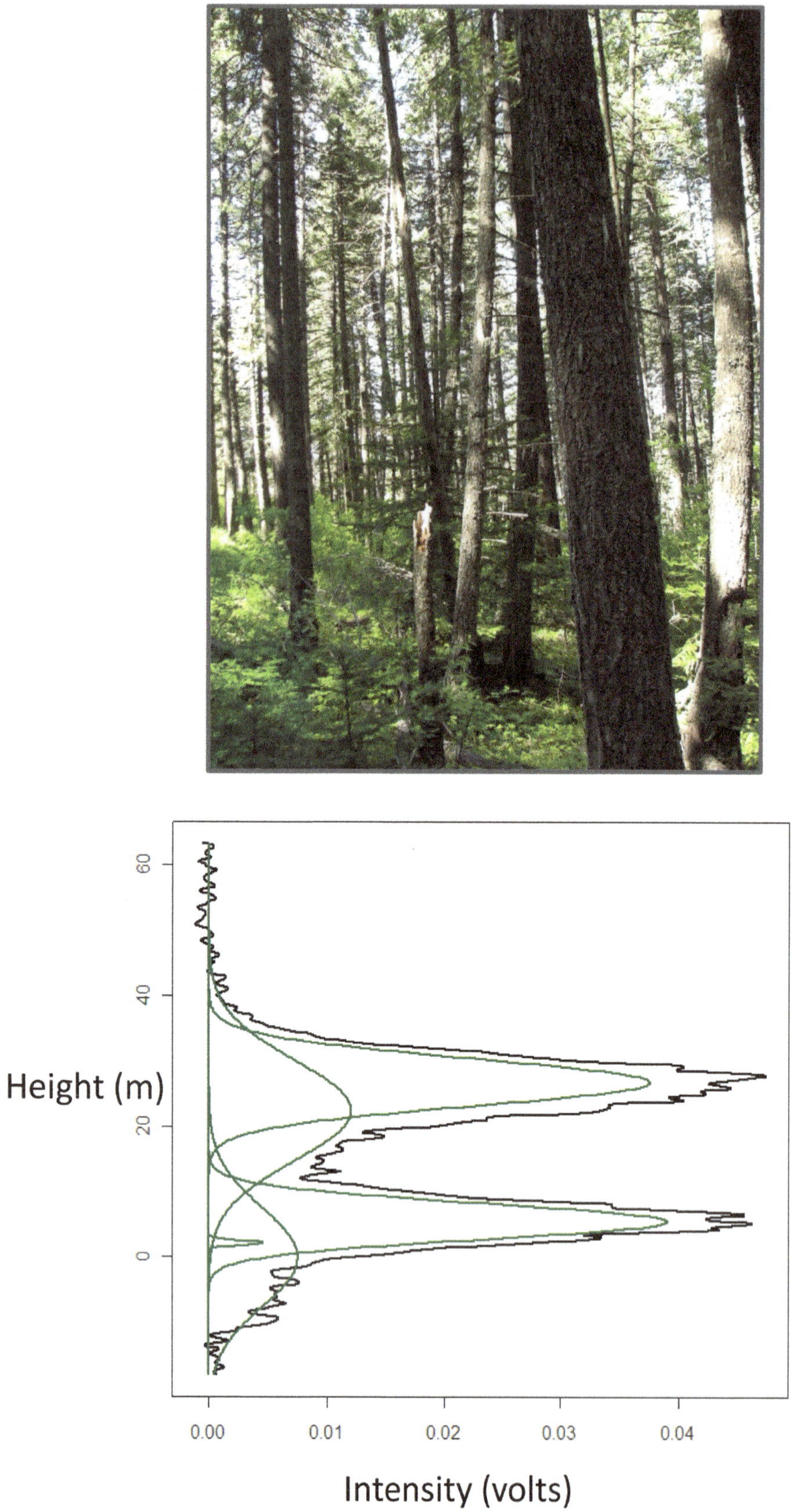

Figure 2. Example GLAS LiDAR waveform signal (bottom panel) and associated woodpecker survey location (top panel). The black line in the graphic represents the actual laser energy signal for the plot location, while the green lines represent the decomposed signal when expressed in best-fit Gaussian curves.

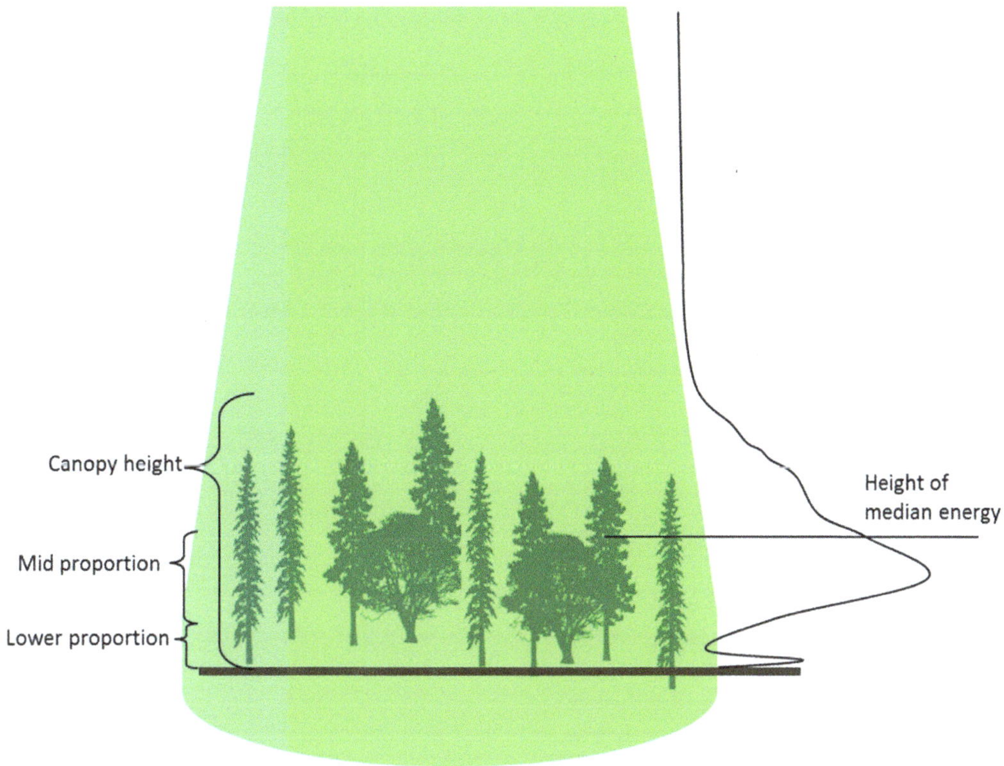

Figure 3. Cartoon depicting locations of vegetation metrics within canopy. The green cone in the figure represents the GLAS laser footprint scale, which subtended an area of roughly 64 m diameter. The black curve on the right of the figure represents a stylized GLAS energy return profile for this particular cartoon arrangement of vegetation and ground.

Table 1. GLAS LiDAR metrics considered for occupancy analysis; only the metrics denoted by double asterisks (**) were used after collinearity analysis.

Metric abbreviation	Metric
**Elev	elevation (m)*
**HLI	heat load index*
**p.lower	proportion of total return energy that represents vegetation between 0 and 3 m above ground level (agl)
**p.mid	proportion of total return energy that represents vegetation between 3 and 10 m agl
**can.height	canopy height computed from Chen (2010b) algorithm
ht.mean	mean canopy height agl
ht.var	variance of vegetation heights
ht.cv	coefficient of variation of vegetation heights
ht.med	median canopy height
ht.mad	median absolute deviation from median of canopy heights
Vdr	vertical distribution ratio (can.height – HOME)/can.height
**veg.density	canopy density – proportion of return energy representing vegetation to total return energy that includes ground returns
**HOME	height of median energy density
**FHD	foliage height diversity (1 m bin); diversity in vegetation distribution. See MacArthur & MacArthur, 1961

These metrics include 6 biotic factors (p.lower, p.mid, can.height, veg.density, HOME, and foliage height diversity (FHD)) and 2 abiotic factors (heat load index (HLI) and elevation.

*Heat load index integrates measures of site slope, aspect, and latitude, computed following [81]. Note that due to difficulties in deriving slope using GLAS data, HLI and elevation were derived from the 10-m National Elevation Database.

Table 2. Airborne LiDAR metrics considered for occupancy analysis; only metrics denoted by double asterisks (**) were used after collinearity analysis.

Metric abbreviation	Metric
**Elev	elevation (m)*
**HLI	heat load index*
p.lower	proportion of total return energy that represents vegetation between 0 and 3 m above ground level (agl)
**p.mid	proportion of total return energy that represents vegetation between 3 and 10 m agl
p.upper	proportion of total return energy that represents vegetation between 10 and 30 m + agl
VDR	Vertical distribution ratio (Goetz *et al.* 2007)
Density	canopy density - proportion of vegetation to total energy return
Hmax	maximum canopy height
Hmedian	median canopy height
Hsd	standard deviation of heights
Hmean	mean canopy height
Crr	canopy Relief Ratio (HMEAN - HMIN)/(HMAX - HMIN)
Hirq	inter-quartile range of heights
**Hmode	height of dominant mode (vegetation density) within return signal
**Hmrange	range between minimum and maximum modes
**Hskew	skewness of height profile across vegetation returns
Hkurt	kurtosis of height profile across vegetation returns
**Hvar	variance of vegetation density weighted height values
Hmad	median absolute deviation from median height
**FHD	foliage height diversity (1 m bin) see MacArthur & MacArthur, 1961

*Heat load index integrates measures of site slope, aspect, and latitude, computed following [81]. Note that while these measures are attainable using airborne lidar, to be consistent with calculations made for GLAS data HLI and elevation were derived from the 10-m National Elevation Database.

Geoid 83 for the vertical datum and NAD83 for the horizontal datum.

The Multiscale Curvature Classification algorithm [49] was used to delineate ground from non-ground coordinate corrected LiDAR data. Returns classified as ground were interpolated into a 1 m resolution DEM and subtracted from non-ground classified points to generate a topographically normalized point cloud. Metrics describing vegetation structural characteristics were derived from all points greater than 1 m in height within 20 m grid cells (see Table 2 for a list of all metrics derived by airborne LiDAR data). The 20 m grid cell resolution was used so that the airborne LiDAR metrics could be compared with and validated against measurements of forest structure collected in the field (see [28]). The resultant 20 m raster grids, each representing a vegetation height metric, were downsampled to 1 m resolution in the vicinity of the GLAS LiDAR footprints to minimize the effect of sampling incomplete pixels around the perimeter of the GLAS footprint. Grid cells within a 65 m diameter circle centered on the woodpecker survey site were aggregated to represent the mean value of each vegetation metric and to coincide with the GLAS footprint.

Woodpecker Surveys

We chose Red-naped Sapsuckers as the focus of our study for multiple reasons. As part of a cavity-excavating guild, sapsuckers are an ecologically important group owing to their status as ecosystem engineers [36]. Sapsuckers and other woodpeckers create keystone structures (i.e. cavities) that many other animals use for nesting and/or roosting [37,50–52]. In addition, cavity excavators in general rely upon a set of forest structural characteristics that may be described via LiDAR. For instance, Red-naped Sapsuckers forage in areas with a sparse understory, and in sections of the canopy that are relatively open to facilitate insect gleaning on trunks [53]. Additionally, the diameter of the nest tree, the fraction of canopy cover, and the distribution of vegetation are all important nest site attributes [54] and these are all potentially quantifiable metrics from both airborne and spaceborne LiDAR data.

Eighty two woodpecker survey sites were subsampled randomly from within the airborne LiDAR flight during 27 May–3 July 2009 from existing song bird point count locations (Fig. 1) and within seventy three GLAS spaceborne LiDAR footprints during 19 May–2 July 2010. A subset of survey sites ($n = 39$) occurred within both spaceborne and airborne LiDAR data samples (Fig. 1). To ensure that we were not double-counting the same birds, we selected sites with a minimum of 340 m spacing (equating to ~12 ha surrounding each survey location) which reflects the mean observed home range of Red-naped Sapsucker [53–55]. Surveys conducted outside of GLAS LiDAR sites were conducted to increase sample size for the airborne LiDAR derived occupancy model.

Woodpecker surveys were conducted following Drever *et al.* [56] using playback recordings to elicit responses from woodpeckers. Two identical playbacks, each comprised of a call and drum, were played for Red-naped Sapsuckers, with a two minute silence in between, and all Red-naped Sapsuckers seen or heard were recorded. Because the purpose of the surveys was to establish breeding bird-habitat relationships, we constrained the dates of the surveys to correspond to the breeding season of this species [55,56]. This timing is important because the use of callbacks is only effective while the birds are breeding and territorial. Each survey point was visited twice during the breeding season because multiple visits are necessary in order to better evaluate detection issues (i.e. when birds are really present but are recorded as absent [57]). Issues of detectability were explicitly incorporated into our analysis (see below). Surveys conducted solely within GLAS footprints were completed in the Idaho Panhandle National Forest (Fig. 1) in 2011 from 19 May–27 June. We selected survey sites by comparing the 2009 airborne LiDAR dataset to an earlier airborne LiDAR dataset flown in 2003 (see [28] for complete description of this multi-temporal analysis) to ensure that harvest operations had not taken place spanning the period from the first GLAS data collection to the date of the airborne LiDAR data acquisition.

Data Analysis

We used a t-test to compare the vegetation metrics between occupied and unoccupied sites, and considered $p \leq 0.05$ to be statistically significant. Our model development was conducted in an exploratory, rather than confirmatory, framework. We utilized this exploratory approach because Walters *et al.* [58] notes that breeding habitat relationships for Red-naped Sapsucker have been conducted where aspen (*Populus tremuloides*) and deciduous forests are important components of the forest, and our study sites contained very little deciduous forest. Preliminary data analysis revealed multi-collinearity among several GLAS and airborne LiDAR derived vegetation height and density metrics (Tables 1 and 2). We computed condition indices and variance decomposition proportions following Belsley [59] to identify the information redundant covariates. After applying conditioning diagnostics, our final set of site covariates included 6 biotic and 2 abiotic variables for the GLAS LiDAR based model (Table 1) and the airborne LiDAR based model (Table 2). We included elevation and heat load index to account for abiotic factors that may influence nest or foraging site selection. The 6 biotic variables are represented by the LiDAR-derived forest structure metrics, while elevation and heat load index were derived from the 10 m National Elevation Dataset [37].

While species presence and absence may be modeled with logistic regression, the approach assumes equal probability of detection across sites and cannot account for detection probabilities less than 1 [57,60,61]. Imperfect detection (i.e. $p<1$) produces biased estimates of latent variables that describe the state processes typically of interest [62,63]. We used a hierarchical occupancy model that explicitly accounts for detection probability that is conditional on the state process controlling species occurrence as implemented in the R package [64] 'unmarked' [65]. Due to the relatively small number of sample sizes for presence/absence modeling for each LiDAR dataset (airborne: n = 82; GLAS: n = 73), we pooled responses from the Moscow Mountain and the Idaho Panhandle National Forest regions for use in a single season occupancy model. Variables from Table 1 were used to model the latent occupancy state of each site. To account for detection probability, we included Julian day and a categorical measure of weather conditions (e.g. wind, cloud cover) where increasing values were associated with an increase in those weather variables.

We computed global, null, and reduced occupancy models for Red-naped Sapsuckers and evaluated relative model performance using Akaike's Information Criterion corrected for small sample size (AICc) [66]. We allowed for all linear combinations of predictor variables, and excluded models where the only predictor variables were the detectability covariates, requiring at least one process variable (vegetation, heat load index, or elevation) to be present. This method generated a list of models that struck a balance between including all possible habitat structure combinations, while excluding ecologically irrelevant models [67].

Global (full) model fit and checks for overdispersion were evaluated using a likelihood ratio test procedure to compute the deviance between the saturated model and the full model. The deviance was then divided by the residual degrees of freedom ($n-p = 72$) to assess fit and potential overdispersion. We performed all first-order subsets regression to determine the top candidate models; while many studies utilize ΔAICc<2, we opted to utilize a more conservative ΔAICc<6 for final model selection evaluation [68,69]. Where model selection uncertainty was high (e.g. weights were <0.3) [66], we utilized a model-averaging approach. Parameter estimates in model averaging represent the sum of the product of the parameter estimates in each model with their associated model weight. As model weight tends towards zero, so too does the value of the parameter estimate. The model-averaging approach generates a new set of parameters that are used to replace the original parameter estimates in the global model. The procedure is recommended when all-subsets modeling is performed [70] and when strong support is lacking for a single model [66,71].

Relative variable importance was computed from model averaged parameter estimate weights to ascertain the contribution of each variable to the averaged model [66]. We believe our approach to model development addresses the concerns raised by Murtaugh [72] and Dahlgren [73], specifically with regards to shrinkage of estimates derived from model averaging that contribute the least amount of information to the model.

Model Cross Validation and Predictive Power

We assessed model performance using the Area Under the Curve (AUC) from the Receiver Operator Characteristic (ROC) curve where presence/absence probability cut-off values were determined using *k*-means cluster analysis by partitioning the predicted probabilities into 2 groups. We generated AUC scores first with unvalidated data, and then generated AUC scores using leave-one-out cross validation (LOOCV) to assess the degree of over-fitting and model sensitivity. AUC values close to 0.5 represent a model whose predictions of presence/absence are no different than random.

Results

We detected Red-naped Sapsucker responses at 48% (39 out of 82) of sites surveyed within area covered by airborne LiDAR and at 28% (28 out of 73) of sites surveyed within area covered by spaceborne LiDAR. Of the 155 total survey sites, 39 were common to both airborne and spaceborne LiDAR surveys. Fifteen of 39 Red-naped Sapsucker responses were reported at the coincident airborne and spaceborne LiDAR sites (44%). There were no statistical differences between the occupied and unoccupied sites (Table 3).

Model Fit and Parameter Estimation

GLAS LiDAR. Global model fit was good (D = 82.8) when compared with the residual degrees of freedom (*rdof* = 63) as indicated by the over-dispersion parameter, $\hat{c}$ ($\hat{c} = 1.3$). Because $\hat{c} \approx 1$, there is no reason to doubt the fit of the global model, nor does over-dispersion appear to be large enough to warrant using a quasi AICc framework [65]. Due to high model selection uncertainty as indicated by AICc weights<0.3 (Table 4), we proceeded with a full model averaging approach for models with ΔAIC≤6. Neither the original global model nor the averaged

Table 3. Means and standard errors of GLAS and airborne LiDAR variables used in sapsucker models.

Metrics	Occupied sites (n = 28)		Unoccupied sites (n = 45)	
GLAS (spaceborne LiDAR; n = 73)	**Mean**	**SE**	**Mean**	**SE**
Elevation (m)	950	21.1	974	17.8
HLI (heat load index)	0.842	0.014	0.871	0.009
p. lower	0.226	0.025	0.205	0.018
p.mid	0.347	0.026	0.389	0.021
Can.height (m)	34.1	2.06	32.2	1.41
Veg.density	0.897	0.014	0.905	0.011
HOME	10.87	1.02	10.96	0.669
FHD	2.092	0.015	2.11	0.012
	Occupied sites (n = 39)		**Unoccupied sites (n = 43)**	
Airborne LiDAR (n = 82)	**Mean**	**SE**	**Mean**	**SE**
Elevation (m)	956	13.9	941	12.8
HLI (heat load index)	0.875	0.008	0.845	0.014
p.mid	0.205	0.019	0.239	0.022
Hmode	2.59	0.736	3.80	0.965
Hmrange	18.4	1.00	16.5	1.00
Hskew	1.51	0.244	1.35	0.340
Hvar	56.9	6.33	47.6	5.09
FHD	1.33	0.014	1.34	0.038

model produced parameter estimates with acceptable confidence as the confidence intervals all included zero (Table 5).

Airborne LiDAR. Global model fit for Red-naped Sapsuckers indicated that the selection and number of parameters were adequate to model the data compared to that of a saturated model. The deviance was 76.6 and the overdispersion parameter (*c-hat*) = 1.05 with 72 residual degrees of freedom ($dof = n - p$). Model selection uncertainty in the airborne LiDAR derived models was high according to AICc model weights so we performed full model averaging including models with AICc≤6 (Tables 6 and 7).

Foliage height diversity (FHD) was highly influential to the averaged model (RVI = 0.95) (Table 8) and positively associated with Red-naped Sapsucker occupancy. Other variables that were important for this species included heat load index, the distance between the lowest and highest vegetation biomass (hmrange), and the distribution of distribution throughout the canopy (hskew). The positive association with hmrange indicated that occupancy increased with a greater vertical distance between the minimum and the maximum nodes of vegetation biomass. Visually, this would be represented by a gap in the vegetation structure at an intermediate height between the biomass mode nearest to the ground (the minimum mode likely representing the shrub layer) and the biomass mode in the canopy (the maximum mode likely representing the canopy). The positive relationship with hskew indicated that there is more vegetation closer to the ground than in the upper canopy. The proportion of vegetation in the mid-story (2.5–10 m) was positively associated with Red-naped Sapsucker presence but contributed the least to the averaged model (RVI = 0.32). Wind was a detectability covariate, and occupancy was negatively associated with high wind.

Model Cross Validation and Predictive Power

Due to the high uncertainty associated with our spaceborne models, we did not proceed with an assessment of model performance based on spaceborne LiDAR. The AUC score for the airborne data indicated that the global model performed better than random, with AUC scores ranging between 0.6 and 0.72 (Fig. 4).

Discussion

This study was novel for two reasons. First, this study addressed whether GLAS spaceborne LiDAR data could be used for wildlife habitat application in the case of one bird species known to be an ecosystem engineer important to creating habitat for other forest species in the Inland Northwest, USA. Secondly, we were able to compare GLAS and airborne LiDAR models developed at the same spatial scale; no studies to date have compared these LiDAR data sets in the context of wildlife habitat assessment. LiDAR data used in this study were averaged over an area of approximately 0.3 ha centered on the survey point. While it is unlikely that the survey points were coincident with the centers of the home ranges of individual birds, the 0.3 ha sampling unit coincided with the size of the GLAS footprint. Thus, we limited our analyses to this spatial scale in order to facilitate comparisons between the two LiDAR data sets as applied to Red-naped Sapsucker habitat.

The model results based upon GLAS spaceborne LiDAR were weak, which may be due to several reasons. First, we had relatively small sample sizes; although we gathered data across three years, fewer birds were detected during 2011. This was a wet summer, and few birds were noted in our expanded GLAS survey sites. Additionally, a scale mismatch may have occurred between the spatial distribution and extent of GLAS footprints and variables

Table 4. Occupancy models for Red-naped Sapsuckers with a ΔAICc≤6 based on GLAS LiDAR.

Model #	Elev (m)	FHD	HOME	Canopy height	HLI	p.lower	p.mid	Vegetation density	Julian Day	sky	df	logLik	AICc	delta	weight
160	−67.05	37.45	−98.89	203.13	−48.58			31.84			8	−66.60	151.45	0.00	0.177
224	−93.36	61.76	−157.13	285.28	−63.16		−9.75	57.39			9	−65.85	152.56	1.11	0.101
382	−68.41		−153.43	156.54	−77.74	−100.44	−2.72		−0.33		9	−65.97	152.79	1.34	0.090
928	−78.24	43.22	−114.44	236.73	−56.85			36.39	−0.36	−0.19	10	−64.69	152.92	1.47	0.085
672	−77.98	43.53	−115.08	236.15	−56.31			37.07		−0.19	9	−66.16	153.17	1.72	0.075
221			−144.38	174.87	−66.23		−78.37	76.93			7	−68.88	153.47	2.02	0.064
176	−147.04	107.85	−110.06	294.86		69.83		62.13			8	−68.05	154.35	2.90	0.041
894	−106.34		−221.14	247.94	−119.19	−147.19	10.6[illegible]		−0.32	−0.20	10	−65.49	154.54	3.08	0.038
206	−68.28		−135.07	132.80			−49.41	79.31			7	−69.64	155.01	3.55	0.030
477			−91.61	127.66	−41.94		−45.09	41.18	−0.29		8	−68.44	155.14	3.68	0.028
733			−177.87	216.25	−82.75		−96.36	94.85		−0.18	8	−68.46	155.16	3.71	0.028
192	−62.66	−15.09	−137.55	140.90	−82.43	−106.1[illegible]		−6.70			9	−67.31	155.47	4.02	0.024
448	−68.42	12.38	−135.16	178.96	−69.03	−64.18		10.00	−0.33		10	−65.97	155.50	4.04	0.023
736	−87.09	70.01	−209.57	296.30	−60.45		−46.08	78.18		−0.21	10	−66.02	155.59	4.14	0.022
223		−33.28	−140.63	197.60	−104.06		−78.12	63.89			8	−68.78	155.81	4.36	0.020
462	−51.77		−103.44	101.11			−38.33	60.88	−0.28		8	−68.79	155.83	4.38	0.020
992	−189.14	121.41	−364.00	606.65	−138.12		−55.46	137.20	−0.32	−0.22	11	−64.77	155.87	4.42	0.019
989			−129.90	157.61	−59.85		−70.45	69.10	−0.29	−0.19	9	−67.55	155.96	4.51	0.019
512	−89.08	41.11	−166.08	248.49	−72.57	−44.97	−1.72	37.27	−0.37		11	−64.95	156.23	4.78	0.016
479		−16.27	−132.75	175.36	−74.68		−77.53	63.72	−0.29		9	−67.75	156.36	4.90	0.015
16	−58.55	40.27	−52.28	88.17							6	−71.58	156.44	4.98	0.015
640	−76.42	2.10	−166.48	180.24	−84.73	−102.70	−3.51			−0.21	10	−66.75	157.04	5.59	0.011
704	−56.05	0.02	−107.61	137.96	−63.41	−64.87		3.77		−0.21	10	−66.79	157.13	5.68	0.010
718	−53.67		−107.68	105.14			−39.86	63.27		−0.15	8	−69.47	157.19	5.74	0.010
768	−62.97	30.52	−102.80	185.63	−49.88	−13.95	−1.48	28.47		−0.19	11	−65.43	157.19	5.74	0.010
238	−50.93		−102.06	107.28		7.77	−41.17	63.69			8	−69.53	157.30	5.85	0.009

Table 5. Model averaged parameter estimates, standard errors, and 95% confidence intervals using GLAS LiDAR data.

Parameter		Estimate	Std. Error	Lower CI	Upper CI
Occupancy	elev	69.6	116	−157	296
	FHD	110	291	−461	681
	HLI	56.5	196	−327	440
	hmode	−59.4	87.4	−231	112
	hmrange	82.8	179	−268	434
	hskew	112	316	−507	731
	hvar	−50.7	103	−253	152
	p.mid	19.7	74.5	−126	166
Detection	wind	−0.26	0.20	−0.64	0.13

Table 7. Original global model parameter estimates, standard errors, and 95% confidence intervals using airborne LiDAR data.

Parameter		Estimate	Std. Error	Lower CI	Upper CI
Occupancy	Elev*	78.3	248	61.7	112
	HLI*	30.3	91.3	16.9	47.9
	p.mid	17.5	98.0	−17.6	41.2
	FHD*	114	301	92.9	146
	Hmode*	−63.2	166	−80.2	−53.0
	Hmrange*	92.5	343	48.0	138
	Hskew*	92.5	268	61.2	141
	Hvar*	−53.7	212	−79.1	−27.9
Detection	wind	−0.25	0.20	−0.65	0.12

*Confidence intervals do not include zero.

that influence Red-naped Sapsucker habitat selection, which is likely to have affected both the GLAS results as well as the performance of the airborne LiDAR models. Indeed, red-naped Sapsucker may respond to habitat features at multiple spatial scales [58], and recently Sadoti and Vierling [74] noted that features at the home range scale (~9.1 ha) were most important in a multi-scale analysis of selection. Multiple studies have been conducted on Red-naped Sapsucker breeding habitat in aspen forests [74,75], but features that might influence Red-naped Sapsuckers in conifer-dominated forests and the associated important spatial scales are relatively unknown [76]. Other studies have noted that woodpeckers select habitat features at multiple spatial scales, and these spatial scales range from nest tree characteristics to landscape features [74,75,77,78]. We were constrained by the spatial distribution and extent of GLAS footprints, and were thus unable to explore the effects of multiple scales. Furthermore, factors within the GLAS footprints that were not detectable might have heavily influenced Red-naped Sapsucker occupancy (e.g. the presence of fungal conks; [75]).

The most important structural variables that influenced Red-naped Sapsucker selection based on airborne LiDAR data included foliage height diversity, the distance between major components in the canopy profile, and the amount of vegetation near the ground. When considered jointly, these three metrics describe a vegetation profile that is uniform throughout its height except for a larger contribution of vegetation near the ground. Red-naped Sapsuckers have been observed to inhabit montane conifer forests, and Walters [53] notes that increased foraging preference was observed in areas of the canopy with fewer branches. Conifer stands with an even distribution of foliage along the height of the forest may represent branch structure that is more open and accessible for foraging. Additionally, a larger contribution of vegetation near the ground, likely representing shrub growth and could support insects. The presence of shrub growth likely supports an increased insect population, and Red-naped Sapsuckers will utilize a combination of flycatching and gleaning to forage during the breeding season [58,74].

The presence of Red-naped Sapsuckers was also positively associated with increasing elevation and heat load index, although their effect sizes were the smallest. Higher heat indices indicate a preference for sites with warmer combinations of ground slope and aspect. Studies have noted tree cavity

Table 6. All subsets models for Red-naped Sapsuckers with ΔAIC≤6 based on airborne LiDAR data.

Model #	Elev	FHD	HLI	hmode	hmrange	hskew	hvar	p.mid	wind	df	logLik	AICc	delta	weight
384	70.50	115.97	27.05	−66.44	82.84	83.16	−54.48		−0.25	10	−79.57	182.24	0.00	0.44
512	78.33	113.98	30.32	−63.20	92.53	92.54	−53.74	17.49	−0.25	11	−79.47	184.70	2.47	0.13
448	73.00	90.95	24.19	−49.88	55.74	111.96		23.54	−0.30	10	−81.13	185.35	3.12	0.09
120	21.96	107.30	65.65		73.64	88.37	−21.12			8	−84.19	186.36	4.12	0.06
188	108.37	193.01		−73.42	140.18	327.12		64.45		8	−84.34	186.66	4.42	0.05
7		−28.76	153.00							4	−89.12	186.76	4.52	0.05
63		339.22	196.97	37.61	145.38	371.68				7	−85.96	187.43	5.20	0.03
263		−34.82	186.06						−0.19	5	−88.53	187.84	5.61	0.03
56	34.05	170.41	111.76		106.84	144.86				7	−86.22	187.96	5.72	0.03
135		−29.03	133.33					5.20		5	−88.60	187.99	5.76	0.02
37			142.39			33.49				4	−89.76	188.04	5.81	0.02
149			95.41		−54.03			−58.84		5	−88.64	188.08	5.84	0.02
124	63.46	76.63		−51.78	63.56	51.64	−34.86			8	−85.08	188.13	5.89	0.02

Table 8. Relative variable importance (RVI) from model averaging.

Parameter	RVI
FHD	0.95
HLI	0.93
hmrange	0.88
hskew	0.88
elev	0.82
hmode	0.77
wind	0.69
hvar	0.65
p.mid	0.32
julDay	–
sky	–

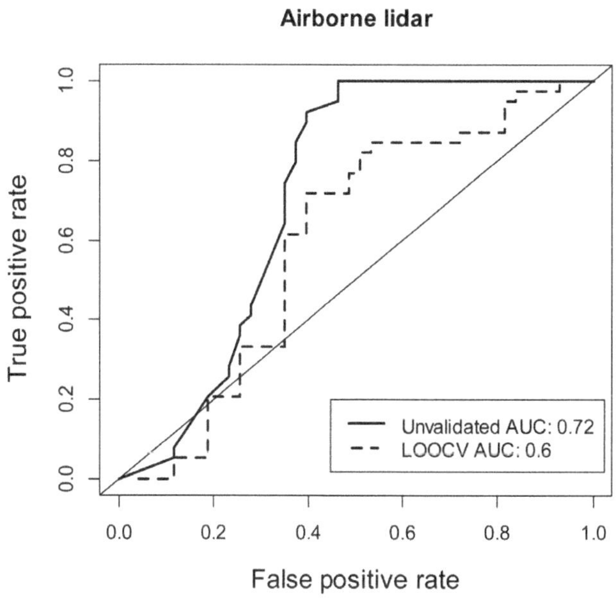

Figure 4. Receiver operator curves with unvalidated and leave-one-out (LOOCV) cross validation for airborne LiDAR.

entrances located on southern aspects in British Columbia [53], and Colorado and Wyoming [79], to no apparent directional preference in Montana [54]. However, while Martinuzzi *et al.* [18] found that slope and aspect could affect the presence of forest snags near the study area, we are not aware of any studies that indicate slope or aspect preference for the nest tree location within the landscape.

Our results demonstrated that small-footprint airborne LiDAR data were able to provide ecologically meaningful information relative to occupancy for Red-naped Sapsuckers when aggregated and analyzed at a scale similar to that sampled by the large-footprint GLAS satellite LiDAR. This finding indicates that the forest structural metrics calculable from airborne LiDAR were likely more suitable for modeling habitat of this bird than GLAS-based structural metrics at this spatial scale. However, because not all samples contained coincident airborne and satellite lidar, this finding requires further testing at additional field sites and wildlife species, and underscores the need for denser spatial sampling in future satellite-based LiDAR missions for application to mapping and modeling wildlife distributions (see [30]). While the large footprint and non-contiguous nature of GLAS data poses significant challenges for wildlife distribution modeling, these data do provide useful information on ecosystem vertical structure, particularly in areas of gentle terrain (e.g. [80]). We therefore encourage the remote sensing and wildlife communities to join efforts so that additional progress can be made to incorporate LiDAR datasets collected from both airborne and past and future satellite platforms (e.g. GLAS, and the planned IceSAT2 mission) with the prospect of improving wildlife biodiversity and single-species modeling for more locations across the globe.

Acknowledgments

This study benefited from discussions with Drs. Fred Hummerich, Dale Goble, Amy Neuenschwander, Laura Duncanson, Sebastian Martinuzzi, Jody Vogeler, and members of the University of Idaho Geospatial Laboratory for Environmental Dynamics. We thank two anonymous reviewers for their constructive comments on eariler versions of the manuscript. We also acknowledge Potlatch Corporation, Bennett Lumber Products, and the Idaho Department of Lands for their gracious permission to conduct research on their property.

Author Contributions

Conceived and designed the experiments: LV KV PA. Performed the experiments: PA. Analyzed the data: LV KV PA AH. Contributed reagents/materials/analysis tools: LV KV PA AH. Wrote the paper: LV KV PA AH.

References

1. Innes JL, Koch B (1998) Forest biodiversity and its assessment by remote sensing. Global Ecology and Biogeography 7: 397–419.
2. Turner W, Spector S, Gardiner N, Fladeland M, Sterling E, et al. (2003) Remote sensing for biodiversity science and conservation. Trends in Ecology & Evolution 18: 306–314.
3. Scott JM, Davis F, Csuti B, Noss R, Butterfield B, et al. (1993) GAP analysis - a geographic approach to protection of biological diversity. Wildlife Monographs 123: 1–41.
4. Rodrigues ASL, Andelman SJ, Bakarr MI, Boitani L, Brooks TM, et al. (2004) Effectiveness of the global protected area network in representing species diversity. Nature 428: 640–643.
5. Nagendra H (2001) Using remote sensing to assess biodiversity. International Journal of Remote Sensing 22: 2377–2400.
6. Scott JM, Peterson CR, Karl JW, Strand E, Svancara LK, et al. (2002) A Gap Analysis of Idaho: Final Report. Moscow: Idaho Cooperative Fish and Wildlife Research Unit.
7. MacArthur R, MacArthur JW (1961) On bird species-diversity. Ecology 42: 594–598.
8. Vierling KT, Vierling LA, Gould WA, Martinuzzi S, Clawges RM (2008) LiDAR: shedding new light on habitat characterization and modeling. Frontiers in Ecology and the Environment 6: 90–98.
9. Elith J, Leathwick JR (2009) Species Distribution Models: Ecological Explanation and Prediction Across Space and Time. Annual Review of Ecology, Evolution and Systematics. 40: 677–697.
10. Hinsley SA, Hill RA, Gaveau DLA, Bellamy PE (2002) Quantifying woodland structure and habitat quality for birds using airborne laser scanning. Functional Ecology 16: 851–857.
11. Hill RA, Hinsley SA, Gaveau DLA, Bellamy PE (2004) Predicting habitat quality for Great Tits (Parus major) with airborne laser scanning data. International Journal of Remote Sensing 25: 4851–4855.
12. Goetz S, Steinberg D, Dubayah R, Blair B (2007) Laser remote sensing of canopy habitat heterogeneity as a predictor of bird species richness in an eastern temperate forest, USA. Remote Sensing of Environment 108: 254–263.

13. Boelman NT, Asner GP, Hart PJ, Martin RE (2007) Multi-trophic invasion resistance in Hawaii: Bioacoustics, field surveys, and airborne remote sensing. Ecological Applications 17: 2137–2144.
14. Clawges R, Vierling K, Vierling L, Rowell E (2008) The use of airborne LiDAR to assess avian species diversity, density, and occurrence in a pine/aspen forest. Remote Sensing of Environment 112: 2064–2073.
15. Goetz SJ, Steinberg D, Betts MG, Holmes RT, Doran PJ, et al. (2010) LiDAR remote sensing variables predict breeding habitat of a Neotropical migrant bird. Ecology 91: 1569–1576.
16. Lesak AA, Radeloff VC, Hawbaker TJ, Pidgeon AM, Gobakken T, et al. (2011) Modeling forest songbird species richness using LiDAR-derived measures of forest structure. Remote Sensing of Environment 115: 2823–2835.
17. Vogeler JC, Hudak AT, Vierling LA, Vierling KT (2013) Lidar-derived canopy architecture predicts Brown Creeper occupancy of two western coniferous forests. The Condor 115: 1–9.
18. Martinuzzi S, Vierling LA, Gould WA, Falkowski MJ, Evans JS, et al. (2009) Mapping snags and understory shrubs for a LiDAR-based assessment of wildlife habitat suitability. Remote Sensing of Environment 113: 2533–2546.
19. Nelson R, Keller C, Ratnaswamy M (2005) Locating and estimating the extent of Delmarva fox squirrel habitat using an airborne LiDAR profiler. Remote Sensing of Environment 96: 292–301.
20. Muller J, Brandl R (2009) Assessing biodiversity by remote sensing in mountainous terrain: the potential of LiDAR to predict forest beetle assemblages. Journal of Applied Ecology 46: 897–905.
21. Vierling KT, Bässler C, Brandl R, Vierling LA, Weiß I, et al. (2011) Spinning a laser web: predicting spider community characteristics using LiDAR. Ecological Applications 21: 577–588.
22. Lefsky MA, Cohen WB, Parker GG, Harding DJ (2002) Lidar remote sensing for ecosystem studies. BioScience 52: 19–30.
23. Dubayah RO, Drake JB (2000) Lidar remote sensing for forestry. Journal of Forestry 98: 44–46.
24. Falkowski M, Smith AM, Hudak AT, Gessler PE, Vierling LA, et al. (2006) Automated estimation of individual conifer tree height and crown diameter via two-dimensional wavelet analysis of lidar data. Canadian Journal of Remote Sensing 32: 153–161.
25. Falkowski MJ, Evans JS, Martinuzzi S, Gessler PE, Hudak AT (2009) Characterizing forest succession with lidar data: An evaluation for the Inland Northwest, USA. Remote Sensing of Environment 113: 946–956.
26. Martinuzzi S, Gould WA, Vierling LA, Hudak AT, Nelson RF, et al. (2013) Quantifying Tropical Dry Forest Type and Succession: Substantial Improvement with LiDAR. Biotropica 45: 135–146.
27. Gonzalez P, Asner GP, Battles JJ, Lefsky MA, Waring KM, et al. (2010) Forest carbon densities and uncertainties from Lidar, QuickBird, and field measurements in California. Remote Sensing of Environment 114: 1561–1575.
28. Hudak AT, Strand EK, Vierling LA, Byrne JC, Eitel JUH, et al. (2012) Quantifying aboveground forest carbon pools and fluxes from repeat LiDAR surveys. Remote Sensing of Environment 123: 25–40.
29. Seavy NE, Viers JH, Wood JK (2009) Riparian bird response to vegetation structure: a multiscale analysis using LiDAR measurements of canopy height. Ecological Applications 19: 1848–1857.
30. Bergen KM, Goetz SJ, Dubayah RO, Henebry GM, Hunsaker CT, et al. (2009) Remote sensing of vegetation 3-D structure for biodiversity and habitat: Review and implications for LiDAR and radar spaceborne missions. Journal of Geophysical Research-Biogeosciences 114: G00e06, DOI: 10.1029/2008JG000883.
31. Schutz BE, Zwally HJ, Shuman CA, Hancock D, DiMarzio JP (2005) Overview of the ICESat Mission. Geophysical Research Letters 32: L21S01, DOI: 10.1029/2005GL024009.
32. Lefsky MA (2010) A global forest canopy height map from the Moderate Resolution Imaging Spectroradiometer and the Geoscience Laser Altimeter System. Geophysical Research Letters 37: L14501, DOI: 10.1029/2010GL043622.
33. Harding DJ, Carabajal CC (2005) ICESat waveform measurements of within-footprint topographic relief and vegetation vertical structure. Geophysical Research Letters 32: L21S10, DOI: 10.1029/2005GL023471.
34. Lefsky MA, Harding DJ, Keller M, Cohen WB, Carabajal CC, et al. (2005) Estimates of forest canopy height and aboveground biomass using ICESat. Geophysical Research Letters 32: L22S02, DOI: 10.1029/2005GL023971.
35. Nelson R, Boudreau J, Gregoire TG, Margolis H, Naesset E, et al. (2009) Estimating Quebec provincial forest resources using ICESat/GLAS. Canadian Journal of Forest Research 39: 862–881.
36. Jones CG, Lawton JH, Shachak M (1994) Organisms as Ecosystem Engineers. Oikos 69: 373–386.
37. Daily GC, Ehrlich PR, Haddad NM (1993) Double keystone bird in a keystone species complex. Proceedings of the National Academy of Sciences of the United States of America 90: 592–594.
38. Aitken KEH, Martin K (2008) Resource selection plasticity and community responses to experimental reduction of a critical resource. Ecology 89: 971–980.
39. Bull EL, Holthausen RS (1993) Habitat Use and Management of Pileated Woodpeckers in Northeastern Oregon. The Journal of Wildlife Management 57: 335–345.
40. Drever MC, Martin K (2010) Response of woodpeckers to changes in forest health and harvest: Implications for conservation of avian biodiversity. Forest Ecology and Management 259: 958–966.
41. Ripper D, Bednarz JC, Varland DE (2007) Landscape Use by Hairy Woodpeckers in Managed Forests of Northwestern Washington. Journal of Wildlife Management 71: 2612–2623.
42. Abshire JB, Sun XL, Riris H, Sirota JM, McGarry JF, et al. (2005) Geoscience Laser Altimeter System (GLAS) on the ICESat mission: On-orbit measurement performance. Geophysical Research Letters 32: L21S02, DOI:10.1029/2005GL024028.
43. Chen Q (2010) Retrieving vegetation height of forests and woodlands over mountainous areas in the Pacific Coast region using satellite laser altimetry. Remote Sensing of Environment 114: 1610–1627.
44. NASA (2012) National Snow and Ice Data Center - ICESat/GLAS Data. Available: http://nsidc.org/. Accessed 5 October 2010.
45. Brenner A, Zwally H, Bentley C, Csatho B, Harding D, et al. (2003) Derivation of range and range distributions from laser pulse waveform analysis for surface elevations, roughness, slope, and vegetation heights. Algorithm Theoertical Basis Document Version 4.1, NASA Goddard Space Flight Center.
46. Chen Q (2010) Assessment of terrain elevation derived from satellite laser altimetry over mountainous forest areas using airborne LiDAR data. ISPRS Journal of Photogrammetry and Remote Sensing 65: 111–122.
47. Lefsky MA, Keller M, Pang Y, de Camargo PB, Hunter MO (2007) Revised method for forest canopy height estimation from Geoscience Laser Altimeter System waveforms. Journal of Applied Remote Sensing 1: 1–18.
48. USDA (2010) Geospatial Data Gateway. Available: http://gcmd.nasa.gov/records/USDA_Geo_Gateway.html. Accessed 15 June 2010.
49. Evans JS, Hudak AT (2007) A multiscale curvature algorithm for classifying discrete return LiDAR in forested environments. IEEE Transactions on Geoscience and Remote Sensing 45: 1029–1038.
50. Cooke HA, Hannon SJ (2011) Do aggregated harvests with structural retention conserve the cavity web of old upland forest in the boreal plains? Forest Ecology and Management 261: 662–674.
51. Gentry DJ, Vierling KT (2008) Reuse of Woodpecker Cavities in the Breeding and Non-Breeding Seasons in Old Burn Habitats in the Black Hills, South Dakota. American Midland Naturalist 160: 413–429.
52. Martin K, Aitken KEH, Wiebe KL (2004) Nest Sites and Nest Webs for Cavity-Nesting Communities in Interior British Columbia, Canada: Nest Characteristics and Niche Partitioning. The Condor 106: 5–19.
53. Walters EL (1996) Habitat and space use of the Red-naped Sapsucker, ~Sphyrapicus nuchalis~, in the Hat Creek valley, south-central British Columbia [Master's Thesis]. Victoria, BC: Univ. of Victoria.
54. McClelland BR, McClelland PT (2000) Red-naped sapsucker nest trees in northern rocky mountain old-growth forest. The Wilson Bulletin 112: 44–50.
55. McClelland BR (1977) Relationships between hole-nesting birds, forest snags, and decay in western larch-Douglas fir forests of the northern Rocky Mountains [PhD dissertation]. Missoula: Univ. of Montana.
56. Drever MC, Aitken KEH, Norris AR, Martin K (2008) Woodpeckers as reliable indicators of bird richness, forest health and harvest. Biological Conservation 141: 624–634.
57. MacKenzie DI, Nichols JD, Lachman GB, Droege S, Royle JA, et al. (2002) Estimating site occupancy rates when detection probabilities are less than one. Ecology 83: 2248–2255.
58. Walters EL, Miller EH, Lowther PE (2002) Red-naped Sapsucker (Sphyrapicus nuchalis) In: Poole A, editor. The Birds of North America Online. Ithaca: Cornell Lab of Ornithology. Available: http://bna.birds.cornell.edu/bna/species/663b; doi:10.2173/bna.663. Accessed 10 January 2012.
59. Belsley D (1991) Conditioning diagnostics : collinearity and weak data in regression. New York: John Wiley. 396 p.
60. Royle JA, Nichols JD (2003) Estimating abundance from repeated presence–absence data or point counts. Ecology 84: 777–790.
61. McCallum DA (2005) A conceptual guide to detection probability for point counts and other count-based survey methods. USDA Forest Service General Technical Report PSW–GTR–191: 754–761.
62. Gu WD, Swihart RK (2004) Absent or undetected? Effects of non-detection of species occurrence on wildlife-habitat models. Biological Conservation 116: 195–203.
63. Mazerolle MJ, Desrochers A, Rochefort L (2005) Landscape characteristics influence pond occupancy by frogs after accounting for detectability. Ecological Applications 15: 824–834.
64. R Development Core Team (2012) R: A language and environment for statistical computing. Vienna, Austria: R Foundation for Statistical Computing.
65. Fiske IJ, Chandler RB (2011) Unmarked: An R Package for Fitting Hierarchical Models of Wildlife Occurrence and Abundance. Journal of Statistical Software 43: 1–23.
66. Burnham K, Anderson D (2002) Model selection and multimodel inference: a practical information-theoretic approach: Springer Verlag. 488 p.
67. Dochtermann NA, Jenkins SH (2011) Developing multiple hypotheses in behavioral ecology. Behavioral Ecology and Sociobiology 65: 37–45.
68. Richards SA (2005) Testing ecological theory using the information-theoretic approach: Examples and cautionary results. Ecology 86: 2805–2814.
69. Richards SA (2008) Dealing with overdispersed count data in applied ecology. Journal of Applied Ecology 45: 218–227.
70. Lukacs P, Burnham K, Anderson D (2010) Model selection bias and Freedman's paradox. Annals of the Institute of Statistical Mathematics 62: 117–125.

71. Burnham KP, Anderson DR, Huyvaert KP (2011) AIC model selection and multimodel inference in behavioral ecology: some background, observations, and comparisons. Behavioral Ecology and Sociobiology 65: 23–35.
72. Murtaugh PA (2009) Performance of several variable-selection methods applied to real ecological data. Ecology Letters 12: 1061–1068.
73. Dahlgren JP (2010) Alternative regression methods are not considered in Murtaugh (2009) or by ecologists in general. Ecology Letters 13: E7–E9.
74. Sadoti G, Vierling KT (2010) Nonideal habitat selection by a North American cavity excavator: pecking up the wrong tree? Canadian Journal of Zoology 88: 527–535.
75. Lawler JJ, Edwards TC (2006) A variance-decomposition approach to investigating multiscale habitat associations. Condor 108: 47–58.
76. Wagner MA (2011) Habitat selection by Red-breasted Sapsucker (Sphyrapicus ruber) in southeast Alaska old-growth forest [Master's thesis]. Humboldt: Humboldt State University. 58 p.
77. Lawler J, Edwards T (2002) Landscape patterns as habitat predictors: building and testing models for cavity-nesting birds in the Uinta Mountains of Utah, USA. Landscape Ecology 17: 233–245.
78. Vierling KT, Lentile LB, Nielsen-Pincus N (2008) Preburn characteristics and woodpecker use of burned coniferous forests. Journal of Wildlife Management 72: 422–427.
79. Crockett AB, Hadow HH (1975) Nest site selection by Williamson's and Red-naped sapsuckers. Condor 77: 365–368.
80. Lefsky MA (2010) A global forest canopy height map from the Moderate Resolution Imaging Spectroradiometer and the Geoscience Laser Altimeter System. Geophysical Research Letters 37: Article Number: L15401 DOI: 10.1029/2010GL043622.
81. McCune B, Keon D (2002) Equations for potential annual direct incident radiation and heat load. Journal of Vegetation Science 13(4): 603–606.

2

Phenology and Productivity of C_3 and C_4 Grasslands in Hawaii

Stephanie Pau[1]*, Christopher J. Still[2]

1 Department of Geography, Florida State University, Tallahassee, Florida, United States of America, 2 Forest Ecosystems and Society, Oregon State University, Corvallis, Oregon, United States of America

Abstract

Grasslands account for a large proportion of global terrestrial productivity and play a critical role in carbon and water cycling. Within grasslands, photosynthetic pathway is an important functional trait yielding different rates of productivity along environmental gradients. Recently, C_3-C_4 sorting along spatial environmental gradients has been reassessed by controlling for confounding traits in phylogenetically structured comparisons. C_3 and C_4 grasses should sort along temporal environmental gradients as well, resulting in differing phenologies and growing season lengths. Here we use 10 years of satellite data (NDVI) to examine the phenology and greenness (as a proxy for productivity) of C_3 and C_4 grass habitats, which reflect differences in both environment and plant physiology. We perform phylogenetically structured comparisons based on 3,595 digitized herbarium collections of 152 grass species across the Hawaiian Islands. Our results show that the clade identity of grasses captures differences in their habitats better than photosynthetic pathway. Growing season length (GSL) and associated productivity (GSP) were not significantly different when considering photosynthetic type alone, but were indeed different when considering photosynthetic type nested within clade. The relationship between GSL and GSP differed most strongly between C_3 clade habitats, and not between C_3-C_4 habitats. Our results suggest that accounting for the interaction between phylogeny and photosynthetic pathway can help improve predictions of productivity, as commonly used C_3-C_4 classifications are very broad and appear to mask important diversity in grassland ecosystem functions.

Editor: Harald Auge, Helmholtz Centre for Environmental Research - UFZ, Germany

Funding: This work was initiated while SP was a Postdoctoral Associate at the National Center for Ecological Analysis and Synthesis, a Center funded by NSF (Grant #EF-0553768), the University of California, Santa Barbara, and the State of California. SP completed revisions with the support of Florida State University's Committee on Faculty Research Support (COFRS) Award. CJS acknowledges the support of a NASA New Investigator Program award that initially supported this research. The funders had no role in study design, data collection and analysis, decision to publish, or preparation of the manuscript.

Competing Interests: The authors have declared that no competing interests exist.

* Email: spau@fsu.edu

Introduction

A critically important problem in understanding ecosystem responses to global change is the relationship between growing season length and productivity. This is because growing season length is commonly used to predict net ecosystem exchange and to inform terrestrial biosphere models of vegetation dynamics and carbon exchange [1,2]. The length of the growing season should be a primary factor controlling the terrestrial carbon cycle by setting the length of time available for photosynthesis and plant growth [3–5]. For example, an extension of the growing season either through earlier spring onset or later autumn senescence has been shown to be positively related to primary productivity and increased carbon assimilation across several different vegetation types [6] and refs. above but also see [7–8] for examples of no or negative relationships. However most studies examine temperate deciduous or evergreen forests, while examples from grasslands and tropical sites are limited.

Grasslands and tropical savannas comprise approximately 29% of the global extent of terrestrial biomes and this corresponds to some 33% of global terrestrial NPP [9]. Photosynthetic pathway is a dominant plant functional trait that has consequences for the global carbon cycle because of the distribution of C_3 and C_4 plants and their different rates of productivity [10]. In general, the efficiency of C_4 photosynthesis should be greater than C_3 photosynthesis under conditions of high temperatures, high light, and greater aridity [11,12]. The classic turnover of C_3-C_4 grasses along altitudinal gradients has been well documented [13–19]. This turnover illustrates the physiological advantage of C_4 grasses in warmer and drier environments, which confers dominance in species composition and relative cover. The occurrence of C_4 grasses in more marginal environments, however, should have consequences for their productivity; thus they may not be more productive than C_3 grasses even with an advantage in resource-use efficiencies under resource-poor conditions [15].

C_3 and C_4 grasses should also partition their activity along temporal environmental gradients resulting in different phenologies and growing season lengths. Differences in phenology have been demonstrated in both mixed C_3-C_4 grasslands [20,21] as well as across large environmental gradients [15]. Studies of C_3 and C_4 grasslands in the Great Plains of North America have demonstrated that, compared to C_3 grasses, C_4 grasses are active later in the growing season when temperatures warm and water becomes limiting in their respective environment [21–25]. The Great Plains region has been the focus of many subsequent studies using remote sensing to discriminate C_3-C_4 regions based on their distinct seasonality and responses to climate variability [26–29].

Yet there is a growing body of work that has re-examined C_3-C_4 comparisons, benefiting from the recent development of well-resolved grass phylogenies [30]. The vast majority of grasses (Poaceae) belong to either the 'BEP' or 'PACMAD' clade, which last shared a common ancestor an estimated 50–80 Mya [31]. 'BEP' is an acronym for the Bambusoideae, Ehrhartoideae, and Pooideae lineages, which contain only C_3 species. 'PACMAD' refers to Panicoideae, Aristidoideae, Chloridoideae, Micrairioideae, Arundinoidae, and Danthonioideae lineages, which contain all of the C_4 species and some C_3 species [30]. Thus, there is considerable diversity within C_3 and C_4 groups associated with different lineages. Comparing C_3 and C_4 grasses restricted to the PACMAD clade enables our understanding of ecological differences previously confounded by different evolutionary histories [32–37]. Furthermore C_3 BEP and PACMAD taxa may exhibit ecological differences relevant to community structure and ecosystem functions.

Previous phylogenetic modeling of Hawaiian grasses has demonstrated that BEP and PACMAD clades capture habitat distinctions not strictly associated with photosynthetic pathway [33]. This work suggested that there are larger differences in aridity between the habitats of C_3 and C_4 grasses compared to temperature differences [33,35,37], a finding largely driven by the much higher precipitation and woody cover of C_3 PACMAD grass environments compared to other grass groupings [38]. Subsequent ecological niche modeling described the changes from C_4 grass-dominated habitats in lower elevation, warm and dry regions, to C_3 PACMADs in mid-elevation, cooler and wetter regions, and finally to C_3 BEPs at the highest elevations in cooler and drier conditions [37]. It was furthermore shown that C_3 PACMADs tend to occur in habitats that receive early season (winter) precipitation, whereas the habitat sorting of C_3 BEPs and C_4 grasses was associated with temperature differences. This work and others add to the growing evidence that there are important ecological differences among lineages that were not captured by previous C_3-C_4 comparisons. Likewise, phylogenetically structured comparisons should highlight when and where photosynthetic pathway may in fact be responsible for habitat sorting and differences in resource strategies.

In this study we examine the phenology and inferred productivity of C_3-C_4 grass habitats using herbarium collection localities across the Hawaiian Islands and a timeseries of satellite images. C_3 and C_4 grasses dominate distinct regions along spatial gradients, thus our examination of the phenology of these habitats reflects both the dominant plant functional type and the environment in which they occur. We compare the habitats of C_3 and C_4 grass species within the PACMAD clade to try to isolate the effects of photosynthetic pathway on the timing, magnitude, and estimated productivity (integrated NDVI) of these grasslands. We also examine habitats of C_3 grass species from both PACMAD and BEP clades to highlight the functional diversity that is commonly grouped. Specifically, we ask the following questions: Do the habitats of C_3 and C_4 grass species within and between clades differ in their phenology (start-of-season, end-of-season, and growing season length) and productivity? Is the relationship between growing season length and productivity of grass species' habitats within and between clades different? The Hawaiian Islands provide an ideal setting for this work because of the considerable variation in both species diversity and climatic gradients within a small geographic region. The majority of grasses in Hawaii are introduced, and thus their distribution should reflect recent ecological sorting and not insular evolutionary history or patterns of extinction.

Methods

Species and environmental data

We used a dataset of 3,595 digitized geo-referenced herbarium records for Poaceae created by [33]. Records from across the main seven Hawaiian Islands (Kauai, Oahu, Molokai, Lanai, Maui, Kahoolawe, and Hawaii) were obtained from the Smithsonian Flora of the Hawaiian Islands Website and The Bishop Museum Herbarium. The dataset represents 152 grass species (there are multiple collections for most species), each assigned to either the 'BEP' or 'PACMAD' clade that together represent most species of Poaceae. Each species' life history was further categorized as annual or perennial based on [39]. Because individual occurrence records may not indicate a large enough expanse of grasses for remote sensing analysis, herbarium point localities were excluded if they did not fall within a grassland landcover class according to the Hawaii GAP Analysis. The Hawaii GAP Analysis is a land cover dataset derived from 30-m resolution Landsat imagery from 1999–2003 and edited using ancillary data and expert knowledge (US Geological Survey, Gap Analysis Program (GAP) August 2011. National Land Cover, Version 2). As a consequence, our C_3 PACMAD samples represent species that occur open grassland environments, yet C_3 PACMADs tend to occur in regions with the greater tree cover [37,38].

Additionally, C_3 PACMADs present a challenge in using remotely sensed data to capture grass phenology because they are globally uncommon. Because of this, we verified that at least ten C_3 PACMAD species in our herbarium records were collected recently based on collection dates. Five species have been collected at recently as 2000 (*Cortaderia jubata*, *Dichanthelium hillebrandianum*, *Oplimensus compositus*, *Rytidosperma pilosum*, and *Schizostachyum glaucifolium*); in addition, another five have been collected since 1990 (*Cortaderia selloana*, *Dichanthelium cynodon*, *Isachne distichophylla*, *Oplimenus hirtella*, *Sacciolepsis indica*). These species are furthermore all considered common and/or occurring densely in open areas in Hawaii, with the exception of *Schizostachum* [39].

Remote sensing data

We used the NASA Terra Moderate Resolution Imaging Spectroradiometer (MODIS) NDVI (MOD13Q1) Collection 5, which provides 16-day composite data at 250-m spatial resolution. Data were screened using the 'QC_Day' scientific data set for only 'good quality' pixels (i.e., not contaminated by clouds or aerosols) or 'check other QA' in the VI Quality dataset. Pixels falling into the latter category were discarded if flagged for clouds (adjacent cloud, mixed clouds and possible shadow), aerosols (high and climatology aerosols), or possible shadow. NDVI has been shown to be less sensitive to view-angle differences compared to the other commonly used index of plant greenness, the Enhanced Vegetation Index (EVI) [40].

An NDVI time series was created from one pixel centered over each collection locality for a 10-year period using 16-day MODIS data (February 2000–February 2010). Each time series was first smoothed using a centered moving-window average of five data points. For each smoothed time series (for each collection locality), piece-wise logistic functions were fit to each ascending and descending phase of NDVI, following [41]. The points of local maxima or minima in the rate of change in the curvature of the fitted logistic models were used to identify the 'Start-of-Season' (SOS) and 'End-of-Season' (EOS) for each year [41]. 'Growing Season Length' (GSL) is the length of time between SOS and EOS, and 'Growing Season Productivity' (GSP) is the time-integrated NDVI between the SOS and EOS (integrated across days - NDVI is a ratio and is unitless). An attempt was made to use each of the 3,595 herbaria point localities for logistic fits. However, in a given year, collection localities were excluded if the fitting procedure failed to converge, for example, because too few data points remained after the QC_Day screening.

On average 107 logistic fits were made each year for C_3 BEPs, 19 for C_3 PACMADs, and 284 for C_4 PACMADs, which in part reflects the number of collections for each group (there were fewer C_3 PACMADs in the herbarium records) as well as potential biases in tree-covered and cloudy regions that were screened more heavily. Of these successful logistic fits, SOS and EOS dates were then averaged across all species falling into each photosynthetic type-clade combination (3 groups: C_3 PACMAD, C_3 BEP, and C_4 PACMAD), for each year (10 years). Annuals and perennials within each photosynthetic type-clade combination were also determined (except there were no annual C_3 PACMADS), which together with the 3 photosynthetic type-clade combinations produced a total of 50 samples for statistical analyses.

Statistical analyses

First, linear models were used to examine the effect of photosynthetic pathway on SOS, EOS, and GSL irrespective of clade (all C_3 vs. C_4 grasses). Then, photosynthetic pathway nested within clades (PACMAD:C_3, PACMAD:C_4, BEP:C_3) were used to assess differences in SOS, EOS, and GSL. Using photosynthetic type nested within clade tests for differences in means among photosynthetic type and differences in means among photosynthetic type-clade combinations. Tukey post-hoc tests were used to examine independent contrasts between each group and to test for differences more likely due to photosynthetic type by controlling for clade identity (C_3 PACMAD vs. C_4 PACMAD). Linear models were also used to examine the relationship between GSL and GSP. Four models were compared using a corrected Akaike's Information Criterion (AICc) for small sample sizes and with greater penalties for extra model parameters [42]. The simplest model considered only GSL as a predictor of GSP (Model A; see Table 1). Then the interactions between GSP and clade (Model B), between GSP and photosynthetic type (Model C), and between GSP and photosynthetic pathway nested within clade (Model D) were included. These models were further compared using Akaike weights (w_i), which indicate a measure of model selection uncertainty, i.e., the probability that a model is 'best' among a set of candidate models.

Results and Discussion

Ignoring phylogeny (i.e., comparing only C_3 and C_4 grass groupings), SOS was later in C_4 habitats compared to C_3 habitats ($F = 9.42$, $p < 0.01$, $df = 48$; Figure 1a). The SOS for C_4 habitats occurred during the end of September (mean day of year or DOY = 269 based on logistic models) compared to the beginning of September for C_3 habitats (mean DOY = 245). In Hawaii, climate seasonality is governed primarily by precipitation, with November-April generally considered the wet season and May-October the dry season [39,43]. Therefore, the SOS for both C_3 and C_4 grasslands appear to fall within the dry season, but C_4 habitats green-up later in the dry season when plants should be more drought-stressed. C_4 grasses in temperate climates have been shown to be active and more productive later in the growing season as temperatures increase and water becomes limiting in their respective environments [20,25,44]. Indeed, this temporal offset forms the basis for predicting C_3-C_4 mixtures in temperate grasslands [26–29,45,46]. The physiological advantage of greater water-use efficiency [47] and dominance of arid, warm environments (reviewed in [12]) by C_4 grasses is well recognized and furthermore shown to be robust in phylogenetically structured screening experiments [36].

When we considered phylogenetic structure by accounting for photosynthetic pathway nested within clade, results revealed that the temporal offset in SOS between photosynthetic types reflected differences between C_4 PACMAD habitats and C_3 BEP habitats ($F = 4.66$, $p < 0.02$, $df = 47$; posthoc Tukey's $p < 0.05$; Figure 1a). C_4 habitats had a significantly later SOS compared to C_3 BEP habitats by about 23 days on average ($p = 0.03$). There was no statistically significant SOS difference between C_3 and C_4 PACMADs (posthoc Tukey's $p = 0.17$). Thus SOS differences were due to BEP-PACMAD identity rather than strictly photosynthetic pathway, otherwise this difference should be apparent in the C_3 and C_4 PACMAD comparison. Our data show large variability in SOS for C_3 PACMAD habitats indicating a diversity of strategies and growing environments for this group. Statistical significance notwithstanding, C_4 habitats started their growing season an estimated 20 days later into the dry season on average compared to C_3 PACMAD habitats. C_3 PACMAD SOS was only about 3 days later than C_3 BEPs on average, and this difference was not statistically significant ($p = 0.99$).

EOS showed the greatest variability among phenological metrics. Although there was a significant difference when considering photosynthetic type alone ($F = 4.85$, $p < 0.05$, $df = 48$), there was no significant difference when photosynthetic type was nested within clade ($F = 2.81$, $p = 0.07$, $df = 47$; post-hoc Tukey's $p > 0.05$ for all comparisons; Figure 1b). C_3 and C_4 habitats brown-down during the transition into the dry season in mid-April/beginning of May (mean EOS DOY for $C_3 = 124$; mean EOS DOY for $C_4 = 107$). C_4 habitats had an earlier EOS compared to C_3 habitats when clade identity was not considered, and this difference appears to be due to the later EOS for BEP taxa (Figure 1b). Environmental controls on EOS or leaf senescence are not well understood (Menzel et al. 2006; Taylor et al. 2007). In other analyses of phenological shifts associated with

Table 1. Comparison of models predicting growing season productivity (GSP; estimated with integrated NDVI) using growing season length (GSL) and its interaction with clade and photosynthetic type.

Model parameters	ΔAICc	Akaike weights (w_i)	k
Model A: GSL	23.823	<0.001	3
Model B: GSL*clade	**1.770**	**0.292**	**5**
Model C: GSL*photosynthetic type	25.003	<0.001	5
Model D: GSL*clade:photosynthetic type	**0**	**0.708**	**5**

Equivalent best models (when ΔAICc is less than or equal to 2) are highlighted in bold and show the importance of clade identity. 'k' = number of model parameters.

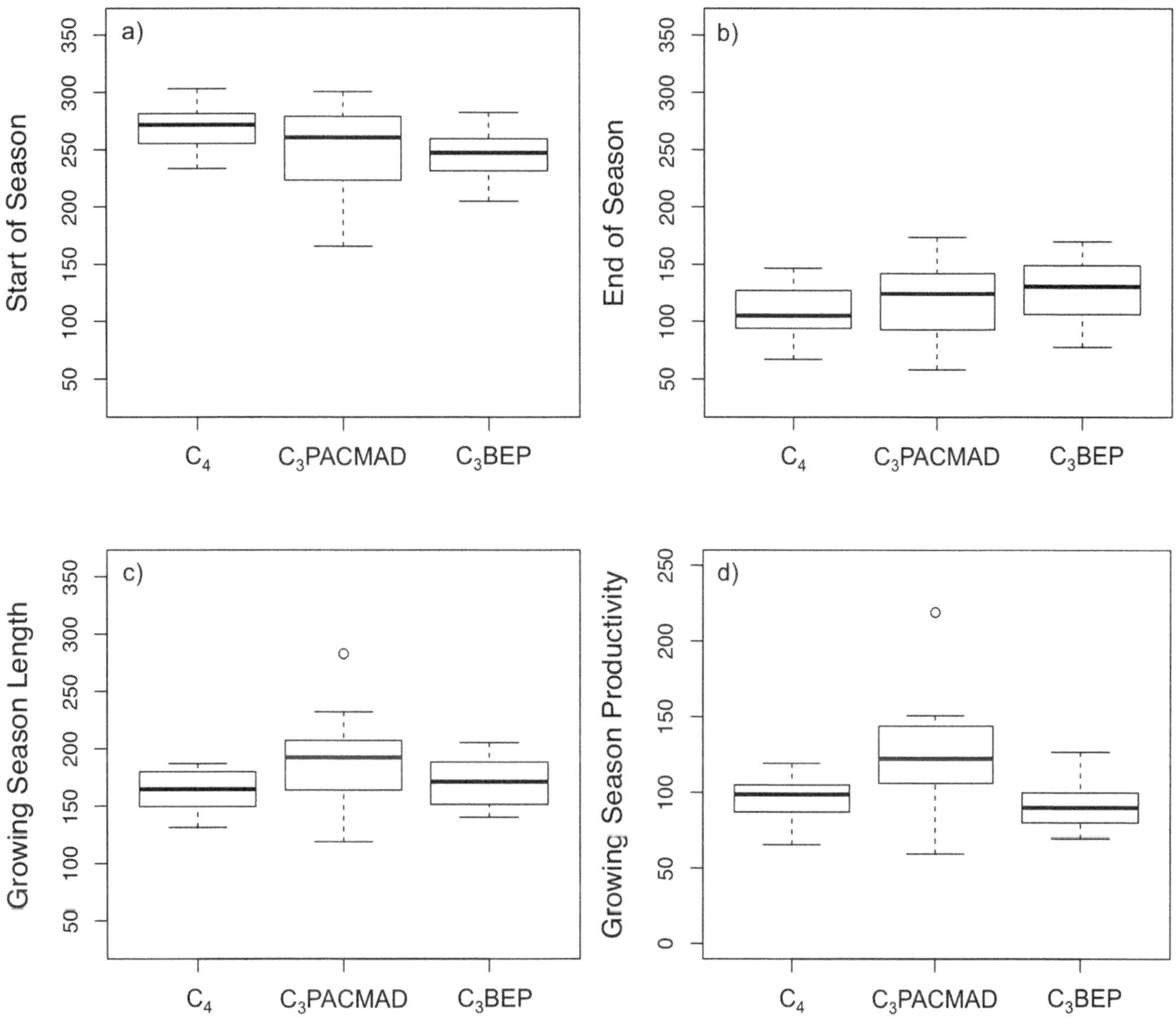

Figure 1. Differences in C_3 and C_4 start-of-season (SOS) (a), end-of-season (EOS) (b), growing season length (c), and growing season productivity (d). Y-axis for panels (a)–(c) is the Julian day-of-year or DOY; y-axis for panel (d) is integrated NDVI, which is unitless, based on logistic models using 10-year timeseries of MODIS NDVI (see Methods). The growing season in Hawaii crosses the calendar year so that SOS begins at a later DOY than EOS. Most Poaceae species fall within the 'BEP' or 'PACMAD' clade. All C_4 grasses are in the 'PACMAD' clade.

climate change, EOS dates have exhibited more variability in comparison to SOS dates, [48–52].

GSL was not significantly different when considering photosynthetic type alone ($F = 3.18$, $p = 0.08$, df = 48), but was indeed different when considering photosynthetic type nested within clade ($F = 4.44$, $p<0.02$, df = 47; Figure 1c). C_3 PACMAD habitats had a longer GSL by about 30 days on average compared to C_4 habitats in the same clade (posthoc Tukey's $p<0.05$); no other comparisons were significantly different. Thus the data show that physiological and/or habitat differences governing GSL are likely masked when C_3 taxa are grouped across clades. But when clades are accounted for, the difference in GSL due to photosynthetic pathway becomes more apparent (i.e., when comparing C_3 and C_4 PACMAD GSL – Figure 1c). Similarly, GSP was not significantly different with respect to just photosynthetic type ($F = 0.10$, $p = 0.32$, df = 48), but was significantly different when considering photosynthetic type nested within clade ($F = 9.59$, $P<0.001$, df = 47). C_3 PACMAD habitats were significantly more productive than C_3 BEP and C_4 habitats (post-hoc Tukey's $p<0.001$ for both comparisons; Figure 1d). The longer GSL and greater GSP of C_3 PACMAD habitats reflect the difference in their growing environment—they prefer or are possibly restricted to more favorable habitats for plant growth—wetter than C_4 habitats and warmer than C_3 BEP habitats [37].

GSL explained a large proportion of the variability in GSP—as GSL increased, there was a corresponding increase in GSP (Figure 1c,d). There were two equivalent best-fit models predicting yearly differences in GSP and these were the two models that included clade as a model parameter (Models B and D). Akaike weights showed more support for the model with photosynthetic type nested within clade (Model D) over the model with clade alone (Model B), and there was strong support for both models over other candidate models that excluded clade. Using photosynthetic type alone to predict GSP resulted in the worst-fit model.

The slope of the relationship between GSL and GSP differed for each photosynthetic type-clade combination (Model D; $F = 108.4$,

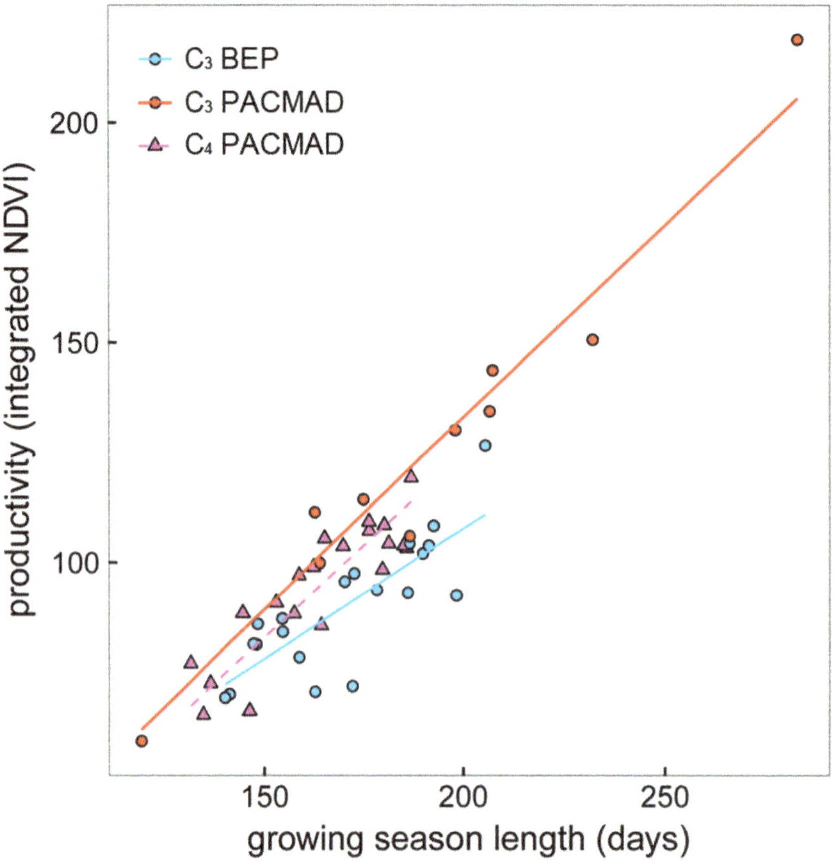

Figure 2. The slope of the relationship between GSL and GSP differed for each photosynthetic type-clade combination (Model D, Table 2). C_3 PACMAD habitats exhibited higher rates of greenness than C_4 PACMAD or C_3 BEP habitats for a given growing season length (slope coefficients = 0.87, 0.83, and 0.59 respectively). Differences within the C_3 functional group were larger than between photosynthetic pathways, i.e., there was a larger clade effect than photosynthetic pathway.

$p<0.01$, df = 45; Figure 2 and Table 2). Productivity increased more per unit of increase in GSL in C_3 PACMAD habitats than other grass habitats. Results show that controlling for GSL, C_3 PACMAD habitats were still more productive than C_4 or C_3 BEP habitats. Interestingly, there was a larger difference within the C_3 functional group than between photosynthetic pathways, i.e., there was a larger clade effect (BEP vs. PACMAD) than photosynthetic pathway effect (C_3 vs. C_4). There was not a significant difference in the GSL of habitats where C_3 BEPs and C_3 PACMADs occur (Figure 1c), but these habitats had the largest difference in GSP. In comparison to C_4 regions, BEP environments had a longer GSL but were less productive (Figure 2), although these differences were not statistically significant. This pattern is consistent with a resource strategy of longer growing seasons at a cost of lower productivity [53,54]. Particularly in resource-poor environments, such as the cold and dry habitats of BEP taxa, species with longer growing seasons are often associated with lower rates of productivity as a strategy to balance energy requirements for tissue growth. BEPs may have traits that have adapted to these marginal environments by adopting a slow and conservative

Table 2. The relationship between growing season length (GSL) and productivity (GSP) (Model D in Table 1) was marginally different between C_3 PACMAD and C_4 grasses (GSL*clade:photosynthetic type), and significantly different between clades irrespective of photosynthetic type (GSL*clade). See Figure 2.

Coefficients	Estimate	Std. Error	t-value	p-value
intercept	−10.80	16.16	−0.67	0.507
GSL	0.59	0.09	6.28	<0.001
clade	−31.11	19.12	−1.63	0.111
GSL*clade	0.28	0.11	2.62	0.012
GSL*clade:photosynthetic type	−0.04	0.02	−2.02	0.049

approach to energy-use. Conversely, plants in high-resource environments tend to produce short-lived leaves with high photosynthetic rates at a cost of being less resistant to environmental stresses and having to produce new leaf tissue [54]. This tradeoff between leaf lifespan and rates of photosynthesis has also been demonstrated in tropical species [55], where the growing season is potentially year-round. Although these strategies may explain C_3 BEP and C_4 differences, C_3 PACMADs did not appear to fit this tradeoff, yet they exhibited greater rates of productivity even when controlling for growing season length.

Notably, life history differences (annual vs. perennial) were not significant in explaining any phenological metric except for GSP ($F = 4.63$, $p < 0.05$, $df = 48$), with perennial species being more productive based on integrated NDVI compared to annual species (posthoc Tukey's $p < 0.05$). However the slope of the relationship between GSL and GSP was not significantly different when considering life history ($p = 0.07$).

Our proxy for productivity (integrated NDVI) only considers aboveground greenness. It is possible that C_4 and C_3 BEP grasses may be less productive aboveground because they are allocating resources to roots, a strategy that would make sense in water- or nutrient-limited environments [56]. A further limitation of the study is the influence of tree cover on the phenology of understory grasses. In our case, although we only used pixels occurring in grassland vegetation classes, C_3 PACMADs preferentially occur in regions with the greater tree cover, which represents a microclimate that is cooler and shadier [37,38]—consequently these samples were excluded and our C_3 PACMADs are represented by species that occur only in open environments.

Conclusions

Our results provide a working hypothesis for understanding C_3 and C_4 grass habitats and growing environments. We show that the clade identity of grasses captures differences in their habitats better than photosynthetic pathway. SOS differences that appear to be associated with the habitats of C_3 and C_4 grasses are in fact associated with habitats of different clades. EOS was not significantly different among any comparisons. GSL and GSP are indeed constrained by photosynthetic pathway; however, the relationships between GSL and GSP were not as expected given the well-recognized habitat sorting of C_3-C_4 photosynthetic pathways. Importantly, the relationship between growing season length and associated productivity differed most strongly between C_3 clade habitats and not between C_3-C_4 habitats. C_3 PACMAD habitats, which in Hawaii represent highly favorable growing environments, had the longest growing season and exhibited the largest variability in all phenological metrics. Taken together, the characteristics of C_3 PACMAD habitats shown here may possibly indicate greater trait diversity within this lineage of globally rare grasses and an advantage in adapting to novel climates [57]. Our results highlight the large functional diversity within Poaceae, one of the largest flowering plant families, and how grass species may respond to future global change. We furthermore identify a need for ecosystem and vegetation models of plant productivity to refine relationships between GSL and GSP. Although our work is limited to grasses and their habitats, in theory, different species or plant functional groups should possess distinct growth strategies resulting in complex relationships between phenology and productivity [58].

Acknowledgments

We thank E.J. Edwards for developing the herbarium dataset and J. Regetz for assistance with the timeseries analysis. We additionally thank the National Center for Ecological Analysis & Synthesis (NCEAS) and Florida State University.

Author Contributions

Conceived and designed the experiments: SP CJS. Analyzed the data: SP. Wrote the paper: SP CJS.

References

1. White M, Running S, Thornton P (1999) The impact of growing-season length variability on carbon assimilation and evapotranspiration over 88 years in the eastern US deciduous forest. Int J Biometeorol 42: 139–145.
2. Richardson AD, Anderson RS, Arain MA, Barr AG, Bohrer G, et al. (2012) Terrestrial biosphere models need better representation of vegetation phenology: results from the North American Carbon Program Site Synthesis. Glob Chang Biol 18: 566–584.
3. Myneni RB, Keeling CD, Tucker CJ, Asrar G, Nemani RR (1997) Increased plant growth in the northern latitudes from 1981 to 1991. Nature 386: 698–701.
4. Baldocchi D, Falge E, Gu L, Olson R, Hollinger D, et al. (2001) FLUXNET: A new tool to study the temporal and spatial variability of ecosystem-scale carbon dioxide, water vapor, and energy flux densities. Bull Amer Meteor Soc 82: 2415–2434.
5. Churkina G, Schimel D, Braswell BH, Xiao A (2005) Spatial analysis of growing season length control over net ecosystem exchange. Glob Chang Biol: 1777–1787.
6. Piao S, Friedlingstein P, Ciais P, Viovy N, Demarty J (2007) Growing season extension and its impact on terrestrial carbon cycle in the Northern Hemisphere over the past 2 decades. Global Biogeochem Cycles 21: GB3018.
7. Dunn AL, Barford CC, Wofsy SC, Goulden ML, Daube B (2007) A long-term record of carbon exchange in aboreal black spruce forest: means, responses to interannual variability, and decadal trends. Glob Chang Biol 13: 577–590.
8. Hu J, Moore DJP, Monson RK (2010) Longer growing seasons lead to less carbon sequestration by a subalpine forest. Glob Chang Biol 16: 771–783.
9. Saugier B, Roy J, Mooney HA (2001) Estimations of global terrestrial productivity: Converging on a single number? In: Roy J, Saugier B, Mooney HA, editors. Terrestrial Global Productivity. San Diego: Academic Press. pp. 543–557.
10. Still CJ, Berry JA, Collatz GJ, DeFries RS (2003) Global distribution of C 3 and C 4 vegetation: Carbon cycle implications. Global Biogeochem Cycles 17.
11. Ehleringer JR, Cerling TE, Helliker BR (1997) C4 photosynthesis, atmospheric CO2, and climate. Oecologia 112: 285–299.
12. Sage RF, Monson RK (1999) C4 Plant Biology. San Diego: Academic Press.
13. Newell C (1968) A phytosociological study of the major vegetation types in Hawai'i Volcanoes National Park, Hawai'i University of Hawai'i.
14. Chazdon RL (1978) Ecological Aspects of the Distribution of C4 Grasses in Selected Habitats of Costa Rica. Biotropica 10: 265–269.
15. Tieszen LL, Senyimba MM, Imbamba SK, Troughton JH (1979) The distribution of C_3 and C_4 grasses and carbon isotope discrimination along an altitudinal and moisture gradient in Kenya. Oecologia 37:337–350.
16. Rundel PW (1980) The ecological distribution of C4 and C3 grasses in the Hawaiian Islands. Oecologia 45: 354–359.
17. Cabido M, Pons E, Cantero JJ, Lewis JP, Anton A (2007) Photosynthetic pathway variation among C 4 grasses along a precipitation gradient in Argentina. Journal of Biogeography 24: 197–204.
18. Cavagnaro JB (1988) Distribution of C3 and C4 grasses at different altitudes in a temperate arid region of Argentina. Oecologia 76: 273–277.
19. Angelo CL, Daehler CC (2013) Upward expansion of fire-adapted grasses along a warming tropical elevation gradient. Ecography 36: 551–559.
20. Fargione J, Tilman D (2005) Niche differences in phenology and rooting depth promote coexistence with a dominant C4 bunchgrass. Oecologia 143: 598–606.
21. Ode DJ, Tieszen LL, Lerman JC (1980) The seasonal contribution of C3 and C4 plant species to primary production in a mized prairie. Ecology 61: 1304–1311.
22. Kemp PR, Williams GJ III (1980) A physiological basis for niche separation between agropyron Smithii (C3) and Bouteloua gracilis (C4). Ecology 61: 846–858.
23. Still CJ, Berry J a, Ribas-Carbo M, Helliker BR (2003) The contribution of C3 and C4 plants to the carbon cycle of a tallgrass prairie: an isotopic approach. Oecologia 136: 347–359.
24. Epstein HE, Lauenroth WK, Burke IC, Coffin DP (1997) Productivity patterns of C3 and C4 functional types in the U.S. Great Plains. Ecology 78: 722–731.
25. Tieszen LL, Reed B, Bliss NB, Wylie B, DeJong D (1997) NDVI, C_3 and C_4 production and distribution in Great Plains grassland land cover classes. Ecol Appl 7: 59–78.
26. Goodin DG, Henebry GM (1997) Monitoring ecological disturbance in tallgrass prairie using seasonal NDVI trajectories and a discriminant function mixture model. Remote Sens Environ 61: 270–278.

27. Davidson A, Csillag F (2003) A comparison of three approaches for predicting C4 species cover of northern mixed grass prairie. Remote Sens Environ 86: 70–82.
28. Foody GM, Dash J (2010) Estimating the relative abundance of C3 and C4 grasses in the Great Plains from multi-temporal MTCI data: issues of compositing period and spatial generalizability. Int J Remote Sens 31: 351–362.
29. Wang C, Jamison B, Spicci A (2010) Trajectory-based warm season grassland mapping in Missouri prairies with multi-temporal ASTER imagery. Remote Sens Environ: 531–539.
30. Grass Phylogeny Working Group II (2012) New grass phylogeny resolves deep evolutionary relationships and discovers C4 origins. New Phytol 193: 304–312.
31. Vicentini A, Barber JC, Aliscioni SS, Giussani LM, Kellogg E a (2008) The age of the grasses and clusters of origins of C 4 photosynthesis. Glob Chang Biol 14: 2963–2977.
32. Edwards EJ, Still CJ, Donoghue MJ (2007) The relevance of phylogeny to studies of global change. Trends Ecol Evol 22: 243–249.
33. Edwards EJ, Still CJ (2008) Climate, phylogeny and the ecological distribution of C4 grasses. Ecol Lett 11: 266–276.
34. Osborne CP, Freckleton RP (2009) Ecological selection pressures for C4 photosynthesis in the grasses. Proc Biol Sci 276: 1753–1760.
35. Edwards EJ, Smith S a (2010) Phylogenetic analyses reveal the shady history of C4 grasses. Proc Natl Acad Sci U S A 107: 2532–2537.
36. Taylor SH, Hulme SP, Rees M, Ripley BS, Woodward FI, et al. (2010) Ecophysiological traits in C3 and C4 grasses: a phylogenetically controlled screening experiment. New Phytol 185: 780–791.
37. Pau S, Edwards EJ, Still CJ (2013) Improving our understanding of environmental controls on the distribution of C3 and C4 grasses. Glob Chang Biol 19: 184–196.
38. Still CJ, Pau S, Edwards EJ (2014) Land surface skin temperature captures thermal environments of C3 and C4 grasses. Glob Ecol Biogeogr 23: 286–296.
39. Wagner WL, Herbst DR, Sohmer S (1999) Manual of the Flowering Plants of Hawai'i. Revised. Honolulu: University of Hawai'i Press.
40. Sims DA, Rahman AF, Cordova VD, El-Masri BZ, Baldocchi DD, et al. (2006) On the use of MODIS EVI to assess gross primary productivity of North American ecosystems. J Geophys Res 111: 1–16.
41. Zhang X, Friedl MA, Schaaf CB, Strahler AH, Hodges JCF, et al. (2003) Monitoring vegetation phenology using MODIS. Remote Sens Environ 84: 471–475.
42. Burnham KP, Anderson DR (2010) Model selection and multimodel inference: a practical information-theoretic approach. Second Edi. New York: Springer.
43. Giambelluca TW, Chen Q, Frazier AG, Price JP, Chen Y-L, et al. (2013) Online Rainfall Atlas of Hawai'i. Bull Am Meteorol Soc 94: 313–316.
44. Reed BC, Brown JF, Vanderzee D, Loveland TR, Merchant W, et al. (2009) Measuring phenological variability from satellite imagery. Inf Syst 5: 703–714.
45. Foody G, Dash J (2007) Discriminating and mapping the C3 and C4 composition of grasslands in the northern Great Plains, USA. Ecol Inform 2: 89–93.
46. Powell RL, Yoo E-H, Still CJ (2012) Vegetation and soil carbon-13 isoscapes for South America: integrating remote sensing and ecosystem isotope measurements. Ecosphere 3: art109.
47. Pearcy RW, Ehleringer JR (1984) Comparative ecophysiology of C3 and C4. plants. Plant Cell Environ 7: 1–13.
48. Menzel A, Sparks TH, Estrella N, Koch E, Aasa A, et al. (2006) European phenological response to climate change matches the warming pattern. Glob Chang Biol 12: 1969–1976.
49. Menzel A, Fabian P (1999) Growing season extended in Europe. Nature 397: 659.
50. Zhou L, Tucker CJ, Kaufmann RK, Slayback D, Shabanov NV, et al. (2001) Variations in northern vegetation activity inferred from satellite data of vegetation index during 1981 to 1999. J Geophys Res 106: 20069–290083.
51. Taylor G, Tallis MJ, Giardina CP, Percy KE, Miglietta F, et al. (2007) Future atmospheric CO2 leads to delayed autumnal senescence. Glob Chang Biol 14: 264–275.
52. Jeong SJ, Ho CH, Gim HJ, Brown ME (2011) Phenology shifts at start vs. end of growing season in temperate vegetation over the Northern Hemisphere for the period 1982–2008. Glob Chang Biol 17: 2385–2399.
53. Mooney HA (1972) The carbon balance of plants. Annu Rev Ecol Syst 3: 315–346.
54. Chapin FS (1980) The mineral nutrition of wild plants. Annu Rev Ecol Syst 11: 233–260.
55. Reich PB, Uhl C, Waiters MB, Ellsworth DS (1991) Leaf lifespan as a determinant of leaf structure and function among 23 amazonian tree species. Oecologia: 16–24.
56. Nippert JB, Wieme R a, Ocheltree TW, Craine JM (2012) Root characteristics of C4 grasses limit reliance on deep soil water in tallgrass prairie. Plant Soil 355: 385–394.
57. Holt RD (2009) Bringing the Hutchinsonian niche into the 21st century: Ecological and evolutionary perspectives. Proceedings of the National Academy of Sciences 106: 19659–19665.
58. Reich PB (2014) The world-wide "fast-slow" plant economics spectrum: a traits manifesto. Journal of Ecology 102: 275–301.

3

Spatial Scale, Means and Gradients of Hydrographic Variables Define Pelagic Seascapes of Bluefin and Bullet Tuna Spawning Distribution

Diego Alvarez-Berastegui[1]*, Lorenzo Ciannelli[2], Alberto Aparicio-Gonzalez[3], Patricia Reglero[3], Manuel Hidalgo[3], Jose Luis López-Jurado[3], Joaquín Tintoré[1,4], Francisco Alemany[3]

1 Balearic Islands Coastal Observing and Forecasting System (SOCIB), Palma de Mallorca, Balearic Islands, Spain, **2** College of Earth, Ocean, and Atmospheric Sciences (CEOAS), Oregon State University, Corvallis, Oregon, United States of America, **3** Instituto Español de Oceanografía - Centre Oceanogràfic de les Balears (COB-IEO), Palma de Mallorca, Balearic Islands, Spain, **4** Institute for Advanced Studies (IMEDEA), Consejo Superior de Investigaciones Científicas (CSIC) and the University of the Balearic Islands (UIB). Esporles, Spain

Abstract

Seascape ecology is an emerging discipline focused on understanding how features of the marine habitat influence the spatial distribution of marine species. However, there is still a gap in the development of concepts and techniques for its application in the marine pelagic realm, where there are no clear boundaries delimitating habitats. Here we demonstrate that pelagic seascape metrics defined as a combination of hydrographic variables and their spatial gradients calculated at an appropriate spatial scale, improve our ability to model pelagic fish distribution. We apply the analysis to study the spawning locations of two tuna species: Atlantic bluefin and bullet tuna. These two species represent a gradient in life history strategies. Bluefin tuna has a large body size and is a long-distant migrant, while bullet tuna has a small body size and lives year-round in coastal waters within the Mediterranean Sea. The results show that the models performance incorporating the proposed seascape metrics increases significantly when compared with models that do not consider these metrics. This improvement is more important for Atlantic bluefin, whose spawning ecology is dependent on the local oceanographic scenario, than it is for bullet tuna, which is less influenced by the hydrographic conditions. Our study advances our understanding of how species perceive their habitat and confirms that the spatial scale at which the seascape metrics provide information is related to the spawning ecology and life history strategy of each species.

Editor: Athanassios C. Tsikliras, Aristotle University of Thessaloniki, Greece

Funding: The development of this study was supported by the Project "PERSEUS" financed by the Seventh Framework Program for Research (FP7) under theme "Oceans Tomorrow", grant Agreement No. 287600 (http://www.perseus-fp7.eu/), the Spanish Foundation for Science and Technology, under the project "BALEARES" Grant number: CTM2009-07944/MAR (http://www.fecyt.es/fecyt/home.do) and the Science, Engineering and Education for Sustainability NSF-Wide Investment (SEES) Research Coordination Network, Grant 1140207 "Sustainability of Marine Renewable Resources in Subarctic Systems Under Incumbent Environmental Variability and Human Exploitation" (http://grantome.com/grant/NSF/OCE-1140207). Hydrographic data used in the study were provided by IBAMAR (http://www.ba.ieo.es/ibamar) a public data base founded by the Council for Economy and Competitiveness of the Balearic Islands Government (http://www.caib.es/govern/organigrama/area.do?coduo = 6&lang = es), grant number AAEE0022/08. The research was developed under the framework of the BLUEFIN TUNA Project driven by the Balearic Island Observing and Forecasting System (www.socib.es) and the Spanish Institute of Oceanography (www.ieo.es). The funders had no role in study design, data collection and analysis, decision to publish, or preparation of the manuscript.

Competing Interests: The authors have declared that no competing interests exist.

* Email: dalvarez@socib.es

Introduction

Seascape ecology represents an emerging field in the study of how the habitat structure shapes the spatial distribution of marine species and influences key ecological processes [1],[2]. This discipline initiated applying techniques and metrics from the traditional landscape ecology to characterize and quantify spatial structure of benthic habitats, observed as a mosaic of patches of different habitat classes [1],[3],[4], [5],[6]. Nevertheless, there is still a gap in the development of concepts and techniques providing metrics to characterize the spatial structure of the seascape in the pelagic environments, where there are no clear boundaries delimitating different habitats [1],[2],[6]. In the framework of landscape ecology spatial gradients have been recently proposed as more appropriated metric than traditional categorical patch mosaic based metrics to characterize continuous habitats [7]. Accordingly, a location in a pelagic seascape would be better characterized by the combination of the value of a particular hydrographic variable and its spatial gradient.

Several studies have applied gradients of hydrographic parameters to characterize the spatial distribution of marine species during various life history stages, as nursery, foraging or spawning [8],[9],[10],[11],[12]. It is likely that the scale at which an individual perceive a change in the environment (i.e., a gradient) varies according to life history, ontogeny and to the hydrographic variable in exam. For instance, while large-scale gradients associated with the North Pacific transition zone drive the location of many pelagic predators including albacore tuna (*Thunnus alalunga*) during their feeding migratory stages [13], once off the west coast of the US, albacore tuna distribution is associated with smaller scale features linked to upwelling fronts [14]. In the Mediterranean Sea during spawning, bluefin tuna distribution is regulated by oceanographic variables that can change at relatively small scales [15],[16]. In spite of the expected importance of gradient scales, to our knowledge there are no studies that have evaluated the effect of changing the spatial scales at which environmental gradients are calculated to model the spatial distribution of fish. The goal of our study is to examine the distribution of large pelagic predators during spawning by explicitly considering mean, gradients and scale of gradients of hydrographic variables.

Atlantic bluefin tuna (*Thunnus thynnus*) and bullet tuna (*Auxis rochei rochei*) are two species of pelagic predators showing different spawning strategies. We target these species in the Balearic Sea (Figure 1), known as a recurrent spawning area for large pelagic species in the Western Mediterranean [17]. Bluefin tuna is a highly migratory species; the Eastern population enters in the Mediterranean Sea from the North Atlantic at the end of spring and early summer [18],[19]. Their spawning activity at the Balearic Sea is linked to the regional oceanography with spawning grounds located in the vicinity of frontal structures formed when the recent Atlantic water mass encounters the more saline resident surface Atlantic waters [15]. The area is characterized by highly dynamic processes that promote a seascape shaped by filaments, fronts and eddies whose location varies between years [20],[21]. Bullet tuna, by contrast, is smaller and more frequent in near coastal areas [22]. The spawning of bullet tuna is associated with the geography. Young larvae are found mainly in coastal areas and with little influence of the local oceanography in comparison with bluefin [16].

We expect that, when selecting spawning locations, a large-bodied and long-distance migratory pelagic fish, such as the bluefin tuna explores its environment at larger spatial scales than bullet tuna, a small-bodied and non-migratory pelagic fish. Therefore, we expect that pelagic seascape metrics based on the combination of hydrographic parameter values and their gradients calculated at appropriate spatial scales provide relevant information for bluefin tuna but not for bullet tuna, where we expect a greater reliance on geographic and hydrographic parameters, calculated at comparatively small-scale.

In this work we analyze the influence of the pelagic environment by depicting the spatial scales at which gradients of hydrographic variables are linked to the spawning ecology of these tuna species. We investigate the two most relevant hydrographic variables describing their spawning spatial distribution: salinity and geostrophic currents velocity [16], already determined in previous studies. Our analytical approach has two steps. Firstly, we identify the scale at which each hydrographic variable (Figure 2) maximizes the performance of a model fitted to larval distribution. Second, we evaluate whether components of the seascape (i.e., mean and gradients of oceanographic variables) are interactively affecting the spatial distribution of tuna larvae. By performing this analysis on two species we evaluate how fish with contrasting life history strategies perceive their environments when deciding for spawning locations.

Materials and Methods

Data acquisition

Bluefin and bullet tuna larvae were collected during ichthyoplankton surveys using Bongo nets from 2001 to 2005. The surveys were conducted by the Instituto Español de Oceanografía (www.ieo.es), a Spanish Government marine research institution. The sampling scheme was communicated and approved by the Spanish Directorate of Fisheries before the sampling was conducted. No specific ethical approval was required and the survey of biological data was conducted using Bongo nets,which are accepted standard

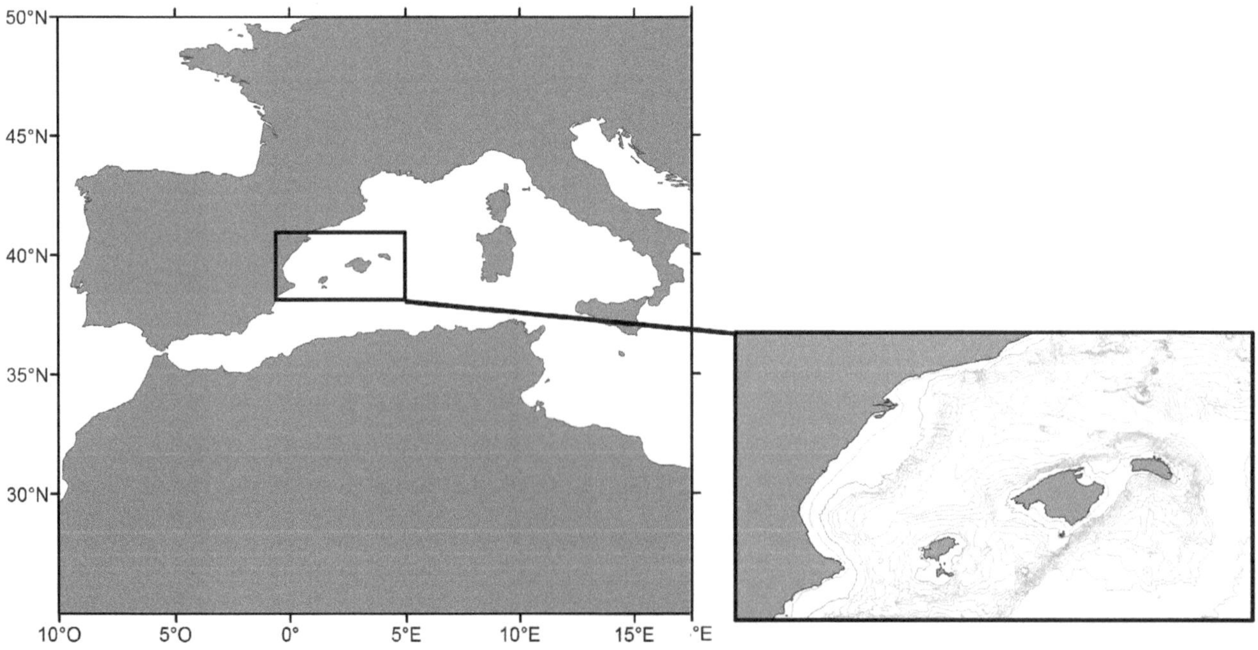

Figure 1. Location of the Balearic Islands, Western Mediterranean.

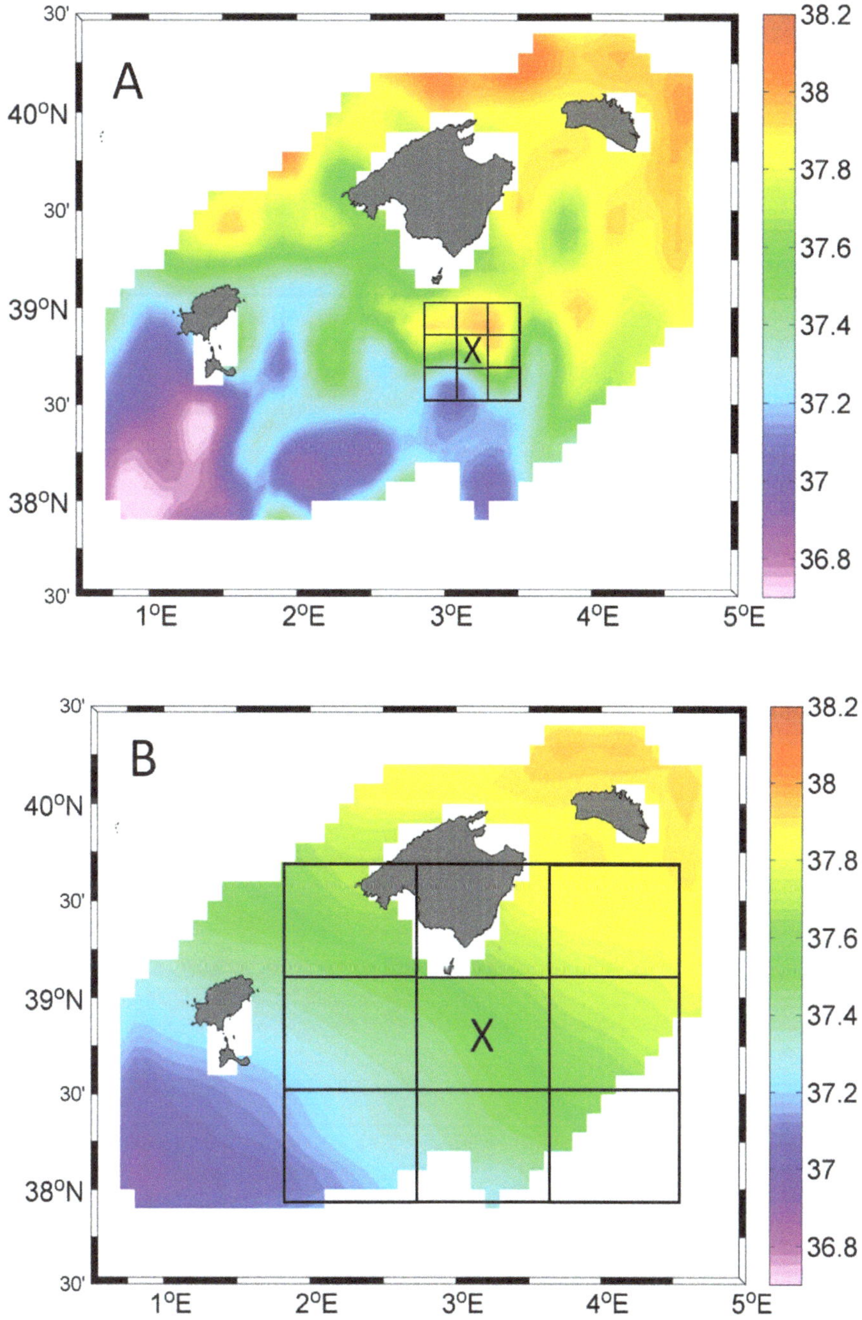

Figure 2. Sea surface salinity field in 2003. Spatial means of sea surface salinity processed at two different scales A) 0.15 degrees and B) 0.75 degrees. Spatial means were interpolated following an objective analysis onto a regular grid by using minimum error variance methods (Bretherton, 1976). The squares in each figure show the polygons used for the calculation of the spatial gradients at the two scales at station X.

techniques for this type of surveys, used worldwide for the collection of plankton samples, including billfish and tuna larvae [15],[17], [23], [24]. The nets were towed at low speeds, around 2 knots, during 8–10 minutes, and plankton samples were immediately preserved with 4% formalin buffered with borax onboard.

Around 200 stations were sampled every year, in a regular sampling grid of 10×10 nm located between 37.85°–40.35° N and 0.77°–4.91°E, covering a total area of 101360 km2 (= 280×362 km) around the Balearic archipelago. The sampling was conducted during June–July coinciding with the spawning period of bullet and bluefin tuna (see [15] for details of the sampling procedures). Tuna larvae were identified to the species level and measured in standard length. All larvae identified as yolk sac and preflexion stages (<4.5 mm) were classified as "stage 1".

Mean and maximum age of larvae under this size are 6 days and 11 days old respectively, accounting for growth [25] and hatching time [26]. Displacement from spawning areas during these intervals in the study area are below 25 kilometers for the mean ages, which are in the range of 1.4 sampling stations, and 46 kilometers for the maximum ages, what is in the range of 2.6 sampling stations. These values have been calculated following methods in [27]. Considering these values, abundances of stage 1 larvae have been defined to get a proxy of spawning locations as in previous research in the study area [28].

Vertical profiles of conductivity, temperature and pressure data were recorded at all stations, by means of Sbe911 CTD. Sea surface salinity at each station (SAL) was calculated as the mean salinity over the mixed layer depth. Geostrophic velocities (GVEL) were calculated at sea surface from the first-derivative of the sea surface height between adjacent points, which was obtained by vertical integration of the specific volume, using 600 m as the level of no motion [21].

These two variables (SAL and GVEL) were selected, since they have been demonstrated to be the two most relevant environmental variables describing the spawning spatial distribution of tuna [16]. Sea surface temperature was also included in the models since in this area it is a secondary but relevant variable mainly related to the phenology of the spawning process [29]. However, the spatial gradient was not explored because sea surface temperature during the summer changes relatively fast due to solar irradiance [21].

Processing of spatial gradients along continuous spatial scales

Spatial gradients from the sea surface salinity field (gSAL) and geostrophic velocity field (gGVEL) within the sampled region were calculated at six spatial scales, from 0.15° to 0.90° with a spatial increment of 0.15°. The minimum (0.15°) and the maximum scale (0.90°) were in the range of the smallest (from 0.13° to 0.27°) and largest (up to 0.92°) mesoscale oceanographic structures in the area [21]. For the computation of the gradients, nine square polygons at every scale were delimitated around each sampling position (see examples for scale 0.15° and 0.75° in figure 2A, 2B respectively). The gradient was then computed as the maximum absolute difference between the mean hydrographic variable at the center polygon and each of the eight surrounding polygons standardized to distance [9]. The software for the spatial processing was developed in R language [30].

Identification of spatial scales

Comparison of how models perform along scales allowed identifying the spatial scales at which information provided by gradients is maximized. The effect of gGVEL and gSAL at each scale on the abundance of bullet and bluefin tuna larvae was assessed using nonparametric regression statistical models (generalized additive models, GAMs, [31]). A base model was formulated to describe inter-annual variability (variable YEAR), sampling location (latitude and longitude variables), and the hour of the day on the catch of tuna larvae. Over-dispersed Poisson distribution family and a natural-log link were selected to model larval data. The volume of water filtered was included as an offset (after natural log transformation), to account for the effort expanded in catching the sample (Equation 1). The effects of these variables on the base model have been already analyzed in previous studies [16]. Here, the base model represented the null hypothesis of no gradient effect on tuna larvae distribution, against which all other more complex formulations will be compared.

Equation 1: Base model

$$\begin{aligned}&\textit{Larvae abundance} = \textit{offset}(\textit{log}(m3)) + \textit{factor}(\textit{year})\\&+ sm_1(\textit{long}, \textit{lat}) + sm_2(\textit{hour})\end{aligned}$$

m3 = volume filtered by the bongo nets (m3); long = longitude; lat = latitude; hour = hour of the day expressed from 0 to 1, sm_1 and sm_2 the smoothing functions.

At each spatial scale a GAM model was processed including the gradient of one hydrographic variable (gSAL, gGVEL) as a new additive term (s3) in the standardization model. The number of knots for the new smoother was always set to a maximum of three (i.e. two degrees of freedom) in order to avoid over fitting in the responses.

The identification of characteristic spatial scales (cgSAL, cgGVEL) was assessed with scalogram where the scale of the covariate is plotted against a measure of the model goodness of fit, which in our case were represented by the adjusted R-squared (Rsq, the higher the better), and the Generalized Cross Validation (GCV, the lower the better) [31]. We selected the scale that maximize Rsq and minimize the GCV. Results of the base model (when a seascape covariate was not included) were presented in the same graphics. Note that due to the greater complexity of the gradient model higher Rsq values do not necessarily imply an improvement in relation to the base model, while they do represent a better performance when compared to other gradient models.

Significant differences of Rsq values between models, or GCVs, were obtained from t-test of these parameters obtained from 500 iterations where 10% of the data was excluded. For all cases, alternative hypothesis (difference in means is not equal to 0) was accepted only if the P value was lower than 0.001, with a confidence level of 0.99. When one variable presented similar Rsq and GCV values at various scales, selection was assessed by inspection of the plot showing the response of the abundance in relation to the gradient processed at those scales.

Once the characteristic scale of the gradients was identified, we tackled the questions of whether the information provided by the gradients is different and complementary to the information provided by the hydrographic variables from which they were calculated, and in that case, how the information from these two variables (spatial mean and gradient) should be combined to maximize the goodness of fit of the models and the ecological information they provide. To assess these questions we analyzed the performance of models with different complexity:

i) The base model from equation 1.
ii) Hydrographic models combining the sea surface salinity, geostrophic velocity and sea surface temperature at the sampling station (stSAL, stGVEL, stSST).
iii) Seascape models combining the gradients at characteristic scales (cgSAL, cgGVEL) and the hydrographic variable at the sampling location (stSAL, stGVEL, stSST). Different seascape models were constructed including the two components of the seascape (values at stations and gradients) as additive and interactive terms. An interaction may be ecologically meaningful when a species is selecting its spawning habitat on a specific side of a frontal region, for example. In such case, it is the combination of both the gradient and the mean that provide the suitable conditions for spawning. The performance of different model configurations for each species was assessed by the delta AIC (ΔAIC), calculated as the difference between model AICs and the base model AIC. The AIC in this case is best suited for model comparisons because each model had different number of variables [32]. Rsq, GCV and explained deviances were used to compare how models perform between the two species, as AIC values among models with different dependent variables are not comparable.

Results

Identification of characteristic spatial scales

In all years considered, the recent Atlantic water masses encountered the more saline resident water masses forming an oceanic frontal zone inside the study area. The size of such frontal zone was bigger than other oceanographic phenomena as small eddies and meanders derived from the instabilities along the haline front and the effect of strong bathymetric changes (Figures S3, S4).

The scalogram of gGVEL for bluefin showed that Rsq values gradually improve as the spatial scales increased to a maximum at 0.6° (Rsq = 0.44, Figure 3A), which was chosen as the geostrophic velocity gradient characteristic scale for bluefin tuna. Values of GCV showed a similar pattern of model improvement, being significantly better than the base model at 0.6 degrees (Figure 3B). At this characteristic scale the response of the larvae abundance is positively related to the gradient of geostrophic velocity (Figure 3C).

The scalograms of gSAL for bluefin tuna showed also an increment of Rsq with higher values at 0.6° and 0.75° that also coincide with lower values of GCV (Figure 3D, 3E). Differences of R-sq between these two scales (0.6° and 0.75°) were not significant. The characteristic scale for gSAL was chosen at 0.6° as the model response at this scale presented a less ambiguous effect on larval abundance (Figure 3F). The gSAL at 0.75° spatial gradients displays a dome-shaped response with a less clear ecological interpretation (Figure S1). At this scale GCV was lower than the base model.

In contrast to the results obtained for salinity, the gradients of geostrophic velocities (gGVEL) did not show any single scale that maximizes R-square and minimizes GCV for bullet tuna (Figure 4A, 4B). The Rsq scalogram showed a flat trend with the highest value at 0.15 degrees. The Rsq value at this scale (= 0.18) showed similar values than other scales (values between 0.170 and 0.173) or when compared to the base model (= 0.166). On the contrary the GCV scalogram showed significantly lower values than the base model at 0.45 and 0.6 degrees, scales at which Rsq were not even significantly higher than the base model. Therefore, the contradictory response of the model performance indicators, the flat trend of Rsq scalogram and their very low values (despite the higher complexity of the gradient models in relation to the base model), may indicate that the spatial gradient of geostrophic velocity is not a valid seascape metric for the spawning locations of bullet tuna. Consequently, gGVEL was excluded from further analysis in relation to this species.

The gSAL scalogram of bullet tuna showed a moderate effect of the scale at which gradients were calculated. The shape of the scalogram did not show a peak scale at which model performs better (Figure 4C). Scales from 0.45°, 0.6° and 0.75° for gSAL seemed to maximize Rsq, but not being different from each other. In this case GCV-scalograms showed the lowest at 0.75° (Figure 4D), being significantly lower than the base model, which

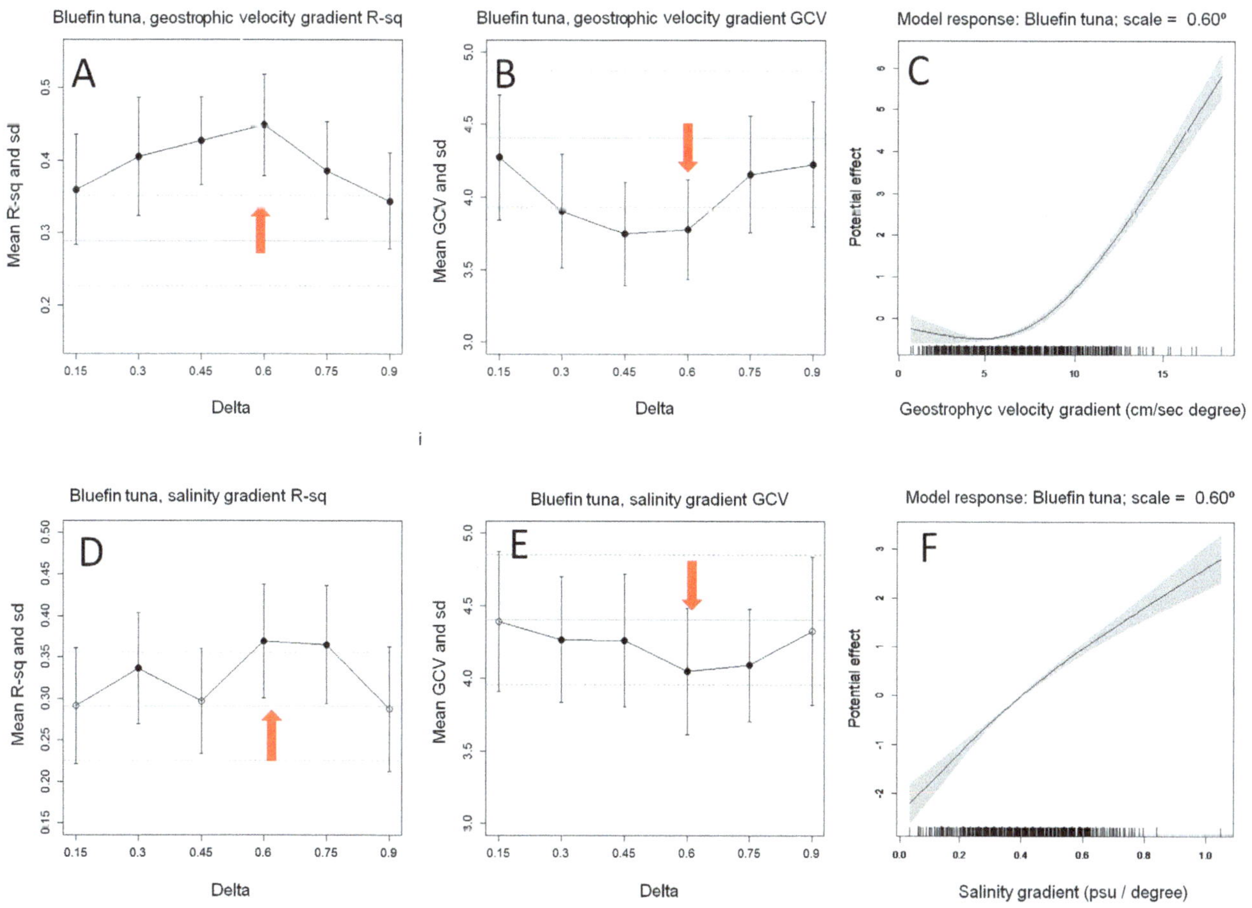

Figure 3. Rsq and GCV Scalograms of bluefin tuna larva abundance models along spatial scales, standard deviations. Horizontal grey lines indicate statistics from the base model (Straight line = mean, dashed line = Sd). Black dots show scales at which values are significantly different from the standardization model. Red arrows indicate the selected characteristic scale.

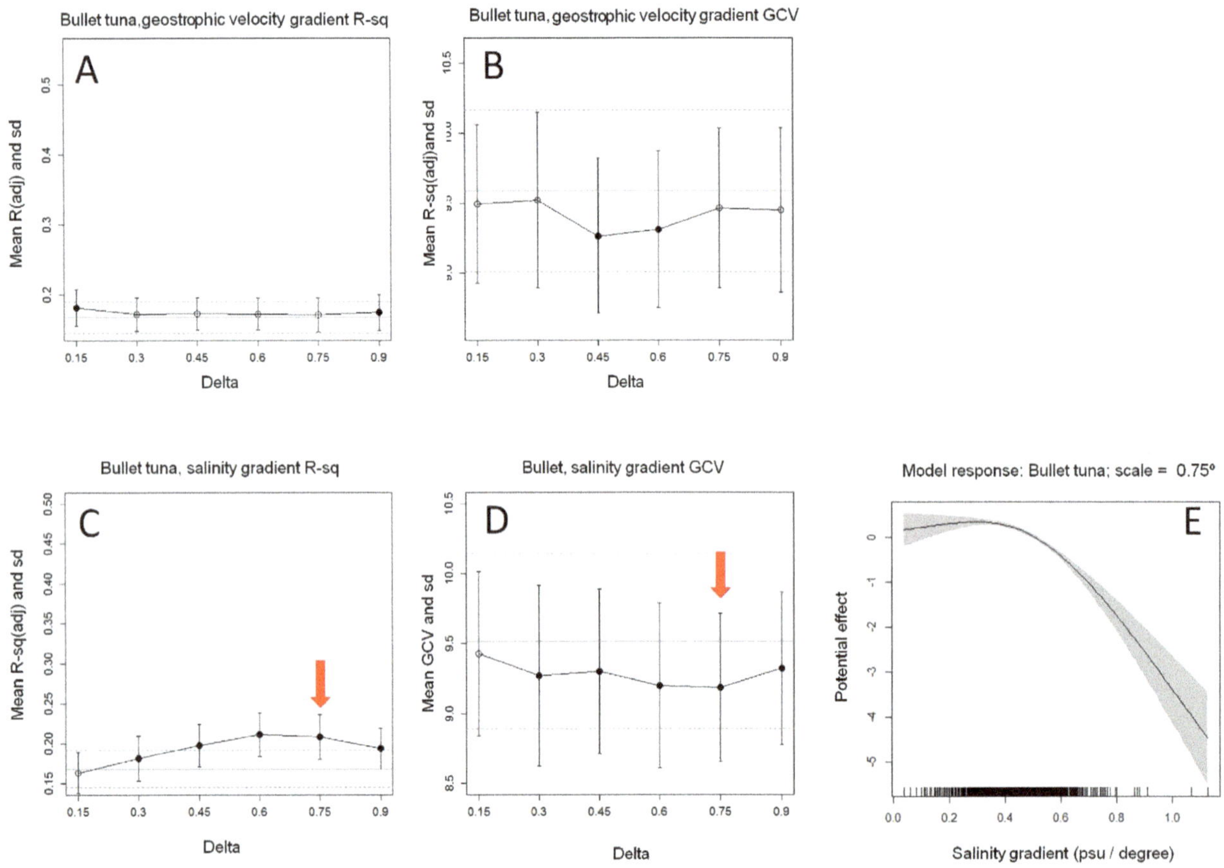

Figure 4. Rsq and GCV scalograms of bullet tuna larva abundance models along spatial scales, standard deviations. Horizontal grey lines indicate statistics from the standardization model (Strait line = mean, dashed line = Sd). Black dots show scales at which values are significantly different from the standardization model. Red arrows indicate selected characteristic scale.

was selected as characteristic scale. The Rsq at this scale was higher (= 0.21) than that of the base model (Rsq = 0.16). At this scale, gradients displayed a negative effect on the bullet tuna larvae abundance (Figure 4E) showing that bullet tuna spawning locations are found with more probability in areas where salinity is spatially homogeneous– a result that contrasted to that obtained for bluefin tuna.

Species-specific seascape characterization

The best model for each species included a gradient and a mean term (Tables 1 and 2). Note however that the hydrographic model already represents an improvement respect to the standardization model (Tables 1 and 2). For bluefin tuna the best seascape model had an improvement of 186% of the Rsq when compared to the base model (Rsq base mode = 0.23; Rsq best seascape model = 0.66, Table 1). The improvement for bullet was 61%, considerably lower compared to bluefin (Rsq base mode = 0.16; Rsq best seascape model = 0.25, Table 2).

Pearson correlation coefficients between hydrographical variables and their gradients at characteristic scales were in all cases below 0.50 and pair plots showed no clear tendencies on the correlations (Figure S2), indicating that selected gradients provided complementary information to that of the hydrographical variable. Models showed a better performance (i.e. lower GCV and higher ΔAIC; Tables 1 and 2) in all the cases when the gradient and the hydrographic value were considered as an interaction term, suggesting dependence in their effect on larvae abundance rather than an additive response. However, larvae abundance of each species responded differently to the interaction of seascape components (Figures 5A, 5B, 5C).

For Bluefin tuna, higher probability of spawning was associated to higher values of geostrophic velocity gradients, but where velocities at station may present either high or low values (Figure 5A), (Figures S7, S8). Considering that a gradient is characterized by an area with high current speed near an area of low current speed, this result indicates that spawning locations were not associated to a particular side of the gradient, but in an area around the location where maximum velocities occurs. The extension of this area would be around a circle of 0.6 degrees of radius (aprox. 65 km in the study area), the characteristic scale at which the gradients were more relevant. In contrast, the interaction of the salinity seascape variables showed high larvae abundances in areas with high salinity gradients and intermediate-high salinity levels (Figure 5B), indicating an effect of the location of the main haline front and a preference for spawning at the high salinity values of that front (Figures S5, S6). The characteristic spatial correlation scale of local oceanographic structures in the area is around 18 nmi (aprox. 0.15 meridian degrees) [21] and therefore surface oceanographic structures at smaller spatial scales are ephemeral. The oceanographic structures at larger spatial scales, relevant for bluefin tuna, are linked to the Med-Atlantic salinity front and last longer.

The functional form of the interaction terms was different for bullet tuna. The relation between the bullet tuna larvae

Table 1. Summary of GAM models of larvae abundance for Atlantic bluefin tuna (*Thunnus thynnus*).

Model group	Model variables	R2	Dev(%)	GCV	AIC	delta AIC
Base model	(latitude, longitude) + filtered volume + hour	0.232	40.8	4,596	3985,54	0
One additive variable models	base model + stGVEL0.15	0,271	43,4	4,412	3827,79	157,75
	base model + gGVEL0.6	0,39	49,4	3,947	3465,03	520,50
	base model + stSAL0.15	0,222	41,8	4,534	3924,70	60,84
	base model + gSAL0.6	0,301	45,1	4,275	3723,53	262,01
Hydrographic model	base model + stGVEL + stSAL + stTEMP	0,472	51,6	3,814	3338,62	646,92
GVEL seascape models	Hydrographic model + stGVEL + gGVEL0.6	0,53	55,8	3,500	3087,14	898,40
	Hydrographic model + (stGVEL,gGVEL0.6)	0,666	59,3	3,251	2881,29	1104,25
SAL seascape models	Hydrographic model + stSAL + gSAL0.6	0,506	54,8	3,571	3145,80	839,74
	Hydrographic model + (stSAL,gSAL0.6)	0,533	57,1	3,417	3009,58	975,96

Interaction terms included in parenthesis.

abundance and the interaction of spatial distribution of sea surface salinity and its gradients is presented in Figure 5C. Spawning locations were associated to areas where the salinity at the station were lower and gradients presented intermediate values, but the interaction plot revealed that spawning also appears in areas of higher salinities associated to very low gradients. Areas defined by this twofold combination were located at both sides of the front avoiding more mixed waters. This spatial distribution was more evident in years 2001, 2003 and 2005 (Figures S9, S10)

Discussion

We have found that the combination of sea surface current velocities, salinities and their gradients calculated at characteristic spatial scales are relevant for the parameterization of the pelagic seascape affecting a key ecological process of bluefin tuna. For bullet tuna only salinity and their gradient provided a valid seascape metric not being relevant the gradients of sea surface current velocities. In agreement with our expectations, the importance of these metrics was much higher for bluefin, a large-bodied, long distance migratory and more dependent on local oceanography than were for bullet tuna a smaller coastal species with shorter migration distance.

Previous studies have documented the links between the frontal activity and the spawning of bluefin tuna [15],[16],[33]. In this study we add to these results by examining the effect of gradients and their interactions with hydrographic mean. These metrics improved our understanding of the conditions for bluefin and bullet tuna spawning when compared to models using just the hydrographical values but not the gradients. Furthermore the identification of characteristic scales of gradients provided a new source of information for the interpretation of how local oceanography determines the selection of the site to spawn.

For bluefin tuna larvae, spatial salinity and geostrophic velocity gradients maximize spatial model performance when calculated at 0.6 degrees. The higher abundance of bluefin tuna larvae in areas with intermediate to high salinities and with high gradients of velocity is consistent among years. Higher abundance occurs around the location of the main frontal area, at the side of higher salinity of the front and where current speed presents high values. Higher salinity water likely has higher resident time near the islands than the less saline water, which may run along the front towards east getting farther from the archipelago. Spawning at the higher salinity side may favor spatial overlap with other larval species that are also located in this water mass [17],[34].

Results for bullet tuna showed that pelagic seascape metrics are not as relevant to explain the spawning distribution as they are for bluefin. In the western Mediterranean, bullet tuna spawning has been associated to near coastal areas [22], being less influenced by the local oceanography than bluefin tuna [16], which is consistent with our results.

Table 2. Summary of GAM models of larvae abundance for bullet tuna (*Auxis rochei rochei*). Interaction terms included in parenthesis.

Model group	Model variables	R2	Dev(%)	GCV	AIC	delta AIC
Base model	(latitude, longitude) + filtered volume + hour	0,158	32,8	9,615	8857,17	0
One additive variable models	base model + GVEL st0.15	0,177	35,3	9,281	8572,91	284,26
	base model + stSAL0.15	0,16	36,5	9,130	8440,06	417,11
	base model + gSAL0.75	0,207	39,6	8,741	8098,28	758,89
Hydrographic model	base model + stGVEL + stSAL + stTEMP	0,192	38,8	8,847	8183,39	673,78
SAL seascape models	Hydrographic model + stSAL + gSAL0.75	0,215	40,6	8,617	7984,18	872,99
	Hydrographic model + (stSAL,gSAL0.75)	0,255	43,1	8,327	7708,82	1148,35

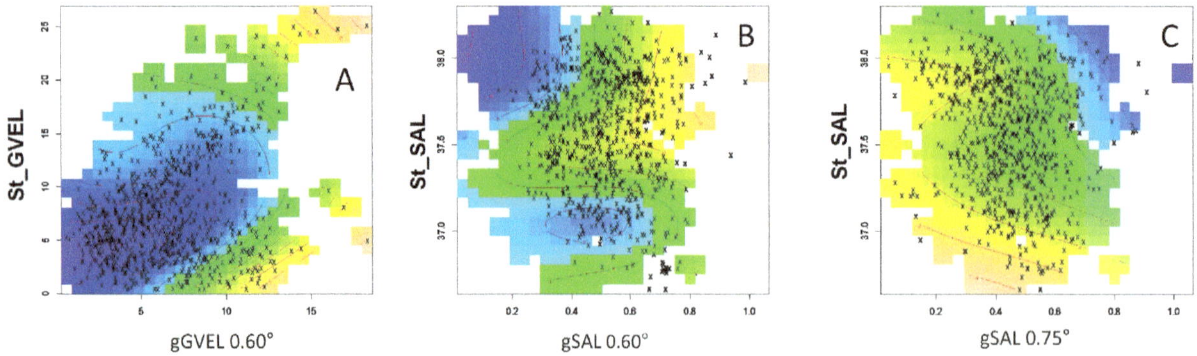

Figure 5. The effect of the interactions of the seascape components on the larval abundance as estimated from the seascape generalized additive model. The effects are shown for bluefin tuna (A–B) and bullet tuna (C). For bluefin tuna: A) the effect of the gGVEL and st_GVEL interaction. B) the effect f the gSAL and st_SAL interaction. For bullet tuna C) the effect of the gSAL and st_SAL interaction. Isolines indicate larval abundances predicted by the model. Peak of abundances are indicated in pink-yellow. Low and very low abundances are indicated in green and blue, respectively.

Despite the lower importance seascape metrics in bullet tuna, the inclusion of salinity gradients provided additional information for the identification of spawning sites. The analysis indicated that bullet tuna spawning areas are mostly found in areas where salinity gradients are low. Bullet tuna was found at both sides of the front but avoiding more mixed waters, located closer to the front. This was verified when observing the spatial distribution of larvae in relation to the salinity seascapes among the different years. For instance, in 2001, 2003 and 2005 high larvae abundances were observed North of the archipelago (high salinity waters with very low gradients), but intermediate abundances, indicating spawning, also occurs in Southern areas (low salinities and intermediate gradients). In 2002 and 2004 higher abundances were linked to low salinity and intermediate gradients shown in the south of the archipelago (Fig. S9 and S10). These results reinforce the theory of bullet tuna spawning occurs in widespread geographic areas, and not only close to the coast and suggest that the location of the main haline front negatively affects the spawning of this species.

Overall results related to bullet tuna point to the fact that, besides the avoidance of areas near strong surface haline gradients, other factors not considered in this study may also be relevant for spawning site selection in this species. It is also relevant that the spatial pattern in relation to the salinity is the opposite to that shown by bluefin tuna, located in areas near the front, suggesting possible avoidance of predators by bullet spawners [35].

The application of seascape metrics derived from salinity and geostrophic currents to characterize the spawning habitat provides new descriptors for environmental variables that improve model quality and predictions. This improvement allows a more precise identification of the relationships between the spatial location of the spawning grounds and the local oceanographic processes. Moreover, our study demonstrates that seascapes must be characterized at specific spatial scales to provide useful information as proposed in previous studies [36] and supporting results on terrestrial landscapes [37],[38] and bottom seascapes [39]. Therefore, the relations between the location of spawning sites and the mesoscale oceanographic processes may prove to be non significant if seascape metrics are not processed at the right spatial scales.

Seascape ecology is an emerging field generally being applied for the analysis of how benthic habitats pattern in coastal areas drives different aspects of marine species ecology [39]. Techniques are applied following categorical approach where the seascape is composed by a number of patches of different type of habitats [40],[7]. However, very little attention has been paid to the techniques and concepts to investigate pelagic seascape ecology due to the complex spatiotemporal dynamics of this system [2]. Thus, the work presented here sheds new light to modeling spatial distribution and investigating key ecological processes of species highly dependent on the variability of the pelagic environment, like spawning ecology of many of the big tuna species are [41]. In areas as the Balearic Sea, for which new operational oceanography platforms provide near real time data of hydrography [42] and also in combination with remote sensing data (e.g. altimetry [43]) and modeling [44],[45] these metrics will improve the species spatial distribution forecast that has proved effective for management [46].

In contrast to seascapes, landscape metrics have a long history in terrestrial ecology, and have improved over time. For instance, the effect caused on the habitat analysis derived from the spatial definition of the input habitat maps or the extent of the study area are common studied topics, [47],[48]. Likewise, calculation of seascape metrics and the final results from their application in ecological studies may be affected by different issues, like the different ways of computing the hydrographic variables and their gradients, or the origin of the input data source like from in situ measurements, remote sensing or hydrodynamic models, each with different sources of uncertainty. A relevant question is how seascapes can provide information for other type of species and ecological processes. Addressing all these challenges and developing comparative studies between different data sources, processing methods, species and ecological processes will allow advancing towards the understanding of how seascape metrics can provide information about how ecological processes and oceanography are linked together.

In summary, pelagic seascapes based on gradients and characteristic scales allow improving spatial distribution models and the identification of essential fish habitat of pelagic species. They also provide a tool for analyzing the links between particular ecological processes and local oceanography going far beyond than stochastic models based on just hydrographic parameters as salinity, temperature or geostrophic velocities. As a consequence these metrics will provide an improvement in all the management approaches and tools pending on the capability of models to identify essential habitats as near real-time spatial management based on habitat predictions [46],[10], pelagic species distribution from deterministic models [49] or the standardization of larvae indices to assess adult stock, [50],[51],[52].

Supporting Information

Figure S1 Model response of bluefin tuna in relation to salinity gradient processed at 0.75 degrees. Fitted line (solid line) and 95% confidence intervals (grey shaded areas) are shown. Whiskers on the x-axis show the locations of measurements.

Figure S2 Correlation between the gradients at the characteristic scales and the hydrographical variables at the sampled station. A) Current velocity and B) salinity for Atlantic Bluefin tuna. C) Salinity for Bullet tuna.

Figure S3 Sea surface salinities in the area during the five years analyzed (2001 to 2005).

Figure S4 Sea surface geostrophic currents in the area during the five years analyzed (2001–2005).

Figure S5 Spatial distribution of bluefin tuna (*Thunnus thynnus*) larvae in relation to the salinity mean calculated at its characteristic scale (0.6 degrees). Relative stage-1 larval abundances are shown in the maps such as dots.

Figure S6 Spatial distribution of bluefin tuna (*Thunnus thynnus*) larvae in relation to the salinity gradient calculated at 0.6 degrees. Relative stage-1 larval abundances are shown in the maps such as dots.

Figure S7 Spatial distribution of bluefin tuna (*Thunnus thynnus*) larvae in relation to the geostrophic velocity mean calculated at its characteristic scale (0.6 degrees). Relative stage-1 larval abundances are shown in the maps such as dots.

Figure S8 Spatial distribution of bluefin tuna (*Thunnus thynnus*) larvae in relation to the geostrophic velocity gradient calculated at the characteristic scale (0.6 degrees). Relative stage-1 larval abundances are shown in the maps such as dots.

Figure S9 Spatial distribution of bullet tuna (*Auxis rochei rochei*) in relation to the salinity mean calculated at 0.75 degrees. Relative stage-1 larval abundances are shown in the maps such as dots.

Figure S10 Spatial distribution of bullet tuna (*Auxis rochei rochei*) in relation to the salinity gradient calculated at 0.75 degrees. Relative stage-1 larval abundances are shown in the maps such as dots.

Acknowledgments

We thank the scientists and crew participating in the TUNIBAL cruises. We also thank P. Tugores and L. Rueda for their help in software development, and R. Balbín for their comments about methodological approaches.

Author Contributions

Conceived and designed the experiments: DAB LC AAG PR FA MH JLLJ JT. Performed the experiments: DAB LC FA. Analyzed the data: DAB LC AAG PR FA MH. Contributed reagents/materials/analysis tools: DAB LC AAG PR FA MH JLLJ JT. Wrote the paper: DAB LC AAG PR FA MH JLLJ JT. Sample collection: FA JLLJ PR. Laboratory sample processing: FA JLLJ AAG. Software development: DAB LC AAG. Numerical modelling design: DAB LC PR. Environmental variables calculation: DAB LC AAG PR FA MH JLLJ JT. Relevant literature review in different fields: DAB LC AAG PR FA MH JLLJ JT. Final manuscript edition: DAB LC AAG PR FA MH JLLJ JT.

References

1. Hinchey EK, Nicholson MC, Zajac RN, Irlandi EA (2008) Preface: marine and coastal applications in landscape ecology. Landscape Ecology 23: 1–5.
2. Pittman SJ, Kneib RT, Simenstad CA (2011) Practicing coastal seascape ecology. Marine Ecology-Progress Series 427: 187–190.
3. Turner MG (1989) Landscape Ecology - the Effect of Pattern on Process. Annual Review of Ecology and Systematics 20: 171–197.
4. Pittman SJ, McAlpine CA, Pittman KM (2004) Linking fish and prawns to their environment: a hierarchical landscape approach. Marine Ecology-Progress Series 283: 233–254.
5. Boström C, Pittman SJ, Simenstad C, Kneib RT (2011) Seascape ecology of coastal biogenic habitats: advances, gaps, and challenges. Marine Ecology Progress Series 427: 191–217.
6. Wedding LM, Lepczyk CA, Pittman SJ, Friedlander AM, Jorgensen S (2011) Quantifying seascape structure: extending terrestrial spatial pattern metrics to the marine realm. Marine Ecology-Progress Series 427: 219–232.
7. Cushman SA, Gutzweiler K, Evans JS, McGarigal K (2010) The gradient paradigm: a conceptual and analytical framework for landscape ecology. In: Huettmann F, Cushman S, editors. Spatial complexity, informatics and wildlife conservation. Tokyo: Springer. pp. 83–110.
8. Mannocci L, Laran S, Monestiez P, Dorémus G, Van Canneyt O, et al. (2013) Predicting top predator habitats in the Southwest Indian Ocean. Ecography 37: 261–278.
9. Worm B, Sandow M, Oschlies A, Lotze HK, Myers RA (2005) Global patterns of predator diversity in the open oceans. Science 309: 1365–1369.
10. Druon JN, Fromentin JM, Aulanier F, Heikkonen J (2011) Potential feeding and spawning habitats of Atlantic bluefin tuna in the Mediterranean Sea. Marine Ecology-Progress Series 439: 223–240.
11. Louzao M, Pinaud D, Peron C, Delord K, Wiegand T, et al. (2011) Conserving pelagic habitats: seascape modelling of an oceanic top predator. Journal of Applied Ecology 48: 121–132.
12. Hidalgo M, Gusdal Y, Dingsor GE, Hjermann D, Ottersen G, Stige LC, et al. (2012) A combination of hydrodynamical and statistical modelling reveals non-stationary climate effects on fish larvae distributions. Proceedings of the Royal Society B: Biological Sciences 279: 275–283.
13. Block BA, Jonsen ID, Jorgensen SJ, Winship AJ, Shaffer SA, et al. (2011) Tracking apex marine predator movements in a dynamic ocean. Nature 475: 86–90.
14. Phillips AJ, Ciannelli L, Brodeur RD, Pearcy WG, Childers J (2014) Spatio-temporal associations of albacore CPUEs in the Northeastern Pacific with regional SST and climate environmental variables. ICES, Journal of Marine Science. In press.
15. Alemany F, Quintanilla L, Velez-Belchi P, Garcia A, Cortés D, et al. (2010) Characterization of the spawning habitat of Atlantic bluefin tuna and related species in the Balearic Sea (western Mediterranean). Progress in Oceanography 86: 21–38.
16. Reglero P, Ciannelli L, Álvarez-Berastegui D, Balbín R, López-Jurado JL, et al. (2012) Geographically and environmentally driven spawning distributions of tuna species in the western Mediterranean Sea. Marine Ecology Progress Series 463: 273–284.
17. Torres AP, Reglero P, Balbin R, Urtizberea A, Alemany F (2011) Coexistence of larvae of tuna species and other fish in the surface mixed layer in the NW Mediterranean. Journal of plankton research 33: 1793–1812.
18. Block BA, Teo SL, Walli A, Boustany A, Stokesbury MJ, et al. (2005) Electronic tagging and population structure of Atlantic bluefin tuna. Nature 434: 1121–1127.
19. Rooker JR, Secor DH, De Metrio G, Schloesser R, Block BA, et al. (2008) Natal homing and connectivity in Atlantic bluefin tuna populations. Science 322: 742–744.
20. La Violette PE, Tintoré J, Font J (1990) The surface circulation of the Balearic Sea. Journal of Geophysical Research 95: 1559–1568.
21. Balbín R, López-Jurado JL, Flexas MM, Reglero P, Vélez-Velchí P, et al. (2013) Interannual variability of the early summer circulation around the Balearic Islands: driving factors and potential effects on the marine ecosystem. Journal of Marine Systems In Press, DOI: 10.1016/j.jmarsys.2013.12.007.

22. Sabatés A, Recasens L (2001) Seasonal distribution and spawning of small tunas (*Auxis rochei* and *Sarda sarda*) in the northwestern Mediterranean. Scientia Marina 65: 95–100.
23. Muhling BA, Lamkin JT, Roffer MA (2010) Predicting the occurrence of Atlantic bluefin tuna (*Thunnus thynnus*) larvae in the northern Gulf of Mexico: building a classification model from archival data. Fisheries Oceanography 19: 526–539.
24. Rooker JR, Simms JR, Wells RD, Holt SA, Holt GJ, et al. (2012) Distribution and habitat associations of billfish and swordfish larvae across mesoscale features in the Gulf of Mexico. PloS one 7: e34180.
25. de la Gándara F, Ortega A, Blanco E, Viguri FJ, Reglero P (2013) La flexión de la notocorda en larvas de atún rojo,*Thunnus thynnus* (L, 1758) cultivadas a diferentes temperaturas. In: Sociedad Española de Acuicultura, editors. XIV Congreso Nacional de Acuicultura. Gijón: Universidad Laboral de Gijón. pp. 181–182. http://www.repositorio.ieo.es/e-ieo/bitstream/handle/10508/1559/5905A.pdf?sequence=2&isAllowed=y. Accessed 23 July 2014.
26. Gordoa A, Carreras G (2014) Determination of Temporal Spawning Patterns and Hatching Time in Response to Temperature of Atlantic Bluefin Tuna (*Thunnus thynnus*) in the Western Mediterranean. PloS one 9: e90691.
27. Reglero P, Balbín R, Ortega A, Alvarez-Berastegui D, Gordoa A, et al. (2013) First attempt to assess the viability of bluefin tuna spawning events in offshore cages located in an a priori favourable larval habitat. Scientia Marina 77: 585–594.
28. Reglero P, Urtizberea A, Torres AP, Alemany F, Fiksen Ö (2011) Cannibalism among size classes of larvae may be a substantial mortality component in tuna. Marine Ecology Progress Series 433: 205–219.
29. Blank JM, Morrissette JM, Landeira-Fernandez AM, Blackwell SB, Williams TD, et al. (2004) In situ cardiac performance of Pacific bluefin tuna hearts in response to acute temperature change. Journal of Experimental Biology 207: 881–890.
30. RDevelopment CORE (2011) TEAM. 2008. R: A language and environment for statistical computing. R Foundation for Statistical Computing, Vienna, Austria. ISBN 3-900051.
31. Wood, Simon N. (2006) Generalized additive models: an introduction with R. CRC, Boca Raton, Forida: Chapman & Hall.
32. Burnham, Kenneth P. and Anderson, David R. (2002) Model selection and multimodel inference: a practical information-theoretic approach. New York: Springer-Verlag. 488 p.
33. Muhling BA, Reglero P, Ciannelli L, Alvarez-Berastegui D, Alemany F, et al. (2013) Comparison between environmental characteristics of larval bluefin tuna *Thunnus thynnus* habitat in the Gulf of Mexico and western Mediterranean Sea. Marine Ecology Progress Series 486: 257–276.
34. Rodriguez JM, Alvarez I, Lopez-Jurado JL, Garcia A, Balbín R, et al. (2013) Environmental forcing and the larval fish community associated to the Atlantic bluefin tuna spawning habitat of the Balearic region (Western Mediterranean), in early summer 2005. Deep Sea Research Part I: Oceanographic Research Papers 77: 11–22.
35. Bakun A (2013) Ocean eddies, predator pits and bluefin tuna: implications of an inferred 'low risk-limited payoff' reproductive scheme of a (former) archetypical top predator test. Fish and Fisheries 14: 424–438.
36. Steele JH (1989) The ocean landscape. Landscape Ecology 3: 185–192.
37. Wiens JA (1989) Spatial Scaling in Ecology. Functional ecology 3: 385–397.
38. Wu JG, Li HB (2006) Perspectives and methods of scaling. In: Jianguo Wu, K.. Bruce ones, Harbin Li, Orie L.. Loucks, editors. Scaling and Uncertainty Analysis in Ecology. Netherlands: Springer. pp. 17–44.
39. Bostrom C, Pittman SJ, Simenstad C, Kneib RT (2011) Seascape ecology of coastal biogenic habitats: advances, gaps, and challenges. Marine Ecology-Progress Series 427: 191–217.
40. Forman, R. T T. (1995) Land mosaics: the ecology of landscapes and regions. Cambridge, UK: Cambridge University Press.
41. Reglero P., Tittensor D.P., Álvarez-Berastegui D., Aparicio-González A., Worm B. (2014) Worldwide distributions of tuna larvae: revisiting hypotheses on environmental requirements for spawning habitats. Marine Ecology Progress Series 501: 207–224.
42. Tintore J, Vizoso G, Casas B, Heslop E, Pascual A, et al. (2013) SOCIB: The Balearic Islands Coastal Ocean Observing and Forecasting System, Responding to Science, Technology and Society Needs. Marine Technology Society Journal 47: 101–117.
43. Pascual A, Bouffard J, Ruiz S, Buongiorno Nardelli B, et al. (2013) Recent improvements in mesoscale characterization of the western Mediterranean Sea: synergy between satellite altimetry and other observational approaches. Scientia Marina 77: 19–36.
44. Juzza M, Mourre B, Renault L, Tintoré J (2014) Assessment and intercomparison of numerical simulations in the Western Mediterranean Sea. Geophysical Research Abstracts, Vol 16, EGU2014-6893, 2014, 27 April, 2014 Vienna, Austria. http://adsabs.harvard.edu/abs/2014EGUGA..16.6893J. Accessed 23 July 2014
45. Juza M, Renault L, Ruiz S, Tintoré J (2013) Origin and pathways of Winter Intermediate Water in the Northwestern Mediterranean Sea using observations and numerical simulation. Journal of Geophysical Research: Oceans 118: 6621–6633.
46. Hobday AJ, Hartmann K (2006) Near real-time spatial management based on habitat predictions for a longline bycatch species. Fisheries Management and Ecology 13: 365–380.
47. Cushman SA, McGarigal K, Neel MC (2008) Parsimony in landscape metrics: strength, universality, and consistency. Ecological Indicators 8: 691–703.
48. Wu JJ (2013) Landscape ecology. In: Rik Leemans, editors. Ecological Systems. New York: Springer. pp. 179–200.
49. Lehodey P, Senina I, Murtugudde R (2008) A spatial ecosystem and populations dynamics model (SEAPODYM)-Modeling of tuna and tuna-like populations. Progress in Oceanography 78: 304–318.
50. Ingram Jr GW, Richards WJ, Porch CE, Restrepo V, Lamkin JT, et al. (2008) Annual indices of bluefin tuna (*Thunnus thynnus*) spawning biomass in the Gulf of Mexico developed using delta-lognormal and multivariate models. ICCAT Working Document SCRS/2008/086.
51. Ingram Jr GW, Alemany F, Alvarez-Berastegui D, García A (2013) Development of indices of larval bluefin tuna (*Thunnus thynnus*) in the Western Mediterranean Sea. ICCAT,Collect Vol Sci Pap 69: 1057–1076.
52. Muhling BA, Lee SK, Lamkin JT, Liu Y (2011) Predicting the effects of climate change on bluefin tuna (*Thunnus thynnus*) spawning habitat in the Gulf of Mexico. ICES Journal of Marine Science: Journal du Conseil 68: 1051–1062.

4

Evaluating the Spatio-Temporal Factors that Structure Network Parameters of Plant-Herbivore Interactions

Antonio López-Carretero[1], Cecilia Díaz-Castelazo[1]*, Karina Boege[2], Víctor Rico-Gray[3]

1 Instituto de Ecología A.C., Xalapa, Veracruz, México, **2** Instituto de Ecología, Universidad Nacional Autónoma de México, México D.F., México, **3** Instituto de Neuroetología, Universidad Veracruzana, Xalapa, Veracruz, México

Abstract

Despite the dynamic nature of ecological interactions, most studies on species networks offer static representations of their structure, constraining our understanding of the ecological mechanisms involved in their spatio-temporal stability. This is the first study to evaluate plant-herbivore interaction networks on a small spatio-temporal scale. Specifically, we simultaneously assessed the effect of host plant availability, habitat complexity and seasonality on the structure of plant-herbivore networks in a coastal tropical ecosystem. Our results revealed that changes in the host plant community resulting from seasonality and habitat structure are reflected not only in the herbivore community, but also in the emergent properties (network parameters) of the plant-herbivore interaction network such as connectance, selectiveness and modularity. Habitat conditions and periods that are most stressful favored the presence of less selective and susceptible herbivore species, resulting in increased connectance within networks. In contrast, the high degree of selectivennes (i.e. interaction specialization) and modularity of the networks under less stressful conditions was promoted by the diversification in resource use by herbivores. By analyzing networks at a small spatio-temporal scale we identified the ecological factors structuring this network such as habitat complexity and seasonality. Our research offers new evidence on the role of abiotic and biotic factors in the variation of the properties of species interaction networks.

Editor: Anna R. Armitage, Texas A&M University at Galveston, United States of America

Funding: Research was supported by grant (project) number 106575 from CONACYT to Cecilia Díaz-Castelazo, CONACYT fellowship number 302184 to Antonio López-Carretero, and Universidad Veracruzana, I010/152/2014 C 133/2014 provided by CONACYT. We acknowledge the financial support provided by Instituto de Ecología A.C. (INECOL) through the Postgraduate Secretary to Antonio López-Carretero. CONACYT URL is http://www.conacyt.gob.mx/Paginas/InicioNueva.aspx, INECOL URL is http://www.inecol.mx/posgrado/. The funders had no role in study design, data collection and analysis, decision to publish, or preparation of the manuscript.

Competing Interests: The authors have declared that no competing interests exist.

* Email: diazcastelazogm@gmail.com

Introduction

Approximately 40% of terrestrial biodiversity is involved in plant-herbivore species interaction networks [1]; these interactions are one of the main conduits for upward energy flow to higher trophic levels [2], and are mainly found in tropical ecosystems [3], [4]. Plant-herbivore interactions are considered antagonistic since the loss of tissue of the host plants caused by herbivore feeding may have detrimental consequences in their fitness or survival [5]. Unlike mutualistic interactions and networks, antagonistic networks tend to be more specialized and have an essentially modular structure [6], [7], characterized by cohesive groups of species that interact intensively (more than expected by chance) and maintain few interactions with other groups of species [8], [9].

Despite the advances in the identification and description of structural patterns of ecological networks, relatively little is known about the ecological and phylogenetic mechanisms that give rise to such patterns [10]. For plant-herbivore trophic networks it has been recently demonstrated that their structural attributes are influenced by species body size, local abundance, phylogeny [7], and by disturbance, since deforestation, habitat fragmentation and secondary plant succession all reduce the specialization and structural complexity of plant-herbivore networks owing to the resulting alteration of the food resources of herbivorous insects [11], [12], [13].

Furthermore, both seasonal changes in precipitation and structural complexity of plant communities are intimately related with changes in microclimate (e.g. temperature and humidity) and with the availability and quality of host plants for herbivores [14], [15], [16], [17]. Thus, both factors can have an important effect on the structure, composition and specialization of herbivorous insect communities [15], [18], [19]. In general, plant communities with high heterogeneity and species diversity offer a wide variety of food and habitats for herbivores and can reduce stressing factors such as temperature and humidity, which affect the performance of invertebrate herbivores [20], [21]. Such conditions promote a more diverse and specialized community of insects than in less structured plant communities with fewer host species and more stressful conditions [4], [21]. Although the effects of seasonality and variation in the quality of food resources on herbivore communities have been well documented for different ecosystems [14], [18], [22], the influence of these factors on interaction networks has not been evaluated.

Addressing the spatio-temporal assessment of ecological networks represents a new frontier in community ecology [23]. This approach can provide evidence of the influence of biotic and abiotic factors in the variation of species interaction networks [24],

[25]. However, this information is particularly scarce for antagonistic interactions [7], [26]. In the present study, we analyzed plant-herbivore networks (lepidopteran larvae feeding on plants) at a small spatio-temporal scale, which has the advantage of identifying the ecological processes that influence the structure of such networks occurring within the same region (coastal habitats spaced as far as 0.4 ha) [27]. In particular we were interested in answering the following questions: i) What is the seasonal variation in the structural parameters of the plant-herbivore network in habitats that differ in their complexity? ii) How does resource availability vary across habitats and seasons? and, iii) Which is the simultaneous contribution of host availability, habitat structure and seasonality for the spatio-temporal structuring of this antagonistic network? In antagonistic systems coevolutionary processes will tend to reinforce constraints against grand generalists (only a small proportion of herbivorous animals are true generalists), thus compartments should often be apparent in such networks [9]. Since host use by phytophagous insects tend to specialization [28], compartmentation or modularity [8] and may be influenced by resource availability [6], [29] we hypothesized that: i) strong seasonal effects would be detectable at the structural parameters of plant-herbivore networks at the studied site, with higher selectiveness (network-level specialization) and modularity during the wet season and forest habitats, ii) host availability and interaction frequency would vary among habitats and temporal periods, thus we may expect iii) higher connectance, interaction evenness and lower selectiveness during dry season periods and open habitats.

This study is unique in the sense that, for the first time, the impacts of resource availability, habitat structural complexity and seasonality on the structure (parameters) of an antagonistic network were assessed.

Methods

Study site

Field work was carried out at La Mancha Coastal Research Center (Centro de Investigaciones Costeras La Mancha, CICOLMA) located on the coast of the state of Veracruz in Mexico (96°24′48″W, 19°40′33″N y 96°22′25″W, 19°31′49″N). Permit to perform the present research was issued by Instituto de Ecología, A.C. (INECOL), since CICOLMA is one of INECOL's field station and private conservation reserve. CICOLMA is a very small fraction of the 1336 Ramsar Site called "La Mancha el Llano" declared in 2005. Thus, the regulatory body concerned with protection of wildlife is given to INECOL by Mexican federal organisms in the matter and framed by The Ramsar Convention on Wetlands (intergovernmental treaty). The study site, the CICOLMA reserve area managed by INECOL covers an area of 70 ha, which includes several native vegetation types (including wetlands and dune vegetation), transformed forestry/agricultural sites, experimental areas and a field station. The major vegetation types here are tropical deciduous forest, tropical dry forest, sand dune scrub, mangrove forest, freshwater marsh and flooded deciduous forest [30], [31].

The climate of the study site is warm sub-humid and total annual precipitation ranges from 899 to 1829 mm, ca. 78% of the total annual precipitation falls during the rainy season (June - September). In this area, 837 plant species have been recorded, 50% of which are herbaceous plants and the rest are shrubs, trees and vines [30]. Currently, there is no list of lepidopteran species for CICOLMA. The field studies presented here did not involve endangered or protected species.

Recording plant-herbivore interactions

The study was carried out at five sites with vegetation types representative of the plant communities in the study region and with differing degrees of structural complexity: 1) pioneer dune vegetation (PIO), 2) coastal dune scrub (DUN), 3) recently established tropical lowland subdeciduous forest (SFY) 4) tropical lowland subdeciduous forest in an advanced stage of succession (SFO), 5) tropical lowland flood forest with a wetland ecotone (FFW).

Structural complexity and vegetation strata differ specially between the open (PIO, DUN) and the forest habitats (SFY, SFO, FFW). The open habitats have moderate vegetation cover due to increased sand movement at the dunes, with dominant herbaceous stratum (although two strata are common at these habitats).These habitats are exposed to abiotic stressful conditions because of high temperatures, radiation, salt water influence and mechanical damage of vegetation during storms and hurricanes [32], [33]. In contrast, the forested habitats (SFY, SFO, FFW) are more stable and with milder conditions because of shade; vegetation structure at these habitats is more complex since three distinct arboreal strata occur here, the taller elements may exceed 20 m; the medium strata with trees of 6 to 15 m high, and finally, a well-developed understory [30].

At each of these vegetation types we placed 14 quadrats (4×4 m) where plant-herbivore interactions were recorded, along with the attributes of each plant community. In monthly censuses during the dry (March, April, May, June) and the wet/rainy (July, August, September, and October) season of 2012, we collected all folivorous caterpillar species (excluding leaf miners) associated with host herbs, lianas, shrubs and trees. In order to search for caterpillars on adult trees, a subsample of four branches was taken per individual tree. All caterpillars were recorded and taken to the laboratory where they were fed until pupated. No ethical approval was required for manipulating lepidopteran specimens, since after caterpillars pupated and adults emerged, only unidentified species (at larval stage) were mounted as adults for taxonomical identification (and only one individual of the unidentified species was mounted). Data underlying the findings of the present study are available within this manuscript, but lepidopteran specimens found at the present study are being processed for subsequent molecular analysis (Instituto de Ecología UNAM, México, by K. Boege) after which specimen identity will be publicly available in reference to this manuscript. Host plants were identified and deposited in the XAL Herbarium of the Instituto de Ecología, A.C. (INECOL, Xalapa, México).

Structural network parameters

To assess spatio-temporal changes in the structural network parameters, for each of the five sampled vegetation types we constructed four matrices of the plant-caterpillar interactions throughout the dry and wet season: dry season 1 (DS1: March, April), dry season 2 (DS2: May, June), rainy season 1 (RS1: July, August) rainy season 2 (RS2: September, October). Previous work at the study site suggests that there is short-scale temporal variation in the structure of insect-plant –mutualistic- interaction networks derived from changes in abiotic properties (temperature and precipitation) [24]. In the present study, the monthly data of plant-herbivore interactions recorded at the field was fused bimestrially in order to have robust temporal information across vegetation types, since at some habitats, monthly census during dry season provide scarce interaction records; furthermore, these bimesters markedly reflect temporal biotic and abiotic changes at the study site: the beginning of the dry season period DS1 mark the end of cold fronts ("nortes", see below), DS2 months corresponds to peak maximum temperatures, RS1 months corresponds to peak rainfall and RS2 correspond to the end of the rainy season. We did not make census from November to February because these months correspond to "nortes", a season of winter cold fronts, characterized by strong winds and a considerable decrease in plant cover [34].

We obtained a total of 20 matrices of species interactions (that result from multiplying 5 vegetation types and 4 bimestrial censuses), which we refer to as sub-networks. We calculated the following structural parameters for each sub-network: connectance (C), Interaction evenness (IE), vulnerability (V), generality (G), modularity (M), number of modules (NM) and network specialization or selectiveness (selective use of the lower trophic level by the higher trophic level) of the entire bipartite network (H^2), given that H^2 describes to which extent observed interactions deviate from those that would be expected given the species marginal sums [27]. 1) Connectance was defined as the fraction of recorded interactions relative to the total number of possible interactions. 2) Interaction evenness indicates how homogeneously the plant-herbivore interactions are distributed throughout the network. It is calculated similarly to Shannon's Index, but does not take into account the absence of interactions (i.e., cells that contain a zero in the interaction matrix). Evenness approaches a value of 1 when the number of interactions between herbivore and host plant species is uniformly distributed. 3) Generality represents the proportion of herbivorous species to host species. 4) Vulnerability was calculated as the proportion of host species to herbivorous species. 5) The number of modules corresponds to the number of subgroups of the network that are not connected to other groups: modules are cohesive groups of highly connected nodes that are loosely connected to other nodes in the network [35], [36]. 6) Modularity is a measure of how host and herbivorous species tend to organize into subgroups of species that interact more frequently among themselves than with other members of the network. Because the degree of modularity determines how dense the connection between different groups of elements in an ecological system is, systems that largely differ in the degree of modularity often differ in their ecological and evolutionary [35]. This parameter was calculated with the M index (range: 0–1) in MODULAR [36] based on Newman and Girvan's algorithm. Its level of significance was calculated using Monte Carlo tests with 1000 randomizations. In order to carry out robust comparisons among sub-networks not affected by neither sampling effort nor network size, we provided a relative value of Modularity (Mr) estimated from the z-score values of the 1000 random replicates of the modularity analysis described above, as: $Mr = M\text{-}Mz/Mz$ where M is the value of the real sub-network data matrix, while Mz refers to the average value of modularity of its random replicates, 7) Network specialization (H^2) is based on the deviation between the observed and expected number of interactions in each network, assuming that all species interact with other species in proportion to their total observed frequency. Given that this metric is not affected by neither sampling effort nor network size, it is possible to make robust, reliable comparisons between different networks [27]. H^2 values, ranging from 0 (low selectiveness) to 1 (high selectiveness). The network indices were calculated for each census and site using the BIPARTITE 3.0.1 package run in R [37].

Attributes of the plant community

In each vegetation type and census, the following plant community attributes were quantified. As a measure of vegetation structure, total plant species richness (TPR) was registered, along with an estimation of canopy cover provided by all plant species (included host and non-host species) found in each quadrat (total plant cover = TPC). As an indicator of food availability to herbivores, we measured plant cover of host species (i.e., exclusively those species in which caterpillars were found; host plant cover = HPC). Species richness of host plants used by herbivorous lepidopterans (host plant richness = HPR) was also registered. These variables were selected because they influence herbivore food availability, habitat structure and the abiotic factors that affect plant-herbivore interactions.

Data analysis

Effect of vegetation type and seasonality on network parameters and plant community attributes. The effect of vegetation type and census on plant community attributes (TPR, TPC, HPC) and structural parameters of each network (C, G, V, IE, M, Mr, NM, H^2) were evaluated using repeated measures ANOVAs, using the sub-network parameters as replicates for each vegetation type and census. Using a standard ANOVA in this case is not appropriate because it fails to model the correlation between the repeated measures (census and vegetation types): the data violate the ANOVA assumption of independence. Therefore the repeated measures ANOVA is suitable for analysis of auto correlated samples. To assess the influence of vegetation type, the model included census as the error term (within subject) and vegetation type as the explanatory variable (between subject). To evaluate the influence of census, the model included vegetation type as the error term (within subject) and census as the explanatory variable.

Spatio-temporal ordination of the sub-networks and their relation with plant community attributes. Plant-herbivore interactions were spatio-temporally analyzed at 20 sub-networks (i.e. 20 frequency matrices). Structural network parameters were subsequently estimated for each sub-network. In order to describe the dissimilarity among these matrices, we performed a NMDS analysis (Non-metric Multi-Dimensional Scaling). The resulting NMDS ordination (biplot) shows the structural parameters of all sub-networks: when sub-networks are close to a specific parameter, these have small dissimilarities in their values for that parameter. This is one of the most effective methods for the ordination of ecological data and the identification of underlying gradients, because it does not assume a linear relationship among the variables [38]. Because this analysis offers more than one solution, we carried out an iterative process to find the model with smallest stress value using the metaMDS function in the MASS routine run in R [39].

To evaluate if spatio-temporal dissimilarity among sub-network parameters was related to plant community attributes, we analyzed the effect of TPR, TPC, HPC (continuous variables) (see "Attributes of the plant community" within this section) on the ordination of plant-herbivore sub-networks using the envfit function from the VEGAN library of the statistical package R [39]. This function fits the vectors or continuous values of the environmental variables (TPR, TPC, HPC) to the NMDS ordination axes, generating a measure of correlation (r) and a significance value based on the probability that 1000 random permutations of environmental variables would have a better fit than the real environmental variables.

To test the null hypothesis of a similar composition of herbivore and host plants species among vegetation types and sampling censuses, we used a PERMANOVA (Permutational Multivariate Analysis of Variance), and the Bray-Curtis Index [40]. The PERMANOVA was performed using the *adonis* function of the VEGAN package run in R. This function runs a multivariate analysis using the distance between matrices (dissimilarities between objects or sampling sites), and the probability of significance P is obtained through permutations [40]. For these analyses, we used the following groups: season (censuses of the dry season vs. censuses of the rainy season) and structural complexity of the habitat (open habitats vs. forest).

To understand if spatio-temporal changes in herbivore species composition (ordination dissimilarity) could affect network structure, we analyzed the effect of parameters C, H^2, M y NM (quantitative variables or vectors) over the ordination (biplot) of an NMDS analysis of the community of lepidopteran herbivorous larvae found foraging on plants at the 20 sub-networks using the *envfit* function from the VEGAN package of the statistical software R. This function provides a correlation measure and is statistical significance (after comparing correlation values with 1000 random permutations).

Results

Effect of vegetation type and seasonality on network structure and plant community attributes

We recorded 654 larvae belonging to 176 morphospecies of Lepidoptera and associated with 56 host plant species. During the dry season there were 237 larvae belonging to 41 morphospecies of Lepidoptera feeding on 33 species of plants, whereas during the rainy season there were 417 larvae belonging to 66 species of Lepidoptera feeding on 47 species of host plants. The composition of the herbivore and host plant communities differed significantly between the studied open and forested habitats (PERMANOVA, herbivores: Pseudo-F = 2.667; P = 0.004, plants: Pseudo-F = 4.043; P = 0.004). For both herbivores (lepidopteran larvae) and hosts plants, species turnover was higher between groups (vegetation types) than within them (Tukey test, herbivores: P = 0.787, plants: P = 0.719). The mean values of C, M, NM and H^2 of the network varied significantly across censuses and among vegetation types (Table 1A). The network parameter V changed across time, while G and IE were not different between vegetation types or throughout the year (Tables 1A).

Post hoc tests indicated that the values for M, Mr, NM and H^2 of the forest habitats (SFY, SFO, FFW) were significantly higher than those with open vegetation (PIO, DUN) (Figure 1A, Table 2). Additionally, the networks of open habitats had the highest C, with values almost twice than those of forest habitats. The network indices did not vary significantly among the three types of forest vegetation, or between the two types of open vegetation (Figure 1A).

The first census of the dry season had the highest connectance of the entire study, with values twice as high as during the rest of the year (Figure 1B). In contrast, species networks of the last three censuses were characterized by high M, Mr (Table 2), NM and H^2 values that were twice as much of those estimated for the networks of the first census. None of these indices were different among the last three censuses (Figure 1B). It is important to notice that all sub-networks (but one) were significantly modular (Table 2).

The HPC and HPR varied significantly among vegetation types and across censuses, while total floristic richness differed only among vegetation types (Table 1B, Figure 1). The TPC did not differ statistically between vegetation types or censuses. In particular HPC in PIO was two times greater than in FFW and SFO and three times greater than in DUN and SFY. In contrast, the DUN had the lowest host plant cover of all the vegetation types studied, even with respect to the PIO (Figure 1C). The TPR and HPR in the forest habitats were twice than in open habitats (Figure 1C).

During the dry season, HPC was two times that of the wet season. Although TPR did not differ statistically across time, the HPR was about three times lower during DS1 relative to the rest of the year (Table 1B, Figure 1D). By comparing TPR and HPR values, we found that in all vegetation types and throughout all censuses, herbivores used approximately one quarter of the available plant species (Figure 1 C–D).

Spatio-temporal ordination of the sub-networks and their relation with plant community attributes

In the NMDS of plant-herbivore sub-networks, axis 1 was associated with the slope defined by seasonality (Figure 2). During the first census of the dry season the sub-networks clustered independently of the vegetation type because of their high C values. In addition sub-networks exhibited a clear separation between the dry and wet seasons, characterized by their high degree of H^2, M and NM (Figure 2). The sub-networks in the open vegetation habitats were defined by high values of C during the dry season. However, during the rainy season the sub-networks in these habitats were characterized by high values of IE, G and M (Figure 2). During the dry season, forest vegetation sub-networks had high values of IE and V, while the sub-networks of the rainy season had similarly high values of H^2, M and NM (Figure 2).

Some attributes of plant communities influenced the structure of plant-herbivore sub-networks: vector TPR (r = 0.485, P = 0.003) and vector HPC (r = 0.274, P = 0.046). The correlation of HPC was greater for the ordination of the sub-networks in habitats with open vegetation and during the dry season (Figure 2). TPR had the highest degree of correlation with sub-network ordination, and its correlation was stronger for the ordination of the sub-networks of the forest vegetation types and rainy season (Figure 2).

The NMDS exploring the relation among sub-network parameters and herbivore species composition showed an important spatio-temporal turnover of lepidopteran herbivore species (Figure 3). Independently of the bimester (seasonality), we found a clear dissimilarity among lepidopteran species foraging in plants at open habitats (PIO, DUN) and forest habitats (FFW, SFY, SFO). Vector fitting at NMDS showed that parameters NM, M and C were significantly correlated to dissimilarity in lepidopteran species composition (r = 0.464, P = 0.007; r = 0.389, P = 0.033; r = 0.364, P = 0.024, respectively). The correlation of M and NM was greater for the ordination of the sub-networks in habitats with forest vegetation, while the highest correlation of C occurred at open habitats (Figure 3). Thus, there are differences in lepidopteran species composition between sites that transcend to network structure (parameters): changes in lepidopteran species composition affects network general topology and connectance, but not specialization (selectiveness) (r = 0.236, P = 0.123).

Discussion

It has recently been demonstrated that the structure of plant-herbivore interaction networks is influenced by ecological, phylogenetic [7] and anthropogenic attributes [12], [13], [41]. However, although the patterns of plant-herbivore interactions can be influenced by finer and more immediate environmental factors such as temporal variation in the quality and availability of food [18], [42] or habitat structure [4], [43], the impact of these factors on the properties of species interaction networks had not been evaluated.

Our results show for the first time how the structural parameters of plant-herbivore networks are spatially and temporally dynamic, influenced by seasonality, habitat complexity and food resource availability. The spatio-temporal changes in host plant communities resulting from abiotic factors (climate and micro environmental variation) were not only reflected in the species composition of

Table 1. Results of the repeated measures analysis of variance (ANOVA) for the effect of census and vegetation type on structural network parameters (A) and the plant community attributes (B).

(A) Network parameters	*DF*	*SS*	*F*	*P*
Generality (G)				
Census	3	0.005	0.626	0.614
Vegetation type	4	0.004	0.363	0.831
Vulnerability (V)				
Census	3	1.961	4.167	***0.033***
Vegetation type	4	0.645	1.028	0.434
I. Evenness (IE)				
Census	3	0.034	1.808	0.203
Vegetation type	4	0.461	1.803	0.198
Connectance (C)				
Census	3	0.099	29.218	*<0.001*
Vegetation type	4	0.066	14.698	*<0.001*
Modularity (M)				
Census	3	0.125	11.568	***0.001***
Vegetation type	4	0.128	8.896	***0.009***
No. Modules (NM)				
Census	3	0.125	11.568	***0.001***
Vegetation type	4	0.128	8.896	***<0.001***
Network Specialization (H_2)				
Census	3	5.842	22.023	***<0.001***
Vegetation type	4	1.483	4.195	***0.026***
(B) Attributes of the plant community				
Total Plant Cover (TPC)				
Census	3	642.717	0.912	0.466
Vegetation type	4	1895.716	2.011	0.161
Total Plant Richness (TPR)				
Census	3	48.133	1.079	0.397
Vegetation type	4	1678.228	28.231	***<0.001***
Host Plant Cover (HPC)				
Census	3	0.342	10.462	***0.001***
Vegetation type	4	0.892	20.456	***<0.001***
Host Plant Richness (HPR)				
Census	3	222.64	11.478	***0.001***
Vegetation type	4	176.95	6.776	***0.005***

Significant values of P (≤0.05) are in bold and italics.

herbivore community as previously reported [14], but also in the emergent properties of the antagonistic network such as its connectance, specialization (selectiveness) and modularity.

In contrast with mutualistic plant-insect networks, which maintain their structural stability over time [44], [45], [46], [47], antagonistic networks (i.e., plant-herbivore) tend to maintain their modular structure [4], [8]. Indeed, in this study plant-herbivore networks were characterized by many specialist interactions and a marked compartmentalization, as has been reported for other plant-insect herbivore networks [6], [7], [8], [48]. To our knowledge this is the first study to report structural seasonal and spatial changes in a plant-herbivore interaction network. Our interaction sampling is robust; since our study habitats are relatively close to each other (spaced as far as 0.4 ha) and the area of each habitat is reduced, we could established permanent quadrats (n = 14) that covered most environmental gradient; thus we have a considerably good sampling effort in reference to the interactions studied. In general, in order to estimate sampling completeness within interaction networks, the number (richness) of interactions accumulated as sampling effort increased is frequently used [49]. Since in our study the assemblage of lepidopteran herbivores showed a high proportion of rare species, the result is similar to other tropical communities of herbivorous insects [3], [50] where asymptotic interaction richness is not reached [15], [29]. This non-asymptotic pattern of accumulation curves occurs as well for lepidopteran species richness hosted even by plant species of the same genus or family [3], [51].

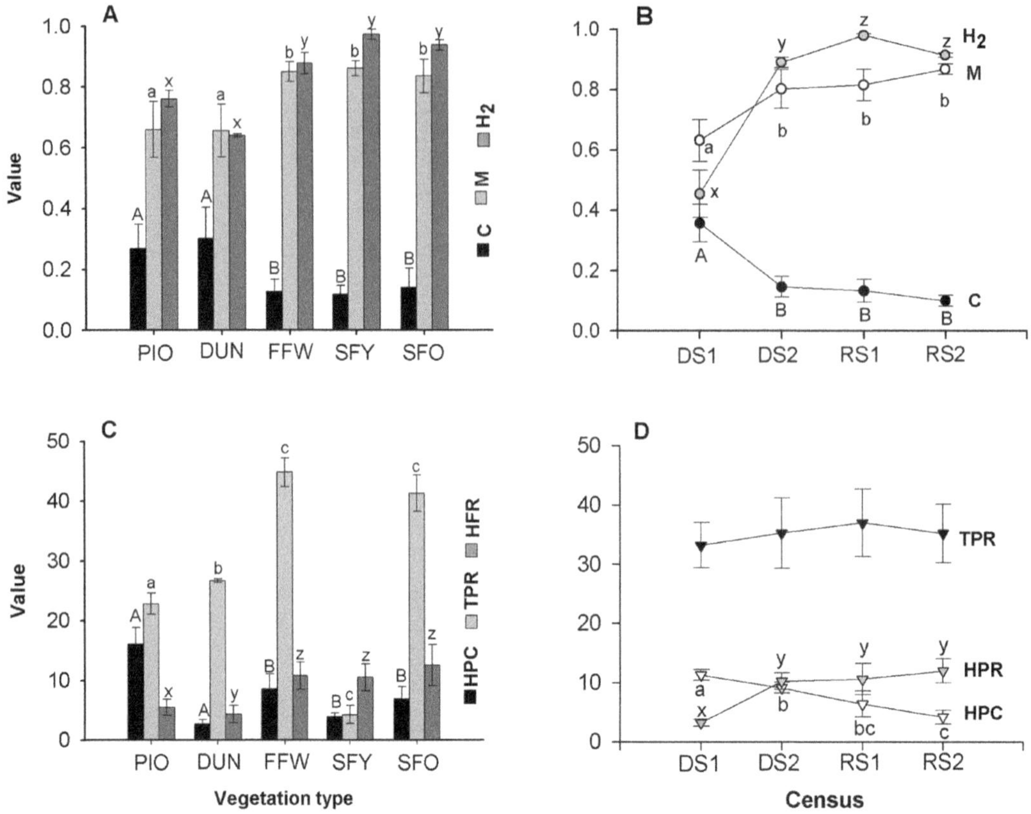

Figure 1. Vegetation type and seasonality effect on network parameters and plant community attributes. Vegetation types (PIO: pioneer dune vegetation, DUN: coastal dune scrub, FFW: flood forest with a wetland ecotone, SFY: recently established tropical lowland subdeciduous forest, SFO: tropical lowland subdeciduous forest in an advanced stage of succession). Census (DS1: Dry season 1, DS2: Dry season 2, RS1: Rainy season 1, RS2: Rainy season 2). Network parameters (H^2: network specialization, M: Modularity, C: Connectance). Plant community attributes (TPR: Total plant richness, HPR: Host plant richness, HPC: Host plant cover). (A) Network parameters by vegetation type. (B) Network parameters by census. (C) Plant community attributes by vegetation type. (D) Plant community attributes by census. Comparisons are valid only within each network parameter (panels A and B, for vegetation types and census, respectively) and within each plant community attribute (panels C and D, for vegetation types and census, respectively). Bars indicate standard error of the mean. Different letters represent significant differences resulting from post-hoc test (P<0.05).

Table 2. Relative modularity (Mr) based in z-score values estimated for the vegetation types and censuses (mean ±SD) in plant-herbivore sub-networks.

Vegetation type	Modularity (Mr)
PIO	1.887±0.303[a]
DUN	1.907±0.194[a]
FFW	2.611±0.111[b]
SFY	2.696±0.191[b]
SFO	2.669±0.140[b]
Census	
DS1	2.049±0.227[a]
DS2	2.325±0.322[ab]
RS1	2.440±0.242[bc]
RS2	2.686±0.108[c]

Different letters among vegetation types and censuses represent significant differences in post-hoc test at P≤0.05. See Methods section for vegetation types and censuses abbreviations.

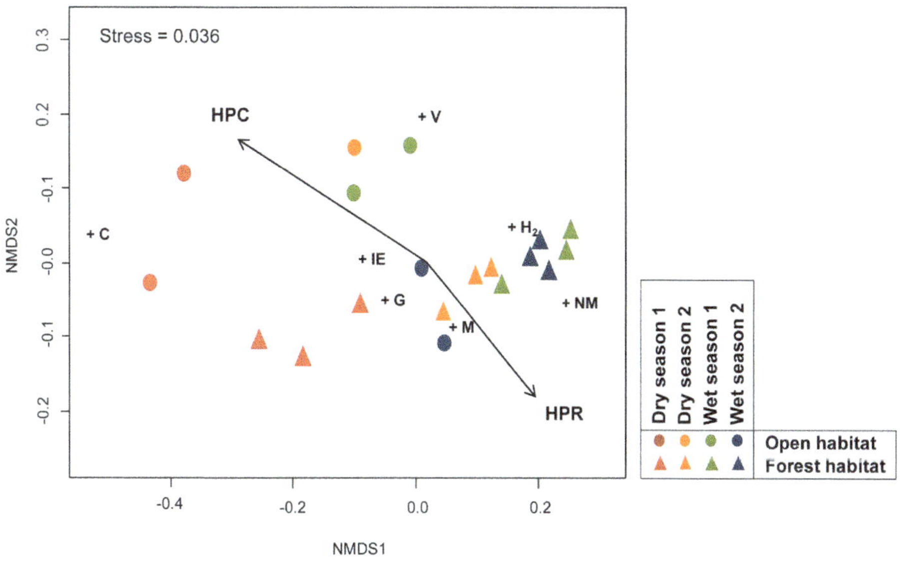

Figure 2. NMDS ordination of network parameters showing the spatio-temporal variation of plant-herbivore sub-networks. The plus signs (+) indicate the network parameters upon which the sub-network ordination was performed (H^2: network specialization, M: Modularity, C: Connectance, IE: interaction evenness, V: vulnerability, G: generality). The vectors on the ordination represent the gradient in HPR (Host plant richness) and HPC (Host plant cover) for all sub-networks. The arrow points to the direction in which the linear change in HPR and HPC was strongest, and the length of the arrow is proportional to the correlation between these variables and the sub-network ordination. Arrows were plotted only for variables with significance of $P \leq 0.05$. Factors fit in the ordination (season and habitat type) are not included. For habitats abbreviations see Figure 1.

Although connectance is a simple parameter sensitive to network size [52], the inverse relation trend generally found between connectance and network size stabilizes after a "threshold" above 50 species conforming the network [53]. Since at the present study our smaller sub-networks have much more than 50 nodes, we consider that connectance can be comparatively used to describe spatio-temporal changes in plant-herbivore interaction "saturation". In general, selectiveness (H^2) metric is not affected by neither sampling effort nor network size [27], and the relative modularity (Mr) provided by the z-score values of "M" metric make our spatio-temporal comparisons robust. In our study, temporal variation in the structure of plant-herbivore networks was associated with changes in climatic factors (e.g., precipitation and temperature) during the transition from the dry to the rainy season, and was accompanied by a decrease in connectance (lower interaction saturation in the community) and an increase in the selectiveness/modularity of networks. These temporal changes in the configuration of species networks could be associated with a high susceptibility of monophagous or oligophagous lepidopteran herbivores that need to synchronize their life cycles to plant phenology [14], [18], [22]. Given that at the present study, no traits of plant species are provided, the H^2 pattern among sub-networks suggest that, independently of the number of species constituting each sub-network, there is higher selectiveness of lepidopteran larvae (not necessarily trophic specialization) during the rainy season when offer of plant species (and foliage availability) is higher; this pattern may reflect changes in host choice by adult females, as a function of temporal changes in plant community [54], [55], [56]. Further studies including nutritional or defensive traits of plants used by herbivores at these networks or phylogenetic signaling of these interactions would shed light on the process generating these patterns.

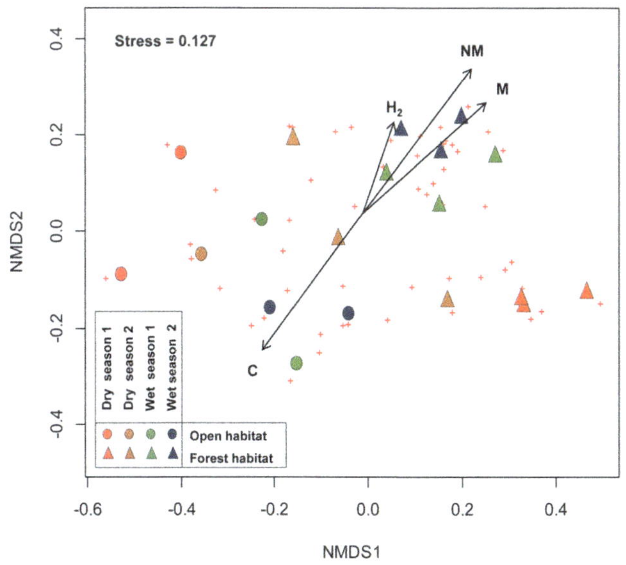

Figure 3. NMDS ordination of lepidopteran herbivore species composition showing the spatio-temporal variation in sub-networks. The plus signs (+) indicate different herbivore species upon which the sub network ordination was performed. The vectors on the ordination represent the gradient in network parameter values across all sub-networks (for habitats abbreviations see Figure 1): H^2: network specialization, M: Modularity, NM: Number of modules, C: Connectance. The arrow points to the direction in which the linear change in network parameters was strongest, and the length of the arrow is proportional to the correlation between these variables and the ordination of herbivore species composition.

The ordination of the most specialized and highly modular sub-networks was highly correlated with floristic richness of forest habitats and the rainy season. The high degree of specialization and modularity of the networks in the richest floristic contexts could be related to a greater diversification by the herbivores in their use of plant resources: a greater availability of niches at the lower trophic level is likely to promote a greater niche differentiation (and therefore greater specialization) in the higher trophic level. The forest sites (tropical lowland subdeciduous forest) were not only floristically the richest of the studied sites, but also offered more canopy cover from trees and shrubs, as well as more plants with different life forms [30]. Greater plant richness and structural complexity of the vegetation provides a wide variety of microhabitats and food that can result in lower competition by food resources among herbivores [4], [57]. In turn, this could translate into greater specialization and modularity in the network patterns. There are two possible reasons that might promote specialization of plant-herbivore networks in more floristically diverse contexts: i) the advantages conferred by specialist foraging behavior in the face of the difficulties herbivores have choosing suitable hosts [58] and/or ii) specialization in host species that provide herbivores enemy-free space in more structurally complex habitats [59], [60]. These avenues remain unexplored in a plant-herbivore network context.

The highest sub-network connectance was found under the most stressful abiotic conditions: Open vegetation and the dry season create conditions that could cause greater water stress in plants as a result of high temperatures and solar radiation. In turn, this could affect their nutrition (water and nitrogen content) and a defense (toughness and phenolic compounds content) attributes rendering the plants unsuitable for many lepidopteran herbivores [14], [18], [19], [42], [61]. For example, the coastal dune scrub and pioneer dune vegetation have the most extreme temperatures (up to 60°C in the scrub) and solar radiation [31], [32] of all our study sites. Moreover, the pioneer dune vegetation is subject to continual disturbance (burial under moving sand, strong winds with a high saline concentration, abrasion from blown sand, flooding and mechanical damage caused by tropical storms and hurricanes) [33], which makes host plants in these habitats highly unpredictable resource for herbivores [62]. In this context, both the unpredictable availability and low nutritional quality of the host plants in open habitats and present during the dry season can impose strong selective pressure on the herbivores [42], favoring species that are less specialized and less susceptible to variations in host resources [63], [64], [65]. In fact, in open vegetation habitats and during the dry season we found a predominance of herbivore species with relatively low host specificity. A generalist habit could make these species less susceptible to fluctuations in resource availability and quality under stressing conditions given that they have more food options.

At the present study most generalist herbivores were associated with host plants with large plant cover, mainly pioneer species adapted to stressful natural habitat conditions such as *Chamaecrista chamaecristoides*, *Crotalaria incana* and *Porophyllum punctatum* [33], [66]. These host species may have provided protection and food under the stressful abiotic conditions and reduced nutritional quality of the plants. Greater plant cover is a structural attribute that provides greater food availability to herbivores that can take advantage of this cover, and would allow them to compensate for the decrease in leaf nutrition quality [67], [68], and reduce larval desiccation under the high temperatures that characterize these environments.

In antagonistic networks, connectance - whether promoted by the range of host plants eaten by herbivores or by resource availability- can mitigate the loss of species when environmental conditions are unstable or stressful [12]. Therefore, the increase in connectance found in our study may have buffered the negative effects of natural disturbance (in stressful habitats and conditions) in plant-herbivore networks. Furthermore, the high network specialization was a structural attribute characteristic of forest habitats with stable host plant population size, compared with open habitats where specialist insects exhibited less spatio-temporal variation [69]. These results are evidence that the structure of antagonistic networks, just as it occurs in mutualistic networks, tends to reduce species vulnerability to extinction, which promotes community stability [70], [71].

In the studied plant-herbivore sub-networks, higher connectance was detected during the dry season in the pioneer habitat, and this seems similar to the changes that other trophic networks undergo when subjected to anthropic disturbances. The way in which the loss of arboreal cover reduced network complexity has been reported as a process of spatio-temporal homogenization of species and their interactions [12], [41], [72]. Moreover, both habitat fragmentation and early secondary succession can decrease the size of the network, resulting in systems that are more connected and less modular owing to the possible loss of specialists [12], [13], [72], [73]. At the study site, the pioneer habitat and the dry season may, represent a significant decrease in resource availability, particularly in the abundance of rare plants and their specialist herbivores. Additionally, the lack of arboreal cover in this habitat increases stressful abiotic conditions for herbivore larvae.

An important property of our plant-herbivore network was the spatio-temporal prevalence of vulnerability and interaction evenness. Regardless of habitat type or season, the ratio of herbivore to plant species was the lowest and no dominant interactions were recorded. Our results are similar to those of other trophic networks for habitats with differing degrees of fragmentation and where there was no evidence of variation in the structural attributes of the network [11]. In contrast, in parasitoid-herbivore trophic networks the spatio-temporal variation in interaction evenness is greater where deforested habitats are dominated by a reduced number of interactions [72]. In our study, the spatio-temporal prevalence of these indices could be a consequence of the high specialization plant-herbivore interactions [12], since the great majority of lepidopteran herbivores in tropical ecosystems are rare species that are low in abundance [14], even in disturbed tropical habitats [74]. Low vulnerability could also be a consequence of the loss of higher trophic levels as occurs in very small habitat fragments [12] [75].

In recent years, there have been efforts to assess the spatio-temporal variation of ecological networks [44], [45], [47], [76]. To our knowledge, this is the first study to address how plant-herbivore trophic networks vary in space and time at small scales, and to evaluate simultaneously the role of habitat complexity, resource availability and seasonality in such variation. This approach opens a new frontier in the study of plant-herbivore interactions in an effort to understand the processes that drive the modular and specialized patterns observed in antagonistic interaction networks.

Acknowledgments

We especially wish to acknowledge Michelle Ramos-Robles for logistical support in the field and caterpillars rearing. We would like to thank also Arturo Bonet Ceballos, María de Jesús Peralta Méndez, Gibrán Pérez Toledo for their support during fieldwork. The suggestions made by Víctor Parra-Tabla at several stages of our work enriched greatly this manuscript. We thank Wesley Dáttilo for his support during modularity analysis. We thank Carlos Durán and Adolfo Ibarra for taxonomic identification of plants and lepidopterans, respectively.

Author Contributions

Conceived and designed the experiments: ALC CDC KB VRG. Performed the experiments: ALC. Analyzed the data: ALC CDC. Contributed reagents/materials/analysis tools: CDC VRG KB. Wrote the paper: ALC CDC KB VRG.

References

1. Price PW (2002) Resource-driven terrestrial interaction webs. Ecol Res 17: 241–247.
2. Futuyma D, Agrawal AA (2009) Macroevolution and the biological diversity of plants and herbivores. Proc Natl Acad Sci USA 84: 8054–18061.
3. Novotny V, Basset Y (2005) Host specificity of insect herbivores in tropical forests. Proc Biol Sci 272: 1083–1090.
4. Lewinsohn TM, Novotny V, Basset Y (2005) Insects on plants: Diversity of herbivore assemblages revisited. Annu Rev Ecol Evol Syst 36: 597–620.
5. Coley PD, Barone JA (1996) Herbivory and plant defenses in tropical forests. Annu Rev Ecol Syst 27: 305–335.
6. Thébault E, Fontaine C (2010) Stability of ecological communities and the architecture of mutualistic and trophic networks. Science 329: 853–856.
7. Cagnolo L, Salvo A, Valladares G (2011) Network topology: patterns and mechanisms in plant-herbivore and host-parasitoid food webs. J AnimEcol 12: 1–10.
8. Prado PI, Lewinsohn TM (2004) Compartments in insect-plant associations and their consequences for community structure. J Anim Ecol 73: 1168–1178.
9. Lewinsohn TM, Prado PI, Jordano P, Bascompte J, Olesen JM (2006) Structure in plant-animal interaction assemblages. Oikos 113: 174–84.
10. Vazquez DP, Bluthgen N, Cagnolo L, Chacoff NP (2009) The determinants of patterns in mutualistic networks: a review. Ann Bot 103: 1445–1457.
11. Kaartinen R, Roslin T (2011) Shrinking by numbers: landscape context affects the species composition but not the quantitative structure of local food webs. J Anim Ecol 80: 622–631.
12. Valladares G, Cagnolo L, Salvo A (2012) Forest fragmentation leads to food web contraction. Oikos 121: 299–305.
13. Villa-Galaviz E, Boege K, del-Val E (2012) Resilience in plant-herbivore networks during secondary succession. PLoS ONE 7(12): e53009.
14. Janzen DH (1993) Caterpillar seasonality in a Costa Rican dry forest. In: Stamp NE, Casey TM, editors. Caterpillars. Ecological and evolutionary constraints on foraging. Chapman and Hall, New York. pp. 448–477.
15. Basset Y (1996) Local communities of arboreal herbivores in Papua New Guinea: predictors of insect variables. Ecology 77: 1909–1916.
16. Poorter L, de Plassche MV, Willems S, Boot RGA (2004) Leaf traits and herbivory rates of tropical tree species differing in successional status. Plant Biol (Stuttg) 6: 746–754.
17. Boege K (2005) Herbivore attack in *Casearia nitida* influenced by plant ontogenetic variation in foliage quality and plant architecture. Oecologia 143: 117–125.
18. Murakami M, Ichie T, Hirao T (2008) Beta-diversity of lepidopteran larval communities in a Japanese temperate forest: effects of phenology and tree species. Ecol Res 23: 179–187.
19. Trotter III, Cobb RT, Whitham TG (2008) Arthropod community diversity and trophic structure: a comparison between extremes of plant stress. Ecol Entomol 33: 1–11.
20. Brehm G, Homeier J, Fiedler K (2003) Beta diversity of geometrid moths (Lepidoptera: Geometridae) in an Andean montane rainforest. Divers Distrib 9: 351–366.
21. Jeffries JM, Marquis JR, Forkner RE (2006) Forest age influences oak insect herbivore community structure, richness and density. Ecol Appl 16: 901–912.
22. Janzen DH (1988) Ecological Characterization of a Costa Rican Dry Forest Caterpillar Fauna. Biotropica 20: 120–135.
23. Bascompte J, Jordano P (2007) Plant-animal mutualistic networks: the architecture of biodiversity. Annu Rev Ecol Evol Syst 38: 567–593.
24. Rico-Gray V, Díaz-Castelazo C, Ramirez-Hernandez A, Guimaraes PR, Holland JN (2012) Abiotic factors shape temporal variation in the structure of an ant-plant network. Arthropod-Plant Interact 6: 289–295.
25. Sánchez-Galván IR, Díaz-Castelazo C, Rico-Gray V (2012) Effect of Hurricane Karl on a plant–ant network occurring in coastal Veracruz, Mexico. J Trop Ecol 28: 603–609.
26. de Ruiter PC, Wolters V, Moore JC, Winemiller KO (2005) Food web ecology: playing Jenga and beyond. Science 309: 68–71.
27. Blüthgen N, Menzel F, Bluthgen N (2006) Measuring specialization in species interaction networks. BMC Ecology 6: 9.
28. Thompson JN (1999) What we know and do not know about coevolution: insect herbivores and plants as a test case. In: Herbivores: Between Plants and Predators, H Olff, V.K Brown and R.H Drent, editors. Oxford, Blackwell Science. pp. 7–30.
29. Novotny V, Miller SE, Leps J, Bito D, Janda M, et al. (2004) No tree an island: the plant-caterpillar food web of secondary rainforest in New Guinea. Ecol Lett 7: 1090–1100.
30. Castillo-Campos G, Travieso-Bello AC (2006) La flora. In: Moreno-Casasola P, editor. Entornos Veracruzanos: la costa de La Mancha. Instituto de Ecología AC, Xalapa. pp. 171–204.
31. Moreno-Casasola P, Travieso AC (2006) Las playas y dunas. In: Moreno-Casasola P, editors. Entornos Veracruzanos: la costa de La Mancha. Instituto de Ecología AC, Xalapa. pp. 205–220.
32. Moreno-Casasola P (1982) Ecología de la vegetación de dunas costeras: factores fisicos. Biotica 7: 577–602.
33. Pérez-Maqueo O (1995) Análisis del efecto de los disturbios en la dinámica de la playa de El Morro de La Mancha, Ver. México DF: UNAM. 65 p.
34. Travieso-Bello AC, Campos (2006) Los componentes del paisaje. In: Moreno-Casasola P, editors. Entornos Veracruzanos: la costa de La Mancha. Instituto de Ecología AC, Xalapa. pp. 140–150.
35. Olesen JM, Bascompte J, Dupont YL, Jordano P (2007) The modularity of pollination networks. Proc Natl Acad Sci USA 104 (50): 19891–19896.
36. Marquitti FMD, Guimaraes PR, Pires MM, Bittencourt LF (2013) MODULAR: Software for the Autonomous Computation of Modularity in Large Network Sets. Ecography arXiv: 1304.2917.
37. R Core Team (2013). R: A language and environment for statistical computing. R Foundation for Statistical Computing, Vienna, Austria. URL: http://www.R-project.org/.
38. Quinn GP, Keough MJ (2002) Experimental design and data analysis for biologists. Cambridge University Press. 537 pp.
39. Dixon P (2009) VEGAN, a package of R functions for community ecology. J Veg Sci: 927–930.
40. Anderson MJ (2001) A new method for non-parametric multivariate analysis of variance. Austral Ecol 26: 32–46.
41. Laliberté E, Tylianakis JM (2010) Deforestation homogenizes tropical parasitoid-host networks. Ecology 91: 1740–1747.
42. Barber NA, Marquis RJ (2011) Leaf quality, predators, and stochastic processes in the assembly of a diverse herbivore community. Ecology 92: 699–708.
43. Lepš J, Novotny V, Basset Y (2001) Habitat and successional status of plants in relation to the communities of their leaf-chewing herbivores in Papua New Guinea. J Ecol 89: 186–199.
44. Alarcón R, Waser NM, Ollerton J (2008) Year-to-year variation in the topology of a plant-pollinator interaction network. Oikos 117: 1796–1807.
45. Olesen JM, Bascompte J, Elberling H, Jordano P (2008) Temporal dynamics in a pollination network. Ecology 89: 1573–1582.
46. Dupont YL, Olesen JM (2009) Ecological modules and roles of species in heathland plant–insect flower visitor networks. J Anim Ecol 78: 346–353.
47. Dupont YL, Padrón B, Olesen JM, Petanidou T (2009) Spatio-temporal variation in the structure of pollination networks. Oikos 118: 1261–1269.
48. Fontaine C, Thébault E, Dajoz I (2009) Are insect pollinators more generalist than insect herbivores? Proc Biol Sci 276: 3027–3030.
49. Chacoff NP, Vázquez DP, Lomáscolo SB, Stevani EL, Dorado J, et al. (2012) Evaluating sampling completeness in a desert plant–pollinator network J An Ecol 81(1): 190–200.
50. Price PW, Abrahamson WG, Hunter MD, Melika G (2004) Using gall wasps on oaks to test broad ecological concepts. Conserv Biol 18: 1405–1416.
51. Novotny V, Basset Y (2000) Rare species in communities of tropical insect herbivores: pondering the mystery of singletons. Oikos 89: 564–572.
52. Heleno R, Devoto M, Pocok M (2012). Connectance of species interaction networks and conservation value: is it any good to be well connected? Ecol Indic 14: 7–10.
53. Dormann CF, Fründ J, Blüthgen N, Gruber B (2009) Indices, graphs and null models: analyzing bipartite ecological networks. The Open Ecology J 2: 7–24.
54. Rausher DM, Papaj DR (1983) Host plant selection by *Battus philenor* butterflies: evidence for individual differences in foraging behavior. Anim Behav 31: 341–334.
55. Rausher MD, Odendaal FJ (1987) Switching and the pattern of host use by *Battus philenor* butterflies. Ecology 68: 869–877.
56. Nylin S (1988) Host plant specialization and seasonality in a polyphagous butterfly, *Polygonia C-Album* (Nymphalidae). Oikos 53: 381–386.
57. Lewinsohn TM, Roslin T (2008) Four ways towards tropical herbivore megadiversity. Ecol Letts 11: 398–416.
58. Bernays EA (2001) Neural limitations in phytophagous insects: implications for diet breadth and evolution of host affiliation. Annu Rev Entomol 46: 703–727.
59. Jeffries MJ, Lawton JH (1984) Enemy free space and the structure of ecological communities. Biol J Linn Soc Lond 23: 269–286.
60. Bernays E, Graham M (1988) On the evolution of host specificity in phytophagous arthropods. Ecology 69: 886–892.
61. Staley JT, Mortimer SR, Masters GJ, Morecroft MD, Brown VK, et al. (2006) Drought stress differentially affects leaf-mining species. Ecol Entomol 31: 460–469.
62. Rico-Gray V, Castro G (1996) Effect of an ant-aphid-plant interaction on the reproductive fitness of *Paullinia fuscescens* (Sapindaceae). Southwest Nat 41: 434–440.
63. Gaston KJ, Lawton JH (1988) Patterns in body size, population dynamics, and regional distribution of bracken herbivores. Am Nat 132: 662–680.
64. Redfearn A, Pimm SL (1988) Population variability and polyphagy in herbivorous insect communities. Ecol Monogr 58: 39–55.
65. Hughes JB (2000) The scale of resource specialization and the distribution and abundance of lycaenid butterflies. Oecologia 123: 375–383.
66. Martínez ML (1994) Sobrevivencia y establecimiento de plántulas de una especie colonizadora de dunas costeras: *Chamaecrista chamaecristoides*. México DF: UNAM. México. 64 p.
67. Bezemer TM, Jones TH (1998) Plant-insect herbivore interactions in elevated atmospheric CO2: quantitative analyses and guild effects. Oikos 82: 212–222.
68. Whittaker JB (2001) Insects and plants in a changing atmosphere. J Ecol 89: 507–518.

69. Lepš J, Spitzer K, Jaroš J (1998) Food plants, species composition and variability of the moth community in undisturbed forest. Oikos 81: 538–548.
70. Dunne JA, Williams RJ, Martinez ND (2002) Food-web structure and network theory: the role of connectance and size. Proc Natl Acad Sci USA 99: 12917–12922.
71. Thébault E, Huber V, Loreau M (2007) Cascading extinctions and ecosystem functioning: contrasting effects of diversity depending on food web structure. Oikos 116: 163–173.
72. Tylianakis JM, Tscharntke T, Lewis OT (2007) Habitat modification alters the structure of tropical host-parasitoid food webs. Nature 445: 202–205.
73. Melián C, Bascompte J (2002) Complex networks: two ways to be robust? Ecol Lett 5: 705–708.
74. López-Carretero A (2010) Composición y diversidad de lepidópteros en la cronosecuencia sucesional del bosque tropical caducifolio, consecuencias sobre la herbivoría de *Casearia nitida*. México DF: UNAM. 86 p.
75. Holt RD, Lawton JH, Polis GA, Martinez ND (1999) Trophic rank and the species-area relationship. Ecology 80: 1495–1504.
76. Díaz-Castelazo C, Guimarães PR, Jordano P, Thompson JN, Marquis RJ, et al. (2010) Changes of a mutualistic network over time: reanalysis over a 10-year period. Ecology 9: 793–801.

Prioritizing Tiger Conservation through Landscape Genetics and Habitat Linkages

Bibek Yumnam[1], Yadvendradev V. Jhala[1]*, Qamar Qureshi[1], Jesus E. Maldonado[2,3], Rajesh Gopal[4], Swati Saini[1], Y. Srinivas[1], Robert C. Fleischer[2]

1 Wildlife Institute of India, Chandrabani, Dehradun 248001, India, **2** Smithsonian Conservation Biology Institute, National Zoological Park, 3001 Connecticut Avenue, Washington, D.C. 20008, United States of America, **3** Department of Vertebrate Zoology, National Museum of Natural History, Smithsonian Institution, Washington, D.C. 20013, United States of America, **4** National Tiger Conservation Authority, Bikaneer House, Shah Jahan Road, New Delhi 110011, India

Abstract

Even with global support for tiger (*Panthera tigris*) conservation their survival is threatened by poaching, habitat loss and isolation. Currently about 3,000 wild tigers persist in small fragmented populations within seven percent of their historic range. Identifying and securing habitat linkages that connect source populations for maintaining landscape-level gene flow is an important long-term conservation strategy for endangered carnivores. However, habitat corridors that link regional tiger populations are often lost to development projects due to lack of objective evidence on their importance. Here, we use individual based genetic analysis in combination with landscape permeability models to identify and prioritize movement corridors across seven tiger populations within the Central Indian Landscape. By using a panel of 11 microsatellites we identified 169 individual tigers from 587 scat and 17 tissue samples. We detected four genetic clusters within Central India with limited gene flow among three of them. Bayesian and likelihood analyses identified 17 tigers as having recent immigrant ancestry. Spatially explicit tiger occupancy obtained from extensive landscape-scale surveys across 76,913 km^2 of forest habitat was found to be only 21,290 km^2. After accounting for detection bias, the covariates that best explained tiger occupancy were large, remote, dense forest patches; large ungulate abundance, and low human footprint. We used tiger occupancy probability to parameterize habitat permeability for modeling habitat linkages using least-cost and circuit theory pathway analyses. Pairwise genetic differences (F_{ST}) between populations were better explained by modeled linkage costs ($r>0.5$, $p<0.05$) compared to Euclidean distances, which was in consonance with observed habitat fragmentation. The results of our study highlight that many corridors may still be functional as there is evidence of contemporary migration. Conservation efforts should provide legal status to corridors, use smart green infrastructure to mitigate development impacts, and restore habitats where connectivity has been lost.

Editor: Alfred L. Roca, University of Illinois at Urbana-Champaign, United States of America

Funding: Funding provided by the National Tiger Conservation Authority: YVJ QQ, Fulbright Fellowship: BY, Wildlife Institute of India: YVJ BY. The funders had no role in study design, data collection and analysis, decision to publish, or preparation of the manuscript.

Competing Interests: Co-author JEM is a PLOS ONE Editorial Board member. This does not alter the authors' adherence to PLOS ONE policies on sharing data and materials.

* Email: jhalay@wii.gov.in

Introduction

By virtue of being at the top of the food chain, large carnivores occur at low densities, have large home ranges and therefore require vast areas to harbor viable populations [1]. Since historical times, large carnivores have been in conflict with humans for food and resources, often resulting in their demise [2]. Habitat destruction, excessive hunting by humans and the use of body parts for traditional medicine have extirpated many populations [3,4,5] while severely reducing, fragmenting, and isolating most others to varying degrees [6,7]. Small and isolated populations are prone to local extirpation [8,9]. Managing such populations in a metapopulation framework [10,11] by connecting them through habitat corridors [12,13] so that individuals have the opportunity to disperse, establish residency and reproduce, reduces the overall risk of extinction [1,14]. Much of the global conservation policy on endangered species is centered on land allocation schemes for securing source populations [15], promoting and maintaining connectivity between fragmented populations [16,17]. Land is one of the most prized resources, and a major challenge to this conservation approach is the difficulty in convincing governments and policy makers on its allocation for conservation purposes. This problem is further compounded when objective criteria for delineating corridor habitats or documenting their functionality based on rigorous scientific data are lacking. As a result, conservation relies heavily on expert opinion and models of corridor connectivity that have little empirical validation.

The tiger (*Panthera tigris*) acts as a flagship species for the conservation of forested ecosystems throughout its range in Asia [18]. Conserving the tiger typifies the prospects and challenges inherent in the current paradigm of fragmented small populations and landscape based conservation models in large carnivores [19]. Extant tiger populations are confined to less than seven percent of their historical range in patchily distributed habitats across a range of twelve regional tiger conservation landscapes (TCLs) in southern and north-eastern Asia [20]. Six global priority TCLs of long-term tiger conservation significance are present in the Indian subcontinent. These Indian TCLs are important for global tiger recovery as they harbor over 50% of the estimated global population of ~3,000 wild tigers [21,22], and>60% of the global genetic variation in the species [23]. The high genetic variation seen in Indian tigers could be attributed to historically high population sizes, numbering about 50,000 individuals until *c.* 200 years ago, and habitat contiguity that permitted genetic exchange between the various regional tiger populations in the area [23]. Due to change in land ownership and forest use policy in the mid nineteenth century during British rule and again during the early years of India's independence a century later, much of the forest was cleared for timber and agricultural needs [24,25]. This change in land use combined with organized trophy hunting and bounty driven extermination resulted in severe decline, fragmentation and isolation of tiger populations throughout India [25,26]. Tiger conservation and subsequent population recovery in India began during the 1970s with the creation of a number of protected areas (Tiger Reserves) under the Project Tiger network in 1973 [18], and aided by comprehensive wildlife legislation (Wildlife Protection Act 1972, [27]). Under Project Tiger, the tiger was used as a flagship and umbrella species for conserving the biodiversity of India's forested ecosystems. However, even though extensive areas have been added to the protected area network, the future of tigers is under severe threat from commercial poaching, and extensive habitat fragmentation within the last two decades [15,18,21]. The rise in human-wildlife conflict and issues dealing with land rights of forest-based dwellers, as people are present both inside and outside tiger reserves, further vexes grass-roots conservation especially at a landscape-scale, and negatively impacts tiger dispersal capability and survival within TCLs [28]. These factors have precipitated the systematic decline in tiger and prey populations from both outside and inside reserves, as attested by the recent local extirpations in few areas [5,29].

Currently in India, the once contiguous tiger population is now fragmented with source populations primarily restricted to tiger reserves. A tiger reserve is legally mandated to designate a critical core area wherein human habitation and resource extraction is not permitted (Wildlife Protection Act 1972, amendment 2005 [30]). This core is surrounded by a buffer zone, which is essentially a multiple use area, wherein conservation objectives are to be given precedence over other land uses. Breeding populations of tigers are mostly located in the core area of tiger reserves, while the buffers usually serve as population sinks [22,28,31]. The size of these tiger reserves vary between 344 km^2 to 3,150 km^2 (average 1,321 km^2), with tiger densities ranging from about 0.1 to 20 individuals per 100 km^2 [22,31,32]. For a demographically viable tiger population, a minimum of 20 to 25 breeding units are believed to be essential [15,32,33]. As such, many extant tiger populations are by themselves inadequate for long-term persistence [33,34], either because of habitats harboring a low number of breeding tigers, small size of the protected area and/or ecological isolation from other populations. High spatial genetic structuring and small population size observed in today's Indian tiger populations [35] dictates preserving them in a metapopulation framework wherein individual populations are connected through a permeable habitat matrix and can occasionally exchange individuals [36,37]. This would result in re-colonization of suitable habitat patches where tigers have become locally extinct and 'rescue' declining local populations from extinction by immigrants [37,38]. Understanding and managing the metapopulation framework of extant tiger populations is an important strategy for ensuring their long-term conservation. This approach entails strict preservation of source populations in protected areas and informed conservation strategies across tiger landscapes.

Due to the relatively high *K* selected life history traits of the tiger in comparison to other large cats, dispersal and immigration play a vital role in long-term viability of tiger populations [32]. Incidentally, it was likely due to the 'rescue effect' by immigrants from high-density populations and intact habitat corridors in the vicinity of Chitwan National Park, Nepal, which enabled the tiger population in the park during the 1930s to recover, even after heavy trophy hunting, to pre-decline levels in only three years [32]. In recent times, tigers have successfully recolonized Rajaji National Park, India, in the Shivalik-Gangetic Plain landscape, from connected source populations further east, within a decade of having completely disappeared from the area [39]. Small tiger populations that become isolated are likely to face extinction due to demographic stochasticity, inbreeding depression [40] and deterministic factors such as poaching [32,33], as witnessed in the small and isolated Indian tiger reserves of Sariska and Panna which recently suffered from local extinction events, although tigers were later re-introduced [5,29]. Habitat connectivity is integral to sustaining regional populations of tigers, as they require contiguous forest connectivity for dispersal and genetic exchange between populations [41]. Currently, within the six tiger occupied landscapes of India, habitat contiguity varies extensively, with the best being within the Western Ghats and the North East, while fragmentation is highest in the Shivalik-Gangetic Plain and the Central Indian Landscapes [42]. Most of the connecting habitats in these landscapes are not within the legal domain of protected areas and are often lost to burgeoning development demands of a growing economy and attrition by human consumptive uses. In India, the transfer of forest-land to other land uses requires approval from the Federal Government as outlined in the Forest (Conservation) Act 1980 [43]. Since Federal Government approvals are usually sought on a case-by-case basis, and rarely are the cumulative impacts of projects or landscape scale conservation significance of forest patches factored into decision making, such permissions are frequently granted [44]. However, when the Supreme Court of India and Federal Government Committees were presented with concrete scientific evidence on the significance of conserving these forest patches, development projects even of national interests have been stalled [45,46,47]. Unfortunately, scientific data rarely exist to substantiate the landscape-level conservation significance of forest patches that constitute habitat corridors, and crucial areas are often lost. Studies on spatial dispersal and gene flow to detect population units and migration between patches can provide a quantitative and formal assessment of corridor function and identify priority populations for conservation action.

Assessing gene flow in species across populations in complex fragmented habitats is critical to understand how landscapes structure genetic variation and maintain metapopulation connectivity. Unfortunately, the traditional validation of habitat connectivity through the direct observation of individual animal movement is logistically difficult as it would entail following the fates of many radio-collared or camera trapped individuals over a regional scale and spanning multiple generations. As a result, alternative genetic assignment methods based on individual clustering approaches [48,49] have gained popularity [7,50]. The integration of metapopulation genetic models with spatial analytic tools in a landscape genetics framework provides a quantitative approach for understanding the role of geography, habitat and land-use features either as barriers or facilitators to gene flow among natural populations [51]. Though initially restricted to analyses correlating with linear distances [51,52], the developing field of landscape genetics has now advanced to include more complicated connectivity modeling incorporating ways in which habitats are actually traversed in nature. The use of landscape heterogeneity patterns and habitat permeability obtained from Geographical Information Systems (GIS) layers to model habitat connectivity by least-cost pathway (LCP) analysis [53] and circuit theory based isolation-by-resistance (IBR) model [54,55] that permit gene-flow between populations provide an objective criteria for delineating and prioritizing habitat corridors. There is a small but rapidly growing body of literature investigating the relationship between genetic and corridor connectivity, with both LCP and IBR models finding promise in gene flow studies on taxa with lower dispersal capabilities and that readily form visible metapopulations such as amphibians (tiger salamanders, *Ambystoma sp.* [56,57]), to wide-ranging carnivore species (cougar, *Puma concolor* [58]; bobcat, *Lynx rufus* [59]; wolverine, *Gulo gulo* [60]; black bear, *Ursus americanus* [61]). Where available, researchers have incorporated information from animal habitat use and movement behavior in the cost parameterization schemes to approximate realistic paths of least resistance, as in Reding *et al.* [59]. However, such data are not readily obtainable, and hence the vast majority of studies rely on expert opinion and *a priori* assumptions on animal presence to assign cost schemes and parameterize landscape resistance to gene flow. Although informative, the parameterization schemes used in landscape resistance surfaces to model movement paths and the assignment of cost schemes to grids in GIS rasters could easily introduce biases which may be more reflective of habitats as perceived by humans rather than by animals [62,63]. Incorporating information obtained from fine-scale species and landscape-specific ground data on suitable habitat, cover, prey availability, disturbance and threats in considering the attribute of surrounding cells, is one way which could help reduce subjectivity involved when assessing resistance or cost of a cell and the likelihood of path usage [63].

In this study, we investigate patterns of landscape heterogeneity and spatial genetic structuring to identify barriers and minimal habitat corridors for gene flow between populations within the fragmented tiger habitats in Central India. The Central Indian landscape is a globally recognized area for tiger conservation, with significant potential for long-term persistence of the species [15,31]. The area supports one of the largest global concentrations of tiger populations (~20% of an estimated 1,700 adult Indian tigers, [31]) in patchily connected habitats. Although the populations were historically connected, rapid infrastructural development and urbanization in recent years threaten to form permanent barriers to dispersing tigers by isolating tenuously connected small populations, thereby effectively reducing long-term metapopulation persistence. Recent population and spatial genetic studies have observed low genetic structure among populations indicative of gene flow [64,65] and long-range dispersal which are affected by increasing urbanization in the area [66]. Although tigers can move huge distances in undisturbed habitats [67], the complex fragmented habitat mosaic in the area, interspersed with high density human settlements and increasingly urbanized centers, have generally been thought to limit long-range dispersal [41]. Dispersal in tigers, like in lions (*Panthera leo* [19]) is male biased, as female offspring tend to reside and breed close to their maternal ranges, while male offspring disperse long distances and establish home ranges far from their natal areas [41]. This study explores a strategy that utilizes genetic assignment methods to detect population genetic structuring and determine which populations are in migratory contact, extensive occupancy modeling and GIS analysis to delineate structural connectivity between populations, and a correlation process between landscape connectivity versus population pairwise genetic distances to determine which of the movement cost schemes and modeled corridors best explain the observed genetic structuring in the area.

We extensively and intensively collected scat and a few tissue samples across seven tiger reserves in the Central Indian Landscape and first identify tiger individuals by genotyping the DNA extracts using eleven autosomal microsatellite loci. Next we assessed spatial genetic structuring and gene flow in the identified individuals through individual clustering methods. We use likelihood based [49] and Bayesian [48,68] assignment methods to detect first and second generation migrants between the identified genetic population clusters. Since resident tigers do not occur outside of forested habitat, we surveyed all of the forested area (76,913 km^2) within 185,100 km^2 of Central India. Based on our understanding of tiger ecology, we predicted *a priori* that tigers should occur in vast, undisturbed, productive forest patches, with high density of large wild ungulate prey, which would be negatively impacted by human disturbances [20,31,41,67]. We tested these *a priori* hypotheses by spatially explicit modeling of tiger occupancy that accounted for imperfect detections, using covariates obtained by remote sensing and ground surveys covering all forest patches within our study area. We then used this spatially explicit information of tiger occupancy as a resource selection probability function [69,70] to model habitat corridors joining tiger populations using LCP [71] and circuit theory [72] analyses in a GIS setting. With genetic data we tested if the observed population structure and dispersal between populations is in concordance with ground reality of tiger occupancy and existing habitat connectivity. Our comprehensive study highlights the importance of particular tiger source populations and intervening forest corridors for maintaining metapopulation structure within Central India. It provides a basis to formulate conservation policy and assist informed decision making for land-use planning at the landscape scale.

Materials and Methods

Ethics Statement

The majority of field sampling was conducted non-invasively from tiger scat, without animal capture and handling. Permits for collection of tiger scat samples were obtained from the National Tiger Conservation Authority and the State Forest Departments. Capture and radio collaring of tigers required the approval of the Ministry of Environment and Forests, Government of India and the Chief Wildlife Warden, Madhya Pradesh State, under the Wildlife (Protection) Act 1972. The permissions define the conditions required for capture of tigers, which include an approved protocol and participation by a Park Official and supervision by a qualified veterinarian in the capture and collaring exercise. Both these permissions were obtained and strictly adhered to. Capture operations were conducted by trained veterinarians and wildlife biologists as per the protocols of the Wildlife Institute of India and the National Tiger Conservation Authority. A tiger tissue sample was obtained from Satpura Tiger Reserve where the tiger died due to natural causes (was killed by another tiger in a territorial strife). This research project was conceived and radio collaring reported in this paper was done prior to the formation of an animal ethics committee at the Wildlife Institute of India.

Study Area

The present study was carried out in the global priority tiger conservation landscape of Central India within the states of Madhya Pradesh, Maharashtra, and Chhattisgarh (Figure 1). A forested area of 76,913 km^2 (20.1–23.5°N and 76.5–81.5°E) covering the seven tiger reserves of Melghat, Satpura, Pench, Kanha, Tadoba, Achanakmar, and Bandhavgarh along with their buffer zones, corridor habitats and adjoining forested habitats were sampled. Sampled sites covered different types of tiger habitats found in Central India ranging from the tropical moist-deciduous Sal (*Shorea robusta*) forests in Kanha and Bandhavgarh to tropical dry-deciduous teak (*Tectona grandis*) dominated forests in Pench, Tadoba and Melghat. The topography varied from about 200 meters above sea level (m a.s.l.) in the low-lying hills to the *dadar* plateaus and meadows in Kanha (500 m a.s.l.) and the rugged Satpura ranges (highest elevation 1,352 m a.s.l.). The rainfall, primarily restricted to the monsoon season (late June to September end) ranged between 1,000 to 2,200 mm per year. The large mammal fauna found in the region included tigers (*Panthera tigris*), leopards (*P. pardus*), sloth bears (*Melursus ursinus*), dholes (*Cuon alpinus*), gaurs (*Bos gaurus*), wild pigs (*Sus scrofa*), sambar deer (*Rusa unicolor*), chital deer (*Axis axis*), barking deer (*Muntiacus muntjak*), swamp deer (*Rucervus duvaucelii*) and nilgai antelopes (*Boselaphus tragocamelus*).

Genetic Sampling and Laboratory Work

Blood from 16 radio-collared tigers, one tissue sample from a dead individual and 587 putative tiger scat (faeces) samples were collected between 2006 and 2011 from the seven Tiger Reserves, and at a few intervening forest corridors in the area (Figure 1 and Table 1). The number of samples obtained was largely proportional to the population size of tigers in that region. However, due to logistical constraints the Tadoba population was under sampled. Scats were stored, either dry with silica or in 75% ethanol, and kept at ambient temperature, prior to laboratory analysis. For each scat, a Global Positioning System (GPS) reading was taken and transferred into a GIS. Scat DNA extractions were performed in a room dedicated to low-copy DNA extraction, using the guanidine isothiocyanate - silica extraction protocol [73]. For every extraction, negative controls composed of reagent only without the scat sample were included to monitor contamination. Extractions from blood and tissue samples were carried out using the DNeasy blood and tissue kit (QIAGEN, Germany).

Tiger scats were identified from among the field collected scats through PCR and *Bam*HI restriction enzyme digestion of the mitochondrial DNA cytochrome *b* gene (Figure S1; see Methods S1 and Tables S10 and S11 for details on protocols and reference species used for PCR). Unambiguously assigned tiger scats and blood and tissue samples were individually identified using a panel of eleven microsatellite loci, derived from domestic cat [74] and tiger [75,76]. The loci consisted of three dinucleotide (Fca304, Fca954, 6Hdz700), three trinucleotide (Pati01, Pati09, Pati15) and five tetranucleotide repeat markers (Fca441, F85, F53, F124,Pati18), variously labeled at the 5′ end of each forward primer with 6–FAM, PET, VIC and NED dyes. Gender identification in individual tigers was carried out by amplifying the Y chromosome linked SRY gene and an X chromosome microsatellite locus Fca651. PCR amplifications were carried out in 10 μl reactions with a multiple panel of 3 to 4 loci using the Multiplex PCR kit (QIAGEN) according to the manufacturer's instructions. Amplified products were resolved on the ABI 3130 Genetic Analyzer and GENEMAPPER 3.7 (Applied Biosystems, USA) was used to score allele sizes. To limit genotyping error due to allelic dropout (nonamplification of one allele in a heterozygote), multiple PCR replicates were conducted as in Navidi *et al.* [77]. Heterozygotes were confirmed with at least two independent replicates and homozygotes with five replicates. The genotype data were checked on MICROCHECKER [78] for identifying and correcting genotyping errors such as those that arose from stuttering patterns, null alleles and small-allele dominance.

Identification of Individuals and Descriptive Statistics

In order to investigate the power of the eleven microsatellite loci to distinguish among closely related individuals in the same population, the conservative sibling probability of identity (*PI sib*) statistic [79,80] was computed in GIMLET 1.1 [81]. Unique multilocus genotypes were identified using the *Identity* analysis option in CERVUS 3.0 [82,83]. Samples that showed mismatches at up to two loci were re-examined for possible genotyping errors and allelic drop-out, and again amplified thrice in order to confirm the multilocus genotypes before assigning them as unique individuals. Multiple replicates of the same individual from the same locality were discarded and only unique multilocus genotypes were used for all further analyses. We used CERVUS 3.0, GENEPOP 4.1 [84,85], GENALEX 6.3 [86] and FSTAT 2.9.3 [87] to calculate the following descriptive statistics: (i) number of alleles per locus, (ii) observed and expected heterozygosity, (iii) tests for deviation from Hardy-Weinberg Equilibrium (HWE), (iv) significance values for linkage disequilibrium (LD) among all pairs of loci, and (v) estimates of population pair-wise F_{ST} [88] and R_{ST} [89] values. Loci with null alleles were flagged by MICROCHECKER, and we tested for deviation from HWE using both null allele adjusted and unadjusted (observed allele frequencies) genotypes. We also tested for associations between genetic variation and estimated population size through Spearman's rank correlation using the pspearman package [90] in R (http://cran.r-project.org).

Population Genetic Structure

We used two types of individual-based analyses to assess genetic differences among individuals and assignment patterns of tigers to populations. First, a Principal Coordinates analysis (PCoA) based on pair-wise Phi_{PT} genetic distances [91] was carried out in GENALEX and the scatter of population-wise individual assignments was plotted on the first three PCo axes using NCSS (www.ncss.com) to understand broad spatial patterns of populations structure in the landscape. Next, we used the Bayesian individual clustering approach in STRUCTURE 2.3.3 [48] to detect population structure among sampled localities in the area by assigning sampled individuals into a number of clusters (K) based on the multilocus genotype data alone. The clustering process ensures that for identified population clusters, deviations from Hardy Weinberg and linkage equilibrium are minimized. We analyzed our data in STRUCTURE by using the admixed model and correlated allele frequencies option to carry out thirty independent simulations at each K ($K = 1$ to 10), with a burn-in length of one million Monte Carlo Markov Chain (MCMC) steps and data collection phase of five million MCMC iterations. These run times were sufficient to ensure convergence of the Markov chains, and all runs were carried out both with (locprior = 1) and without (locprior = 0) using prior sampling locality information. The true K or most likely number of population clusters in the dataset was inferred from (i) the *ad hoc* parameter of log-likelihood change in probability of individual assignments to K clusters (*Ln P(K)*, [48]), and (ii) the second order rate of change in the likelihood of K values (*delta K*, [92]). Both these values were computed from the STRUCTURE output using the program STRUCTURE HARVESTER v0.6.91 [93]. We also carried out

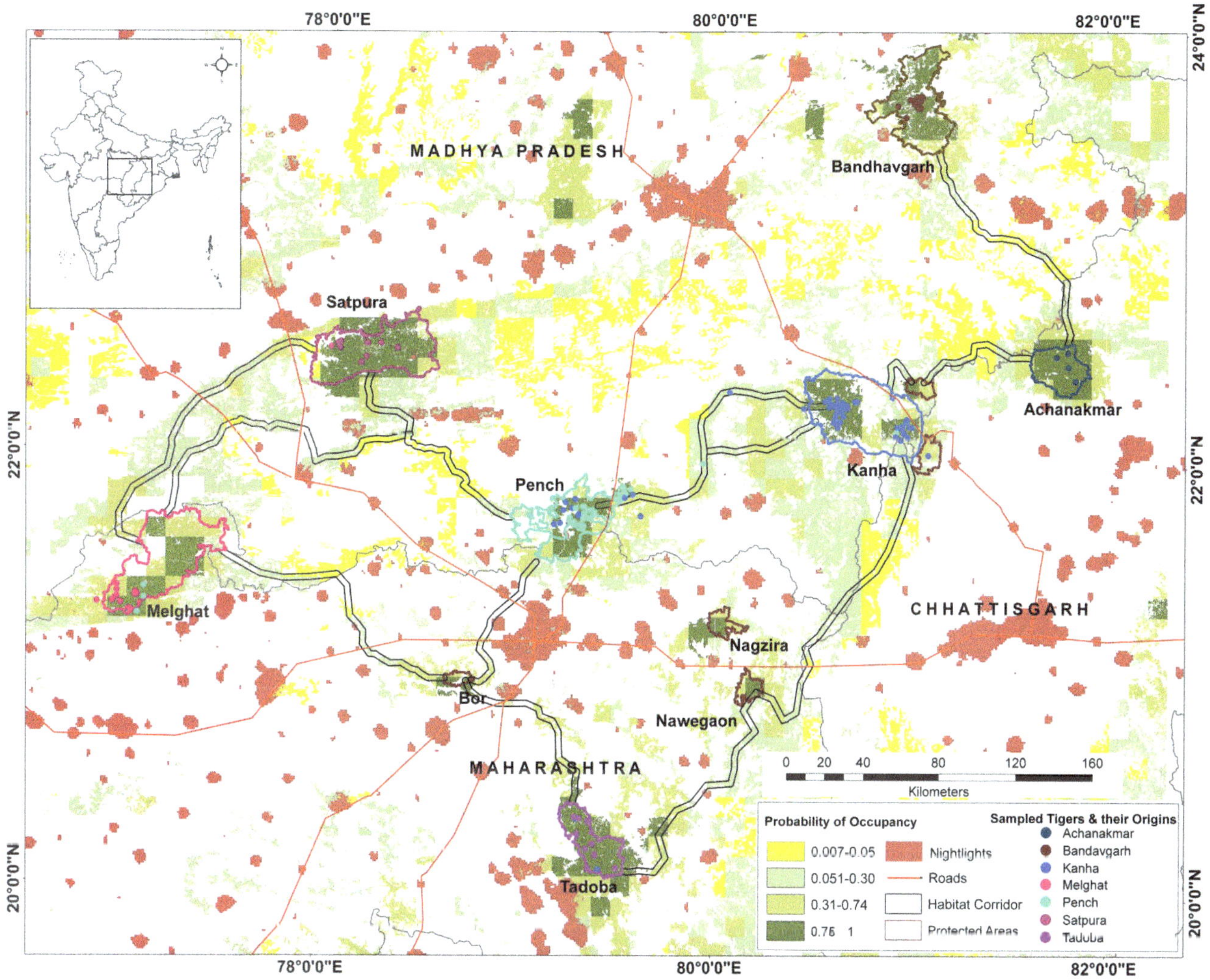

Figure 1. Study area map showing sampled sites, genotype locations and habitat connectivity. The study area of Central India spanning the states of Madhya Pradesh, Maharashtra and Chhattisgarh, showing tiger habitat (forest cover) coded with tiger occupancy probability, protected areas, human habitation (night lights), major roads and least-cost habitat corridors connecting tiger reserves. Individually genotyped tigers (n = 165) are shown as color coded dots at their sampled locations with their colors matching their genetically assigned population.

an Analysis of Molecular Variance (AMOVA, [91]) in GENALEX to compare the population clusters identified by the STRUCTURE analysis. Genetic variances were partitioned at two levels, viz. among all the STRUCTURE identified population groups (*K* clusters) and among subpopulations within each group.

Detection of Migrants

We used three Bayesian approaches implemented in STRUCTURE 2.3.3 [48], GENECLASS 2.0 [49] and BAYESASS 1.3 [68], for identification of migrant and admixed individuals. STRUCTURE was used to calculate the posterior probability of whether individuals are residents of their sampled population or migrants from other areas by incorporating the previously identified population cluster information with *a priori* designation of the migration rate (MIGPRIOR = 0.05). All other parameters and run times followed previous population clustering runs (as detailed above in the preceding methods section on analyzing population genetic structure). We detected no biases in *a priori* assignment of the migration rate as the selection of particular MIGPRIOR values (0.001 to 0.1) did not substantially influence the STRUCTURE output, therefore only results for MIGPRIOR = 0.05 are presented here.

Next, we used the likelihood-based estimator L_h/L_{max} in combination with the resampling algorithm of Paetkau *et al.* [94], implemented in the 'detect migrants' function in GENECLASS to exclusively identify first generation migrants, i.e. individuals assigned to a different population other than the sampled population. The Paetkau *et al.* [94] routine was selected on basis of its superior simulation scheme which closely mimics natural processes and results in accurate type I error rates. L_h/L_{max}, is the ratio of L_h, the likelihood of a given individual being assigned to its sampled population to L_{max}, the greatest likelihood among all sampled populations [94]. We employed the Bayesian criterion of Rannala and Mountain [95] in combination with the re-sampling algorithm of Paetkau *et al.* [94] using a simulated set of 10,000 area-specific genotypes, to determine migrant thresholds (type I error α levels of 0.01 and 0.05) of L_h/L_{max}. We considered a minimum log likelihood L_h/L_{max} ratio of 2.0, which corresponds to a 100 times probability of being cross-assigned, as the threshold level for determining putative migrant status of an individual [7].

Table 1. Area-wise estimates showing population extents, occupied habitats, sampling effort and number of tiger individuals identified.

Population	Naïve Occupancy (km^2)	Critical Core Area (km^2)	Tiger Reserve Size (km^2)	total scats/cyt *b* PCR amplified/ tiger scats/STR genotypes	other samples - blood (b), tissue (t)	all samples/all tiger STR genotypes/ individuals	estimated population size (SE range) *	% population sampled	identified tiger individuals		
									male	female	unknown sex
Melghat (M)	2,376	1,501	2,769	66/47/44/35	0	66/35/15	35 (30–39)	43	6	8	1
Satpura (S)	1,554	1,339	2,133	46/21/17/17	1 (t)	47/18/11	43 (42–46)	26	3	6	2
Pench (P)	2,174	669	1,921	137/94/86/82	3 (b)	140/85/51	65 (53–78)	78	25	24	2
Kanha Pench corridor (KPC)	1,013	*NA*	*NA*	26/24/5/5	0	26/5/5	*NA*	*NA*	2	2	1
Kanha (K)	1,850	917	2,052	159/67/65/56	12 (b)	171/68/50	60 (45–75)	83	22	27	1
Achanakmar (A)	904	626	914	23/16/5/5	0	23/5/4	12 (11–13)	33	2	2	0
Tadoba (T)	3,519	626	1728	32/20/17/17	0	32/17/11	69 (66–74)	16	5	3	3
Bandhavgarh (B)	1,700	717	1,537	98/41/36/33	1 (b)	99/34/22	59 (47–71)	37	9	13	0
Total	**15,090**	**6395**	**13,054**	**587/330/275/250**	**17**	**604/267/169**	**343**	**49**	**74**	**85**	**10**

* in Jhala *et al.* [31], *NA* - not available.

Third, the non-equilibrium Bayesian assignment test of BAYESASS was used to trace each individual's immigrant ancestry within the last two to three generations. Unlike STRUCTURE and GENECLASS analyses which require loci to follow Hardy-Weinberg and linkage equilibrium, BAYESASS is robust to violations from Hardy-Weinberg equilibrium as it measures contemporary gene flow within the last few generations based only on multilocus allele sharing among individuals [68]. To ensure convergence, a total run length of eight million MCMC iterations was performed, of which the first two million runs were the burn-in phase and the remaining six million runs comprised the data collection phase with the MCMC chain being sampled every 2,000 steps. Individual assignments and immigrant ancestries were calculated at a migration rate prior of 0.05, and keeping all other settings at default. Varying the prior rate (0.01 to 0.15) did not affect the results.

Lastly, to corroborate the results of the above assignment based migrant decisions, a likelihood based parentage analysis was carried out in CERVUS 3.0 [82,96] to identify likely parent-offspring relationships between putative migrants and an individual in the cross-assigned source population based on log of the odds (LOD) scores. The LOD score that is the natural logarithm of the overall likelihood ratio for each candidate parent is calculated by multiplying together the likelihood ratios at each locus. A positive LOD score means that the candidate parent is likely to be the true parent, whereas a negative value means that the candidate parent is less likely to be the true parent. We estimated LOD scores for strict (95%) and relaxed (80%) confidence limits as 7.0 and 4.9 respectively, which were calculated from a simulated set of 10,000 offspring and 300 candidate parent genotypes, assuming 25% of candidate parents were sampled, 93% of loci were typed with typing error of 0.01 to 0.10.

Designation of migrant status to an individual was contingent upon - (i) significant assignment of first generation migrant status in GENECLASS ($P<0.01$, $Lh/Lmax \geq 2.0$); (ii) observance of> 50% migrant or cross-assignment probability in STRUCTURE and BAYESASS; (iii) high assignment probability to first generation immigrant ancestry state (gen1>50%) in both STRUCTURE and BAYESASS; and (iv) high membership ($Q>$ 0.8) to a single non-home cluster in the STRUCTURE analysis without prior population information. Further, in most cases of putative migrants, the successful parentage assignment corroborated the migrant status of the individual tiger. We considered a conservative approach by identifying individuals as putative migrants only if all three programs suggested evidence of immigrant ancestry.

Estimation of Contemporary and Historical Migration Rates

In order to study gene flow across different timescales, we used the programs BAYESASS 1.3 [68] and MIGRATE 3.3.2 [97,98] to compare migration rates over contemporary and historical timescales, respectively. Although the two programs use different approaches to derive estimates of gene flow, both programs generate parameters from which a comparative estimate of the proportion of genetic migrants in the population per generation (m) can be inferred. In BAYESASS, a Bayesian approach incorporating an MCMC sampling scheme is used to estimate migration rates between pairs of populations over the last few (approximately <5) generations back. With an estimated generation time of four to five years in tigers [99], this period corresponds to a timescale of nearly 20–25 years ago. MIGRATE on the other hand, uses the coalescent to estimate the relative mutation-scaled effective population size, theta, θ_{Ne} ($4N_e\mu$; where

N_e is the effective population size and μ is the mutation rate) and asymmetric mutation-scaled immigration rate M (m/μ). The mutation-scaled immigration rate M, which is the immigration rate m divided by the mutation rate μ, is a measure of the importance of immigration events over mutation in contributing to variation in the population [97]. The relative effective population size, theta, is the number of individuals representing an idealized (Wright-Fisher) population that will result in the same amount of genetic drift as in the actual population. The number of migrants per generation, $4N_em$, is the product of theta and M. MIGRATE assumes mutation-migration-drift equilibrium with values of M and theta constant over time and parameter estimates in MIGRATE date back nearly $4Ne$ generations into the past (approximately thousands of years ago). Hence, these migration rates provide estimates of gene flow that post and pre-date the estimated time (approximately 600 years ago) when humans began to significantly alter the habitats in which these tigers currently live.

For both the BAYESASS and MIGRATE runs, we used the STRUCTURE defined population clusters to estimate pairwise migration rates. BAYESASS runs were performed as described in the preceding methods section on detecting migrants. A total of 8×10^6 MCMC iterations were carried out, by discarding the first 2×10^6 steps as burn-in and sampling at every 2,000 iteration intervals of the remaining 6×10^6 MCMC chain. Runs were carried out with a migration rate prior of 0.05 while other parameters were kept at default settings. The average result from three independent BAYESASS runs is presented. Estimates of historical gene flow and effective population size were carried out in MIGRATE by using the Bayesian approach and the Brownian motion model as an approximation for the step-wise microsatellite mutation model. Following initial trial runs, the Bayesian search criteria for the MCMC sampler was set at 10 replicates of one long chain of 50,000 steps with every 100 steps of the chain being recorded, producing a total of 5×10^7 visited parameter values. The initial 10^7 steps of the MCMC run were discarded as burn-in, and the remaining 4×10^7 runs were sampled. To increase efficiency of the sampler, we used four chain-heating temperatures of 1, 1.5, 3 and 10,000, which allows a more efficient exploration of the genealogy space. We used parameter estimates from the initial run to calculate starting values of theta, for use as new parameters during subsequent runs. Parameter estimates from the final run were similar to the results of the initial runs. All MIGRATE runs were carried out on the Bioportal cluster computing facility at the University of Oslo, Norway (https://www.bioportal.uio.no/; accessed 12 May 2013).

Detection of Genetic Bottleneck

To detect past occurrences of genetic bottleneck in the sampled populations, we evaluated three summary statistics - (i) Wilcoxon's sign rank test and (ii) mode-shift test, implemented in the program BOTTLENECK 1.2.02 [100], and (iii) M ratio test [101] implemented in ARLEQUIN 3.1 [102]. We were not interested in quantifying population expansion/decline or dating the time of and therefore avoided using other considerably lengthy Bayesian procedures [103,104]. Wilcoxon's test detects bottlenecks based on the probability of heterozygosity excess over that expected at mutation-drift equilibrium in a population. It is most effective at detecting historic bottlenecks occurring over approximately 2–4 N_e generations in the past. The mode-shift test is more suited for detecting bottlenecks within the last few dozen generations [105,106]. This test is based on the premise that a stable population at mutation-drift equilibrium will have a large proportion of alleles at low frequency and a smaller proportion at intermediate frequencies and few at high frequencies. The resulting allele proportions yield an L-shaped distribution. In bottlenecked populations the mode is shifted because of the rapid loss of alleles present at low frequency. We ran 10,000 simulations using the program BOTTLENECK in 5 populations (excluding KPC, ATR and TATR) under both the stepwise mutation model (SMM) and the two-phase mutation model (TPM) with 95% single step mutations and 5% multi-step mutations and a variance of 12 as recommended by Piry *et al.* [100]. P-values from the Wilcoxon's test were used to examine bottlenecks at each timescale and were assessed at the alpha 0.05 level. The M ratio ($M = k/r+1$) was calculated from the mean ratio of the number of alleles (k) and the allele size range (r). Assuming loci follow a generalized stepwise mutation model, the loss of rare alleles would diminish the value of k at a faster rate than r thereby a drop in the M ratio below a threshold of 0.68 would be suggestive of populations that experienced a recent bottleneck [101].

Field Data Collection for Occupancy Analysis

(A) Tiger Sign Surveys. The entire study area was divided into 10×10 km grids. Each grid that contained potential tiger habitat (all types of forest cover) was surveyed by replicate search paths for tiger sign. The number of surveys per grid ranged from 3 to 35, and was proportional to the amount of tiger habitat within each grid. Areas under agriculture, industry, and human habitation that were known to be non-habitat for tigers were not surveyed in an occupancy framework. Each survey consisted of a 5 km search for tiger signs with approximately one survey for every five km^2 of habitat. Surveys were not random, but instead conducted along features that were likely to have tiger sign (e.g. dirt roads, dry water courses, and animal trails) so as to maximize detections [22]. Surveys were conducted by the local guard and a local assistant who had intimate knowledge of the forest and were trained to observe and record tiger sign in pre-designed datasheets. All encounters of tiger pugmark track sets and scats were recorded. These were distinguished from those of other carnivores based on criteria described by Jhala *et al.* [107] and Karanth and Nichols [108]. A total of 79,000 km of search effort was invested in 15,800 replicate surveys between December 2009-February 2010 (cold and dry season) across the entire study landscape to adhere with the assumption of occupancy closure [70] and have minimal influence of weather (rainfall) on sign detections and distribution. A total of 1,851 grids were sampled.

(B) Prey Assessment. Within each forest beat, one or two permanent line transects of two to four km in length were delineated. Each transect was walked on two or three subsequent mornings (06:30 to 08:30 hrs) by two observers to record encounter rates of wild ungulates and domestic livestock. Data on number of each species seen and the length of transect were recorded to compute encounter rates of each species. In disturbed forests (outside of protected areas) wild ungulate densities were low, animals were shy, and difficult to record using line transects. Therefore, at every 400 m along the line transect we also sampled a plot of 20×2 m to record ungulate dung. Dung was visually distinguished to species [107] and dung density for each species, wild ungulates as a group, and domestic livestock was computed separately. Encounter rates of ungulates and dung density were used as indices of ungulate abundance. The number of transects within each 100 km^2 grid ranged from 1 to 24, and were proportional to the quantum of tiger habitat within that grid. The total effort invested in transect surveys was 71,168 km of walking on 26,688 occasions.

(C) Human Disturbance. At every 400 m along transects established for ungulate assessment a plot of 15 m radius was sampled to assess indices of human impact. Presence of (a) human/livestock trails within the plot, and (b) sighting of humans and livestock from the plot were recorded [107]. The number of plots within a 100 km^2 grid ranged from 5 to 120. The total number of plots sampled across the landscape was 51,073.

Remotely Sensed Variables

Remotely sensed data that depict landscape characteristics and human impacts were obtained from various sources and extracted at the 10×10 km grid resolution. Forest cover was obtained from the Forest Survey of India [109] that is based on IRS 1D LISS III satellite with 4 multispectral band data at 23.5 m resolution. Normalized Differential Vegetation Index (NDVI) information were derived from 1 km^2 Advanced Very High Resolution Radiometer (AVHRR) data, acquired from the National Aeronautics and Space Administration's (NASA) Television Infrared Observation Satellite (TIROS) (http://science.nasa.gov/missions/tiros/; accessed 23 Dec 2010). Road and drainage information were obtained from Digital Chart of the World (http://statisk.umb.no/ikf/gis/dcw/; accessed 20 Dec 2010). Protected area shape files were obtained from the wildlife database at the Wildlife Institute of India, National Tiger Conservation Authority and State Forest Departments of India. The Shuttle Radar Topography Mission has produced the most complete, high-resolution digital elevation model of the earth [110]. Within each 1 km^2 grid, this information was used for computing average elevation and the coefficient of variation of elevation used as a measure of terrain ruggedness. Night light data was obtained from the United States Air Force Defense Meteorological Satellite Program (DMSP) and National Oceanic and Atmospheric Administration's (NOAA) Operational Linescan System (OLS) (http://www.ngdc.noaa.gov/dmsp/sensors/ols.html; accessed 18 Dec 2010). Density of roads (length of paved road per km^2), and Euclidean distances to roads, protected areas and night lights were computed in ArcGIS 9.3 (www.esri.com) software.

Occupancy Modeling

Though sampling was done at the level of the forest beat (an administrative unit of about 15 km^2) so as to ensure an even distribution of sampling effort across the landscape, analysis was done at the scale of 1,851 grids, each of size 10×10 km. This grid size was chosen since it was larger than the average home range size of a tiger [111,112] and the size was relevant for subsequent administrative and managerial inputs. Sign surveys of 5 km independent spatial replicates within each grid [113] were modeled to address the issue of imperfect detections of tiger sign using the program PRESENCE v6.3 [114]. Detection of tiger signs was likely to depend on the abundance of tigers within a grid [115]. We first modeled the detection process by (i) keeping detection ($\widehat{P}$) constant across surveys, (ii) $\widehat{P}$ varying across surveys and (iii) $\widehat{P}$ as a function of tiger abundance in that grid, wherein we used average encounter rate of tiger sign as an index that surrogated tiger abundance [31,115]. The model that best explained the detection process based on Akaike Information Criteria (AIC) was then used in all subsequent models of tiger occupancy [70].

Tiger site occupancy was *a priori* expected to be positively influenced by (a) prey abundance, and (b) amount and quality of tiger habitat, and negatively influenced by (c) human disturbance [22,115,116]. We tested these hypotheses by modeling variables representing these factors as covariates using the logit link function in PRESENCE [70,114]. We initially generated data on 23 site covariates that represented landscape and habitat features (forest area, core forest area, forest patch size, NDVI, elevation, ruggedness, drainage density, rainfall, distance to protected area), prey availability (chital, sambar, wild pig and gaur encounter rates on line transect walks, and wild ungulate dung density), human disturbance (distance to night lights, distance to roads, humans and livestock encountered on transect walks, human/livestock trails within sampled plots, and livestock dung density) that could potentially explain tiger occupancy. These covariates were examined with exploratory data analyses for their interrelationships and relationship to tiger presence (by scatter plots, box plots, and correlation analysis) and based on this a subset of 16 variables was selected for inclusion as site covariates for occupancy modeling (see Table S1 for univariate coefficients). Since many of the covariates had high correlation coefficients between them, their contribution to explaining tiger occupancy would be redundant. To account for this collinearity and to reduce the dimensionality of the covariate matrix we extracted Principal Components (PCs) from 16 variables [117]. The varimax rotated PCs were further used as independent variables in a logit link function to model tiger occupancy in the program PRESENCE ([70], available for download from http://www.proteus.co.nz/). Model selection was done using AIC, and model fit was assessed by comparing the actual detection histories with simulations generated from 50,000 parametric bootstrap runs of the target model in PRESENCE. The over dispersion parameter $\hat{C}$ close to one suggests that the model provides an adequate description of the data, values of $\hat{C}$ greater than one suggests more variation in the data than expected by the model, and values less than one suggest less variation in the data. The standard errors of model estimates were adjusted by the square root of $\hat{C}$ as recommended by MacKenzie *et al.* [70]. Models were built using PCs that represented prey abundance, human disturbance and habitat quality; these were evaluated against the null model and the full model by their delta AIC values. A total of six models were evaluated for modeling tiger occupancy and coefficient estimates for all models with delta AIC <4 were averaged based on model weights [70] to arrive at occupancy probability (Ψ) in each grid.

Tiger Population Extents and Occupied Habitats

We used two approaches to estimate population extents and area of occupied habitats; (i) a more conservative approach wherein we considered only those grids that detected tiger sign as being occupied (the naïve estimate) and (ii) model inferred occupancy that corrected for detection bias and covariates in PRESENCE. Herein, landscape scale occupancy was computed by sum of cell occupancy probability values and divided by the total number of cells. Tiger habitat (forested area) in each grid was weighted by the tiger occupancy probability of that grid and summed across all grids to arrive at occupied tiger habitat for the landscape [115]. All adjacent tiger sign detected cells were joined and were considered to be occupied by a single tiger population.

Habitat Corridor Modeling

Grid based tiger occupancy probability (Ψ) obtained from PRESENCE was used as a measure of habitat suitability for tigers [69,70]. A cost surface for tiger habitat suitability across grids was generated as 1-Ψ. For the non-tiger habitat (human land uses) where tiger occupancy values were not available, we considered them permeable to tigers at higher costs than forested habitats, although human habitations (townships) were considered impermeable to tigers (see Methods S2 for details). These costs were used as a resistance layer for modeling habitat connectivity using LCP [63] and circuit theory [118] analyses. Based on these cost surfaces the resulting models would optimize connectivity by selecting high quality tiger habitat and minimize gaps formed by non-tiger habitat. Least-cost pathways (LCP) were modeled using PATHMATRIX [71], and resistance pathways were modeled using CIRCUITSCAPE [72]. Core areas of tiger reserves were considered as "sources" or areas of high potential from which tiger movement across paths of least resistance were modeled across the landscape. PATHMATRIX models several potential routes in a radiating manner from the "source" to connect to another adjacent "source". It then selects a single "least cost" pathway as the best alternative. CIRCUITSCAPE models connectivity through habitat swaths considering resistance to movement based on pixel cost and corridor width [118]. It provides one to several potential alternatives for joining sources and helps in identifying bottlenecks within the corridors. Since the Central Indian landscape is highly fragmented and human dominated with clearly defined and limited forested habitat, we could overlay LCP on high resolution Google Earth images and align them to match geographic features within occupancy grids, to delineate realistic corridors. These least cost corridors buffered by 1.5 km (LCC) were considered the minimal essential corridors joining two tiger reserves.

Genetic Structure, Migrants and Corridor Cost

The genetic structure between populations is a consequence of the amount of genetic drift to which each local population has been subjected, due to its local effective size, and/or due to its degree of demographic, geographic and ecological isolation [119]. Since all the Central Indian tigers likely belonged to a large, mostly contiguous population till a few hundred years ago [23], pairwise genetic distances between populations should reflect levels of differentiation and barriers to gene flow, i.e., the cost of movement between these populations. The proportion of migrants between population pairs would reflect gene flow in current or recent times while F_{ST} values would indicate genetic differences over historical, more long-term time scales. We therefore expected to have more migrants detected between geographically closer populations that had lower movement costs between them. This exploratory analysis would corroborate the short-term mechanisms (migration events) that result in long-term (F_{ST}) genetic differences between populations due to tiger movement across the landscape.

Pairwise F_{ST}, R_{ST} and Phi_{PT} genetic distance estimates, obtained from AMOVA analysis in GENALEX, were linearized using the formula $F_{ST}/(1 - F_{ST})$, as given by Rousset [120]. In order to determine which spatial model best explained genetic structuring, matrices of linearized pairwise genetic distances were correlated against matrices representing geographic distances (GGD), log-transformed geographic distances ($\log_{10}$ GGD), least-cost pathway distances (LCPD) least-cost corridor distances (LCCD), and resistance distances (RD). In addition, the effectiveness of the modeled corridors and spatial distance matrices was compared using partial correlations that allowed one model to be tested, while controlling for other competing models [55]. The biologically realistic model not only exhibited the highest significant, positive correlation, but also displayed significant positive partial correlations after controlling for each of the competing models. Mantel [121] and partial Mantel [122] tests were carried out with 10,000 randomizations in the program zt [123] to evaluate the significance of the correlations.

As patterns of isolation-by-distance (IBD) are known to bias tests of hierarchical structuring and vice-versa, we used the population clusters as a covariate in a partial Mantel test, to model the partial correlations between pairwise genetic distances and spatial distances, while controlling the effect of population clusters (following Meirmans [124]). Partial correlations between matrices representing pairwise genetic distances and spatial distance matrices were calculated with a third matrix describing whether population comparisons were made between (1) or within (0) the STRUCTURE identified clusters. A non-significant or negative partial correlation of genetic with geographic distance, after controlling for population clusters, would rule out patterns of underlying IBD in the observed genetic structure.

Results

Identified Individuals and Descriptive Statistics

Out of 587 scat samples, 330 scats were successfully PCR amplified using the felid specific mitochondrial DNA cytochrome *b* (mtDNA cyt *b*) primers, of which 275 were identified as tiger scat based on *Bam*HI restriction enzyme digestion profile of the PCR products (Table 1 and Figure S1). From the 275 tiger scats, only 250 samples yielded microsatellite genotype data at a minimum of seven loci to be considered for *Identity* analysis in CERVUS. We identified 169 individuals with 81 recaptures from a total of 267 microsatellite loci genotypes (250 scats, 16 blood and 1 tissue). The total number of individuals identified here constitutes roughly 49% of the estimated total tiger population in the entire sampled area (Table 1). Sex identification yielded 74 males and 85 females, with the sex ratio being nearly symmetric in most localities, except in the lower sampled areas. On average, individual multilocus genotypes were 93.2% complete. We included 97% (164 out of 169) of samples, those that had complete or near complete genotypes with a maximum of two missing loci, for further analyses. Two Tadoba individuals with three missing loci (73% complete) and three individuals from the Kanha Pench corridor with four missing loci (64% complete) were also retained since sample size from these areas was small and the genotypes represented unique tigers. The panel of eleven microsatellite loci that was used for individual discrimination had very low cumulative sibling probability of identity (*PI-sib*) of 1.5×10^{-4} (Table S1), indicating very high power to discriminate individuals. Even in the samples with the least amount of genotype information (four missing loci), the cumulative *PI-sib* value (1.6×10^{-3}) of the samples was sufficiently low to determine unique individuals.

No significant evidence of linkage disequilibrium (LD) among all pair-wise loci combinations was observed, when all sampling locations were pooled ($p>0.05$ at 1,000 permutations). Except for deviation at three loci, Fca441, Pati09 and Pati18 ($p<0.001$), all other loci were in Hardy-Weinberg equilibrium (HWE) in the pooled population after adjusting the critical P-value using the Bonferroni correction procedure (Table S1). Loci not in HWE or with null allele frequencies>5% appear to be random with respect to population (Table S2). Pati09 showed significant deviation from HWE across three populations, while Pati01 and Fca441 showed significant deviation from HWE in two populations. 6Hdz700, F53 and Fca304 each deviated from HWE in one population. Such departures from HWE could indicate the possibility of genetic structuring among populations and likely presence of related individuals in the data. The deviation from HWE in few populations could also be explained by the presence of null alleles in the data. While MICROCHECKER tests did not show any evidence of scoring errors due to stuttering or small allele dominance in the dataset, five loci (Pati01, Pati15, Pati18, Fca304 and F53) with high null allele frequencies>9% were detected (Tables S1 and S2). Three loci deviating from HWE were likely due to the presence of null alleles in the data - Fca304 in Bandhavgarh, and Pati01 and F53 in the Melghat-Satpura-Tadoba cluster, as they showed significant deviation from HWE only with observed allele frequencies and not with null allele corrected frequencies (not shown). Although Pati15 and Pati18 also contained null alleles at high frequencies (>9%) in Kanha, they did not deviate from HWE in any population. Since these patterns were random with respect to population (e.g., no departures from HWE in Kanha, and null allele frequency $<$ 5% in Pench), it was more likely that these deviations could be due to the presence of population structuring in the data. Hence we retained all eleven loci in subsequent analyses.

Most tigers genotyped in this study showed high heterozygosity (Figure S2). The minimum number of heterozygotes observed across loci for an individual tiger's multilocus genotype was two and three heterozygous loci, for one and two tigers respectively. The most polymorphic individuals (n = 5) were heterozygous at all eleven loci. Individuals that were completely homozygous at all eleven loci, were not observed. Nearly 90% of the tigers (i.e. 152 out of 169 individuals) had heterozygous genotypes at five to nine loci, and 12 tigers were heterozygous at ten to eleven loci. Genetic diversity estimates showed a mean number of alleles per locus to be 9.1±2.2 with heterozygosity *He* to be 75.4%±3.9 and *Ho* to be 70.1%±5.9 in the pooled population (Table 2). A summary of population-wise genetic variation, revealed significantly higher allelic diversity and heterozygosity levels in the larger populations of Pench and Kanha compared to all other localities (Table 2 and Table S2). The number of population diagnostic private alleles was high in Pench (n = 8) and Kanha (n = 6), while individual frequency for a private allele was highest in Bandhavgarh compared to all other localities (Table S3). Estimated population size was positively and linearly correlated to genetic diversity statistics (Number of alleles, $rho = 0.857, p = 0.011$; polymorphism information content, $rho = 0.738, p = 0.0458$; Shannon's index of diversity, $r = 0.714$, $p = 0.058$). Partial correlations between estimated population size and the allele diversity indices, after controlling for the effect of sample size, were positive and highly significant (Number of alleles, $r = 0.82$, $p = 0.012$; polymorphism information content, $r = 0.659$, $p = 0.054$; Shannon's index of diversity, $r = 0.714$, $p = 0.036$). Heterozygosity values were not correlated with population size (*Ho*: $rho = -262$, $p = 0.428$; *He*: $rho = 0.357$, $p = 0.389$). Correlation of estimated population size with actual sampled size was highly significant ($r = 0.964$, $p = 0.001$), meaning that our samples were in proportion to the size of the population.

Population Genetic Structure

According to the results of the Principal Coordinates Analysis (PCoA) based on Phi_{PT} genetic distance estimator, individuals in the area were clustered into roughly four groups with varying degrees of population partitioning (Figure 2A). The three coordinate axes accounted for 60% of the variation in the dataset. Tigers were observed to cluster in four major groups. Tigers from Kanha, Pench and Melghat formed three distinct clusters that partially overlapped each other, while Bandhavgarh tigers formed a discrete cluster with minimal overlap. Tigers from Satpura, Tadoba, and Achanakmar were scattered within the clusters formed by Kanha-Pench-Melghat.

Calculation of delta K from the output of the STRUCTURE runs using prior population information (locprior = 1), produced the largest modal value of the statistic at $K = 4$, suggesting pronounced population subdivision at $K = 4$ (Figure S3). On the other hand, the log-likelihood $L(K)$ value reached an inflection point at $K = 4$ before gradually plateauing at $K = 6$ to 7 and finally leveling off at $K = 8$ (Figure S3). The variance in $L(K)$ increased at higher values of K, as reported previously with other studies [92,125]. The disparity in population structuring patterns between delta K and log-likelihood values occasionally occurs in cases

Table 2. Summary statistics of mean genetic variation and bottleneck tests across sampled populations.

Population	Observed heterozygosity (St. Dev.)	Expected heterozygosity (St. Dev.)	Number of alleles (St. Dev.)	Allelic size range (St. Dev.)	*M* ratio (St. Dev.)	Wilcoxon's heterozygosity excess test	Mode-shift test
M	0.654 (0.234)	0.674 (0.110)	4.7 (0.6)	17.5 (4.3)	0.75 (0.21)	ns	L-shaped
S	0.673 (0.230)	0.639 (0.146)	4.2 (0.9)	16.7 (4.6)	0.70 (0.23)	ns	L-shaped
T	0.705 (0.195)	0.704 (0.094)	4.9 (1.5)	16.5 (5.9)	0.80 (0.15)	ns	L-shaped
M-S-T	0.691 (0.141)	0.774 (0.036)	7.6 (1.6)	24.9 (7.6)	0.86 (0.16)	ns	L-shaped
P	0.750 (0.070)	0.734 (0.056)	7.6 (1.5)	26.8 (11.5)	0.83 (0.18)	ns	L-shaped
K	0.685 (0.061)	0.674 (0.106)	7.2 (1.9)	22.5 (5.2)	0.86 (0.13)	ns	L-shaped
K-A	0.683 (0.085)	0.713 (0.051)	7.9 (1.6)	26.5 (10.6)	0.86 (0.20)	ns	L-shaped
B	0.641 (0.216)	0.607 (0.148)	4.5 (1.0)	18.6 (6.2)	0.66 (0.20)	ns	L-shaped
All	0.701 (0.059)	0.754 (0.039)	9.1 (2.2)	28.8 (10.5)	0.90 (0.14)	ns	L-shaped

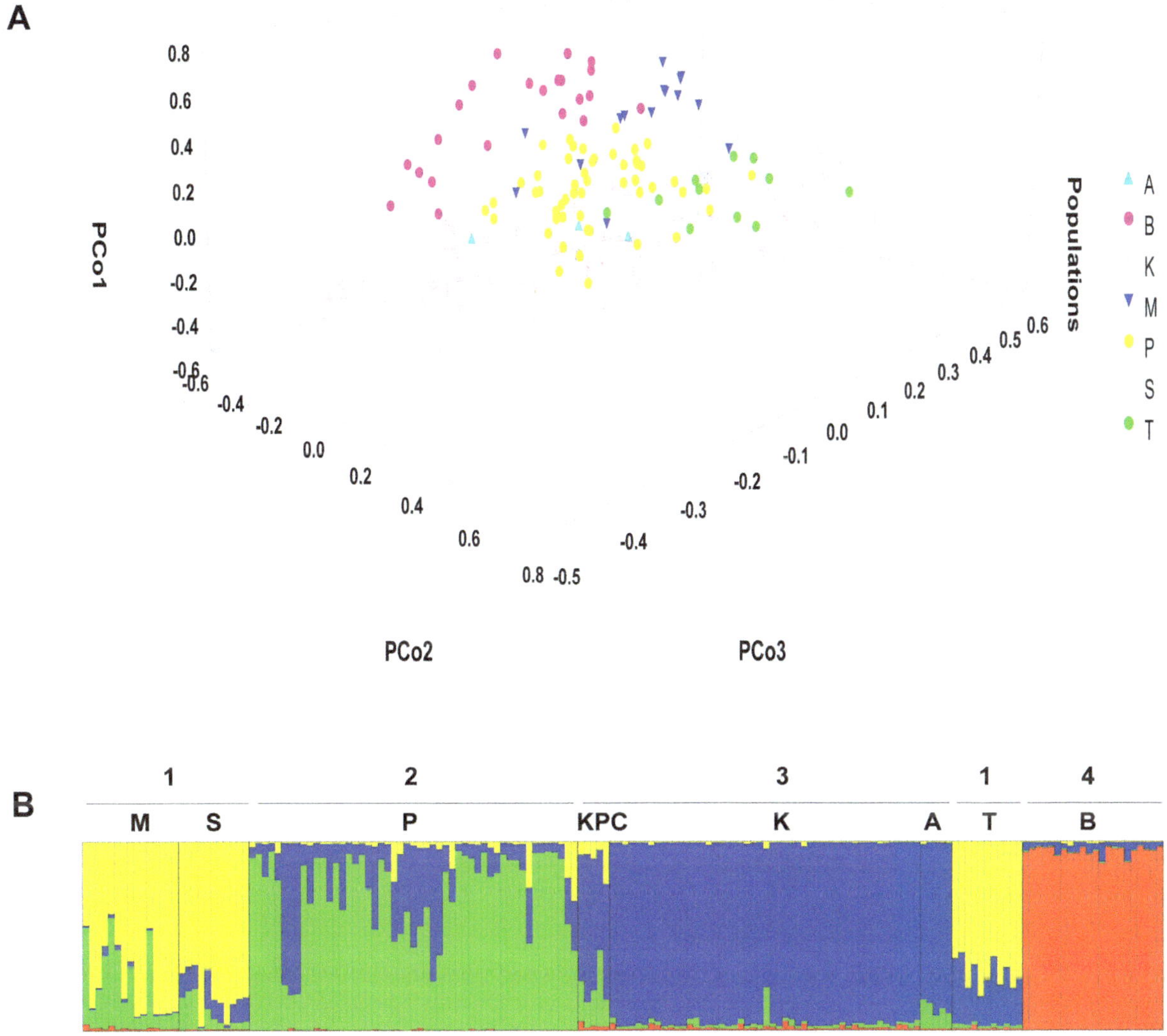

Figure 2. Results of individual clustering analyses. (**A**) Three dimensional plot showing partitioning between different populations as obtained from PCo analysis based on Phi_{PT} co-dominant genetic distance among individuals. (**B**) Summary barplot of STRUCTURE run at $K=4$ showing population assignments for each individual. Four distinct population clusters are observed. Sampled populations are Melghat (M), Satpura (S), Pench (P), Kanha-Pench corridor (KPC), Kanha (K), Achanakmar (A), Tadoba (T) and Bandhavgarh (B).

where the F_{ST} values are significant [126]. Examination of individual Q summary barplots (Figure S3) yielded identical clustering patterns at all runs between assumed $K=2$ to 4 (carried out both with and without prior sampling location information), and distinct population saturation, indicative of population subdivision was evident at $K=4$ in conformance with the delta K approach. Based on the above, the four cluster solution ($K=4$) best describes the levels of genetic subdivision in our sample of the Central Indian tiger population (Figure 2B).

The four population clusters identified by STRUCTURE were similar to the clusters observed using PCoA plots, and in consonance with geographic configuration as well (Figure 1). Melghat, Satpura and Tadoba represented populations with patchy connectivity and formed a unique cluster in the western and southern limits of the landscape, with few individuals being cross-assigned to the Pench and Kanha clusters. To the east of Melghat and Satpura, Pench formed a unique cluster. A few individuals in Pench had cross-assignments with Kanha suggesting gene flow between these two population clusters substantiated by a functional habitat corridor between Pench and Kanha. The next cluster in the eastern part of the landscape was represented by tiger populations in Kanha and Achanakmar. Individuals from the forested corridor between Pench and Kanha were also assigned to this cluster, but most had mixed assignments to both populations indicating that this was not a distinct but rather an admixed population. The last cluster was represented by Bandhavgarh, in the north-eastern part of the landscape, which formed a distinct isolated population where all individuals were assigned to the sampled locality.

Partitioning of genetic variation in the AMOVA test indicated low but significant differentiation ($p<0.01$) between the STRUCTURE identified clusters (Table S4). The major portion of genetic variance was found within populations (88%) with 7% among the population clusters and 5% among populations within clusters. Exact tests showed significant genetic variance on all three levels ($p<0.01$). Both F_{ST} and Phi_{PT} values showed highly significant structuring ($p\leq0.001$), and had relatively similar trends in magnitude with low sampling variance. In contrast, R_{ST} estimates showed no variation between groups, and had unreliably high sampling variances and mean square error estimates. The results of pair-wise F_{ST} and R_{ST} calculations indicated significant ($p < 0.05$) and varied (F_{ST} 0.049 to 0.241; R_{ST} 0.000 to 0.330) genetic structuring between all sampled populations in Central India (Table S5). Within cluster F_{ST} estimates were mostly lower, ranging from moderately low (0.048 to 0.062) to high (0.079 to 0.102), compared to pairwise estimates between different clusters (0.127 to 0.217).

Migrants

(A) STRUCTURE results. Eight putative migrants were identified in STRUCTURE (Table 3). Of these, four individuals (D954, D955, D958, D1399) were identified with>80% migrant and cross-assignment probability to a single non-home cluster in the STRUCTURE analysis, carried out without prior population information. The remaining four samples represented individuals (D1843, D1892, D1297, D2058) that had weaker migrant probability ($P>0.5$ to <0.7) and showed variable Q (0.289 to 0.758), and with the majority of samples having cluster memberships to more than one non-home locality.

(B) GENECLASS results. GENECLASS also identified eight individuals (Table 3) as putative migrants ($P<0.01$), with high log likelihood of cross-assignment ($L_h/L_{max}>2.0$). Lowering the likelihood threshold ($L_h/L_{max}<2.0$) yielded further six putative migrants ($P<0.05$).

(C) BAYESASS results. A total of fifteen individuals with likely immigrant or admixed ancestry were detected in the BAYESASS analysis (Table 3). Of the fifteen total migrants, eleven individuals had high migrant cross-assignment probabilities ($P>0.8$), and four individuals had intermediate migrant assignments ($P = 0.509$ to 0.617). In general, the posterior probabilities of migrant assignment were higher in BAYESASS compared to the STRUCTURE analysis. Additionally, BAYESASS identified six other individuals (D1075, D1381, D1383, D1393, D1400, D1987) as potential migrants ($P = 0.509$ to 0.926) that were not assigned as migrants by either STRUCTURE or GENECLASS.

(D) CERVUS results. Identification of offspring-candidate parent pairs in CERVUS yielded parentage relationships in thirteen out of seventeen putative migrants (Table 3). No evidence of likely parentage (or sibling) relationships in the offspring population was observed. The cross-assigned population in ten of the thirteen individuals matched the parentage assignment in CERVUS, which serves to further corroborate the results of the migrant assignments. Only three mismatches (D1075, D1393 and D1987) were observed between the parent populations identified by CERVUS and the population assignments depicted by the migrant analysis, but this could be due to low information in the data as opposed to incorrect migrant assignment. Except for negative LOD value in two pairs (D955–D1182, LOD = −0.36; D1400–D1168, LOD = −2.15), LOD scores were positive in the remaining eleven putative parent-offspring pairs. The relationship between a potential migrant sampled in Pench (D958) with a candidate parent from Kanha (D1205) was identified with>80% confidence in assignment (LOD = 5.32). LOD scores in remaining offspring-parent pairs were below the 80% confidence limit (<4.9).

The detection of migrants by the above methods yielded a total of seventeen individuals with putative immigrant ancestry (Table 3 and Figure 3). Identical migrant assignment across all three programs was observed in seven individuals (D954, D955, D958, D1297, D1399, D1843, D1892), while there was equivocal assignment in the remaining ten individuals. Sex identification revealed 12 out of 75 males (16%) and 5 out of 84 females (6%) as individuals with immigrant ancestry in the entire area. Figure 3 shows the posterior distributions of individuals assigned to nonimmigrant (gen0), first (gen1) or second generation immigrant (gen2) ancestry states in GENECLASS, STRUCTURE and BAYESASS. All GENECLASS migrants with $L_h/L_{max}>2.0$ were classified as 100% first generation migrants. Two individuals (D955, D958) with>90% gen1 assignment and three individuals (D954, D1399, D1843) with relatively high gen1 assignment probability (0.5 to 0.7) were considered as migrants. Five individuals (D525, D1892, D2058, D1297 and D1987) showed moderate levels of migrant assignment and immigrant ancestry patterns are indicative of admixed status. The assignment status of seven more individuals (D1075, D1381, D1383, D1393, D1400, D1140 and D2154) was equivocal. While STRUCTURE could not assign them as migrants, they were identified as potential second generation migrants or admixed individuals in BAYESASS.

Contemporary and Historical Migration Rates

The mean posterior distributions of pairwise immigration rates depicting contemporary gene flow estimates in BAYESASS are shown in Tables 4 and 5. Most populations have low migrant proportions with the exception of migration from Pench to Melghat ($m = 0.09$) and Kanha to Pench ($m = 0.07$) where the rates were more than 5% (Table 4). Gene flow between Melghat and Pench was likely asymmetric and there appears to be a source-sink relationship because the expected proportion of migrants into the Pench population from Melghat is much smaller ($m = 0.015$). Asymmetric migration was also visible between the Pench and Kanha population clusters, as the proportion of migrants from Pench to Kanha was negligible ($m = 0.006$). The Bandhavgarh population was devoid of migrants as suggested by the lack of gene flow between other populations ($m<0.01$). As in the locality-wise analysis, similar asymmetry and rates of migration were obtained between the various STRUCTURE defined population clusters (Table 5).

Results from the MIGRATE analysis showed low estimates of relative effective population size (theta, θ) and historical mutation scaled immigration rate (M) suggesting very low overall migrant proportions in the area over the long-term (Table 6). Theta estimates were low to moderate and ranged from 0.57 (Bandhavgarh), 0.77 (Kanha-Achanakmar, and Melghat-Satpura-Tadoba) to 1.5 (Pench). Estimates of M ranged from a high of 6.17 (Pench to Bandhavgarh) to a low of 0.9 (Bandhavgarh to Melghat-Satpura-Tadoba), revealing limited to no migration among populations in the landscape (Table 6). Of the twelve pairwise population comparisons, ten pairs showed asymmetric migration patterns, with the higher value of M representing immigration from the population with the larger theta value (Kanha and Pench clusters), to the population with the smaller theta value (Melghat-Satpura-Tadoba and Bandhavgarh clusters). The number of migrants per generation ranged from almost zero (Bandhavgarh to Melghat-Satpura-Tadoba cluster) to nine (Pench to Bandhavgarh). The Pench and Kanha clusters represented the largest source populations for immigrants in the area. Though marginally higher migration from Pench to Kanha was visible compared to migration in the other direction, both population clusters had overall symmetric gene flow.

Genetic Bottleneck

Except for Bandhvagarh, which had a low M ratio of 0.66, all other populations showed M values above the critical threshold of 0.68 (Table 2). Wilcoxon's sign rank test of heterozygosity excess were not significant ($p>>0.05$) regardless of the mutation models used and all localities tested showed normal L-shaped allele distributions in the mode-shift test, indicative of stable non-bottlenecked populations (Table 2). Only, the Pench and Kanha

Table 3. Summary of migrant assignments based on STRUCTURE, GENECLASS (migrants based on $**\alpha_{0.01}$ and $*\alpha_{0.05}$ type I error levels) and BAYESASS analyses.

tiger ID	sex	sampled locality	GENECLASS assigned population	GENECLASS *Lh/Lmax*	STRUCTURE *Q* assignment clusters (MST/P/KA/B; no prior population information, K = 4)	STRUCTURE migrant probability (gen1, gen2)	BAYESASS assigned population	BAYESASS migrant probability (gen1, gen2)	CERVUS assigned parent/population/pair LOD score	Final migrant status
D955	♂	*PTR*	*KTR*	*3.234***	*0.035/0.009/0.889/0.067*	*0.996 (0.978, 0.018)*	*KTR*	*0.999 (0.990, 0.009)*	*D1182/KTR/-0.360*	Migrant
D958	♀	*PTR*	*KTR*	*2.959***	*0.033/0.016/0.936/0.015*	*0.982 (0.964, 0.018)*	*KTR*	*0.997 (0.987, 0.010)*	*D1205/KTR/5.320*	Migrant
D954	♂	*PTR*	*KTR*	*3.883***	*0.009/0.019/0.963/0.009*	*0.863 (0.774, 0.089)*	*KTR*	*0.962 (0.825, 0.137)*	*D402/KTR/0.760*	Migrant
D1399	♂	*PTR*	*KTR*	*3.222***	*0.087/0.022/0.885/0.007*	*0.868 (0.552, 0.316)*	*KTR*	*0.998 (0.702, 0.296)*	*NE*	Migrant
D1843	♂	*MTR*	*PTR*	*2.343***	*0.275/0.527/0.186/0.013*	*0.590 (0.502, 0.088)*	*PTR*	*0.986 (0.842, 0.144)*	*D1381/PTR/4.210*	Migrant
D1297	♀	*KPC*	*PTR*	*6.444***	*0.075/0.683/0.215/0.026*	*0.610 (0.130, 0.480)*	*PTR*	*0.559 (0.124, 0.436)*	*D1043/PTR/2.490*	Admixed
D1987[†]	♂	STR	PTR	2.407**	0.481/0.325/0.180/0.014	0.311 (0.005, 0.306)	PTR	0.550 (0.006, 0.544)	D2057/MTR/0.300	Admixed
D525	♂	TATR	KTR	2.562**	0.664/0.023/0.300/0.013	0.170 (0.031, 0.139)	KTR	0.617 (0.086, 0.531)	NE	Admixed
D2154	♀	TATR	PTR	<2.0*	0.460/0.158/0.358/0.024	0.007 (0.001, 0.006)	TATR	0.188 (0.056, 0.132)	NE	Admixed
D1892	♂	MTR	PTR	<2.0*	0.658/0.289/0.042/0.011	0.584 (0.113, 0.471)	PTR	0.996 (0.011, 0.985)	D578/PTR/0.070	Admixed
D2058	♀	MTR	MTR	0	0.147/0.758/0.019/0.076	0.508 (0.173, 0.335)	PTR	0.895 (0.036, 0.859)	NE	Admixed
D1140	♀	KTR	PTR	<2.0*	0.070/0.556/0.344/0.030	0.124 (0.016, 0.108)	PTR	0.241 (0.035, 0.206)	D1043/PTR/3.540	Admixed
D1075	♂	PTR	KTR	<2.0*	0.065/0.237/0.663/0.034	0.096 (0.012, 0.084)	KTR	0.509 (0.161, 0.348)	D1926/STR/2.473	Admixed
D1381	♂	PTR	KTR	<2.0*	0.213/0.466/0.316/0.004	0.190 (0.000, 0.190)	KTR	0.832 (0.012, 0.820)	D1185/KTR/0.351	Admixed
D1383	♂	PTR	PTR	0	0.068/0.510/0.420/0.003	0.159 (0.001, 0.158)	KTR	0.926 (0.009, 0.917)	D1114/KTR/4.471	Admixed
D1393	♂	PTR	KTR	<2.0*	0.089/0.083/0.807/0.021	0.188 (0.013, 0.175)	KTR	0.872 (0.264, 0.608)	D578/PTR/1.066	Admixed
D1400	♂	PTR	PTR	0	0.146/0.110/0.738/0.006	0.229 (0.015, 0.214)	KTR	0.829 (0.201, 0.628)	D1168/KTR/-2.155	Admixed

Q assignments depict individual membership to each of the four STRUCTURE identified population clusters of Melghat-Satpura-Tadoba (MST), Pench (P), Kanha-Achanakmar (KA), and Bandhavgarh (B). Migrant probability refers to the *P* value of each individual to its cross- assigned population. The assignment likelihood of each migrant to a likely candidate parent and its source population is shown by CERVUS log (LOD) scores. Italicized individuals depict potential migrants identified unanimously by all three programs. Localities depicted include Melghat (MTR), Satpura (STR), Pench (PTR), Kanha-Pench corridor (KPC), Kanha (KTR) and Tadoba (TATR).

[†]Tigers reported dead during the study period, M-Male, F-Female, NE- not established.

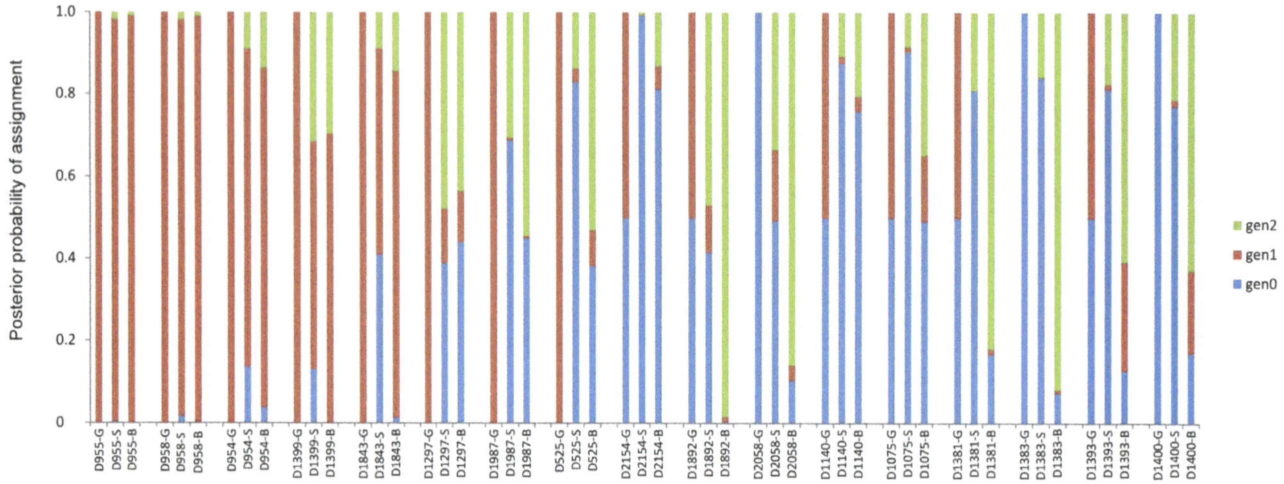

Figure 3. Individual ancestry states of putative migrants. Posterior distributions of individual assignment to nonimmigrant (gen0), and first (gen1) and second generation immigrant (gen2) ancestry states. Suffixes after indvidual names indicate assignment probabilties as obtained in GENECLASS (G), STRUCTURE (S) and BAYESASS (B).

populations showed significant ($p<0.05$) evidence of heterozygosity deficiency (Table S6), which could suggest recent events of expansion in these two populations.

Tiger Occupancy, Habitat, Prey and Human Disturbance

Ten Principal Components (PCs) had Eigen values greater than one and together explained 86.6% of the variability of the original variables (Table 7). The factor loadings of the components permitted unambiguous ecological interpretation of PCs (Table 7). The detection probability of tiger sign was 21.98% (SE 0.04). Detection of tigers was best explained by index of tiger abundance. The closest competing model for detection differed by a ΔAIC value of 1207 (Table 8). Tiger occupancy was best explained by a model that included PC1 to PC9 and all the covariates conformed to the *a priori* predictions of their influence on tiger occupancy. The best model differed from the full model by ΔAIC of 0.45 and from the null model by ΔAIC of 1682. Tiger occupancy in a grid was best explained by (a) PC1 that represented availability of large undisturbed good canopy forests, (b) PC2 and PC4 that represented grids that had low human and livestock disturbances, (c) PC3, PC6 and PC9 that represented high ungulate prey especially in the form of chital, sambar, wild pig and gaur, (d) PC5 representing lower elevation, (e) PC7 representing grids within or in the proximity of legally protected areas and (f) PC8 that gave greater loadings for grids with higher rainfall (Tables 7 and 8). The sign of the coefficients of the best model for each PC was the same when models were run using original variables in a univariate model (Table S7). The goodness-of-fit test for observed data against 50,000 model based bootstrap samples failed to show lack of model fit (χ^2 P value $=0.11$), $\hat{C}$ statistic was estimated at 1.4 and standard errors shown in Table 9 are corrected for over-dispersion by a factor of 1.18. Out of the 185,100 km^2 area of the sampled grids, 76,913 km^2 was forested or potential tiger habitat. The naïve estimate of grid occupancy was 17.5% while the model-averaged estimate of occupancy was 20.87% giving a 3.4% increment over the naïve estimate. The naïve estimate of tiger

Table 4. Locality-wise contemporary migration rates, *m*, estimated using BAYESASS, showing means (± standard deviation) of the posterior distributions along with the 95% confidence intervals in parentheses.

	M	S	P	K	T	B
M	**0.856±0.042 (0.773, 0.937)**	0.014±0.015 (0.000, 0.054)	0.092±0.038 (0.026, 0.173)	0.015±0.016 (0.000, 0.057)	0.011±0.013 (0.000, 0.046)	0.012±0.014 (0.000, 0.051)
S	0.014±0.017 (0.000, 0.059)	**0.899±0.045 (0.805, 0.975)**	0.037±0.031 (0.000, 0.115)	0.026±0.026 (0.000, 0.099)	0.012±0.015 (0.000, 0.049)	0.012±0.015 (0.000, 0.054)
P	0.004±0.006 (0.000, 0.020)	0.025±0.014 (0.003, 0.059)	**0.897±0.029 (0.838, 0.952)**	**0.065±0.025 (0.022, 0.120)**	0.005±0.006 (0.000, 0.024)	0.003±0.004 (0.000, 0.016)
K	0.008±0.007 (0.000, 0.026)	0.003±0.004 (0.000, 0.014)	0.005±0.007 (0.000, 0.024)	**0.976±0.015 (0.940, 0.996)**	0.007±0.008 (0.000, 0.030)	0.002±0.004 (0.000, 0.014)
T	0.025±0.032 (0.000, 0.114)	0.012±0.015 (0.000, 0.054)	0.015±0.021 (0.000, 0.068)	0.021±0.029 (0.000, 0.102)	**0.917±0.068 (0.756, 0.998)**	0.010±0.014 (0.000, 0.048)
B	0.003±0.006 (0.000, 0.017)	0.003±0.006 (0.000, 0.020)	0.003±0.006 (0.000, 0.020)	0.003±0.006 (0.000, 0.019)	0.003±0.006 (0.000, 0.019)	**0.986±0.014 (0.952, 0.999)**

The populations into which individuals are migrating are listed in the rows, while the sources of the migrants are listed in the columns. Values along the diagonal are proportions of individuals derived from the source populations each generation. Migration rates ≥0.05 are in bold. Individuals from Achanakmar ($n=4$) and the Kanha-Pench corridor ($n=5$) were excluded due to low sample sizes, and only included in the cluster based test (provided in Table 5).

Table 5. Cluster-wise contemporary migration rates, *m*, showing means (± standard deviation) of the posterior distributions along with the 95% confidence intervals in parentheses.

	MST	P	K	B
MST	**0.915±0.027 (0.860, 0.962)**	**0.064±0.025 (0.020, 0.117)**	0.015±0.016 (0.000, 0.058)	0.006±0.007 (0.000, 0.026)
P	0.015±0.016 (0.000, 0.054)	**0.905±0.029 (0.846, 0.958)**	**0.076±0.026 (0.029, 0.129)**	0.004±0.005 (0.000, 0.019)
K	0.009±0.009 (0.000, 0.033)	0.006±0.008 (0.000, 0.029)	**0.981±0.014 (0.950, 0.999)**	0.003±0.004 (0.000, 0.017)
B	0.005±0.007 (0.000, 0.028)	0.005±0.007 (0.000, 0.027)	0.005±0.008 (0.000, 0.028)	**0.986±0.014 (0.950, 0.999)**

The population clusters into which individuals are migrating are listed in the rows, while the sources of the migrants are listed in the columns. Values along the diagonal are proportions of individuals derived from the source populations each generation. Migration rates ≥0.05 are in bold.

occupied habitat was 19,845 km^2 while the model inferred occupancy of forests was 21,290 km^2 giving an increment of 1.8% in occupancy estimate. The habitat variables of forest area, forest core area, rainfall, NDVI, and terrain ruggedness had significantly higher values in tiger occupied grids compared to unoccupied cells. All wild tiger prey indices were higher, while livestock abundance indices were lower in tiger occupied grids. Relevant human impact indices had significantly lower values in tiger occupied grids (Table 9 and Figure S4).

Tiger Population Extents and Occupied Habitats

Tiger populations within the landscape were primarily located in and around tiger reserves (Figure 1). The Pench-Kanha-Achanakmar tiger population was located in the largest patch of contiguous forest comprising 16,063 km^2 with intermittent tiger presence recorded throughout this patch even outside the legal reserve boundaries. The Satpura-Melghat forest patch was 12,720 km^2, while the forest patches that contained Tadoba and Bandhavgarh Tiger reserves were smaller 2,088 km^2 and 1,902 km^2 respectively and connected to larger adjacent patches by fragmented forests (Figures 1 and 4). Melghat Tiger Reserve had the largest critical core area that is legally mandated to be made free of human habitation (Table 1). The total protected area in the landscape under the tiger reserve system was 13,054 km^2 with 6,395 km^2 as core area. Tiger population extent was largest for the Tadoba population at 3,519 km^2, while the smallest area occupied (904 km^2) was recorded in Achanakmar (Table 1).

Corridor models – maps and corridor cost values between source pairs

The least-cost corridor plot joining tiger reserves (Figure 1) shows that the longest corridor was between Bandhavgarh and Melghat; and the shortest corridor was between Kanha and Achanakmar (Table S8). The maximum number of barriers in the form of crossings of national highways was five for the corridor connecting Bandhavgarh to Melghat. CIRCUITSCAPE results detected bottlenecks in connectivity in corridors connecting Kanha with Pench, Tadoba with Kanha and Tadoba with Melghat (Figure 4). Alternative habitat connectivity besides the least cost corridor was detected by CIRCUITSCAPE to exist between Bandhavgarh and Melghat along the sparse ridge forests on the southern banks of the Narmada river as well as between Tadoba and Melghat with patchy connectivity observed through Bor wildlife sanctuary. Habitat connectivity between Achanakmar and Bandhavgarh was diffused with no clearly defined flow pathways being observed (Figure 4).

Genetic structure, migrants and corridor costs

Mantel's *r* correlations between pairwise genetic distances and spatial distance metrics were positive, and showed a similar trend across all three genetic distance estimators (Table S9). The highest correlations were observed for IBR and LCC distances with genetic distance while geographical distances showed the lowest correlations with genetic distance. Linearized R_{ST} had weak non-significant correlations with spatial distances, compared to F_{ST} and Phi_{PT} estimates. A significant linear relationship was observed between population pairwise $F_{ST}/(1-F_{ST})$ genetic distances vs.

Table 6. Means of posterior distributions of mutation scaled immigration rate, *M*, along with the 95% confidence limits (before comma) and mean number of migrants per generation (after comma) estimated from Bayesian runs in MIGRATE.

	MST	P	KA	B
MST	0.77 [0–1.93]	4.43 [0–7.2],	1.83 [0–3.87],	0.9 [0–2.67],
		6.6	1.4	0.5
P	1.43 [0–3.53],	1.5 [0.533–3.0]	1.83 [0–4.4],	2.83 [0–5.4],
	1.1		1.4	1.6
KA	2.97 [0–6.13],	2.17 [0–5.4],	0.77 [0–2.0]	2.43 [0–4.8],
	2.3	3.3		1.4
B	3.9 [0–19.4],	6.17 [0–25],	4.03 [0–21.8],	0.57 [0–1.53]
	3	9.3	3.1	

Population clusters listed in rows depict the populations into which individuals migrate, while the source populations of individuals are shown in the columns. The values along the diagonal are estimates of the relative effective population size, θ.

Table 7. Principal component (PC) loadings of covariates relevant for modeling tiger occupancy, Eigen values of the components, the percent variation of the original data explained by the PC, their ecological interpretation and *a priori* effect on tiger occupancy.

Covariates	PC1 (Forest)	PC2 (Human-Livestock)	PC3 (Chital, Sambar, Dung)	PC4 (Livestock Dung)	PC5 (Elevation)	PC6 (Gaur-Sambar)	PC7 (Legal Protection)	PC8 (Rainfall)	PC9 (Wild Pig)	PC10 (Distance to Urbanization)
Elevation (m)	−0.109	0.035	0.019	−0.032	0.962	−0.021	0.004	−0.059	0.042	0.116
Rainfall (mm)	−0.276	−0.033	0.009	0.005	−0.079	−0.048	−0.027	0.871	0.098	0.180
Mean NDVI	−0.795	0.017	0.037	−0.010	0.035	0.102	−0.002	0.195	0.037	−0.017
Forest Area of Grid	−0.796	0.041	−0.116	0.063	0.113	−0.067	−0.095	0.131	0.060	0.288
Core Forest Area in a Grid	−0.820	0.027	−0.059	0.053	0.008	−0.191	−0.120	−0.034	−0.015	0.069
Chital Encounters per km	0.032	0.031	−0.645	0.126	0.106	0.081	−0.037	0.167	−0.506	−0.075
Gaur Encounters per km	−0.100	0.058	−0.144	0.023	0.019	−0.950	−0.065	0.039	−0.033	0.023
Sambar Encounters per km	−0.072	0.062	−0.776	0.087	0.065	−0.268	−0.125	0.117	−0.086	−0.105
Wild Pig Encounters per km	0.053	−0.007	−0.153	−0.052	−0.068	−0.055	−0.050	−0.127	−0.929	−0.009
Wild Ungulate Dung Density	−0.084	−0.008	−0.782	−0.230	−0.201	−0.001	−0.014	−0.333	−0.019	0.118
Livestock Dung Density	0.075	−0.078	−0.009	−0.976	0.032	0.026	0.003	−0.011	−0.022	−0.051
Livestock seen on transect walk	0.040	−0.960	0.033	−0.044	−0.030	0.013	0.047	0.017	−0.009	−0.012
Human and Livestock Trails	0.023	−0.921	0.014	−0.028	0.012	0.042	−0.004	−0.013	0.018	−0.035
People seen on transect walk	0.014	−0.963	0.022	−0.026	−0.030	0.018	0.038	0.021	−0.005	−0.017
Distance of Grid to a Protected Area	0.165	−0.068	0.119	−0.004	0.005	0.071	0.968	−0.024	0.058	−0.055
Distance of Grid to Night Lights	−0.237	0.056	0.053	0.054	0.126	−0.021	−0.056	0.160	0.023	0.913
Eigen Value	2.143	2.723	1.712	1.045	1.026	1.047	0.991	1.016	1.149	1.006
% Variance Explained	13.39	17.02	10.7	6.53	6.41	6.55	6.19	6.35	7.18	6.29
Cumulative Variance Explained	13.39	30.41	41.11	47.64	54.06	60.6	66.8	73.15	80.33	86.62
Ecological Attribute	Large Undisturbed Canopied Forests	People and Cattle	Wild Ungulate Prey	Livestock	Elevation	Large Wild Ungulates	Distance to Protected Area	Rainfall	Wild Pig	Distance to Night Light
***A Priori* influence of PC on Tiger Occupancy**	Negative	Positive	Negative	Positive	Negative	Negative	Negative	Positive	Negative	Positive

Table 8. Model selection results for estimating tiger occupancy within the Central Indian Landscape incorporating imperfect detections and covariates of landscape characteristics, prey abundance, and human disturbance represented by 10 Principal Components.

Models for detection	AIC	delta AIC	AIC weight	Model likelihood	no. par.	−2 log likelihood
psi(.),p(TigSign)	7261.85	0	1	1	3	7255.85
psi(.),p(Survey Specific)	8469.56	1207.71	0	0	36	8397.56
psi(.), p(.)	8514.35	1252.50	0	0	2	8510.35
Models for Occupancy using best model for Detection						
psi(PC1+PC2+PC3+PC4+PC5+PC6+PC7+PC8+PC9),p(TigSign)	6831.99	0	0.556	1	12	6807.99
psi(PC1+PC2+PC3+PC4+PC5+PC6+PC7+PC8+PC9+PC10),p(TigSign)	6832.44	0.45	0.444	0.7985	13	6806.44
psi(PC1+PC2+PC3+PC4+PC5+PC6+PC7+PC8),p(TigSign)	6852.66	20.67	0	0	11	6830.66
psi(PC1+PC2+PC3+PC4+PC5+PC7+PC8),p(Tig. Sign)	6863.40	31.41	0	0	10	6843.40
psi(PC1+PC2+PC3+PC5+PC7+PC8),p(Tig. Sign)	6985.28	153.29	0	0	9	6967.28
psi(PC1+PC2+PC3),p(Tig. Sign)	7034.80	202.81	0	0	6	70228.00

Details of the 10 Principal Components are provided in Table 6. Goodness of fit χ^2 P value = 0.11, $\hat{C}$ = 1.4, for best model.
TigSign – Tiger sign (pugmark, scat); Survey Specific – Survey specific detection.

resistance ($r = 0.549$, $p = 0.019$) and least-cost corridor distances ($r = 0.533$, $p = 0.009$). In comparison to Euclidean geographic distance correlations, this corresponded to a 29.5% and 25.7% increase in model fit for the circuit theory based IBR and the LCC models respectively. The proportion of migrants was higher between population pairs that had lower corridor costs (Figure 5). Linearized Phi_{PT} estimates also produced significant positive correlation with both least-cost corridor ($r = 0.416$, $p = 0.035$) and resistance distances ($r = 0.462$, $p = 0.023$). This meant a 46.2% and 31.6% increase over log-transformed Euclidean distance correlations for the circuit theory and corridor models respectively.

Correlations in partial Mantel tests showed a significant, positive relationship between genetic and resistance distances, after controlling for all competing spatial distance metrics (linearized F_{ST}: $r>0.48$, $p<0.02$; linearized Phi_{PT}: $r>0.39$, $p<0.05$). In contrast, after controlling the effect of resistance distance, significant positive correlation was only observed between linearized F_{ST} vs. geographic and least-cost corridor distances, while partial correlations between all other spatial and genetic distance metrics were either non-significant or negative. Although significant positive correlation was observed between F_{ST} and geographic distances in the standard Mantel tests, genetic and geographic distances were uncorrelated in the partial Mantel tests, after controlling the effect of population clusters, thus affirming that isolation-by-distance (IBD) pattern was absent in the data. Only resistance (IBR) and LCC distances retained a significant, positive partial correlation with genetic distances, when controlled for the effect of population clusters (linearized F_{ST}: $r>0.4$, $p<0.03$; linearized Phi_{PT}: $r>0.4$, $p<0.04$).

Discussion

Genetic diversity and population structure

The present study is illustrative of the general strengths and challenges in using non-invasive genetic samples to assess spatial

Table 9. Coefficient estimates for the best model selected for estimating tiger occupancy in the Central Indian Landscape.

Occupancy Covariates	Coefficient β Estimate	SE corrected for Ĉ	Exponent (β)	Sign of variable loading on PC
Constant a1	−1.932	0.134	0.145	NA
PC1 (Forested Habitat)	−0.838	0.1	0.433	Negative
PC2 (Anthropogenic Disturbance)	0.452	0.151	1.571	Negative
PC3 (Chital, Sambar Encounters & Wild Ungulate Dung)	−0.909	0.142	0.403	Negative
PC4 (Cattle Dung)	0.258	0.137	1.294	Negative
PC5 (Elevation)	−0.436	0.107	0.647	Positive
PC7 (Distance to Protected Area)	−1.109	0.111	0.330	Positive
PC8 (Rainfall)	0.343	0.131	1.410	Positive
PC6 (Gaur Encounters)	−0.325	0.105	0.723	Negative
PC9 (Wild Pig Encounters)	−0.391	0.109	0.676	Negative
Detection Covariates				
Constant b1	−1.072	0.075	0.342	NA
Average Encounter Rate of Tiger Sign	0.840	0.058	2.317	NA

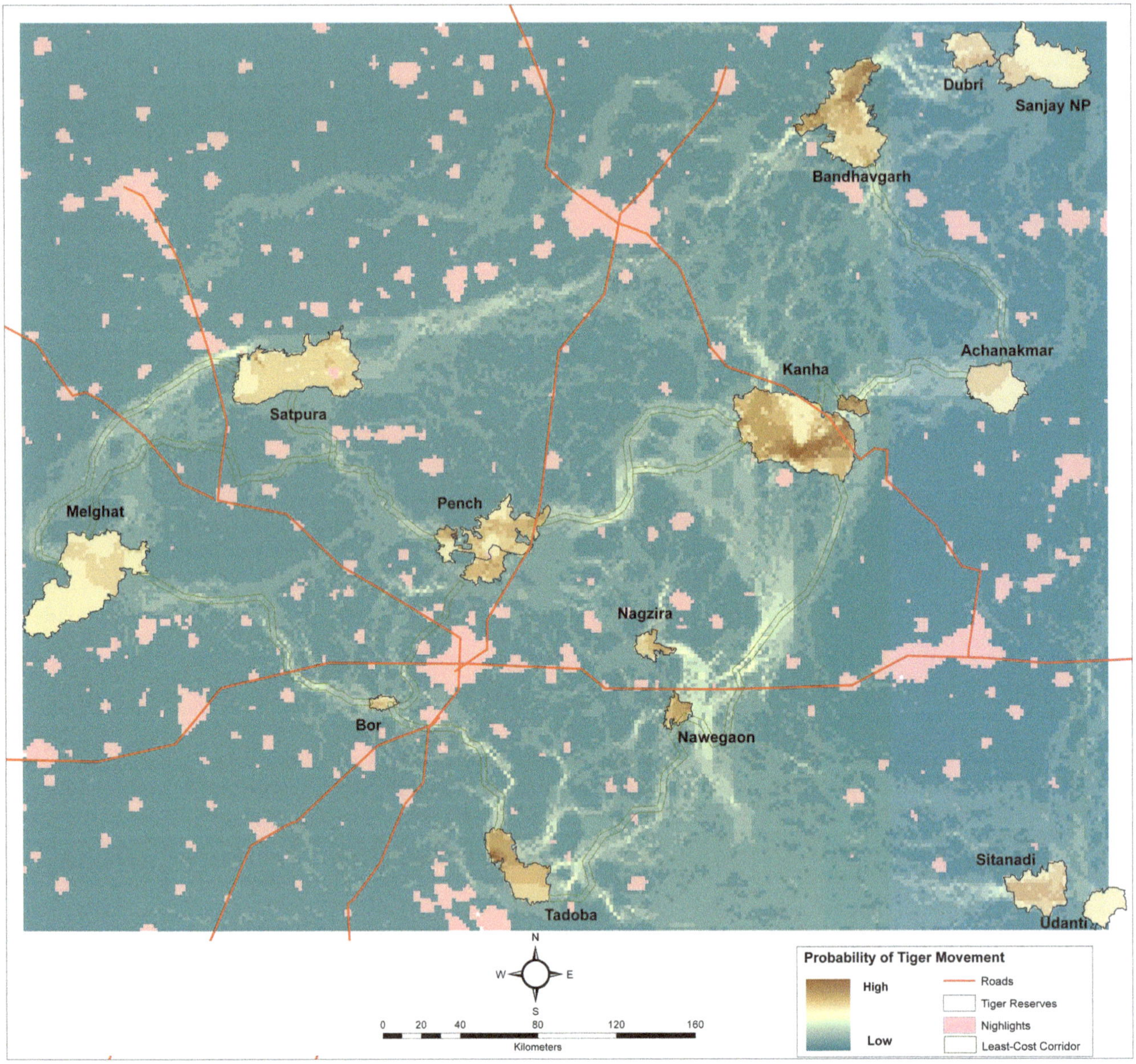

Figure 4. CIRCUITSCAPE model of cumulative current flow used to estimate landscape permeability to tiger movement. Tiger movement modeled as current flow within the Central Indian Landscape using tiger occupancy probability and drainage systems as conductance layers and human settlements as high resistance barriers in CIRCUITSCAPE. Light colors indicate potential habitat corridors. Note the prominent bottlenecks observed in the Kanha-Pench, Kanha-Tadoba, and Tadoba-Melghat habitat corridors.

distribution patterns in an endangered and cryptic species living in fragmented habitats. Although maternally inherited mitochondrial DNA could elucidate patterns of female gene flow, the sequencing of mitochondrial genes was not attempted as the primary objective of the study was to assess the functionality of corridors through the detection of first and second generation migrants. Hence, highly polymorphic, bi-parentally inherited, co-dominant and selectively neutral nuclear loci such as microsatellites were most suited for this purpose. Microsatellite marker amplification success was low from scats, reflecting the often poor quality of samples, typical of tropical environments [127]. Even though standardized methods of sample preservation and DNA extraction were used for the study, finding fresh scats (which may have improved results) was rare despite repeated surveys, due to the low tiger density of many sampled areas, the remoteness of the terrain, and further compounded by the heat and humidity of the region which enhanced degradation of scats. Despite these constraints a relatively sizeable sampling was achieved with the total number of identified individuals ($n = 169$) representing about 49% of the estimated population [31] in the entire landscape (Table 1). The gender proportions observed in our data are similar to the estimates obtained from camera-trap studies in the area [31]. The relatively large sample size in relation to the estimated population enhances the significance of our results.

Most individuals genotyped in this study, possessed heterozygous genotypes at>50% of typed loci, and there was no discernible differences in heterozygosity between migrant and resident individuals (Figure S2). However, unlike resident individ-

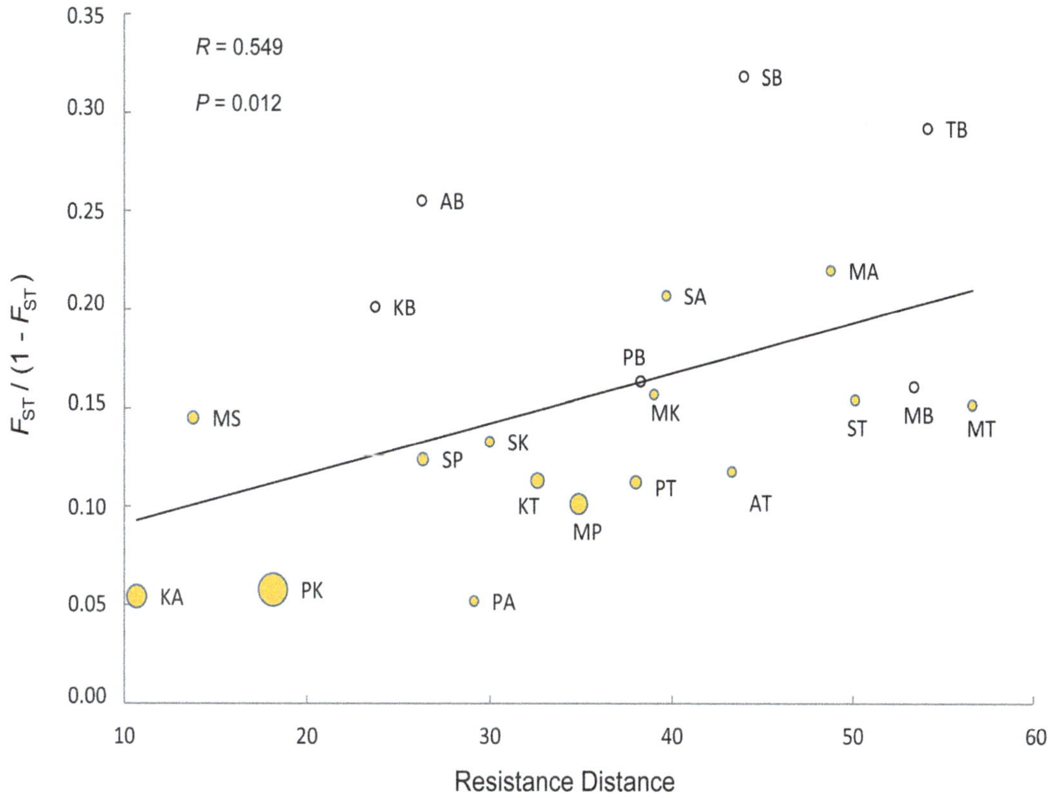

Figure 5. Regression of population pair-wise linearized F_{ST} values with corridor cost. The size of each circle is representative of the proportion of migrants shared between each population pair. Depicted corridors are Kanha-Bandhavgarh (KB), Kanha-Achanakmar (KA), Pench-Kanha (PK), Melghat-Kanha (MK), Satpura-Kanha (SK), Satpura-Pench (SP), Melghat-Pench (MP), Melghat-Satpura (MS), Kanha-Tadoba (KT), Pench-Achanakmar (PA), Melghat-Achanakmar (MA), Achanakmar-Bandhavgarh (AB), Pench-Bandhavgarh (PB), Satpura-Bandhavgarh (SB) and Melghat-Bandhavgarh (MB).

uals, we found no migrants with fewer than five heterozygous loci, or completely heterozygous across all eleven loci. Eleven of the seventeen identified migrant individuals were slightly skewed towards a higher heterozygosity distribution, being heterozygous at eight to ten loci, while the rest of the migrants were polymorphic at five to seven loci. This could be due to the contribution by second generation migrant individuals which constituted the bulk of migrant pool. These individuals are more likely to possess a higher number of heterozygous loci than resident tigers, since they are admixed with parental genotypes from different allelic distributions.

The allelic diversity (9.1±2.2) and heterozygosity ($Ho = 0.701 \pm 0.059$, $He = 0.754 \pm 0.039$) observed in our study is typical of genetic diversity prevalent among tigers in the Indian subcontinent [23,128,129]. Related studies, using different microsatellite markers than the ones used here, have observed levels of average expected heterozygosity (*He*) and mean number of alleles (*A*) to be lower in other subspecies, viz. *P. t. altaica* ($He = 0.26 \pm 0.11$, $k = 2.6 \pm 0.84$ [130]), *P. t. sumatrae* ($He = 0.493 \pm 0.039$, $A = 3.60 \pm 1.48$ [129]), *P. t. jacksoni* ($He = 0.571 \pm 0.027$, $A = 3.90 \pm 1.18$ [129]) and *P. t. corbetti* ($He = 0.670 \pm 0.027$, $A = 6.03 \pm 1.81$ [129]). Interestingly, allele diversity and heterozygosity in the Indian tiger was comparable to other felid species [129], such as the jaguar (*P. onca*, $He = 0.792 \pm 0.0137$, $A = 8.67 \pm 1.72$ [131]), Asian leopard (*P. pardus*, $He = 0.790 \pm 0.0174$, $A = 10.71 \pm 2.31$ [129]), African lion (*P. leo*, $He = 0.610 \pm 0.0348$, $A = 5.0 \pm 1.75$ [129]), South American puma (*Puma concolor*, $He = 0.774 \pm 0.0247$, $A = 7.0 \pm 1.76$ [129]), and cheetah (*Acinonyx jubatus*, $He = 0.708 \pm 0.130$, $A = 6.1 \pm 1.8$ [132]). Although not directly comparable with the present study because of the different markers used, a recent study in the Satpura-Maikal landscape in Central India [64] based on seven microsatellite loci, detected similar levels of heterozygosity ($Ho = 0.65 \pm 0.09$, $He = 0.80 \pm 0.05$), and allelic diversity ($A = 7.76 \pm 1.96$), and very low genetic subdivision (mean $F_{ST} = 0.013 \pm 0.006$). Another study in the southern parts of the Central Indian landscape, focused primarily in peripheral habitats just outside the scope of this study area [66] also observed high mean heterozygosity ($Ho = 0.54$, $He = 0.81$), and allelic diversity ($A = 11.71$) levels at fourteen microsatellite loci. A range-wide study [23] conducted with five microsatellite loci showed that Indian tigers have higher heterozygosity ($Ho = 0.70 \pm 0.16$) and allelic diversity ($A = 12.4 \pm 3.6$) compared to all other tiger subspecies ($Ho = 0.53 \pm 0.07$, $A = 7.2 \pm 1.2$). The study also reported low and non-significant genetic structuring of the Central Indian tigers with the Northern ($F_{ST} = 0.027$, $p = 0.063$) and Southern Indian populations ($F_{ST} = 0.019$, $p = 0.054$). This could be attributed to a historically large effective population size and inter-population connectivity in the region of Central and Peninsular India, explaining why, despite centuries of immense trophy hunting and continued habitat fragmentation, extant tiger populations in the region currently retain close to 60% of the global genetic variation in the species [23].

By using a combination of classical population differentiation and individual clustering approaches, we were able to detect patterns of population sub-structuring in the region. Results from both the PCoA and STRUCTURE analyses suggest the presence of four genetic clusters in the area. The clustered localities occur in close proximity to each other and the overall pattern of genetic structuring observed in the landscape is concordant with existing habitat connectivity and indicative of the role that habitat fragmentation plays in shaping the distribution of allele frequencies across populations. Significant genetic structuring was also detected in the population pairwise F_{ST} and R_{ST} statistics, which may have biased parameter estimations in the STRUCTURE runs carried out without prior population information (locprior = 0) resulting in higher hierarchical levels of population subdivision at $K = 8$. In contrast, the runs carried out with population information (locprior = 1, $K = 4$) did not detect unnecessary genetic structuring, and ignored the prior sampling location information if the ancestry of individuals was uncorrelated [126]. The cluster solution of $K = 4$ appeared optimal for the following reasons. In cases where STRUCTURE detects multiple clustering options with similar probabilities, typically the lowest K value which captures much of the biological complexity in the sample is considered to be the most conservative [126]. Additionally, the presence of related individuals in our sample and the model of correlated allele frequencies used for analysis can lead to overestimation of the true K value [126]. Variation across the four STRUCTURE-defined population clusters was weak, but significant, in the AMOVA test, as the major portion of genetic variance was attributed to within population variation. Though the AMOVA results and F_{ST} statistics indicated significant pairwise structuring across all populations in the area, the pattern shared certain similarities to the population clusters identified by STRUCTURE and the ordination results in PCoA. F_{ST} values were generally lower for localities within the same cluster compared to pairwise estimates between different clusters. We therefore treated the genetic distance estimators including F_{ST} and its analogues as relative measures of population differentiation. The estimation of these parameters requires prior identification of populations and unless the population units are clearly known, such *a priori* designation may not reflect realistic biological patterns as they would only be representing *ad hoc* division of populations [133]. Importantly, the assumptions of demographic and genetic equilibrium along with long time scales under which F_{ST} and its analogues are based may not be suited for estimating genetic differences between populations that have undergone fragmentation or demographic fluctuation events only recently. On the other hand, the results of the model based STRUCTURE clustering approach, which partitioned individuals into relatively distinct clusters based on iterative assignments, made much more sense of the biological realities of the area as the structuring observed here is likely an artifact of recent population fragmentation.

Additionally, the differences observed between F_{ST} and R_{ST} values can provide valuable insights into the balance between genetic drift and mutation events in the studied populations [134]. In this study, pairwise genetic differences between tiger populations in the landscape showed higher F_{ST} values compared to R_{ST} values. The F_{ST} statistic is based on allelic identity and accounts for gene flow between different populations as the basic premise under which it estimates pairwise genetic differences [134]. In contrast, R_{ST} relies on allele size and single stepwise mutations are the primary contributors of genetic variation for this statistic [134]. R_{ST} estimations produced non-significant results across many pairwise comparisons, and had higher sampling variances compared to F_{ST}. This suggests that populations in the area were not isolated long enough for mutations to have caused the genetic differences between populations and the primary cause of genetic structuring in the area is due to recent genetic drift. Some measure of this drift is evident from the number of private alleles found in each population, especially Bandhavgarh, which has the highest individual allele frequency compared to all other localities (Table S3).

Although the STRUCTURE and PCoA results suggest subdivision at four clusters for the Central Indian tiger population, the significant F_{ST} structuring observed between all populations could also be indicative of ongoing fine-scale genetic differentiation in the area. Existing patterns of population structuring are a result of past fragmentation effects and fragmentation is not a static process. Gene flow in the area is currently meager and likely to become even lower, due to continued habitat loss and burgeoning anthropogenic activity in the area. Many localities still retain marginal inter-population connectivity, as evidenced by the presence of individuals having immigrant ancestry and further substantiated by camera trapping and radio-telemetry. However, genetic isolation of almost all populations in the foreseeable future is likely if current patterns of habitat fragmentation persist. In the case of Bandhavgarh the extent of fragmentation appears to be so great that the population may have already become genetically isolated for a longer period than the other reserves. Efforts should be made to revitalize the least cost corridor connecting Bandhavgarh with Achanakmar and subsequently to the gene pool of the main Central Indian complex by a combination of restorative ecology and legal instruments (See conservation implications below).

Migration in the Central Indian tiger population

The Central Indian landscape is a mosaic of habitats where tiger population densities [31] varying from high (Bandhavgarh 14 tigers/100 km^2) to low (Achanakmar 0.1 tigers/100 km^2) are juxtaposed in a matrix encompassing a range of land-use regimes from undisturbed natural forest to high density human settlements (Figure 1). Despite being one of the most fragmented TCL in India [31], some potential for gene flow across populations within the landscape exist because tigers, like most large carnivores, have the ability to disperse across great distances where habitat connectivity is present (e.g., ~150 km linear distance between two reserves in Nepal [67]). However, in the fragmented human-dominated Terai Arc landscape of Nepal, tigers were not found to disperse across expanses of agricultural land (10 to 20 km wide) though they did traverse through stretches of degraded forest [41]. So far, the only published dispersal estimates from the Central Indian landscape are from recent genetic studies that observed first generation migration and long-range dispersal across protected areas located roughly 200 km [65] and 650 km [66] apart, suggesting that tigers may be more resilient in traversing fragmented habitats than previously reported. Extensive annual camera-trapping since 2006 has recorded tigers dispersing between Pench and Kanha, a geographic distance of more than 150 km (Jhala and Qureshi unpublished data).

Identification of migrants by the methods used in our study requires that the populations being investigated have sufficient genetic separation and low levels of migration. We found in our study that the power to detect migrant individuals varied across localities in the region. There was no power to detect migration between localities in the same cluster (such as Kanha-Achanakmar). In these situations it was difficult to tease apart actual migration events from similar allelic patterns that arose from shared population histories and low genetic separation. On the other hand, migrant identification between localities in different population clusters produced robust assignments because of distinct genetic differences. Migrant detection was highest between localities that had discernible genetic separation and had relatively intact habitat corridors with confirmed tiger presence, such as between Kanha and Pench populations where four migrants and seven likely admixed individuals were detected. Other studies by Joshi *et al.* [66] and Sharma *et al.* [65], respectively, detected one and two first generation migrants between Pench and Kanha. The intervening forest patch between Kanha and Pench not only served as a movement corridor but also had some resident tigers as evidenced from camera trap data [135]. A noninvasive genetics study by Sharma *et al.* [64] detected seventeen individuals from this corridor. At the extreme end in Bandhavgarh, no migrants were detected even though distinct genetic structuring of Bandhavgarh from other population clusters provided sufficient power to distinguish migrants. Between the Melghat and Pench populations where three individuals with migrant ancestry were obtained, there was limited but adequate power to detect migrants as genetic separation was distinct but at a lower level than Bandhavgarh. One individual sampled in Satpura was cross-assigned to Pench. The corridor between Pench and Satpura has fragmented forest connectivity and tigers were reported from this area within the last ten years [42]. Two individuals with admixed ancestry were also detected between Kanha and Tadoba, though one of these had mixed assignments with Pench as well suggesting gene flow in the last few generations. Two individuals with admixed ancestry were also detected between Kanha and Tadoba. Areas of degraded forest and low density tiger occupancy such as Nagzira and Nawegaon Wildlife Sanctuaries located strategically between Tadoba and Kanha, and Bor located between Tadoba, Pench and Melghat may serve to provide stepping stone [136] type corridor connectivity (Figure 1). These areas have resident tigers, and it seems likely that the populations are sustained by sporadic dispersing individuals from larger source populations. Joshi *et al.* [66] observed that Tadoba, Nagzira and Melghat form a genetic cluster, and based on asymmetric migration patterns in their study it appears that the tiger population in Nagzira acts as a sink population for the high density source population in Tadoba.

Based on our strict criteria of defining migrants, we designated four individuals, three males (D954, D955 and D1399) and one female (D958) that had high cross-assignments (Q>0.8) to a single non-home cluster, as first generation migrants. Further, a fourth male individual (D1853) that had high cross-assignments, albeit with marginally lower Q value (0.527) compared to the above four migrants, was also designated as a first generation migrant. The immigrant ancestry status in the remaining ten putative migrant individuals was classified as admixed based on - (i) intermediate levels of migrant probability in STRUCTURE; (ii) membership to more than one cluster in the individual Q assignments; and (iii) high probability of assignment to second and not to first generation immigrant ancestry state in STRUCTURE and BAYESASS.

Though one of the first generation migrants was a female, most individuals with immigrant ancestry were males (80%: 4 males out of 5 total migrants), confirming that dispersal is male-biased in tigers [41]. The low number of strict first generation migrants suggests that contemporary migration events are minor in proportion to the number of individuals with likely admixed ancestry. The identification of samples with admixed ancestry suggests that most migration events in this landscape have occurred within a few generations and is evidence that migrant tigers are able to reproduce in the new locality. This genetic evidence is further supported by field observations where a sub-adult male tiger photo-captured in Pench tiger reserve in 2006 was observed to be a territorial breeding male in Kanha tiger reserve in 2010 (Jhala and Qureshi unpublished data). The data and analyses show that tigers likely disperse between Kanha-Pench, Pench-Satpura-Melghat, Kanha-Tadoba, Melghat-Tadoba and Tadoba-Pench.

Gene flow rates and past demography

Both the contemporary and historical analyses detected low estimates of migration rate between populations in the landscape. While this result is superficially similar to the findings of Joshi *et al.* [66], it is in sharp contrast to Sharma *et al.* [65] where they found both high historical and contemporary gene flow, although historical rates were much higher than contemporary rates. As observed in our study and by Joshi *et al.* [66], low levels of contemporary gene flow are expected given that the area is highly fragmented and extant populations occupy habitats that are completely surrounded by heavily modified landscapes, altered by agriculture and high density human settlements, thus making it difficult for tigers to disperse between populations. The highest estimate of contemporary migration was seen from Pench to Melghat, but there appears to be an asymmetric source-sink relationship between the two populations with very low migration from Melghat to Pench. Such a pattern is expected given that Pench has high tiger density (4 tigers/100 km^2) likely serving as a source population, while the population density in Melghat is lower (2 tigers/100 km^2). Asymmetric contemporary rates of gene flow were also apparent between Pench and Kanha, as migration from Kanha (6 tigers/100 km^2) to Pench was much higher compared to movement in the opposite direction from Pench to Kanha. Joshi *et al.* [66] similarly detected high gene flow (>5%) between Kanha and Pench, with Kanha acting as the biggest contemporary source for immigrants. In contrast, Sharma *et al.* [65] state that only Pench is acting as a contemporary source population and contemporary gene flow from Pench to Kanha and Satpura is very high, and has remained relatively stable since historic pre-disturbance times. A comparative evaluation at historical and contemporary time scales in their study [65] showed that gene flow between the Pench–Satpura and Melghat Satpura pairs remains similar, whereas there has been a 47–70% reduction in gene flow between Kanha–Satpura, Pench–Melghat and Kanha–Melghat. Our results showed that historical patterns of migration between the two major population clusters of Kanha and Pench were of equal, albeit low proportion in both directions, and that both Kanha and Pench acted as source populations in contemporary times as well. Although Sharma *et al.* [65] reported higher historic and contemporary gene flow rates compared to our study, both studies similarly observed that Kanha and Pench were important historic source populations and were the main drivers of gene flow in the area.

Contemporary density estimation studies using photographic capture-recapture techniques in the region [31] help explain some of the patterns observed in the gene flow analysis. During 2006, the Kanha population experienced a phase of relative expansion compared to a decline observed later in 2010. The situation was reversed in Pench where the population was relatively low during 2006–2009 compared to a later phase of expansion in 2010 [31,42]. From the genetic sampling done in Pench during both 2007 and 2010, which pre and post-dated this period of population expansion in the locality, a range of individuals representing likely migrant or admixed ancestry to Kanha was obtained. All first generation migrants from Kanha to Pench were detected during 2007 only, when the population in Kanha was in expansion phase. In 2010, seven individuals showing admixed ancestry from Kanha were obtained in Pench. Dispersal and subsequent breeding by immigrants from Kanha to Pench during periods of population expansion in the former appears to explain the patterns of immigrant ancestry detected in the Pench population.

As parameter estimates of migration rate and theta in MIGRATE average across at least 4*Ne* past generations, [137] or several thousands of years ago in absolute time, the low long-term gene flow estimates observed in this study could suggest that populations may have been subject to historical fragmentation and genetic drift. Although ancient population fragmentation is perceived to have occurred in response to the forest clearing activity of agro-pastoralist Neolithic people during the mid-Holocene about 5,000–3,000 years ago [138], the effect is difficult to evaluate as early farmers likely did not clear forests at a scale comparable to recent centuries [139]. However, it is conceivable that the shift from hunter-gatherer to agro-pastoralist lifestyle and successful colonization of new areas would have negatively impacted large mammal populations including tigers and prey species through hunting (and reduction of prey base), as seen during human colonization events in other parts of the world [140,141]. Evidence from an exhaustive range-wide study of tigers has detected a massive decline of about 98% in the number of tigers over the last 200 years in peninsular India [23]. The wanton slaughter of tigers and other wildlife can be glimpsed from historical hunting records of the area, where upwards of 1,000 tigers and 2,000 leopards from this region were killed within just a few decades [25]. The population decline continues apace today, as omnipresent fragmentation and rampant poaching of tigers, other carnivores and prey species have reduced tiger populations such as at Achanakmar to only a few individuals [31], and resulted in their local extirpation in adjacent reserves of Sariska [5] and Panna [29].

Alternatively, the low estimates of historic migration rate raise concerns whether these results are artifacts of sampling or population related. MIGRATE parameter estimations assume coalescent-based models of constant population sizes and mutation-migration-drift equilibrium. Departures from the equilibrium model such as recent and sudden declines in population sizes can negatively bias the posterior parameter distributions of theta and hence migration rate estimations [137]. However, this study and a recent work by Sharma *et al.* [64] did not find significant evidence of past demographic contraction. In our study, only the population of Bandhavgarh had below par *M* ratios (<0.68) suggestive of bottleneck, but the evidence was equivocal since no significant heterozygosity excess or a mode-shift in allele frequencies was detected. Though bottlenecks were not observed in this study, the analyses may be undermined in a few populations because of low sample sizes (<20 individuals), as tests such as the mode-shift in allele frequencies is known to be affected by sample size variances [106]. Also, a demographic decline may not necessarily result in a bottleneck as several factors such as duration of decline, pre-bottleneck diversity, and gene flow between populations can affect the probability of detecting a bottleneck [101,106,142,143]. Furthermore, the significant heterozygosity deficit (symptomatic of recent population expansion) observed in the Pench and Kanha populations could mask signatures of population bottleneck as the addition of new individuals could increase the number of rare alleles which can bias allelic and heterozygosity distributions [106]. Demographic expansion in Pench and Kanha is likely, given the recent history of tiger population recovery in the area in the 1970s–1980s [28,144], or about six to eight generations ago (considering a five year generation time in tigers [99]). This period is well within the window of detection for genetic bottleneck tests of heterozygosity deficiency/excess which assess short term demographic changes only, maximum up to ten generations before present [143]. We could not account for other genetic effects or demographic events which could affect the heterozygosity distributions in these two populations, although our analysis did detect some amount of higher hierarchical STRUCTURE clustering patterns which could be indicative of lineage or extended family structure, since the data definitely contained related individuals. However, even if kin-based segregation did produce a signature of heterozygosity deficit, the specific nature of the effect is difficult to evaluate without in-depth parentage and relatedness analyses, and backed by field investigations, which was beyond the scope of this study. Based on both the theta estimates and absence of bottlenecks, our results suggest that the bigger source populations in the study area have had a relatively stable population history, compared to the smaller populations, as also observed by Sharma *et al.* [65].

Our study and the recent study by Joshi *et al.* [66] had higher power to resolve hierarchical genetic structuring in the area compared to Sharma *et al.* [65] where indistinct structuring was observed. This was most likely due to the higher number of loci used in both studies compared to Sharma *et al.* [65]. Except for Tadoba where sampling was low due to logistic constraints, the small sample sizes from some other sites were due to small tiger populations. Simulation studies by Paetkau *et al.* [94] caution regarding the use of MCMC resampling methods implemented in Rannala and Mountain [95], as they tend to over-estimate migrants from a limited data set. In our case, this translates to the possibility that there may be less migration than we report in some of the smaller populations in the study system. Due to low sample sizes, individuals from Achanakmar (n = 4) and the Kanha-Pench corridor (n = 5) were not analyzed separately, as doing so resulted in overestimation of migration rates. Instead these localities were merged with the Kanha cluster, following the results of STRUCTURE assignments. Low sampling was also a problem in the study of Joshi *et al.* [66] as sites such as Kanha, Pench and Melghat were clearly under-sampled.

Tiger Occupancy, Habitat, Prey and Human Disturbance

Occupancy has often been used as a straightforward and economical state variable in place of abundance to monitor populations [145]. Occupancy of tigers has been estimated in the Mysore-Wayanad Landscape [115] and the Corbett-Rajaji Landscape [116] and across three major tiger landscapes of India [22]. Correcting for detection bias by replicate surveys within sampling units is especially important when detection probabilities are low, sample sizes and survey lengths are small resulting in naïve occupancy estimates that are substantially negatively biased [115]. The recent availability of powerful analytical tools and software [70,114] that permit accounting for bias caused by detections being <1 has promoted the use of correcting for such bias post data collection [115,146] compared to more rigorous designed field methods that minimize such bias in the first place. In this study we use independent spatial replicates, each of 5 km in length, with a minimum effort of one survey for every 5 km^2 of tiger habitat. Thus a 10×10 km^2 grid had anywhere between 3 to 35 replicate surveys depending on the amount of tiger habitat present within that grid. The detection probability of tiger sign was high in comparison to other studies [115] due to longer spatially independent surveys as well as higher density of surveys. Therefore, the difference between naïve and bias corrected estimates of occupancy were comparatively small (3.4%). Several assumptions that are difficult to meet underlie the use of occupancy as a state variable to monitor populations [147]. Further, tiger populations are at risk of poaching and they can be severely reduced or even extirpated from prime habitats [5,29]. Thus, model inferred occupancy from covariates of habitat, prey, and human disturbance can provide misleading inference regarding true occupancy in the case of tigers and other species that can be severely depleted by poaching, and should therefore be interpreted with caution. However, occupancy models serve to provide a good assessment of habitat suitability at a scale relevant for conservation management [70,146]. It is in this context that we primarily use the occupancy probability in this paper so as to model habitat corridors joining tiger populations.

Detection of tiger sign was better modeled by including encounter rate of sign per km walk. Intensity of sign was found to be a reliable index of tiger abundance [22]. Detection of sign was likely to be higher with greater abundance of tigers and this was reflected in our model. Principal Components that represented distance from protected areas, prey abundance and remote canopied forests had the highest coefficients in the logit-link function that modeled tiger occupancy. All the model coefficients were in concordance with the *a priori* predictions based on our understanding of tiger ecology. Since PCs are orthogonal to each other [117] coefficients can be interpreted in terms of their sign and magnitude as the models are free from the effects of multicollinearity. The model coefficients of all site variables when used independently to model tiger occupancy as univariate models also conformed to the *a priori* predicted effects and to the coefficients obtained by the multivariate best model obtained from their PCs by PRESENCE. These results indicate that tigers are a conservation-dependent species, requiring areas having effective legal protection, good forest cover with least human impacts and high prey availability.

Corridor models, genetic structure, migrants and dispersal costs

Although both LCP and circuit theory used tiger occupancy probability as a base layer to parameterize landscape resistance surfaces, LCP defines the optimal minimal route that is required to connect two tiger reserves [53], while the resistance layers obtained from circuit theory analysis provides an understanding of all habitat connectivity currently available between tiger reserves [118]. Circuit theory is especially important to identify bottlenecks in corridors where current is constrained to flow through a narrow channel due to high resistance of the neighboring matrix. Such bottlenecks are highlighted in Figure 4 and exist in almost all corridors connecting tiger reserves. A corridor is as good as its weakest link and if the high resistance habitat matrix surrounding these bottlenecks increase in their extent they could choke the corridor, thus forming a barrier to tiger dispersal.

In our study, the ecological distances that were generated using information from habitat heterogeneity and landscape resistance surfaces (LCCD and RD), correlated significantly with pairwise genetic distances, which validated the effectiveness of the modeled linkages in representing realistic biological scenarios. In particular, IBR models depicted well the ecological costs of movement as they accounted for multiple pathways, irregular patch effects, landscape heterogeneity and wider habitat swaths connecting populations [54]. Although the raw LCP model accounted for habitat heterogeneity while selecting a single pathway as the optimal minimal corridor, it did not correlate significantly with pairwise genetic distances (except F_{ST}). Instead, the LCCs which were delineated by matching and buffering the pathway results from the raw LCP model with existing ground forest cover maps of the area provided a more realistic relationship of the ecological costs with observed genetic differences. In contrast, Euclidean distances (GGD and $\log_{10}$GGD) did not correlate significantly with genetic distances, due to the null model assumptions of spatial homogeneity where habitats are arrayed in an infinite lattice [124], and hence do not conform to the ground realties of fragmentation in the Central Indian landscape. The apparent IBD pattern observed due to the significant correlation of F_{ST} (but not Phi_{PT} or R_{ST}) with geographic distance is an artifact of metapopulation structure and discrete population clusters present in the area. This observation was affirmed by non-significant partial correlations between genetic vs. geographic distances (but not LCCD and RD) using the STRUCTURE identified population clusters as a control. Apart from fragmentation induced spatial heterogeneity, genetic structure is also a result of population history.

The results of our study imply that population subdivision and genetic structure across most localities in the area was strongly associated with habitat features that offer resistance to dispersal at different intensities such as agricultural land, roads, railway lines, high density human settlements and urban infrastructure and not only by geographic distance between populations. Mantel tests showed significant positive correlation of genetic differentiation with least-cost corridors and landscape resistance surfaces associated in traversing corridors, further confirming this observation. The tiger habitats in the region are patchy with some populations still having connectivity, while being reduced or conspicuously absent in others. The best patches of contiguous forested habitat are present in the corridor between Pench and Kanha, which extends eastward to Achanakmar. Likewise, the Satpura and Melghat populations are also connected through swathes of degraded forests, which are interspersed with agricultural land and medium density human settlements. Connectivity between Pench and Satpura is fragile as parts of the linkage are disrupted by mining activities, and broken up in places by agriculture, habitations, major highways and railway lines. The population in Tadoba is linked with Kanha in a stepping stone connectivity through patchily distributed forests. Even though the intervening matrix between Tadoba, Pench and Melghat is heavily human dominated, the populations are tenuously linked by degraded forest patches and tiger occupied habitats such as Bor Wildlife Sanctuary. Bandhavgarh has linkages with forest habitat and tiger reserves further east [31], but seems isolated from tiger populations in the study area by human settlements and agricultural land. Corridors identified herein need to be given legal status, and mitigated with appropriate green infrastructure [148] for development projects within corridor habitats so as to ensure continued gene flow between populations.

Conservation implications

It is indeed surprising that in spite of being highly fragmented [31], the tiger habitats in Central India still exist as a metapopulation with gene flow occurring between most population clusters in contemporary times. Similar conclusions were also reached in other studies [65,66], which observed indistinct genetic structuring and low migration amongst populations, although the genetic differences were not substantial to permit unambiguous identification of migrants. These findings suggest that tigers are able to disperse through suboptimal habitat fragments, than was earlier believed [41,149]. The results of our study underscore the importance of conserving and maintaining corridor connectivity for the continued persistence of tiger populations in the landscape. We found seventeen individuals or about 10% of the sampled tigers with migrant ancestry. As revealed from this study, the sizes of genetic population clusters are clearly beyond the boundaries of protected areas, and have to be managed in a metapopulation framework stressing the need for a landscape conservation policy in place of the current conservation policy focused on protected areas. The functionality of corridors as shown in this study has important implications for the persistence of small populations such as Achanakmar, Satpura and Melghat by the rescue effect of emigrants from the large source populations in the landscape. Apart from poaching, the ecological isolation from other source populations, due to the absence of corridors was one of the prominent causes that led to the local extinction of tigers from the nearby protected areas of Panna and Sariska [5,29]. Now both of these areas have been repopulated through translocation of tigers from nearby reserves. A range of anthropogenic pressures are fragmenting important habitat corridors such as the ones linking Satpura with Pench, Pench with Kanha, Kanha with Tadoba, and Tadoba with Melghat. Dispersing tigers often have to negotiate passage through suboptimal prey-poor habitats and spend extended periods of time in proximity to human habitations, which greatly reduces survival and successful gene flow. The limited migration occurring in the landscape is probably due to tigers avoiding areas of human activity and high mortality due to commercial poaching or retributive killing in response to predation on livestock and humans. The eventual loss of genetic connectivity between populations in the near future, as observed currently in Bandhavgarh, is likely for most populations in the area if present trends of fragmentation continue unabated. In such situations, last resort alternative strategies such as translocation may be considered, as it could easily enrich genetic diversity by moving tigers across each of the four distinct genetic population clusters, to mimic natural patterns of gene flow. However, maintaining and restoring habitat corridors is the preferred conservation strategy to maintain genetic exchange between tiger populations, since corridors would serve the same function for other biota as well, exemplifying the role of the tiger as a flagship species.

Conservation efforts in such a fragmented and dynamic human-dominated landscape presents enormous challenges, but should be attempted in all areas where substantial habitat is still present. Our study gives renewed hope to tiger conservation efforts within the Central Indian Landscape and similar habitats with small fragmented tiger populations. We show that minimal habitat connectivity permits gene flow between populations, which is essential to maintain metapopulation connectivity. Our findings suggest that tigers could negotiate passage through stepping stone dispersal [136] as observed between Kanha-Tadoba, and Pench-Melghat. The most functional corridor was observed between Kanha and Pench, which has evidence of prey and offers the possibility of resident tigers in some of its larger forest patches [31,135]. However, infrastructural development in the form of adding lanes to national highways and widening of railway lines, threaten to form permanent barriers even within this corridor unless proper safeguards and mitigation measures [44] are built into these development projects. Tiger range countries like India are heavily investing in infrastructure development and mining to meet the needs of a growing economy [148]. Identification of minimal habitat corridors is vital for conservation efforts of tigers and other wide-ranging fauna like elephants (*Elephas maximus*), gaurs (*Bos gaurus*), leopards (*Panthera pardus*), and dholes (*Cuon alpinus*). They need to be offered legal protection with smart green infrastructure [148] being the norm of development policies within these corridor habitats. The extinction (nearly two decades ago) of gaur in Bandhavgarh, preceding their recent reintroduction with individuals from Kanha [150], is symptomatic of habitat fragmentation events that have impaired movement with other source populations in the landscape. We demonstrate here an integrated approach to generate reliable information to document metapopulation structure of tigers and the required habitat connectivity [17] to maintain it in Central India. Legal mechanisms to safeguard these minimal corridors could potentially be the eco-sensitive category under the Environment Protection Act (1986) legislation. Currently tigers and other mega-fauna exist as a metapopulation and do exchange genetic material through functional corridors [65,66,128,151]. Such opportunities are likely to be lost rapidly in the wake of new wave of development unless legal sanctity, active restoration and mitigation of development projects become the norm. Conservation policy needs to shift the focus from protected area centered preservation to landscape scale conservation where development policies incorporate a conservation ethic.

Supporting Information

Figure S1 PCR-RFLP identification of tiger scats. (**A**) MtDNA cyt *b* alignment with GenBank and reference sequences showing polymorphism at the particular *Bam*HI restriction enzyme between tiger and leopard. (**B**) Enzyme digested bands of the 187 bp PCR product, targeting this region, showing different profiles in tiger and leopard for species identification.

Figure S2 Frequency of heterozygous genotypes observed at each individual multilocus genotype in all tiger individuals ($n = 169$) in this study.

Figure S3 Results of STRUCTURE analysis. (**A**). Difference in delta K and mean LnP(K) for an estimated number of K populations, in models run with (locprior = 1) and without (locprior = 0) prior sampling location information. (**B**). Summary barplots depicting prior and non-prior STRUCTURE runs (assumed $K = 2$ to 8), of sampled populations in central India showing cluster affiliations according to individual Q values. Cluster saturation at $K = 4$, indicative of four population clusters, is observed in runs carried out both with and without *a priori* location information. At $K>4$, increased sub-structuring is detected, but there is no concordance in clustering between the prior and non-prior runs.

Figure S4 Differences in tiger present ($n = 311$) and tiger absent ($n = 1540$), 10×10 km grids shown as violin plots in the Central Indian Landscape. All variables are normalized by z transformation to make the scales comparable.

Table S1 Information on the 11 microsatellite loci used in this study. Allele diversity statistics, observed (*Ho*) and expected (*He*) heterozygosity, Hardy-Weinberg equilibrium (HWE) tests, null allele frequencies and sibling probability of identity (*PI-sib*) values obtained across 169 tiger individuals. Null allele frequencies>+ 0.05 are italicized.

Table S2 Summary of population cluster-wise genetic diversity statistics at each locus. Depicted are number of alleles (*k*), number of individuals typed (*N*), observed (*Ho*) and expected (*He*) heterozygosity, Hardy-Weinberg equilibrium (HWE) test significance and null allele frequencies (*Null*). The Achanakmar ($n = 4$) and Kanha-Pench corridor ($n = 5$) individuals are included in the Kanha cluster.

Table S3 List of diagnostic alleles present in the sampled populations.

Table S4 AMOVA results.

Table S5 Population pair-wise F_{ST} (below diagonal) and R_{ST} (above diagonal) estimates.

Table S6 Bottleneck test results for loci under different mutation models. P values <0.05 are depicted in italics.

Table S7 Variable attributes used as site covariates in modeling tiger occupancy.

Table S8 Habitat corridors, major roads, corridor cost between tiger reserves.

Table S9 Results of standard and partial Mantel tests for correlation between pairwise genetic and spatial distance metrics. The correlation coefficient (r) and probability (p) are shown using three different genetic distance estimators. Significant values ($p<0.05$) are indicated by an asterisk (*).

Table S10 Details of cyt *b* PCR amplification in reference samples using felid-specific (187 bp) and universal primers (309 bp). + indicates all samples which amplified. NA – not amplified.

Table S11 Information on the pilot test carried out on scats ($n = 65$) for species identification of tiger samples by PCR and *Bam*HI restriction enzyme digestion.

Method S1 Species identification from scat samples.

Method S2 Generating the cost surface for PATHMATRIX and CIRCUITSCAPE.

Acknowledgments

We thank the Ministry of Environment and Forests, and the Chief Wildlife Wardens of Madhya Pradesh, Maharashtra, and Chhattisgarh states, India, for permissions and logistic support. We thank J. Borah, J. Vattakaven, K. Deka, Lalthanpuia, J. Jena, A. Bais, M. Aggarwal, G. George and the research fellows of the 2010 tiger assessment team for assisting with scat collection and laboratory work. The Director and Dean of WII are acknowledged for their support.

Author Contributions

Conceived and designed the experiments: YVJ QQ. Performed the experiments: BY YVJ QQ RG YS. Analyzed the data: BY YVJ QQ SS. Contributed reagents/materials/analysis tools: RG RCF QQ YVJ. Wrote the paper: BY YVJ. Field logistics: RG. Designed the genetic component of the study: BY. Scat sampling and laboratory analysis of genetic data: BY YVJ. Landscape scale occupancy data, field sample collection: YVJ QQ RG. Assisted in laboratory analysis genetic data: YS. Analyzed genetic data: BY. Supervision of genetic analysis: JEM RCF YVJ. Occupancy analysis: YVJ QQ. Corridor analysis: SS YVJ QQ. Critical revision of the manuscript: JEM RCF QQ.

References

1. Purvis A, Agapow PM, Gittleman JL, Mace GM (2000) Nonrandom Extinction and the Loss of Evolutionary History. Science 288: 328–330. doi:10.1126/science.288.5464.328.
2. Gittleman JL, Funk SM, Macdonald DW, Wayne RK, eds (2001) *Carnivore conservation*. Cambridge: Cambridge University Press. 690 p.
3. Woodroffe R (2000) Predators and people: using human densities to interpret declines of large carnivores. Anim Conserv 3: 165–173.
4. Clark W, Curlee AP, Reading RP (1996) Crafting effective solutions to the large carnivore conservation problem. Conserv Biol 10: 940–948.
5. Check E (2006) Conservation biology: The tiger's retreat. Nature 441: 927–930.
6. Crooks K (2002) Relative Sensitivities of Mammalian Carnivores to Habitat Fragmentation. Conserv Biol 2: 488–502. doi:10.1046/j.1523-1739.2002.00386.x.
7. Proctor MF, McLellan BN, Strobeck C, Barclay RMR (2005) Genetic analysis reveals demographic fragmentation of grizzly bears yielding vulnerably small populations. Proc R Soc B 272: 2409–2416. doi:10.1098/rspb.2005.3246.
8. Wilcox BA, Murphy DB (1985) Conservation strategy: The effects of fragmentation on extinction. Am Nat 125: 879–887.
9. Woodroffe R, Ginsberg JR (1998) Edge effects and the extinction of populations inside protected areas. Science 280: 2126–2128. doi:10.1126/science.280.5372.2126.
10. Harrison S (1991) Local extinction in a metapopulation context: an empirical evaluation. Biol J Linn Soc Lond 42: 73–88.
11. Hanski IA (1994) A practical model of metapopulation dynamics. J Anim Ecol 63: 151–162.
12. Beier P, Noss RF (1998) Do habitat corridors provide connectivity? Conserv Biol 12: 1241–1252. doi:10.1111/j.1523-1739.1998.98036.x.
13. Noss RF (1987) Corridors in real landscapes: a reply to Simberloff and Cox. Conserv Biol 1: 159–164. doi:10.1111/j.1523-1739.1987.tb00024.x.
14. Banerjee K, Jhala YV (2012) Demographic parameters of endangered Asiatic lions (*Panthera leo persica*) in Gir Forests, India. J Mammal 93: 1420–1430. doi:http://dx.doi.org/10.1644/11-MAMM-A-231.1.
15. Walston J, Robinson JG, Bennett EL, Breitenmoser U, da Fonseca GAB, et al. (2010) Bringing the tiger back from the brink – the six percent solution. PLoS Biol 8: e1000485. doi:10.1371/journal.pbio.1000485.
16. Margules CR, Pressey RL (2000) Systematic conservation planning. Nature 405: 243–253. doi:10.1038/35012251.
17. Prugh LR, Hodges KE, Sinclair ARE, Brashares JS (2008) Effect of habitat area and isolation on fragmented animal populations. Proc Natl Acad Sci USA 105: 20770–20775. doi:10.1073/pnas.0806080105.
18. Tilson RL, Seal US (1987) Tigers of the world. Noyes Publications, Park Ridge, New Jersey.
19. Dolrenry S, Stenglein J, Hazzah L, Lutz RS, Frank L (2014) A metapopulation approach to African lion (*Panthera leo*) conservation. PLoS One 9: e88081. doi:10.1371/journal.pone.0088081.
20. Dinerstein E, Loucks C, Wikramanayake E, Ginsberg J, Sanderson E, et al. (2007) The fate of wild tigers. BioScience 57: 508–514. doi: 10.1641/B570608.
21. Sanderson E, Forrest J, Loucks C, Ginsberg J, Dinerstein E, et al. (2006) Setting Priorities for the Conservation and Recovery of Wild Tigers: 2005–2015. The Technical Assessment. Wildlife Conservation Society, World Wildlife Fund, Smithsonian Institution, and National Fish and Wildlife Foundation-Save the Tiger Fund, New York and Washington, D.C. 206 p.
22. Jhala YV, Qureshi Q, Gopal R (2010) Can the abundance of tigers be assessed from their signs? J Appl Ecol 48: 14–24. doi:10.1111/j.1365-2664.2010.01901.x.
23. Mondol S, Karanth KU, Ramakrishnan U (2009) Why the Indian Subcontinent Holds the Key to Global Tiger Recovery. PLoS Genet 5: e1000585. doi:10.1371/journal.pgen.1000585.
24. Rangarajan M (1996) Fencing the forest: conservation and ecological change in India's Central Provinces, 1860–1914. Oxford University Press, New Delhi.245 p.
25. Rangarajan M (2006) India's wildlife history: an introduction. Permanent Black, New Delhi. 150 p.
26. Narain S, Panwar HS, Gadgil M, Thapar V (2005) Joining the dots: The report of the Tiger Task Force. Project Tiger, Union Ministry of Environment and Forests, New Delhi. Available: http://projecttiger.nic.in/WriteReadData/PublicationFile/full_report.pdf. Accessed 3 Jan 2013.
27. Anonymous (1972) The Indian Wildlife (Protection) Act, (1972). Ministry of Environment and Forests, Government of India. Nataraj Publishers, Dehradun, India.
28. Karanth KU, Gopal R (2005) An ecology-based policy framework for human-tiger coexistence in India. In: Woodroffe R, Thirgood S and Rabinowitz A, editors. People and Wildlife: Conflict or Coexistence? Cambridge University Press. pp.373–387.
29. Gopal R, Qureshi Q, Bharadwaj M, Singh RKJ, Jhala YV (2010) Evaluating the status of the endangered tiger *Panthera tigris* and its prey in Panna Tiger

Reserve, Madhya Pradesh, India. Oryx 44: 383–398. doi: http//dx.doi.org//10.1017/S0030605310000529.
30. Anonymous (2005) The Indian Wildlife (Protection) Act (1972) Amendment Act, 2005. Ministry of Environment and Forests, Government of India. Nataraj Publishers, Dehradun, India.
31. Jhala YV, Qureshi Q, Gopal R, Sinha PR, editors (2011) Status of the tigers, co-predators, and prey in India, 2010. National Tiger Conservation Authority, Ministry of Environment and Forests, Government of India, New Delhi and Wildlife Institute of India, Dehradun. TR2011/003. 302 p. Available: http://www.wii.gov.in/protected_download/publications/researchreports/2011/tiger/staus_tiger_2010. Accessed 3 Mar 2013.
32. Chapron G, Miquelle DG, Lambert A, Goodrich JM, Legendre S, et al. (2008) The impact on tigers of poaching versus prey depletion. J Appl Ecol 45: 1667–1674. doi:10.1111/j.1365-2664.2008.01538.x.
33. Kenney JS, Smith JLD, Starfield AM, McDougal CW (1995) The long term effects of tiger poaching on population viability. Conserv Biol 9: 1127–1133. doi:10.1046/j.1523-1739.1995.9051116.x-i1.
34. Linkie M, Chapron G, Martyr DJ, Holden J, Leader-Williams N (2006) Assessing the viability of tiger subpopulations in a fragmented landscape. J Appl Ecol 43: 576–586. doi:10.1111/j.1365-2664.2006.01153.x.
35. Mondol S, Bruford MW, Ramakrishnan U (2013) Demographic loss, genetic structure and the conservation implications for Indian tigers. Proc R Soc B 280: 20130496. doi:10.1098/rspb.2013.0496.
36. Hanski I, Gilpin M, eds (1997) Metapopulation biology: ecology, genetics and evolution. Academic Press, San Diego, California. 512 p.
37. Hanski I (1999) Metapopulation ecology. Oxford University Press, New York.313 p.
38. Brown JH, Kodric-Brown A (1977) Turnover rates in insular biogeography: effect of immigration on extinction. Ecology 58: 445–449. doi:http://dx.doi.org/10.2307/1935620.
39. Seidensticker J, Dinerstein E, Goyal SP, Gurung B, Harihar A, et al. (2010) Tiger range collapse and recovery at the base of the Himalayas. In: Macdonald DW and Loveridge AJ, editors. Biology and Conservation of Wild Felids, Oxford University Press, Oxford, UK. pp.305–324.
40. Frankham R, Ballou JD, Briscoe D (2002) Introduction to Conservation Genetics. Cambridge University Press, Cambridge, UK.
41. Smith JLD (1993) The role of dispersal in structuring the Chitwan tiger population. Behaviour 124: 165–195.
42. Jhala YV, Gopal R, Qureshi Q, eds (2008) Status of the tigers, co-predators, and prey in India. National Tiger Conservation Authority, Ministry of Environment and Forests, Government of India, New Delhi and Wildlife Institute of India, Dehradun. TR08/001. 164 p. Available: http://www.wii.gov.in/protected_download/publications/statusof_tigers2008. Accessed 13 Mar 2013.
43. Forest (Conservation) Act (1980) Handbook of Forest (Conservation) Act, 1980 (With Amendments made in 1988), Forest (Conservation) Rules, 2003 (With Amendments made in 2004), Guidelines & Clarifications (Up to June, 2004), Government of India, Ministry of Environment & Forest, New Delhi. 126 p. Available: http://www.arunachalpwd.org/pdf/Hanbook%20on%20Forest%20(Conservation)%20Act,%201980%20(updated%202004).pdf. Accessed 29 Mar 2013.
44. Fernandes A (2012) How coal mining is trashing tigerland. Greenpeace, Bengaluru, India. 115 p. Available: http://www.greenpeace.org/india/Global/india/report/How-Coal-mining-is-Trashing-Tigerland.pdf. Accessed 30 Mar 2013.
45. Pinjarkar V (2011) SC refuses to grant relief in NH7 four-laning along Pench. The Times of India, Nagpur.Available: http://articles.timesofindia.indiatimes.com/2011-04- 06/nagpur/29387911_1_laning-mansar-cuts-tiger-corridor. Accessed 28 Apr 2013.
46. Ramachandran R (2011) Ministry's 'no' to Neutrino Observatory project in Nilgiris. The Hindu. Available: http://www.thehindu.com/2009/11/21/stories/2009112154952000.htm. Accessed 23 Mar 2013.
47. Vattakaven J (2010) Fragmentation threat in Kanha-Pench corridor: Implications of the Gondia-Jabalpur railway line on corridor connectivity and tiger dispersal. WWF-India, New Delhi, 19 p. Available: http://awsassets.wwfindia.org/downloads/fragmentation_threat_in_the_kanha_pench_corridor_1.pdf. Accessed 15 Apr 2013.
48. Pritchard JK, Stephens M, Donelly P (2000) Inferences of population structure using multilocus genotyping data. Genetics 155: 945–959.
49. Piry S, Alapetite A, Cornuet J-M, Paetkau D, Baudouin L, et al. (2004) GENECLASS2: a software for genetic assignment and first generation migrant detection. J Hered 95: 536–539. doi:10.1093/jhered/esh074.
50. Bergl RA, Vigilant L (2007) Genetic analysis reveals population structure and recent migration within the highly fragmented range of the Cross River gorilla (*Gorilla gorilla diehli*). Mol Ecol 16: 501–506. doi:10.1111/j.1365-294X.2006.03159.x.
51. Manel S, Schwartz MK, Luikart G, Taberlet P (2003) Landscape genetics: combining landscape ecology and population genetics. Trends Ecol Evol 18: 189–197. doi:http://dx.doi.org/10.1016/S0169-5347(03)00008-9.
52. Spear SF, Peterson CR, Matocq MD, Storfer A (2005) Landscape genetics of the blotched tiger salamander (*Ambystoma tigrinum melanostictum*). Mol Ecol 15: 2333–2344. doi:10.1111/j.1365-294X.2005.02573.x.
53. Epps CW, Wehausen JD, Palsboll PJ, McCullough DR (2007) Optimizing dispersal and corridor models using landscape genetics. J Appl Ecol 44: 714–724. doi:10.1111/j.1365-2664.2007.01325.x.
54. McRae BH (2006) Isolation by resistance. Evolution 60: 1551–1661. doi:10.1111/j.0014-3820.2006.tb00500.x.
55. McRae BH, Beier P (2007) Circuit theory predicts gene flow in plant and animal populations. Proc Natl Acad Sci USA 104: 19885–19890. doi:10.1073/pnas.0706568104.
56. Wang IJ, Savage WK, Shaffer HB (2009) Landscape genetics and least-cost path analysis reveal unexpected dispersal routes in the California tiger salamander (*Ambystoma californiense*). Mol Ecol 18: 1365–1374. doi:10.1111/j.1365-294X.2009.04122.x.
57. Savage WK, Fremier AK, Shaffer HB (2010) Landscape genetics of alpine Sierra Nevada salamanders reveal extreme population subdivision in space and time. Mol Ecol 19: 3301–3314. doi:10.1111/j.1365-294X.2010.04718.x.
58. Castilho CS, Marins-Sá LG, Benedet RC, Freitas TO (2011) Landscape genetics of mountain lions (*Puma concolor*) in southern Brazil. Mamm Biol 76: 476–483. doi:10.1016/j.mambio.2010.08.002.
59. Reding DM, Cushman SA, Gosselink TE, Clark WR (2013) Linking movement behavior and fine-scale genetic structure to model landscape connectivity for bobcats (*Lynx rufus*). Landsc Ecol 28: 471–486. doi:10.1007/s10980-012-9844-y.
60. Schwartz MK, Copeland JP, Anderson NJ, Squires, JR, InmanRM, et al. (2009) Wolverine gene flow across a narrow climatic niche. Ecology 90: 3222–3232. doi:http://dx.doi.org/10.1890/08-1287.1.
61. Cushman SA, McKelvey KS, Schwartz MK (2008) Use of empirically derived source-destination models to map regional conservation corridors. Conserv Biol 23: 368–376. doi:10.1111/j.1523-1739.2008.01111.x.
62. Spear SF, Balkenhol N, Fortin MJ, McRae BH, Scribner K (2010) Use of resistance surfaces for landscape genetic studies: considerations for parameterization and analysis. Mol Ecol 19: 3576–3591. doi:10.1111/j.1365-294X.2010.04657.x.
63. Sawyer SC, Epps CW, Brashares JS (2011) Placing linkages among fragmented habitats: do least-cost models reflect how animals use landscapes? J Appl Ecol 48: 668–678. doi:10.1111/j.1365-2664.2011.01970.x.
64. Sharma S, Dutta T, Maldonado JE, Wood TC, Panwar HS, et al. (2012) Spatial genetic analysis reveals high connectivity of tiger (*Panthera tigris*) populations in the Satpuda-Maikal landscape of Central India. Ecol Evol 3: 48–60. doi:10.1002/ece3.432.
65. Sharma S, Dutta T, Maldonado JE, Wood TC, Panwar HS, et al. (2013) Forest corridors maintain historical gene flow in a tiger metapopulation in the highlands of central India. Proc R Soc B 280: 20131506. doi:10.1098/rspb.2013.1506.
66. Joshi A, Vaidyanathan S, Mondol S, Edgaonkar A, Ramakrishnan U (2013) Connectivity of Tiger (*Panthera tigris*) Populations in the Human-Influenced Forest Mosaic of Central India. PLoS One 8: e77980. doi:10.1371/journal.-pone.0077980.
67. Sunquist ME, Karanth KU, Sunquist F (1999) Ecology, behaviour and resilience of the tiger and its conservation needs. In: Seidensticker J, Christie S and Jackson P, editors. Riding the tiger: Tiger conservation in human dominated landscapes. Cambridge University Press, Cambridge, UK. pp.5–18.
68. Wilson GA, Rannala B (2003) Bayesian inference of recent migration rates using multilocus genotypes. Genetics 163: 1177–1191.
69. Boyce MS, McDonald LL (1999) Relating populations to habitats using resource selection functions. Trends Ecol Evol 14: 268–272. doi:10.1016/S0169-5347(99)01593-1.
70. MacKenzie DI, Nichols JD, Royles JA, Pollock KH, Bailey LL, et al. (2006) Occupancy estimation and modeling: inferring patterns and dynamics of species occurrence.Academic Press. 324 p.
71. Ray N (2005) PATHMATRIX: a geographical information system tool to compute effective distances among samples. Mol Ecol Notes 5: 177–180. doi:10.1111/j.1471-8286.2004.00843.x.
72. McRae BH, Shah VB (2009) Circuitscape users' guide online. The University of California, Santa Barbara. Available: http://www.circuitscape.org. Accessed 10 Oct 2013.
73. Boom R, Sol CJ, Salimans MM, Jansen CL, Wertheim-van Dillen PM, et al. (1990) Rapid and simple method for purification of nucleic-acids. J Clin Microbiol 28: 495–503.
74. Menotti-Raymond M, David VA, Lyons LA, Schäffer AA, Tomlin JF, et al. (1999) A Genetic Linkage Map of Microsatellites in the Domestic Cat (*Felis catus*). Genomics 57: 9–23. doi:10.1006/geno.1999.5743.
75. Williamson JE, Heubinger RM, Sommer JA, Louis EE, Barber RC (2002) Development and cross-species amplification of 18 microsatellite markers in the Sumatran tiger (*Panthera tigris sumatrae*). Mol Ecol Notes 2: 110–112. doi:10.1046/j.1471-8286.2002.00163.x.
76. Wu J-H, Lei Y-L, Fang S-G, Wan Q-H (2008) Twenty-one novel tri- and tetranucleotide microsatellite loci for the Amur tiger (*Panthera tigris altaica*). Conserv Genet 10: 567–570. doi:10.1007/s10592-008-9571-8.
77. Navidi W, Arnheim N, Waterman MS (1992) A multiple-tubes approach for accurate genotyping of very small DNA samples by using PCR: statistical considerations. Am J Hum Genet 50: 347–359.
78. vanOosterhout C, Hutchinson WF, Wills DP, Shipley P (2004) MICRO-CHECKER: software for identifying and correcting genotyping errors in

microsatellite data. Mol Ecol Notes 4: 535–538. doi:10.1111/j.1471-8286.2004.00684.x.
79. Taberlet P, Luikart G (1999) Non-invasive genetic sampling and individual identification. Biol J Linn Soc Lond 68: 41–55. doi:10.1111/j.1095-8312.1999.tb01157.x.
80. Waits LP, Luikart G, Taberlet P (2001) Estimating the probability of identity among genotypes in natural populations: cautions and guidelines. Mol Ecol 10: 249–256. doi:10.1046/j.1365-294X.2001.01185.x.
81. Valière N (2002) GIMLET: a computer program for analyzing genetic individual identification data. Mol Ecol Notes 2: 377–379. doi:10.1046/j.1471-8286.2002.00228.x-i2.
82. Kalinowski ST, Taper ML, Marshall TC (2007) Revising how the computer program CERVUS accommodates genotyping error increases success in paternity assignment. Mol Ecol 16: 1099–1006. doi:10.1111/j.1365-294X.2007.03089.x.
83. Marshall TC, Slate J, Kruuk LEB, Pemberton JM (1998) Statistical confidence for likelihood-based paternity inference in natural populations. Mol Ecol 7: 639–655. doi:10.1046/j.1365-294x.1998.00374.x.
84. Rousset F (2008) Genepop'007: a complete reimplementation of the Genepop software for Windows and Linux. Mol Ecol Resour 8: 103–106. doi:10.1111/j.1471-8286.2007.01931.x.
85. Raymond M, Rousset F (1995) GENEPOP (version 1.2): population genetics software for exact tests and ecumenicism. J Hered 86: 248–249.
86. Peakall R, Smouse PE (2006) Genalex 6: genetic analysis in Excel. Population genetic software for teaching and research. Mol Ecol Notes 6: 288–295. doi:10.1111/j.1471-8286.2005.01155.x.
87. Goudet J (2001) FSTAT, a program to estimate and test gene diversities and fixation indices (version 2.9.3). Available: http://www2.unil.ch/popgen/softwares/fstat.htm. Accessed 3 Jul 2012.
88. Weir BS, Cockerham CC (1984) Estimating F-statistics for the analysis of population structure. Evolution 38: 1358–1370.
89. Slatkin M (1995) A measure of population subdivision based on microsatellite allele frequencies. Genetics 139: 457–462.
90. Savicky P (2009) pspearman: Spearman's rank correlation test. R package version 0.2–5. Available: http://CRAN.R-project.org/package=pspearman. Accessed 10 Jun 2014.
91. Excoffier L, Smouse PE, Quattro JM (1992) Analysis of molecular variance inferred from metric distances among haplotypes: applications to human mitochondrial DNA restriction data. Genetics 131: 479–491.
92. Evanno G, Regnaut S, Goudet J (2005) Detecting the number of clusters of individuals using the software STRUCTURE: a simulation study. Mol Ecol 14: 2611–2620. doi:10.1111/j.1365-294X.2005.02553.x.
93. Earl DA, vonHoldt BM (2012) STRUCTURE HARVESTER: a website and program for visualizing STRUCTURE output and implementing the Evanno method. Conserv Genet Resour 4: 359–361. doi:10.1007/s12686-011-9548-7. Available: http://taylor0.biology.ucla.edu/struct_harvest/index.php. Accessed 12 Apr 2013.
94. Paetkau D, Slade R, Burden M, Estoup A (2004) Genetic assignment methods for the direct, real-time estimation of migration rate: a simulation–based exploration of accuracy and power. Mol Ecol 13: 55–65. doi:10.1046/j.1365-294X.2004.02008.x.
95. Rannala B, Mountain JL (1997) Detecting immigration by using multilocus genotypes. Proc Natl Acad Sci USA 94: 9197–9201.
96. Slate J, Marshall TC, Pemberton JM (2000) A retrospective assessment of the accuracy of the paternity inference program CERVUS. Mol Ecol 9: 801–808. doi:10.1046/j.1365-294x.2000.00930.x.
97. Beerli P (2008) Migrate-n, version 3.3.2: a maximum likelihood and Bayesian estimator of gene flow using the coalescent. Available: http://popgen.scs.fsu.edu/migrate-n.html. Accessed 4 May 2013.
98. Beerli P, Felsenstein J (2001) Maximum likelihood estimation of a migration matrix and effective population sizes in *n* subpopulations by using a coalescent approach. Proc Natl Acad Sci USA 98: 4563–4568. doi:10.1073/pnas.081068098.
99. Smith JLD, McDougal C (1991) The contribution of variance in lifetime reproduction to effective population size in tigers. Conserv Biol 5: 484–490. doi:10.1111/j.1523-1739.1991.tb00355.x.
100. Piry S, Luikart G, Cornuet J-M (1999) BOTTLENECK: A computer program for detecting recent reductions in the effective population size using allele frequency data. J Hered 90: 502–503.
101. Garza JC, Williamson EG (2001) Detection of reduction in population size using data from microsatellite loci. Mol Ecol 10: 305–318. doi:10.1046/j.1365-294X.2001.01190.x.
102. Excoffier L, Laval G, Schneider S (2005) Arlequin (version 3.0): an integrated package for population genetics data analysis. Evol Bioinform Online 1: 47–50.
103. Beaumont M, Barratt EM, Gotelli D, Kitchener AC, Daniels MJ, et al. (2001) Genetic diversity and introgression in the Scottish wildcat. Mol Ecol 15: 319–336. doi:10.1046/j.1365-294X.2001.01196.x.
104. Storz JF, Beaumont M (2002) Testing for genetic evidence of population expansion and contraction: An empirical analysis of microsatellite DNA variation using a hierarchical Bayesian method. Evolution 56: 154–166. doi:10.1111/j.0014-3820.2002.tb00857.x.
105. Luikart G, Allendorf FW, Cornuet JM, Sherwin WB (1998) Distortion of allele frequency distributions provides a test for recent population bottlenecks. J Hered 89: 238–247. doi:10.1093/jhered/89.3.238.
106. Cornuet JM, Luikart G (1996) Description and power analysis of two tests for detecting recent population bottlenecks from allele frequency data. Genetics 144: 2001–2004.
107. Jhala YV, Qureshi Q, Gopal R, Amin R (2009) Field Guide: Monitoring tigers, co-predators, prey and their habitats. Third edition. Technical publication of National Tiger Conservation Authority, Government of India, New Delhi and Wildlife Institute of India, Dehradun, India. Available: http://www.wii.gov.in/images/field_guide_monitoring_tigers_co_predators.pdf. Accessed 28 Mar 2013.
108. Karanth KU, Nichols JD, eds (2002) Monitoring tigers and their prey. A manual for researchers, managers, and conservationists in Tropical Asia. Center for Wildlife Studies, Bangalore, India.193 p.
109. Forest Survey of India (2009) India: State of Forest Report. Ministry of Environment and Forests, Government of India, Dehradun, India. Available: http://fsi.nic.in/details.php?pgID=sb_61. Accessed 15 March 2013.
110. Rodriguez E, Morris CS, Belz JE, Chapin EC, Martin JM, et al. (2005) An assessment of the SRTM topographic products, Jet Propulsion Laboratory Publications, D31639. 143 p.
111. Sharma RK, Jhala YV, Qureshi Q, Vattakaven J, Gopal R, et al. (2010) Evaluating capture-recapture population and density estimation of tigers in a population with known parameters. Anim Conserv 13: 94–103. doi: 10.1111/j.1469-1795.2009.00305.x.
112. Karanth KU, Sunquist ME (2000) Behavioural correlates of predation by tiger (*Panthera tigris*) leopard (*Panthera pardus*) and dhole (*Cuon alpinus*) in Nagarahole, India. J Zool 250: 255–265.
113. Kendall WL, White GC (2009) A cautionary note on substituting spatial subunits for repeated temporal sampling in studies of site occupancy. J Appl Ecol 46: 1182–1188. doi:10.1111/j.1365-2664.2009.01732.x.
114. Hines JE (2006) PRESENCE2 – Software to estimate patch occupancy and related parameters. Version 2.4. USGS-PWRC. Available: http://www.mbr-pwrc.usgs.gov/software/presence.html. Accessed 20 Feb 2013.
115. Karanth KU, Gopalaswami AM, Kumar NS, Vaidyanathan S, Nichols JD, et al. (2011) Monitoring carnivore populations at the landscape scale: occupancy modeling of tigers from sign surveys. J Appl Ecol 48: 1048–1056. doi:10.1111/j.1365-2664.2011.02002.x.
116. Harihar A, Pandav B (2012) Influence of Connectivity, Wild Prey and Disturbance on Occupancy of Tigers in the Human-Dominated Western Terai Arc Landscape. PLoS One 7: e40105. doi:10.1371/journal.pone.0040105.
117. Jolliffe IT (2002) Principal Component Analysis. Springer, Heidelberg.493 p.
118. McRae BH, Dickson BG, Keitt TH, Shah VB (2008) Using circuit theory to model connectivity in ecology, evolution and conservation. Ecology 89: 2712–2724. doi:http://dx.doi.org/10.1890/07-1861.1.
119. Wright S (1951) The genetic structure of populations. Ann Eugen 15: 323–354.
120. Rousset F (1997) Genetic differentiation and estimation of gene flow from F-statistics under isolation by distance. Genetics 145: 1219–1228.
121. Mantel N (1967) The detection of disease clustering and a generalized regression approach. Cancer Res 27: 209–220.
122. Smouse PE, Long JC, Sokal RR (1986) Multiple regression and correlation extensions of the Mantel test of matrix correspondence. System Zool 35: 627–632.
123. Bonnet E, Van de Peer Y (2002) zt: a software tool for simple and partial Mantel tests. J Stat Softw 7: 1–12.
124. Meirmans PG (2012) The trouble with isolation by distance. Mol Ecol 21: 2839–2846. doi:10.1111/j.1365-294X.2012.05578.x.
125. Rosenberg NA, Burke T, Elo K, Feldman MW, Freidlin PJ, et al. (2001) Empirical evaluation of genetic clustering methods using multilocus genotypes from 20 chicken breeds. Genetics 159: 699–713.
126. Pritchard J, Wen W, Falush D (2009) Documentation for STRUCTURE software: Version 2.3. Department of Human Genetics, University of Chicago, Chicago and Department of Statistics, University of Oxford, Oxford.
127. Goossens B, Salgado-Lynn M (2013) Advances and difficulties of molecular tools for carnivore conservation in the tropics. Raffles Bull Zool 28: 43–53.
128. Reddy PA, Gour DS, Bhavanishankar M, Jaggi K, Hussain SM, et al. (2012) Genetic Evidence of Tiger Population Structure and Migration within an Isolated and Fragmented Landscape in Northwest India. PLoS One 7: e29827. doi:10.1371/journal.pone.0029827.
129. Luo S-J, Kim J-H, Johnson WE, VanderValt J, Martenson J, et al. (2004) Phylogeography and genetic ancestry of tigers (*Panthera tigris*). PLoS Biol 2: e442. doi:10.1371/journal.pbio.0020442.
130. Henry P, Miquelle D, Sugimoto T, McCullough DR, Caccone A, et al. (2009) *In situ* population structure and *ex situ* representation of the endangered Amur tiger. Mol Ecol 18: 3173–3184. doi:10.1111/j.1365-294X.2009.04266.x.
131. Elzirik E, Kim J-H, Menotti-Raymond M, Crawshaw PG Jr, O'Brien SJ, et al. (2001) Phylogeography, population history and conservation genetics of jaguars (*Panthera onca*, Mammalia, Felidae). Mol Ecol 10: 65–79. doi: 10.1046/j.1365-294X.2001.01144.x.
132. Marker LL, Wilkerson AJP, Sarno RJ, Martenson J, Breitenmoser-Wursten C, et al. (2008) Molecular genetic insights on cheetah (*Acinonyx jubatus*) ecology and conservation in Namibia. J Hered 99. 2–13. doi:10.1093/jhered/esm081.
133. Pearse DE, Crandall KA (2004) Beyond F_{ST}: Analysis of population genetic data for conservation. Conserv Genet 5: 585–602. doi:10.1007/s10592-003-1863-4.

134. Hardy OJ, Charbonnel N, Freville H, Heuertz M (2003) Microsatellite allele sizes: a simple test to assess their significance on genetic differentiation. Genetics 163: 1467–1482.
135. WWF India (2012) The return of tigers to Kopijhola, Madhya Pradesh, India: a tigress camera trapped in the Kopijhola forest block signifies its importance as a corridor. WWF–India Satpura Maikal Landscape Program. Available: http://www.wwfindia.org/news_facts/?uNewsID=7600. Accessed 21 Oct 2013.
136. Kimura M (1953) "Stepping Stone" model of population. Annual Report National Institute of Genetics, Japan 3: 62–63.
137. Beerli P (2009). How to use MIGRATE or why are Markov chain Monte Carlo programs so difficult to use. In: Bertorell G, Bruford MW, Hauffe HC, Rizzoli A, Vernesi C, editors. Population Genetics for Animal conservation. Cambridge University Press, Cambridge. pp.42–79.
138. Chandran MDS (1997) On the ecological history of the Western Ghats. Curr Sci 73: 146–155.
139. Misra VN (2001) Prehistoric human colonization of India. J Biosci 26: 491–531. doi:10.1007/BF02704749.
140. Alroy J (2001) A multispecies overkill simulation of the end-Pleistocene megafaunal mass extinction. Science 292: 1893–1896. doi:10.1126/science.1059342.
141. Surovell TA, Waguespack NM, Brantingham PJ (2005) Global archaeological evidence for Proboscidean overkill. Proc Natl Acad Sci USA 102: 6231–6236. doi:10.1073/pnas.0501947102.
142. Girod C, Vitalis R, Leblois R, Freville H (2011) Inferring population decline and expansion from microsatellite data: a simulation-based evaluation of the Msvar method. Genetics 188: 165–179.
143. Peery MZ, Kirby R, Reid BN, Stoelting R, Doucet-Beer E, et al. (2012) Reliability of genetic bottleneck tests for detecting recent population declines. Mol Ecol 21: 3403–3418. doi:10.1111/j.1365-294X.2012.05635.x.
144. Panwar HS (1987) Project Tiger: the reserves, the tigers and their future. In: Tilson RL and Seal US, editors. Tigers of the World: the biology, biopolitics, management and conservation of an endangered species. Noyes Publications, Park Ridge, New Jersey. pp.110–117.
145. MacKenzie DI, Nichols JD (2004) Occupancy as a surrogate for animal abundance. Anim Biodivers Conserv 27: 461–467.
146. Sunarto S, Kelly MJ, Parakkasi K, Klenzendorf S, Septayuda E, et al. (2012) Tigers need cover: Multi-scale occupancy study of the big cat in Sumatran forest and plantation landscapes. PLoS One 7: e30859. doi:10.1371/journal.pone.0030859.
147. Efford MG, Dawson DK (2012) Occupancy in continuous habitat. Ecosphere 3: art32. doi:http://dx.doi.org/10.1890/ES11-00308.1.
148. Quintero J, Roca R, Morgan AJ, Mathur A (2010) Smart Green Infrastructure in Tiger Range Countries: A Multi-Level Approach. Global Tiger Initiative, GTI-SGI working group, technical paper. Accessed: http://www.globaltigerinitiative.org/download/GTI-Smart-Green-Infrastructure-Technical-Paper.pdf. Accessed 28 Jan 2013.
149. Carter NH, Shrestha BK, Karki JB, Pradhan NMP, Liu J (2012) Coexistence between wildlife and humans at fine spatial scales. Proc Natl Acad Sci USA 109: 15360–15365. doi:10.1073/pnas.1210490109.
150. Pabla HS, Carlisle L, Cooper D, Cooke J, Nigam P, et al. (2011) Reintroduction of Gaur (*Bos gaurus gaurus*) in Bandhavgarh Tiger Reserve, India. Technical Report. Madhya Pradesh Forest Department, & Beyond and Wildlife Institute of India, Dehradun, India. 73p. Available: http://www.wii.gov.in/images/publications/researchreports/2011/reintroduction_gaur_report.pdf. Accessed 10 Sept 2013.
151. Dutta T, Sharma S, Maldonado JE, Wood TC, Panwar HS, et al. (2013) Gene flow and demographic history of leopards (*Panthera pardus*) in the Central Indian Highlands. Evol Appl 6: 949–959. doi:10.1111/eva.12078.

6

Spatial Analysis of Factors Influencing Long-Term Stress in the Grizzly Bear (*Ursus arctos*) Population of Alberta, Canada

Mathieu L. Bourbonnais[1]*, Trisalyn A. Nelson[1], Marc R. L. Cattet[2], Chris T. Darimont[3], Gordon B. Stenhouse[4]

1 Spatial Pattern Analysis and Research Lab, Department of Geography, University of Victoria, Victoria, British Columbia, Canada, **2** Canadian Cooperative Wildlife Health Centre, Western College of Veterinary Medicine, University of Saskatchewan, Saskatoon, Saskatchewan, Canada, **3** Applied Conservation Science Lab, Department of Geography, University of Victoria, Victoria, British Columbia, Canada, **4** Foothills Research Institute, Hinton, Alberta, Canada

Abstract

Non-invasive measures for assessing long-term stress in free ranging mammals are an increasingly important approach for understanding physiological responses to landscape conditions. Using a spatially and temporally expansive dataset of hair cortisol concentrations (HCC) generated from a threatened grizzly bear (*Ursus arctos*) population in Alberta, Canada, we quantified how variables representing habitat conditions and anthropogenic disturbance impact long-term stress in grizzly bears. We characterized spatial variability in male and female HCC point data using kernel density estimation and quantified variable influence on spatial patterns of male and female HCC stress surfaces using random forests. Separate models were developed for regions inside and outside of parks and protected areas to account for substantial differences in anthropogenic activity and disturbance within the study area. Variance explained in the random forest models ranged from 55.34% to 74.96% for males and 58.15% to 68.46% for females. Predicted HCC levels were higher for females compared to males. Generally, high spatially continuous female HCC levels were associated with parks and protected areas while low-to-moderate levels were associated with increased anthropogenic disturbance. In contrast, male HCC levels were low in parks and protected areas and low-to-moderate in areas with increased anthropogenic disturbance. Spatial variability in gender-specific HCC levels reveal that the type and intensity of external stressors are not uniform across the landscape and that male and female grizzly bears may be exposed to, or perceive, potential stressors differently. We suggest observed spatial patterns of long-term stress may be the result of the availability and distribution of foods related to disturbance features, potential sexual segregation in available habitat selection, and may not be influenced by sources of mortality which represent acute traumas. In this wildlife system and others, conservation and management efforts can benefit by understanding spatial- and gender-based stress responses to landscape conditions.

Editor: R. Mark Brigham, University of Regina, Canada

Funding: The authors would like to thank the Natural Sciences and Engineering Research Council of Canada, the Yellowstone to Yukon Conservation Initiative, and partners from the Foothills Research Institute Grizzly Bear Program for financial support. The funders had no role in the study design, data collection and analysis, decision to publish, or preparation of the manuscript.

Competing Interests: The authors have declared that no competing interests exist.

* E-mail: mathieub@uvic.ca

Introduction

Spatial patterns of species decline and extinction have been linked to complex interactions among anthropogenic factors, such as habitat loss and fragmentation [1–3], over-exploitation [4], climate change [5,6], and competition with invasive species [7,8]. Whereas many large predators lack the behavioural plasticity necessary to adapt to rapid change [9], at greatest risk are large bodied predators with diminished geographic range, small population size, low fecundity, and which occupy higher trophic levels [4,10,11]. To date, efforts to examine spatial patterns of species decline have focused primarily on changing patterns of species distributions, abundance, and mortality in response to anthropogenic activities and habitat fragmentation [12–14]. Although such studies provide essential understanding about how wild populations have responded to changing environments, they are generally retrospective due to a temporal disconnect between the disturbance event and associated population decline [15]. What effective policy intervention requires are real-time measures of potential stressors with associated spatial methods to reliably understand where individuals within populations might be at most risk of declines.

Recently, measures of the physiological response of wildlife to external stressors are emerging as a viable approach for analyzing contemporary impacts of habitat conditions and disturbance on the health and fitness of individuals and populations [15,16]. Cortisol concentrations measured from saliva, blood, feces, and hair [17–19] have been used as a non-invasive approach to quantify long-term stress responses in animals such as the northern spotted owl [20], squirrel gliders [21], ungulates [22–24], wolves [25], and grizzly bears [26,27]. Vertebrates respond to noxious external stimuli by activation of the hypothalamic-pituitary-adrenal axis with the resultant release of glucocorticoids, such as cortisol, into the blood circulation [28] allowing the organism to respond to the stressor with the goal of maintaining or re-establishing homeostasis [29]. However, persistent or repeated exposure to stressful stimuli, and resultant continued circulation of glucocorticoids, have been found to impair immune system performance, increase susceptibility to disease, and decrease growth and reproductive capacity in some species [15,28,30].

An emblematic species of western North America, the grizzly bear (*Ursus arctos*) has experienced substantial reduction in its historic range due to human settlement and development [31]. Conservation of remaining localized populations is difficult due to conflicting public opinion and land-use [32]. Grizzly bears in Alberta, Canada, have recently been estimated to number fewer than 700 individuals [33], and as a result the provincial population was listed as *Threatened* in 2010 [34]. Grizzly bears in Alberta occupy a landscape heavily impacted by human activities and resource extraction. Industrial activities (e.g., forestry, oil and gas exploration, mining, and agriculture) and extensive road networks are prevalent throughout grizzly bear habitat within the province resulting in a highly fragmented multi-use landscape [35,36]. Although a number of parks and protected areas exist, many are subject to a wide variety of recreational pursuits and high human visitation rates. Anthropogenic land-use and open road access features represent primary causes of grizzly bear mortality as bears have been found to select anthropogenic disturbed habitats to exploit seasonal food availability, which has increased contact with humans [35,37–39]. Further, patterns of land-use have resulted in genetically fragmented sub-populations that may not be viable in the long-term [36,40]. Less clear is how these activities, which vary spatially and in their character, affect the physiology of individuals.

Spatial patterns of anthropogenic land-use, forest conditions, and topography also influence the distribution of available resources (e.g., [38]). Conceptually, habitat quality accounts for the range of conditions that have an impact on the health of an animal occupying the habitat [41]. Grizzly bears in Alberta occupy large home ranges (ranging from approximately 300 km^2 for females to upwards of 1500 km^2 for males) allowing them to seasonally select a diverse assemblage of habitat types necessary to meet their nutritional needs [42,43]. True to their description as opportunistic omnivores, the diet of grizzly bears in Alberta is generally low in protein consisting mostly of green vegetation, fruits, and insects [38,44], although the consumption of ungulates varies seasonally [45]. Accounting for spatial variability in environmental factors related to habitat, such as forest conditions, landcover, vegetation productivity, and elevation, that influence the availability of food resources and have an impact on the health of individuals is an important consideration when assessing the relationship between landscape conditions and physiology [46,47].

We know that human impacts to the landscape, including habitat loss and alteration [33], contribute to grizzly bear mortality [48–50]. Yet, little is understood regarding how human activities and habitat conditions affect bears physiologically. Given small grizzly bear population sizes [36], densities, and reproductive rates [51], research on the interaction between physiological status and landscape conditions is essential. Although physiological status can be represented by a wide range of metrics, we specifically focus on long-term stress in this study because of growing recognition that long-term stress is an important factor linking ecological change with impaired health and population performance in wildlife [52,53]. Accordingly, our goal is to quantify spatial relationships between landscape conditions and long-term stress levels in grizzly bears by statistically integrating a spatially and temporally broad dataset of hair cortisol concentrations (HCC) with data representative of habitat conditions and anthropogenic disturbance. To meet this goal we address the following objectives:

1. Quantify impacts of habitat conditions and anthropogenic landscape disturbance on observed spatial patterns of HCC levels in Alberta grizzly bears.
2. Develop a spatially explicit model to predict HCC levels across grizzly bear habitat based on current landscape conditions.
3. Interpret the spatial distribution of predicted HCC values using data describing the relative importance and security of grizzly bear habitat.

To address these objectives, novel spatial methods are required to integrate HCC data with spatially continuous data representing environmental conditions and to quantify observed relationships. Methods presented here, including marked point pattern analysis using kernel density estimation and non-parametric regression using random forests, provide an effective means for analyzing ecological data and are appropriate for future research on wildlife and HCC.

Study Area

The study was carried out for five grizzly bear management units (BMUs) in Alberta, Canada (Figure 1). Representing an area of nearly 111,000 km^2, the Grande Cache, Yellowhead, Clearwater, Livingstone, and Castle BMUs are divided by major east-west transportation corridors. As such, the BMUs largely represent genetically isolated populations, although some inter-population movement does occur [36]. Due to the geographic extent of the study area, vegetation, topography, and local weather conditions are highly variable. Elevation ranges from 450 m to 3500 m and increases from east to west. In the western mountainous region, habitat types include alpine and sub-alpine ecosystems comprised of fir (*Abies* spp.), pine (*Pinus* spp.), and spruce (*Picea* spp.), and wet-meadow complexes [54,55]. To the east, lower elevation foothills comprised of mixed-wood forests of pine, aspen and poplar (*Populus* spp.), spruce, and balsam fir (*Abies balsamea*) represent a transitional zone between the Rocky Mountains and the prairies [54]. Mean temperatures range from 12°C in the summer to −7.5°C in the winter, and mean annual precipitation is 450–800 mm. Major grizzly bear foods found in the region include herbaceous plant growth such as sweetvetch (*Hedysarum* spp), Canada buffaloberry (*Shepherdia canadensis*), bearberry (*Arctostaphylos uva-ursi*), horsetail (*Equisetum* spp), dandelions (*Taraxacum officinale*), and clover (*Trifolium* spp), as well as a variety of insects and ungulates [38,56].

Resource extraction activities within the BMUs also vary spatially. They include forestry, oil and gas exploration, mining, and agriculture. To service resource extraction activities, an extensive network of roads exists that provide access to grizzly bear habitat resulting in increased human-bear conflict and mortality [49,50,57]. These roads and areas are also widely used for recreation, including hunting, fishing, hiking, and trail-riding with all-terrain vehicles and snowmobiles. A network of parks and protected areas, including Jasper, Banff, and Waterton National Parks, as well as a number of provincial parks and wilderness reserves, which generally exclude resource extraction activities, are also found throughout the BMUs.

Materials and Methods

Ethics Statement

Grizzly bear hair collection was undertaken as part of an initiative to conduct a population inventory program for the species. The population inventory work was carried out at the request of the Government of Alberta (Environment and Sustainable Resource Development) by the Foothills Research

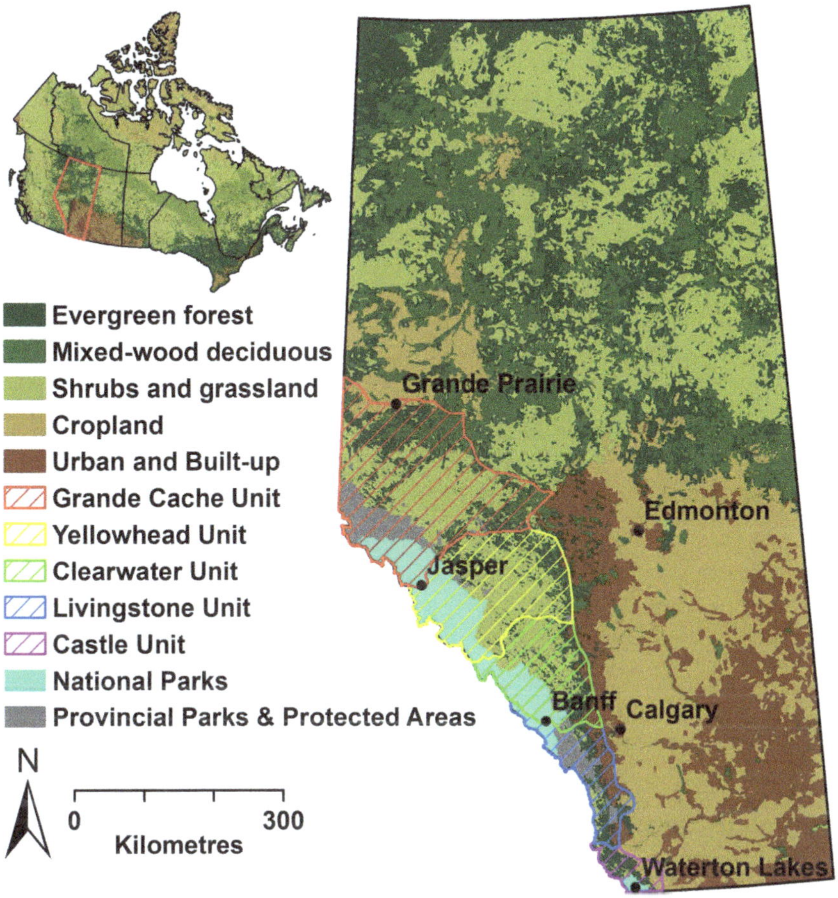

Figure 1. Study area location in Alberta, Canada. Grizzly bear hair samples were collected in each bear management unit during a single summer (Yellowhead - 2004; Clearwater - 2005; Livingstone - 2006; Castle - 2007; Grande Cache - 2008).

Institute's Grizzly Bear Program (FRIGBP) and followed the techniques described by Woods et al. [58] and Proctor et al. [59]. Grizzly bears are not provincially endangered and were not a protected species during the time when this sampling occurred. All samples were collected on provincial and federal lands under the authority of the Government of Alberta (Environment and Sustainable Resource Development and Alberta Parks) and the Government of Canada (Parks Canada). This inventory program was approved by the Alberta Department of Sustainable Resource Development animal care committee and by Parks Canada when sampling occurred in their jurisdiction in each year of data collection. Annual research permits and animal care approvals were obtained from both provincial and federal agencies responsible for permits and licensing of these activities.

Hair Cortisol Concentrations

Cortisol concentrations (picograms per milligram of hair - pg/mg) were measured in grizzly bear hair samples obtained annually from 2004 to 2008. As cortisol accumulates in hair for the duration of its growth, the HCC values used in this analysis represent a stress signal from the period of hair growth during the year preceding hair sample capture. Hair is a stable medium that can be collected non-invasively from free-ranging animals [26]. It can be transported and stored with relative ease (e.g., paper envelope at room temperature) and substances incorporated into growing hair, including cortisol, remain detectable for years to centuries [60]. These attributes make hair cortisol concentrations particularly effective for quantifying long-term stress in far-ranging species such as grizzly bears [26,61]. Hair samples were collected using barbwire hair snags [55,56] randomly placed within 7 km × 7 km grid cells and repositioned at 14 day intervals throughout known grizzly bear habitat in each BMU during the spring and early summer (see [62–66] for methods used to collect grizzly bear hair samples in each BMU). A total of 304 HCC values ($n = 168$ females, $n = 136$ males) were extracted from hair samples collected in the five BMUs (see [26] for details regarding procedures to extract cortisol concentrations from grizzly bear hair). As only two of the 304 HCC values were from the same individual, the HCC data represent an effective characterization of long-term stress in the Alberta grizzly bear population. The range of observed HCC values (0.16–14.94 pg/mg vs. 0.16–23.66 pg/mg) and mean HCC ±95% CI (1.20±0.29 pg/mg vs. 1.52±0.34 pg/mg) were similar between males and females, respectively. Further, comparison between the mean HCC values for males and females did not indicate a statistically significant difference (independent samples t-test with observed values: $p = 0.19$; independent samples t-test with ln-transformed values: $p = 0.13$).

Habitat Conditions

We represented grizzly bear habitat conditions by integrating a variety of spatial data. These variables, characterizing forest conditions, landcover, topography, and vegetation productivity measured using remotely sensed data, represent proxies for food availability (see Table 1 for rationale for inclusion of variables).

Percent crown closure and percent conifer were modelled and scaled from 0% to 100% through classification of Landsat Thematic Mapper (TM) 5 and Landsat 7 Enhanced Thematic Mapper Plus (ETM+) imagery informed by topographic derivatives from a digital elevation model (DEM) [67]. Landcover was classified into eight classes: upland trees, wetland trees, upland herbs, wetland herbs, shrubs, water, barren land, and snow/ice, using Landsat TM 5 and ETM+7 imagery, and topographic derivatives [68,69]. We assessed terrain conditions using a DEM detailing elevation for the study area obtained from the Government of Canada spatial data portal Geobase and resampled to 1 km grid cells. A terrain ruggedness index, providing a measure of terrain complexity and variability [70], and a compound topographic index, which represents potential soil moisture based on slope, catchment area, and upstream water sources [71], were derived from the DEM.

We used indices from the Dynamic Habitat Index (DHI) [72,73], which has been linked to observed spatial patterns of avian species [74], biodiversity gradients [75], and home range size of carnivores [76], to characterize vegetation productivity in the study area. The DHI indices are calculated from remotely sensed imagery and summarize annual trends in monthly images of the fraction of photosynthetically active radiation (fPAR). In this study, fPAR is derived from Advanced Very High Resolution Radiometer (AVHRR) reflectance values with a spatial resolution of 1 km from 2003 to 2007 [72,77]. The DHI is comprised of three indices representing vegetation productivity: cumulative greenness, variation in greenness, and minimum cover [72,78]. Cumulative greenness, which represents total vegetation productivity, is estimated annually by summing monthly fPAR observations. Variation in greenness, representative of seasonal changes in productivity, and consequently the seasonal availability of food resources, is calculated using the coefficient of variation in fPAR values over a year. Highly seasonal landscapes, such as alpine environments, where greenness values vary substantially due to snowpack, receive higher values than regions that are productive year round, such as evergreen forests [72]. The minimum cover is an estimate of the lowest level of vegetative productivity available year round and may influence the persistent use of habitat by herbivorous species. As production of leafy biomass and fruits consumed by grizzly bears is seasonally dependent [45,79], regions with high seasonality and cumulative greenness, as well as high minimum cover, are likely representative of high quality grizzly bear habitat.

Habitat Selection

We characterized spatial patterns of seasonal habitat use by grizzly bears throughout the study area using resource selection functions (RSFs) where the probability of habitat use is ranked from 0 (low) to 10 (high) [43,80]. An RSF models the probability of use of a resource, relative to its availability, based on occurrence patterns of an animal on the landscape [81]. The RSFs developed

Table 1. Variables used to predict HCC levels in grizzly bears.

Abbreviation	Variable	Range	Rationale
Habitat conditions			
cc	Percent crown closure (%)	0–100	Influences forest understory vegetation abundance
pctcon	Percent conifer (%)	0–100	Correlated with herbaceous food abundance
lcover	Landcover (categorical)	1–8	Proxy for presence and abundance of food sources
dhi_cum	Dynamic Habitat Index - cumulative greenness (unitless)	0.33–18.50	Estimate of total vegetation productivity
dhi_cv	Dynamic Habitat Index - coefficient of variation (unitless)	0.19–1.35	Estimate of seasonal change in vegetation productivity
dhi_min	Dynamic Habitat Index - minimum cover (unitless)	0–0.40	Lowest estimated annual vegetation productivity
elev	Elevation (m)	450–3500	Impacts landcover, vegetation cover, and human access
tri	Terrain ruggedness index (unitless)	0–189.33	Impacts human access and grizzly bear mortality
cti	Compound topographic index (unitless)	3.86–18.03	Correlated with herbaceous foods and presence of ungulates
Habitat selection			
rsf_s1	Resource Selection Function - hypophagia (categorical)	0–10	Probability of habitat selection following den emergence
rsf_s2	Resource Selection Function - early hyperphagia (categorical)	0–10	Probability of habitat selection during the summer
rsf_s3	Resource Selection Function - late hyperphagia (categorical)	0–10	Probability of habitat selection during the fall
rsf_max	Resource Selection Function - maximum value (categorical)	0–10	Maximum observed habitat selection across all three seasons
Anthropogenic disturbance			
rd_dd	Roads - distance decay (unitless)	0–1	Impacts human access and contribute to landscape fragmentation
rail_dd	Railways - distance decay (unitless)	0–1	Contribute to grizzly bear mortality
wl_dd	Oil and gas well-sites - distance decay (unitless)	0–1	Concentrated sites of human activity and contribute to habitat fragmentation
ln_den	Secondary linear features - density (km/km^2)	0–7.28	Contribute to habitat fragmentation, density of forest edges, and impacts human access
cblk_l	Forest harvest blocks - ≤ than 15 years old (% cut/km^2)	0–100	Younger seral forests have greater abundance of herbaceous foods
cblk_g	Forest harvest blocks ->than 15 years old (% cut/km^2)	0–100	Food availability decreases as time since disturbance increases
pa	Proportion parks and protected area (unitless)	0–1	Less disturbance compared to surrounding landscape

for this study area are based on third order selection [82], and as such the modelled probability of habitat use by grizzly bears is representative of selection at the patch level (see [43] for RSF model details and accuracy). Separate RSF models were used for male and female bears in three seasons, hypophagia (1 May to 15 June), early hyperphagia (16 June to 31 July), and late hyperphagia (1 August to 15 October), as well as the maximum observed RSF value for all three seasons (Table 1).

Anthropogenic Influence

Anthropogenic features we considered included roads, railways, oil and gas well-sites, cut-lines, power-lines, pipelines, and forest harvest blocks (Table 1). Anthropogenic data were provided by Alberta Sustainable Resource Development and are mapped and updated by the FRIGBP based on appearance of disturbance features in Landsat imagery. We represented roads, railways, and oil and gas well-sites using an exponential distance decay function, e^{-ad} where d is the distance in metres to the feature and a was fixed at 0.002 [80]. The distance decay surfaces decrease linearly from a value of 1 at the location of the anthropogenic feature to a value of 0 at a distance of approximately 2000 m. We represented secondary anthropogenic linear features, such as cut-lines, power-lines, and pipelines, as a cumulative linear density per 1 km^2 grid cell (km/km^2).

We divided forest harvest blocks into two age classes (≤15 years old; >15 years old) to account for regeneration and resultant differences in grizzly bear food availability present within the harvested areas [38,83]. We assessed the influence of forest harvest blocks by calculating the proportion of harvested area in 1 km^2 grid cells. Resultant grid cells ranged from a value of 1 representing an area that has been completely harvested to 0 indicating no harvesting had occurred in the area. Finally, we modelled the influence of the parks and protected areas in the region, which represent a noted contrast in terms of land-use compared to the surrounding landscape, based on proportion parks and protected area within a 10 km radius from 1 km^2 grid cells. Values ranged from 1 for cells completely within parks and protected areas to 0 when no parks and protected areas were in the immediate vicinity.

Conservation Areas and Habitat Security

Core and secondary grizzly bear conservation areas, based on observed patterns of grizzly bear occurrence, resource availability, and road density, have been mapped within the five BMUs [80]. Core conservation areas, which are meant to act as a population source, are defined as regions with high quality habitat (using RSF scores as surrogates) and road densities below 0.6 km/km^2. Secondary conservation areas also contain high quality habitat, however road densities are higher at 1.2 km/km^2 which increases the risk of mortality and decreases the population source capacity of the area. Within these regions, habitat states have been assessed by characterizing the landscape based on habitat quality and mortality risk [84]. In this analysis we incorporate three defined habitat states in 1 km^2 grid cells: secure habitat (low mortality risk), sink habitat (high mortality risk), and non-critical habitat. Combined with the present network of parks and protected areas, the core and secondary conservation units as well as the observed habitat states were used to help interpret geographic relationships in the modelled distribution of HCC values within the five BMUs.

Kernel Density Estimation of Hair Cortisol Concentrations

To integrate the HCC point data with spatially continuous habitat and anthropogenic variables, we converted HCC data from points to a continuous surface using kernel density estimation (KDE). KDE is frequently used in wildlife analyses to summarize spatial variation in the intensity of habitat selection over large areas (e.g., the utilization distribution) based on telemetry point data [85]. KDE is also employed to spatially allocate attribute data (based on a statistically or ecologically significant bandwidth) over an area representative of the sample locations to understand how underlying environmental characteristics influence spatial pattern [86]. The KDE surface provides greater detail regarding the spatial variability of the underlying point pattern, which is the case in the spatial distribution of HCC levels. Kernel density estimation of a marked point pattern is defined by

$$g(u) = \frac{\sum_i k(e(u) - x_i)v_i}{\sum_i k(e(u) - x_i)},$$

where k is a Gaussian kernel, the HCC point data values are given by $v_1,\dots,v_n$ at locations $x_1,\dots,x_n$, u are the smoothed HCC values, and $e(u)$ is an edge correction factor based on the reciprocal of the kernel mass

$$\frac{1}{e(u)} = \int_W k(v-u)dv$$

inside the observation window or spatial extent W [87,88]. The spatial extent was defined separately for males and females in each BMU as the minimum convex polygon (MCP) of HCC sample locations [89]. Kernel values were stored in a 1 km cell matching the spatial resolution of landscape variables. A 9 km kernel bandwidth was defined as it corresponds to the average daily area used by an adult female grizzly bear [43] and was supported by least-squares cross-validation, which identifies the bandwidth that minimizes the summed squared error between observed values and smoothed values.

We assessed the influence of individual HCC points on the stability of the male and female KDE surfaces by generating 99 leave-one-out bootstrap HCC kernel density surfaces for validation [90]. We quantified uncertainty in the HCC surfaces for each BMU by determining the proportion of KDE pixels that fell within a 95% confidence interval (p<0.01) of similar observed values in the bootstrap surfaces [91]. We compared the distribution and central tendency of the values in the KDE HCC surfaces to the original HCC values using Kolmogorov-Smirnov and Mann-Whitney U tests.

Random Forest Models

We used random forest models to quantify the influence of habitat and anthropogenic variables on the spatial distribution of male and female grizzly bear HCC values from the KDE surfaces. A non-parametric recursive regression method, random forests combine multiple regression trees built using bootstrap samples of data [92]. Each individual regression tree is grown to its maximum size using random subsets of predictor variables [93]. Trees are combined by averaging and estimation of response values is performed using the withheld out-of-bag observations [93,94]. The model variance explained is assessed based on the accuracy of the prediction of out-of-bag data. Random forests have been found to be ideally suited to ecological data as they do not require linear relationships, effectively model variable interactions, can handle missing data and correlated variables, are more stable than traditional regression trees to minor changes in input data, and have high predictive power [92–94]. Variable importance in random forest models is quantified using two complementary output metrics. The first is a normalized comparison of the mean square error of model predictions with predictions generated using randomly permuted predictor values from the out-of-bag data [93]. The second is the average total decrease in node impurity attributed to splitting on each variable measured using the residual sum of squares and provides an indication of node prediction accuracy attributed to each variable.

We ran random forest models for both male and female grizzly bears using a random subset of 50% of the available data. The remaining 50% of the data were withheld for model validation. Each model included 1000 trees to allow stabilization of out-of-bag error and 18 variables were randomly selected for consideration at each split. Due to varying landscape conditions and anthropogenic influence inside and outside of parks and protected areas, we produced secondary random forest models with an inside parks and protected areas/outside parks and protected areas distinction to explore potentially differing variable importance.

We used validated random forest models to predict male and female HCC values in 1 km^2 grid cells for the area of the five BMUs outside the confines of the MCPs used in the development of the KDE surfaces. We explored relationships between the 10 most influential variables and the predicted HCC values by summarizing the mean predictor values that corresponded with the lower (0.16–0.45 pg/mg: low HCC), inner (0.46–1.62 pg/mg: moderate HCC), and upper (>1.62 pg/mg: high HCC) quartiles of the input HCC point data. We used the pooled interquartiles of the input male and female HCC values in order to facilitate direct comparison of HCC levels between the sexes. These same HCC value breaks were used to summarize the percentage of pixels within the area of each BMU classified as a low, moderate or high HCC.

To aid interpretation of predicted HCC values, we assessed the associations between HCC values and parks and protected areas, core conservation areas, and secondary conservation areas based on the frequency distribution of male and female HCC values occurring within each of these management units. We also assessed relationships between the predicted male and female HCC values and classified secure, sink, and non-critical habitat types using frequency distributions based on 1 km^2 pixel associations.

Results

Validation of HCC Kernel Density Layers

Comparison of input HCC values and generated KDE HCC surfaces indicated that the KDE represented the range and spatial distribution of HCC values. Greater than 80% of pixels in the HCC KDE surfaces for all five BMUs fell within the 95% confidence interval ($p<0.01$) when compared to values in the 99 leave-one-out bootstrap surfaces. Kolmogorov-Smirnov and Mann-Whitney U tests showed the attributes of HCC KDE surfaces were not significantly different than measured HCC values in each of the BMUs (Table 2).

Male HCC Models

Using random forest metrics, habitat conditions and anthropogenic disturbance variables considered explained 74.28% of the variance in the male HCC data (MSE = 0.17). Validation of the male model using the withheld data returned an r^2 of 0.71. The male total model predicted a mean HCC value of 0.89 pg/mg (±95% CI: 0.01 pg/mg; range: 0.17 pg/mg –3.25 pg/mg). Proportion parks and protected areas was the most influential variable in the male model (Figure 2A). Habitat condition variables with the greatest influence included the three topographic metrics (elevation, terrain ruggedness index, and compound topographic index), the DHI metrics (cumulative greenness, variation in greenness, and minimum cover), as well as crown closure and the late hyperphagia RSF. Influential anthropogenic variables included distance decay to roads, and to a lesser extent distance decay to railways and the density of secondary linear features (Figure 2A). Mean variable values associated with low, moderate, and high predicted HCC values revealed generalized trends in the relationship between variables and HCC values (Table 3). Male HCC values had an inverse relationship with the proportion parks and protected areas. HCC levels were low-to-moderate at high and low elevations, and low stress-levels were associated with rugged terrain and low soil wetness. Observed HCC levels increased as the distance decay to roads decreased and as forest crown closure increased. DHI metrics showed male stress levels were lowest in areas with greater seasonality, low minimum cover and moderate cumulative greenness. The late hyperphagia RSF had the strongest relationship with male HCC values among the habitat selection variables considered. Spatial associations showed male HCC levels increased with increasing incidence of habitat use in the fall season.

Evaluation of the influence of differing landscape conditions inside and outside parks and protected areas on the modelled HCC values altered the importance of predictor variables. The male grizzly bear outside parks and protected areas model explained 55.34% of the variance in the HCC data (MSE = 014; $r^2 = 0.72$). The mean predicted male HCC for the outside parks and protected areas model was 1.01 pg/mg (±95% CI: 0.01 pg/mg; range: 0.21 pg/mg –3.25 pg/mg). Excluding the influence of parks and protected areas increased the importance of density of secondary linear features and proportion of forest harvest blocks greater than 15 years old (Figure 2B). The male inside parks and protected areas model explained 74.96% of the variance (MSE = 0.07, $r^2 = 0.84$), similar to the total model. The inside parks and protected areas model mean predicted male HCC was 0.55 pg/mg (±95% CI: 0.02 pg/mg; range: 0.17 pg/mg – 2.88 pg/mg). While the influence of variables related to habitat quality, such as elevation, terrain ruggedness, crown closure, and DHI metrics, were similar to the total model, the importance of distance decay to roads increased substantially for males when the spatial extent of the model was restricted to parks and protected areas (Figure 2C).

HCC values predicted using the total male random forest model showed spatial patterns of long-term stress in males were generally low in western national parks and increased in non-protected and low elevation regions to the east and south (Figure 3A). The highest HCC values for males were restricted to zones in and around smaller protected areas, while low and moderate HCC values were more frequent and continuously distributed on the landscape (Table 4). Frequency distributions of predicted HCC values associated with parks and protected areas, core conservation areas, and secondary conservation areas revealed similar trends. Low male HCC values had more frequent spatial associations with parks and protected areas (Figure 4A). Regions designated as core conservation areas had a higher frequency of moderate male HCC values, while secondary conservation areas were associated with moderate-to-high male HCC values (Figure 4A). In terms of secure, sink, and non-critical habitats, moderate-to-high male values were more frequently associated with non-critical and sink habitats (Figure 4B).

Female HCC Models

Total female random forest models explained 68.46% (MSE = 0.21; $r^2 = 0.70$) of the variance in female HCC values. The mean

Table 2. HCC kernel density estimation validation results by bear management unit (BMU).

BMU	Proportion data within 95% CI ($p<0.01$)	Kolmogorov-Smirnov	Mann-Whitney U
Castle	0.89	$p = 0.07$	$p = 0.09$
Livingstone	0.86	$p = 0.12$	$p = 0.17$
Clearwater	0.88	$p = 0.09$	$p = 0.08$
Yellowhead	0.88	$p = 0.16$	$p = 0.14$
Grande Cache	0.87	$p = 0.21$	$p = 0.27$

predicted female HCC from the total model was 1.01 pg/mg (±95% CI: 0.02 pg/mg; range: 0.16 pg/mg –5.82 pg/mg). Variable importance was similar to the total male model, as proportion parks and protected areas was again the most influential variable (Figure 2D). Important habitat variables in the total female model included the three topographic metrics, the three DHI metrics, crown closure, the hypophagia RSF, and percent conifer. Important anthropogenic variables included distance decay to roads and railways, as well as the proportion of forest harvest blocks less than 15 years old. Generally, relationships between the mean values of the 10 most influential variables associated with low, moderate, and high predicted female HCC values contrasted those in the male model (Table 3). Female HCC values increased as the proportion parks and protected areas, elevation, and terrain ruggedness increased. Unlike males, low female HCC levels were associated with areas closer to roads, areas which have a higher density of recent forest harvest blocks, and forests with moderate crown closure. DHI metrics showed females had lower HCC levels in areas exhibiting less seasonality and higher minimum cover and cumulative greenness. Hypophagia RSF values had the strongest associations with female HCC levels which increased as the probability of habitat use increased in the spring season.

The female outside parks and protected areas model had a higher variance explained (67.61%; MSE =0.20; $r^2 = 0.71$), compared to the female inside parks and protected areas model (58.15%; MSE =0.15; $r^2 = 0.58$). The mean female HCC was 0.73 pg/mg (±95% CI: 0.02 pg/mg; range: 0.16 pg/mg – 5.82 pg/mg) for the outside parks and protected areas model and 1.86 pg/mg (±95% CI: 0.02 pg/mg; range: 0.40 pg/mg – 3.10 pg/mg) for the inside parks and protected areas model. Compared to the male model, the change in variable importance was more pronounced in the female outside parks and protected areas model as the influence of proportion of forest harvest blocks less than 15 years old, distance decay to railways, and distance decay to oil and gas well-sites all increased (Figure 2E). However, unlike the male inside parks and protected areas model, restricting the female model to landscape conditions within parks and protected areas decreased the influence of all anthropogenic variables while increasing the influence of the DHI and topographic metrics (Figure 2F).

Spatial patterns of the predicted long-term stress levels for female grizzly bears contrasted predicted male HCC patterns. Predicted female HCC values were generally low in the eastern portions of the study area where anthropogenic disturbance is concentrated and increased substantially in the foothills and high elevation parks and protected areas in the west (Figure 3B). As a result, low female HCC values were more frequently associated with secondary conservation areas and sink habitats (Figure 4C; Figure 4D). High, moderate, and low predicted female HCC values were more evenly distributed on the landscape, and high female values had a far greater geographic representation compared to similar male HCC values (Table 4). Similar to males, the Livingstone BMU had the greatest percent area with high female HCC values, while the Grande Cache BMU had the greatest concentration of low male and female HCC values (Table 4).

Discussion

Grizzly bears in Alberta, Canada face many challenges as a result of increasing anthropogenic activities and habitat fragmentation throughout their current range. While knowledge regarding causes of mortality, habitat use, and spatial patterns of habitat fragmentation and loss is considerable, less is known regarding the impacts of landscape conditions on the health of grizzly bears [26]. We found no obvious differences in observed input HCC values between males and females. Similarity in male and female input HCC levels suggests that baseline cortisol levels are robust to differences in the types and timings of life history events. However, the spatial distribution of predicted HCC values, based on random forest models integrating HCC stress surfaces with environmental covariates, differed between males and females. In general, female HCC levels appeared to follow a gradient based on elevation with moderate-to-high values predicted in the mountain parks and core conservation areas, and low-to-moderate values predicted in lower elevation forests, prairies, and aspen parkland. The pattern for male HCC values also appeared to follow a similar gradient. Male HCC values were predicted to be low in the mountain parks and moderate in the prairies and aspen parkland located in the eastern portion of the study area. In contrast to females, focal areas of high HCC values were not evident for males.

Gender based variation in spatial patterns of HCC levels suggest the type and intensity of stressors are not uniform across the landscape. Results suggesting females, and to a lesser extent males, have low-to-moderate long-term stress levels in response to environmental conditions present in regions highly impacted by humans appear counterintuitive. However, disturbance features related to resource extraction activities (e.g., forest harvest blocks, roads, pipelines), which represent the dominant forest disturbances in the study area due to extensive fire suppression [95,96], have been correlated with the presence and abundance of foods consumed by grizzly bears [38,97,98]. Findings presented here support the hypothesis that long-term stress levels in certain wildlife species may be correlated with the availability and abundance of foods. Examining glucocorticoid levels detected in grizzly bear scat in the Yellowhead BMU, Wasser et al. [99] found preliminary evidence that grizzly bear stress levels were generally low in areas with high densities of anthropogenic disturbance and higher in areas with fewer disturbances. They hypothesized that observed patterns of stress may be related to the greater availability and distribution of foods in anthropogenic disturbance

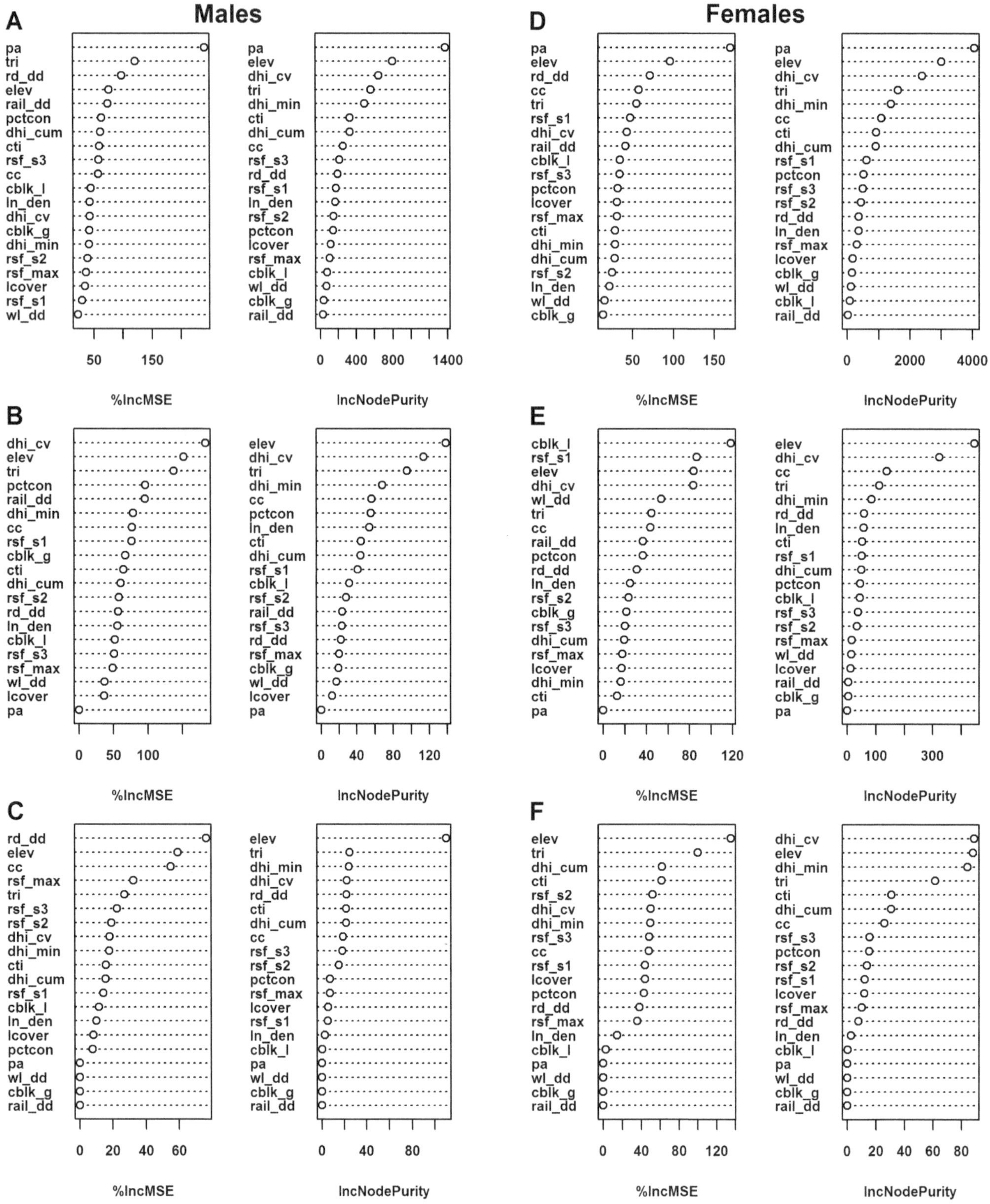

Figure 2. Variable importance metrics for male and female HCC random forest models. Variable importance for the male (A) total model, (B) outside parks and protected areas model, and (C) inside parks and protected areas model, as well as the female (D) total model, (E) outside parks and protected areas model, and (F) inside parks and protected areas model. Variable importance plots on the left of each panel (%IncMSE) represent the accuracy of random forest model predictions based on regression tree splits made using each variable. Plots on the right of each panel (IncNodePurity) indicate how often each variable was used as a split in regression trees aggregated through the random forest. For example, in panels A & D the proportion parks and protected areas was selected often as a tree split in the random forest and had a high predictive HCC value accuracy.

Table 3. Mean values of the 10 most influential variables in the total random models associated with lower, mid, and upper quartiles of the predicted HCC levels in male and female grizzly bears.

Variable	HCC range (pg/mg)		
	0.16–0.45	**0.46–1.62**	**>1.62**
Males			
pa	0.85	0.13	0.17
elev	2164.92	1277.83	1690.15
rd_dd	0.01	0.56	0.36
tri	33.87	10.67	21.32
dhi_cv	0.71	0.43	0.41
dhi_min	0.04	0.11	0.14
dhi_cum	6.02	11.11	10.13
cti	6.62	14.12	10.80
cc	17.22	43.66	48.76
rsf_s3	3	4	6
Females			
pa	0.01	0.12	0.68
elev	940.91	1427.64	1980.69
rd_dd	0.69	0.55	0.12
tri	3.88	12.15	29.49
dhi_cv	0.43	0.53	0.65
dhi_min	0.11	0.11	0.06
dhi_cum	11.47	8.42	6.54
cblk_l	0.03	0.02	0.01
cc	47.10	39.54	30.76
rsf_s1	2	5	6

features. Similar inverse relationships between glucocorticoid levels and food availability and diet have been observed in other grizzly bear populations [27,100], as well as in a number of bird [47] and marine species [101]. For grizzly bears in Alberta, with a diet comprised largely of seasonally dependent herbaceous food sources [45], optimizing nutritional uptake over a short summer season is essential to ensure energy demands are met over the winter. Consequently, areas with more abundant food sources may present a low-stress environment for grizzly bears despite human activities.

While male and female stress levels in highly industrialized areas in the east were comparable, the high mountain parks in the west represented notable broadly dispersed areas of high stress for females and low stress for males. The observed differences in HCC levels may be the result of gender-specific differences in exposure to stressors and/or differences in perception and response to stressors both inside and outside of parks and protected areas. For example, Graham et al. [57] found females were more likely to select habitat associated with roads and to cross roads during the day, while males avoided roads and were more likely to cross them at night. Similarly, females have been found to select edge habitats associated with anthropogenic disturbance while male habitat selection is more frequently associated with naturally occurring edges [97]. As a result, outside of parks and protected areas the higher density of anthropogenic disturbance features, which are generally avoided by males, may explain the slightly higher observed male HCC levels compared to values inside the parks and protected areas where anthropogenic activities are more localized. However, as males occupy very large home ranges (upwards of 1500 km^2) they may be more capable of mitigating the cumulative effects of external environmental stressors compared to females whose home ranges are smaller (approximately 300 km^2) and have less inter-annual variability. This may explain the lower spatial variability in HCC levels observed in males across the study area.

Compared to males, female exposure to external stressors elicits strong stress levels inside parks and protected areas. Despite the importance of parks and protected areas as wildlife refuges, Gibeau et al. [102] classified nearly half of the available habitat in these areas as unsuitable for grizzly bears (i.e., high elevation areas with poor food availability and concentrated areas of human recreation and activity). As a result, the availability of high quality habitat and foods is restricted in these regions and may result in higher observed stress levels in females. We propose the same conditions do not elicit a strong stress response in males due to sexual segregation of high quality habitat selection within parks and protected areas. Females have been found to make greater use of sub-optimal habitats and areas in close proximity to humans to avoid males and reduce the risk of infanticide when they are with cubs [36,57,103–107]. Female habitat selection is also philopatric as female offspring tend to occupy home ranges that take in part of, or are in close proximity, to the home ranges of their mother [59,106,108–110]. If males are dominating or excluding female use of quality habitat in parks and protected areas the resulting disparity in resources available to each gender could help explain observed geographic differences in the stress levels in parks and protected areas.

Further, in an assessment of grizzly bear body condition across the study area, Cattet et al. [111] found bears inside mountainous parks had poorer body condition compared to bears outside parks and protected areas, and suggested observed patterns may be linked to relatively low food availability in high mountain parks. If food availability is linked to long-term stress in grizzly bears the differences in body condition may in fact be the result of observed differences in stress levels inside and outside of parks and protected areas and warrants further study. While the observed spatial associations between high stress levels and increased incidence of habitat selection in the hypophagia and late-hyperphagia RSF layers provide preliminary evidence of potential impacts of competition on long-term stress, future studies explicitly considering impacts of sexual segregation and conspecific competition on stress in grizzly bears would also be of interest.

The observed spatial patterns of long-term stress in areas with high densities of human disturbance and activity represent a potential grizzly bear management opportunity. Numerous studies have proposed restricting human access may help reduce human-induced grizzly bear mortality (e.g., [49,50]). We suggest managing human access to grizzly bear habitat, for example by closing roads following the conclusion of resource extraction activities or restricting access to high quality habitat in parks and protected areas, could lead to health benefits for individuals, particularly females. As external environmental stressors can negatively impact reproduction [30], the ability of females to occupy low-stress environments with abundant food sources may help ensure the long-term viability of populations. However, potential physiological gains afforded by food resources associated with anthropogenic disturbed habitats may be offset if grizzly bears continue to experience high rates of human-induced mortality in these areas (humans are responsible for upwards of 80% of observed grizzly bear mortality in the area [48,49,112]).

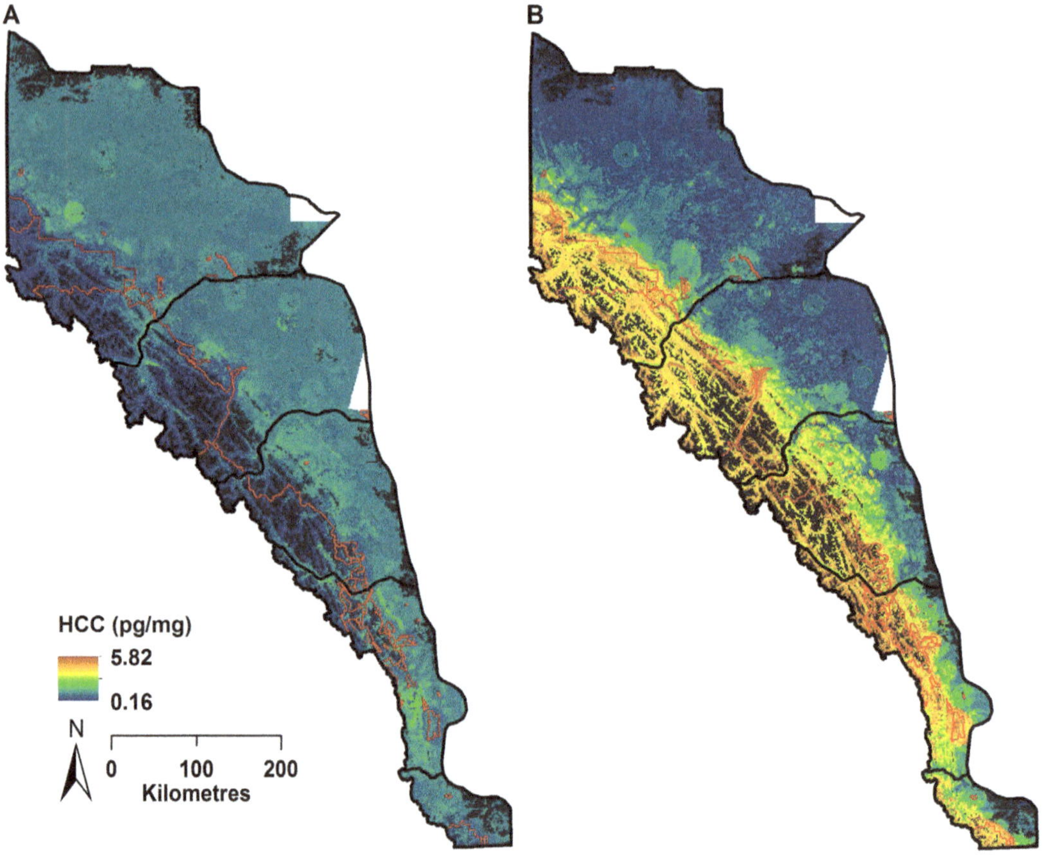

Figure 3. Geographic distribution of the predicted HCC levels from gender-specific total random forest models. Predicted HCC values for (A) male and (B) female grizzly bears. Parks and protected areas are shown in red. Regions of non-habitat (e.g., rock and ice) are shown in grey.

While one might expect habitats associated with high rates of mortality to elicit a strong long-term stress response, human-caused grizzly bear mortalities represent events characterized by acute traumas and not the result of prolonged continued exposure to stressors. Consequently, we hypothesize that observed low-to-moderate stress levels associated with high-risk (e.g., high chance of mortality) high-reward (e.g., food availability) habitat suggests a willingness to risk human contact to optimize foraging opportunities. Moreover, the observed stress profiles might also signal a degree of habituation to such contact that potentially puts grizzly bears at greater risk of mortality in low-stress environments. Future research should consider whether observed long-term stress levels influence the probability of grizzly bear mortality in human influenced landscapes.

Conclusion

We have shown that the geographic dissimilarity in long-term stress levels of grizzly bears appears to be both context-dependent [21], and similar to northern spotted owls [20], gender specific. Similar to black bears [46] and grizzly bears in other regions [27,100] the availability and ability to procure resources appears

Table 4. Percent area of the bear management units (BMU) and study area classified as low, moderate, and high HCC based on the geographic distribution of predicted male and female HCC values.

BMU	Low HCC (0.16–0.45 pg/mg)		Moderate HCC (0.46–1.62 pg/mg)		High HCC (>1.62 pg/mg)	
	Male (%)	Female (%)	Male (%)	Female (%)	Male (%)	Female (%)
Castle	2.93	6.83	95.67	58.90	0.73	33.58
Clearwater	22.17	10.22	75.92	49.34	1.76	40.61
Grande Cache	13.04	51.86	83.47	26.96	0.91	18.60
Livingstone	11.54	0.54	78.53	42.61	8.37	55.29
Yellowhead	13.04	34.37	74.60	27.95	1.30	34.77
Total study area	16.11	33.89	79.90	33.53	1.88	30.47

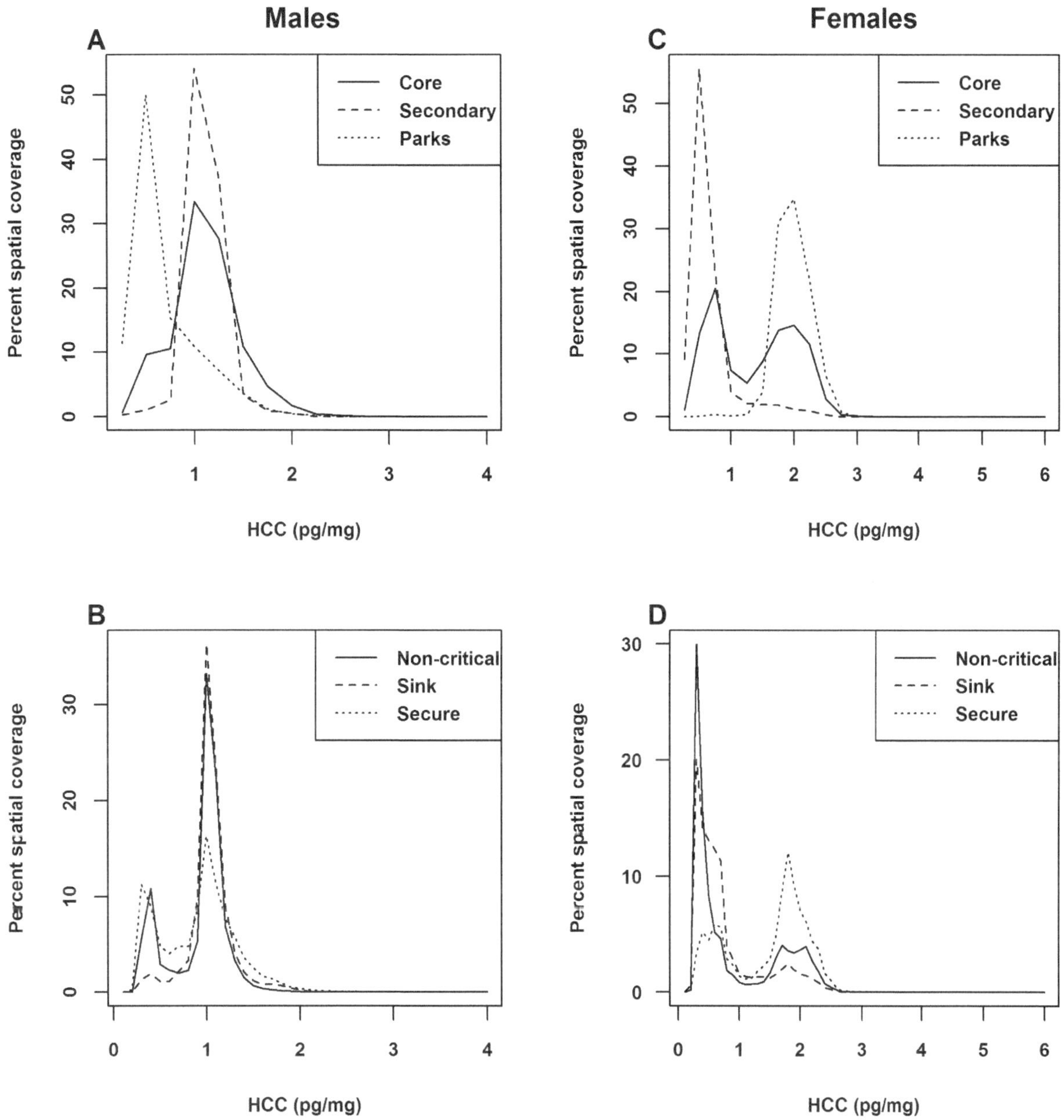

Figure 4. Frequency distributions of predicted HCC values associated with conservation management units and habitat states. Percent spatial coverage of predicted HCC values associated with (A – males; C – females) parks and protected areas, core conservation areas, and secondary conservation areas, as well as (B – males, D – females) secure, sink, and non-critical habitat states.

to impact spatial patterns of the modelled stress levels in Alberta grizzly bears. Our results also reflect findings in species such as squirrel gliders [21], ungulates [22,24], and wolves [25], where the measured stress levels varied according to the density, distribution, and perhaps most importantly, the nature of anthropogenic disturbance and activities throughout their habitat. The methodology we present here extends an aspatial glucocorticoid metric to a spatially local representation of potential long-term stress in grizzly bears based on current environmental conditions. The information provided highlights spatial variability in the long-term stress levels of male and female grizzly bears. Future directions in this and other wildlife systems include modelling the impacts of individual behaviour and the effects of interaction on stress, predicting changes in long-term stress based on future disturbance patterns and climate change, and exploring spatial relationships between stress and body condition. Our methods may also be applicable in spatial analyses of point sampled stress metrics taken from other far ranging wild animals (e.g., polar bears [113]). Finally, while our models are specific to the grizzly bear population in Alberta, many populations in western North America occupy landscapes with similar environmental stressors. Consequently, our findings may offer an indication of similar interactions in other regions. Future species conservation efforts should therefore attempt to better understand gender and spatial based differences in the physiological response of wildlife to landscape conditions.

Acknowledgments

We thank the researchers with the Grizzly Bear Program at the Foothills Research Institute for their careful work collecting grizzly bear hair samples and for providing access to habitat and anthropogenic data used in this analysis. We also wish to thank two anonymous reviewers for their comments and feedback.

Author Contributions

Conceived and designed the experiments: MLB TAN MRLC CTD GBS. Performed the experiments: MLB TAN. Analyzed the data: MLB TAN. Contributed reagents/materials/analysis tools: MLB TAN MRLC CTD GBS. Wrote the paper: MLB TAN MRLC CTD GBS.

References

1. Wilcox BA, Murphy DD (1985) Conservation strategy: The effects of fragmentation on extinction. Am Nat 125: 879–887.
2. Bender DJ, Contreras TA, Fahrig L (1998) Habitat loss and population decline: A meta-analysis of the patch size effect. Ecology 79: 517–533.
3. Brooks TM, Mittermeier RA., Mittermeier CG, Da Fonseca GAB, Rylands AB, et al. (2002) Habitat loss and extinction in the hotspots of biodiversity. Conserv Biol 16: 909–923.
4. Purvis A, Gittleman JL, Cowlishaw G, Mace GM (2000) Predicting extinction risk in declining species. Proc Biol Sci 267: 1947–1952.
5. Travis JMJ (2003) Climate change and habitat destruction: A deadly anthropogenic cocktail. Proc Biol Sci 270: 467–473.
6. Thomas CD, Cameron A, Green RE, Bakkenes M, Beaumont LJ, et al. (2004) Extinction risk from climate change. Nature 427: 145–148.
7. Gurevitch J, Padilla DK (2004) Are invasive species a major cause of extinctions? Trends Ecol Evol 19: 470–474.
8. Didham RK, Tylianakis JM, Gemmell NJ, Rand TA, Ewers RM (2007) Interactive effects of habitat modification and species invasion on native species decline. Trends Ecol Evol 22: 489–496.
9. Weaver JL, Paquet PC, Ruggiero LF (1996) Resilience and conservation of large carnivores in the Rocky Mountains. Conserv Biol 10: 964–976.
10. Cardillo M, Mace GM, Jones KE, Bielby J, Bininda-Emonds ORP, et al. (2005) Multiple causes of high extinction risk in large mammal species. Science 309: 1239–1241.
11. Russell GJ, Brooks TM, McKinney MM, Anderson CG (1998) Present and future taxonomic selectivity in bird and mammal extinctions. Conserv Biol 12: 1365–1376.
12. Elith J, Leathwick JR (2009) Species distribution models: Ecological explanation and prediction across space and time. Annu Rev Ecol Evol Syst 40: 677–697.
13. Guisan A, Thuiller W (2005) Predicting species distribution: Offering more than simple habitat models. Ecol Lett 8: 993–1009.
14. Ewers RM, Didham RK (2006) Confounding factors in the detection of species responses to habitat fragmentation. Biol Rev Camb Philos Soc 81: 117–142.
15. Ellis RD, McWhorter TJ, Maron M (2012) Integrating landscape ecology and conservation physiology. Landsc Ecol 27: 1–12.
16. Romero LM (2004) Physiological stress in ecology: Lessons from biomedical research. Trends Ecol Evol 19: 249–255.
17. Palme R, Rettenbacher S, Touma C, El-Bahr SM, Möstl E (2005) Stress hormones in mammals and birds: Comparative aspects regarding metabolism, excretion, and noninvasive measurement in fecal samples. Ann N Y Acad Sci 1040: 162–171.
18. Sheriff MJ, Dantzer B, Delehanty B, Palme R, Boonstra R (2011) Measuring stress in wildlife: Techniques for quantifying glucocorticoids. Oecologia 166: 869–887.
19. von der Ohe CG, Servheen C (2002) Measuring stress in mammals using fecal glucocorticoids: Opportunities and challenges. Wildl Soc Bull 30: 1215–1225.
20. Wasser SK, Bevis K, King G, Hanson E (1997) Noninvasive physiological measures of disturbance in the northern spotted owl. Conserv Biol 11: 1019–1022.
21. Brearley G, McAlpine C, Bell S, Bradley A (2012) Influence of urban edges on stress in an arboreal mammal: A case study of squirrel gliders in southeast Queensland, Australia. Landsc Ecol 27: 1407–1419.
22. Millspaugh JJ, Woods RJ, Hunt KE, Raedke KL, Brundige GC, et al. (2001) Fecal glucocorticoid assays and the physiological stress response in elk. Wildl Soc Bull 29: 899–907.
23. Ashley NT, Barboza PS, Macbeth BJ, Janz DM, Cattet MRL, et al. (2011) Glucocorticosteroid concentrations in feces and hair of captive caribou and reindeer following adrenocorticotropic hormone challenge. Gen Comp Endocrinol 172: 382–391.
24. Wasser SK, Keim JL, Taper ML, Lele SR (2011) The influences of wolf predation, habitat loss, and human activity on caribou and moose in the Alberta oil sands. Front Ecol Environ 9: 546–551.
25. Creel S, Fox JE, Hardy A, Sands J, Garrott B, et al. (2002) Snowmobile activity and glucocorticoid stress responses in wolves and elk. Conserv Biol 16: 809–814.
26. Macbeth BJ, Cattet MRL, Stenhouse GB, Gibeau ML, Janz DM (2010) Hair cortisol concentration as a noninvasive measure of long-term stress in free-ranging grizzly bears (*Ursus arctos*): Considerations with implications for other wildlife. Can J Zool 88: 935–949.
27. von der Ohe CG, Wasser SK, Hunt KE, Servheen C (2004) Factors associated with fecal glucocorticoids in Alaskan brown bears (*Ursus arctos horribilis*). Physiol Biochem Zool 77: 313–320.
28. Reeder DM, Kramer KM (2005) Stress in free-ranging mammals: Integrating physiology, ecology, and natural history. J Mammal 86: 225–235.
29. McEwen BS, Wingfield JC (2003) The concept of allostasis in biology and biomedicine. Horm Behav 43: 2–15.
30. Wingfield JC, Sapolsky RM (2003) Reproduction and resistance to stress: When and how. J Neuroendocrinol 15: 711–724.
31. Laliberte AS, Ripple WJ (2004) Range contractions of North American carnivores and ungulates. Bioscience 54: 123–138.
32. Kellert SR, Black M, Rush CR, Bath AJ (2002) Human culture and large carnivore conservation in North America. Conserv Biol 10: 977–990.
33. Alberta Sustainable Resource Development, Alberta Conservation Association (2010) Status of the grizzly bear (*Ursus arctos*) in Alberta: Update 2010. Alberta Sustainable Resource Development. Wildlife Status Report No. 37 (Update 2010). Edmonton, AB. 44 p.
34. Clark DA, Slocombe DS (2011) Grizzly bear conservation in the Foothills Model Forest: Appraisal of a collaborative ecosystem management effort. Policy Sci 44: 1–11.
35. Berland A, Nelson T, Stenhouse G, Graham K, Cranston J (2008) The impact of landscape disturbance on grizzly bear habitat use in the Foothills Model Forest, Alberta, Canada. For Ecol Manage 256: 1875–1883.
36. Proctor MF, Paetkau D, McLellan BN, Stenhouse GB, Kendall KC, et al. (2012) Population fragmentation and inter-ecosystem movements of grizzly bears in western Canada and the northern United States. Wildlife Monographs 180: 1–46.
37. McLellan BN, Hovey FW (2001) Habitats selected by grizzly bears in a multiple use landscape. J Wildl Manage 65: 92–99.
38. Nielsen SE, Munro RHM, Bainbridge EL, Stenhouse GB, Boyce MS (2004) Grizzly bears and forestry II. Distribution of grizzly bear foods in clearcuts of west-central Alberta, Canada. For Ecol Manage 199: 67–82.
39. Nielsen SE, Boyce MS, Stenhouse GB (2004) Grizzly bears and forestry I. Selection of clearcuts by grizzly bears in west-central Alberta, Canada. For Ecol Manage 199: 51–65.
40. Proctor MF, McLellan BN, Strobeck C, Barclay RMR (2005) Genetic analysis reveals demographic fragmentation of grizzly bears yielding vulnerably small populations. Proc Biol Sci 272: 2409–2416.
41. Morrison ML (2001) A proposed research emphasis to overcome the limits of wildlife-habitat relationship studies. J Wildl Manage 65: 613–623.
42. McLoughlin PD, Ferguson SH, Messier F (2000) Intraspecific variation in home range overlap with habitat quality: a comparison among brown bear populations. Evol Ecol 14: 39–60.
43. Nielsen SE, Boyce MS, Stenhouse GB, Munro RHM (2002) Modelling grizzly bear habitats in the yellowhead ecosystem of Alberta: Taking autocorrelation seriously. Ursus 13: 45–56.
44. Mowat G, Heard DC (2006) Major components of grizzly bear diet across North America. Can J Zoo 84: 473–489.
45. Munro RHM, Nielsen SE, Price MH, Stenhouse GB, Boyce MS (2006) Seasonal and diel patterns of grizzly bear diet and activity in west-central Alberta. J Mammal 87: 1112–1121.
46. Hellgren EC, Rogers LL, Seal US (1993) Serum chemistry and hematology of black bears: Physiological indices of habitat quality or seasonal patterns? J Mammal 74: 304–315.
47. Kitaysky AS, Piatt JF, Wingfield JC (2007) Stress hormones link food availability and population processes in seabirds. Mar Ecol Prog Ser 352: 245–258.
48. McLellan BN, Hovey FW, Mace RD, Woods JG, Carney DW, et al. (1999) Rates and causes of grizzly bear mortality in the interior mountains of British Columbia, Alberta, Montana, Washington, and Idaho. J Wildl Manage 63: 911–920.
49. Benn B, Herrero S (2002) Grizzly bear mortality and human access in Banff and Yoho National Parks, 1971–98. Ursus 13: 213–221.
50. Nielsen SE, Herrero S, Boyce MS, Mace RD, Benn B, et al. (2004) Modelling the spatial distribution of human-caused grizzly bear mortalities in the Central Rockies ecosystem of Canada. Biol Conserv 120: 101–113.
51. Knight RR, Eberhardt LL (1985) Population dynamics of Yellowstone grizzly bears. Ecology 66: 323–334.
52. Wikelski M, Cooke SJ (2006) Conservation physiology. Trends Ecol Evol 21: 38–46.

53. Sheriff MJ, Krebs CJ, Boonstra R (2010) The ghosts of predators past: Population cycles and the role of maternal programming under fluctuating predation risk. Ecology 91: 2983–2994.
54. Ecological Stratification Working Group (1995) A National Ecological Framework for Canada. Agriculture and Agri-Foods Canada, Research Branch, Centre for Land and Biological Resources Research and Environment Canada, State of Environment Directorate, Ottawa/Hull. 125 p.
55. Stenhouse G, Boulanger J, Lee J, Graham K, Duval J, et al. (2005) Grizzly bear associations along the eastern slopes of Alberta. Ursus 16: 31–40.
56. Nielsen SE, McDermid G, Stenhouse GB, Boyce MS (2010) Dynamic wildlife habitat models: Seasonal foods and mortality risk predict occupancy-abundance and habitat selection in grizzly bears. Biol Conserv 143: 1623–1634.
57. Graham K, Boulanger J, Duval J, Stenhouse G (2010) Spatial and temporal use of roads by grizzly bears in west-central Alberta. Ursus 21: 43–56.
58. Woods JG, Paetkau D, Lewis D, McLellan BN, Proctor M, et al. (1999) Genetic tagging of free-ranging black and brown bears. Wildl Soc Bull 27: 616–627.
59. Proctor MF, Mclellan BN, Strobeck C, Barclay RMR (2004) Gender-specific dispersal distances of grizzly bears estimated by genetic analysis. Can J Zoo 82: 1108–1118.
60. Webb E, Thomson S, Nelson A, White C, Koren G, et al. (2010) Assessing individual systemic stress through cortisol analysis of archaeological hair. J Archaeol Sci 37: 807–812.
61. Meyer JS, Novak MA (2012) Minireview: Hair cortisol: A novel biomarker of hypothalamic-pituitary-adrenocortical activity. Endocrinology 153: 4120–4127.
62. Boulanger J, Stenhouse G, Proctor M, Himmer S, Paetkau D, et al. (2005) 2004 population inventory and density estimates for the Alberta 3B and 4B grizzly bear management area. Alberta Sustainable Resource Development, Edmonton. 28 p.
63. Boulanger J, Stenhouse G, MacHutchon G, Proctor M, Himmer S, et al. (2005) Grizzly bear population and density estimates for the 2005 Alberta (proposed) Unit 4 management area inventory. Alberta Sustainable Resource Development, Edmonton. 31 p.
64. Boulanger J, Stenhouse G, MacHutchon G, Proctor M, Paetkau D, et al. (2007) Grizzly bear population and density estimates for the 2006 Alberta unit 5 management area inventory. Alberta Sustainable Resource Development, Edmonton. 37 p.
65. Boulanger J, MacHutchon G, Stenhouse G, Proctor M, Cranston J, et al. (2008) Grizzly bear population and density estimates for Alberta Bear Management Unit 6 and British Columbia Management Units 4–1, 4–2, and 4–23. Alberta Sustainable Resource Development, Edmonton. 46 p.
66. Boulanger J, MacHutchon G, Stenhouse G, Proctor M, Cranston J, et al. (2009) Grizzly bear population and density estimates for the 2008 DNA inventory of the Grande Cache bear management area (BMA 2). Alberta Sustainable Resource Development, Edmonton. 36 p.
67. Mcdermid GJ (2005) Remote sensing for large-area, multi-jurisdictional habitat mapping. Ph.D. Thesis. University of Waterloo: Canada. 258 p.
68. Franklin SE, Peddle DR, Dechka JA, Stenhouse GB (2002) Evidential reasoning with Landsat TM, DEM and GIS data for landcover classification in support of grizzly bear habitat mapping. Int J Remote Sens 23: 4633–4652.
69. McDermid GJ, Franklin SE, LeDrew EF (2005) Remote sensing for large-area habitat mapping. Prog Phys Geog 29: 449–474.
70. Wilson PE, Gallant JC (2000) Terrain analysis: Principles and applications. New York: John Wiley & Sons. 479 p.
71. Gessler PE, Moore ID, McKenzie NJ, Ryan J (1995) Soil-landscape modelling and spatial prediction of soil attributes. Int J Geogr Inf Syst 9: 421–432.
72. Coops NC, Wulder MA, Duro DC, Han T, Berry S (2008) The development of a Canadian dynamic habitat index using multi-temporal satellite estimates of canopy light absorbance. Ecol Indic 8: 754–766.
73. Berry S, Mackey B, Brown T (2007) Potential applications of remotely sensed vegetation greeness to habitat analysis and the conservation of dispersive fauna. Pac Conserv Biol 13: 120–127.
74. Coops NC, Waring RH, Wulder MA, Pidgeon AM, Radeloff VC (2009) Bird diversity: A predictable function of satellite-derived estimates of seasonal variation in canopy light absorbance across the United States. J Biogeogr 36: 905–918.
75. Andrew ME, Wulder MA, Coops NC, Baillargeon G (2012) Beta-diversity gradients of butterflies along productivity axes. Glob Ecol Biogeogr 21: 352–364.
76. Nilsen EB, Herfindal I, Linnell JDC (2005) Can intra-specific variation in carnivore home-range size be explained using remote-sensing estimates of environmental productivity? Ecoscience 12: 68–75.
77. Fontana FMA, Coops NC, Khlopenkov KV, Trishchenko AP, Riffler M, et al. (2012) Generation of a novel 1 km NDVI data set over Canada, the northern United States, and Greenland based on historical AVHRR data. Remote Sens Environ 121: 171–185.
78. Duro DC, Coops NC, Wulder MA, Han T (2007) Development of a large area biodiversity monitoring system driven by remote sensing. Prog Phys Geog 31: 235–260.
79. Hamer D, Herrero S (1987) Grizzly bear food and habitat in the front ranges of Banff National Park, Alberta. International Conference on Bear Research and Management 7: 199–213.
80. Nielsen SE, Cranston J, Stenhouse GB (2009) Identification of priority areas for grizzly bear conservation and recovery in Alberta, Canada. J Conserv Plan 5: 38–60.
81. Manly BFJ, McDonald LL, Thomas DL, McDonald TL, Erickson WP (2002) Resource selection by animals: Statistical design and analysis for field studies. Second edition. Dordrecht: Kluwer Academic Publishers. 221 p.
82. Johnson DH (1980) The comparison of usage and availability measurements for evaluating resource preference. Ecology 61: 65–71.
83. Stewart BP, Nelson TA, Wulder MA, Nielsen SE, Stenhouse G (2012) Impact of disturbance characteristics and age on grizzly bear habitat selection. Appl Geogr 34: 614–625.
84. Nielsen SE, Stenhouse GB, Boyce MS (2006) A habitat-based framework for grizzly bear conservation in Alberta. Biol Conserv 130: 217–229.
85. Worton BJ (1989) Kernel methods for estimating the utilization distribution in home-range studies. Ecology 70: 164–168.
86. Nelson TA, Boots B, Wulder MA, Carroll AL (2007) The environmental characteristics of mountain pine beetle infestation hot spots. BC J Ecosyst Manage 8: 91–108.
87. Nadaraya EA (1964) On estimating regression. Theor Probab Appl+9: 141–142.
88. Diggle P (1985) A kernel method for smoothing point process data. J R Stat Soc Ser C Appl Stat 34: 138–147.
89. Worton BJ (1987) A review of models of home range for animal movement. Ecol Modell 38: 277–298.
90. Bowman AW, Azzalini A (1997) Applied smoothing techniques for data analysis: The kernel approach with S-Plus illustrations. New York: Oxford University Press. 193 p.
91. Nelson T, Boots B, Wulder MA (2006) Large-area mountain pine beetle infestations: Spatial data representation and accuracy. Forest Chron 82: 243–252.
92. Breiman L (2001) Random forests. Mach Learn 45: 5–32.
93. Cutler D, Edwards TC Jr, Beard KH, Cutler A, Hess KT, et al. (2007) Random forests for classification in ecology. Ecology 88: 2783–2792.
94. Prasad AM, Iverson LR, Liaw A (2006) Newer classification and regression tree techniques: Bagging and random forests for ecological prediction. Ecosystems 9: 181–199.
95. Andison DW (1998) Temporal patterns of age-class distributions on foothills landscapes in Alberta. Ecography 21: 543–550.
96. Rhemtulla JM, Hall RJ, Higgs ES, Macdonald SE (2010) Eighty years of change: vegetation in the montane ecoregions of Jasper National Park, Alberta, Canada. Can J Forest Res 32: 2010–2021.
97. Stewart BP, Nelson TA, Laberee K, Nielsen SE, Wulder MA, et al. (2013) Quantifying grizzly bear selection of natural and anthropogenic edges. J Wildl Manage In press. doi:10.1002/jwmg.535.
98. Roever CL, Boyce MS, Stenhouse GB (2008) Grizzly bears and forestry I: road vegetation and placement as an attractant to grizzly bears. For Ecol Manage 256. 1253 1261.
99. Wasser S, Davenport B, Ramage E, Hunt K, Parker M, et al. (2004). Scat detection dogs in wildlife research and management: application to grizzly and black bears in the Yellowhead Ecosystem, Alberta, Canada. Can J Zool 492: 475–492.
100. Bryan H, Darimont C, Paquet P, Wynne-Edwards K, Smits J (2013) Stress and reproductive hormones reflect inter-specific social and nutritional conditions mediated by resource availability in a bear-salmon system. Plos One: In Press.
101. Trites AW, Donnelly CP (2003) The decline of Steller sea lions *Eumetopias jubatus* in Alaska: a review of the nutritional stress hypothesis. Mamm Rev 33: 3–28.
102. Gibeau ML, Herrero S, McLellan BN, Woods JG (2001) Managing for grizzly bear security areas in Banff National Park and the central Canadian rocky mountains. Ursus 12: 121–129.
103. Elfström M, Zedrosser A, Støen O-G, Swenson JE (2012) Ultimate and proximate mechanisms underlying the occurrence of bears close to human settlements: Review and management implications. Mamm Rev In press. doi:10.1111/j.1365-2907.2012.00223.x.
104. Libal NS, Belant JL, Leopold BD, Wang G, Owen PA (2011) Despotism and risk of infanticide influence grizzly bear den-site selection. PloS One 6: e24133. doi:10.1371/journal.pone.0024133.
105. Rode KD, Farley SD, Robbins CT (2006) Sexual dimorphism, reproductive strategy, and human activities determine resource use by brown bears. Ecology 87: 2636–2646.
106. Støen O-G, Bellemain E, Sæbø S, Swenson JE (2005) Kin-related spatial structure in brown bears *Ursus arctos*. Behav Ecol Sociobiol 59: 191–197.
107. Steyaert SMJG, Kindberg J, Swenson JE, Zedrosser A (2013) Male reproductive strategy explains spatiotemporal segregation in brown bears. J Anim Ecol: In press. doi:10.1111/1365-2656.12055.
108. Dahle B, Støen O-G, Swenson JE (2006) Factors influencing home-range size in subadult brown bears. J Mammal 87: 859–865.
109. Blanchard BM, Knight RR (1991) Movements of yellowstone grizzly bears. Biol Conserv 58: 41–67.
110. Nielsen SE, Shafer AB a, Boyce MS, Stenhouse GB (2013) Does learning or instinct shape habitat selection? PloS One 8: e53721. doi:10.1371/journal.pone.0053721.
111. Cattet MRL, Caulkett NA, Obbard ME, Stenhouse GB (2002) A body-condition index for ursids. Can J Zool 80: 1156–1161.

112. Garshelis DL, Gibeau ML, Herrero S (2005) Grizzly bear demographics in and around Banff Natioanl Park and Kananaskis Country, Alberta. J Wildl Manage 69: 277–297.

113. Bechshøft TØ, Sonne C, Dietz R, Born EW, Novak MA, et al. (2011) Cortisol levels in hair of East Greenland polar bears. Sci Total Environ 409: 831–834.

7

Effects of Natural and Anthropogenic Change on Habitat Use and Movement of Endangered Salt Marsh Harvest Mice

Katherine R. Smith[1,2*¤], Laureen Barthman-Thompson[2], William R. Gould[3], Karen E. Mabry[1]

1 Department of Biology, New Mexico State University, Las Cruces, New Mexico, United States of America, **2** California Department of Fish and Wildlife, Stockton, California, United States of America, **3** Department of Economics, Applied Statistics and International Business, New Mexico State University, Las Cruces, New Mexico, United States of America

Abstract

The northern salt marsh harvest mouse (*Reithrodontomys raviventris halicoetes*) is an endangered species endemic to the San Francisco Bay Estuary. Using a conservation behavior perspective, we examined how salt marsh harvest mice cope with both natural (daily tidal fluctuations) and anthropogenic (modification of tidal regime) changes in natural tidal wetlands and human-created diked wetlands, and investigated the role of behavioral flexibility in utilizing a human-created environment in the Suisun Marsh. We used radio telemetry to determine refuge use at high tide, space use, and movement rates to investigate possible differences in movement behavior in tidal versus diked wetlands. We found that the vast majority of the time salt marsh harvest mice remain in vegetation above the water during high tides. We also found no difference in space used by mice during high tide as compared to before or after high tide in either tidal or diked wetlands. We found no detectable difference in diurnal or nocturnal movement rates in tidal wetlands. However, we did find that diurnal movement rates for mice in diked wetlands were lower than nocturnal movement rates, especially during the new moon. This change in movement behavior in a relatively novel human-created habitat indicates that behavioral flexibility may facilitate the use of human-created environments by salt marsh harvest mice.

Editor: Cédric Sueur, Institut Pluridisciplinaire Hubert Curien, France

Funding: This work was supported financially by the California Department of Water Resources and New Mexico State University. The funders had no role in study design, data collection and analysis, decision to publish, or preparation of the manuscript.

Competing Interests: The authors have declared that no competing interests exist.

* Email: ratsmith@ucdavis.edu

¤ Current address: Wildlife, Fish, and Conservation Biology Department, University of California at Davis, Davis, California, United States of America

Introduction

Animal behavior has the potential to contribute substantially to conservation efforts, but the field of conservation behavior has been slow to develop [1–6]. Only recently have attempts been made to develop unifying principles and overarching frameworks for the study of conservation behavior [7–9]. Because behavior is one of the most flexible traits an animal may possess, it represents a promising mechanism by which species may cope with human-induced rapid environmental change (HIREC) [10–13].

Behavioral flexibility is the ability of an animal to modify its behavior under different environmental conditions [10], [14], [15]. Behavioral responses are often rapid and reversible, and afford animals some degree of control over their external stimuli by choosing their surrounding environment [12]. Appropriate behavioral responses to environmental change are vital for animals that live in naturally heterogeneous environments, and will likely be critical for the persistence of many species in the face of HIREC [12].

Both natural environmental heterogeneity and the effects of HIREC occur across time and space, and this environmental variability affects species occurrence, distribution, abundance, population stability, and individual behavior [16–18]. Some habitat types are relatively stable, while others are highly variable in their natural state. For example, wetland biota is influenced by a variety of both predictable (tides, seasonal flows) and stochastic (flood, drought) changes on a variety of time scales [19]. Anthropogenic changes to wetlands, such as modification of tidal inundation regimes through the construction of earthen dikes/levees, can greatly interfere with these natural patterns of heterogeneity. Because coastal wetlands are both highly variable in their natural state and have been heavily modified by human activities, organisms that live in coastal wetlands present opportunities for the study of both behavioral flexibility in response to natural heterogeneity and behavioral responses to HIREC.

We investigated the response of the endangered northern salt marsh harvest mouse (*Reithrodontomys raviventris halicoetes*) to both natural and anthropogenic environmental heterogeneity. The salt marsh harvest mouse is an ideal species in which to study behavioral flexibility and responses to HIREC, because it is endemic to changeable wetland habitats and currently occurs in both natural tidal (full tidal influence) and anthropogenically-altered (diked) wetlands.

Features of salt marsh harvest mouse biology that are consistent with adaptation to tidal environments become clear when comparisons are made to the sympatric congener the Western harvest mouse (*R. megalotis*), which occurs in tidal marshes but also in a variety of other habitat types. Compared to Western harvest mice, the salt marsh harvest mouse is a stronger swimmer and is more capable of consuming sea water [20]. They are also capable of entering torpor [20], a mechanism thought to aid small mammals in coping with osmotic stress [21], and are more active during warmer daytime temperatures which allows for a lower resting metabolic rate than activity during cool, damp nights [20], [22]. Finally, as a species that naturally resides in a highly changeable environment, the salt marsh harvest mouse may possess a high level of behavioral flexibility that predisposes it to alter its behavior to utilize resources in novel habitat types, such as diked wetlands [23].

Understanding how salt marsh harvest mice use their habitat during daily tidal flooding is especially important because tidal restoration is the primary conservation measure for preserving this species which is endangered due to dramatic habitat loss (~90%) and faces further loss due to predicted sea level rise [24–27]. As the only terrestrial mammals that is entirely restricted to coastal marshes, the salt marsh harvest mouse must somehow avoid drowning during high tides [28]. It has long been believed that salt marsh harvest mice spend the majority of their time in wetlands, and move upland only to escape tide and flood waters [20].

Since the 1950s, researchers have attempted to characterize salt marsh harvest mouse behavior during tidal inundation to determine whether mice move vertically into tall vegetation or horizontally upland to escape the tide [20], [29–31]. In tidal wetlands salt marsh harvest mice typically experience tidal flooding twice a day. In contrast, diked wetlands are flooded continuously during the rainy season (approximately October through March), and dried out for the remainder of the year. Thus, the patchwork of modified and unmodified habitat in the San Francisco Bay Estuary presents the opportunity to study habitat use of mice in both highly changeable natural (tidal) habitat and highly modified (diked, surrounded by levees with water control structures) habitat that is temporally more stable.

Conclusions from previous studies of the behavior of the salt marsh harvest mouse at high tide have been split on the question of how mice deal with daily tidal inundation of their habitat. Some concluded based on trapping data or visual observations that animals move out of tidal wetlands and into upland areas or onto levees to escape rising waters [29], [30], and others concluded that they remain in tall, dense vegetation over water where they can easily move about in the thatch layer [20], [31]. However, although trapping methods allow researchers to determine that a mouse had been at a location at some point during the night, it is impossible to tell from these previous trapping studies where mice were during the time that the habitat was actually flooded. The contrasting conclusions of previous studies suggest that the only way to definitively understand salt marsh harvest mouse habitat use during high tide is through the use of radio telemetry [31].

The objective of this study was to determine how salt marsh harvest mice respond to both natural environmental heterogeneity and HIREC, focusing on the effects of regular tidal inundation. We approached this investigation from a conservation behavior perspective which uses general principals of animal behavior to address conservation issues [32]. We applied two questions from a recently developed conservation behavior framework [7]. One focus of the Berger-Tal et al. framework stresses performing behavior-based management, which considers behavior in conservation decision making protocols [7]. In our case, answering the question *Do salt marsh harvest mice move vertically (into tall vegetation) or horizontally (into upland areas) to escape the high tide?* will allow managers develop restoration and enhancement priorities in the little habitat that remains, and is a research priority explicitly identified in the recovery plan for this species [26]. A second focus of the framework addresses if and how HIREC affects behavior, in the case of this endangered species, *Do mice in anthropogenically altered diked wetlands behave differently than those subject to tidal influence?* [7].

To answer these questions, we tracked mouse movements during high tide events in natural tidal wetlands using radio telemetry to identify the type of refuge used to escape incoming water. During the same periods, we tracked mice in adjacent diked wetlands. This allowed us to determine space use and estimate movement rates in response to high tides, as well as independent of tidal influence in diked habitats. With regard to our first question, we predicted that during tidal flooding, mice would remain in tall vegetation over water. Our predictions for the second question depended on the results of the first. If the prediction that mice remain in tall vegetation during the high tide was confirmed, we expected that movement distances (and thus rates) in natural tidal and diked wetlands would be similar. In contrast, if mice instead moved long distances upland to escape the tide, then we expected that movement rates in natural tidal wetlands would be greater than those in diked wetlands.

Methods

Ethical note

The research reported here was conducted under an approved IACUC protocol from New Mexico State University (2011–013) and a Memorandum of Understanding between the California Department of Fish and Wildlife and the United States Fish and Wildlife Service for handing of endangered species, with direct involvement and supervision by the California Department of Fish and Wildlife. All procedures were consistent with the guidelines for the use of wild mammals in research from both ASAB/ABS and the American Society of Mammalogists.

Study Area

We conducted this study between May and August 2011 in the Suisun Marsh in Solano County, CA, USA (38° 11′ 15.86″ N, 122° 3′ 52.67″ W), which is a large wetland complex in the San Francisco Bay Estuary (Fig. 1a). We established six study areas which were grouped into three blocks, each containing one natural tidal and one diked wetland. Study areas were established on public lands with the permission of the California Department of Fish and Wildlife and the Suisun Resource Conservation District. Paired wetlands (one tidal and one diked) within each block were 100 to 600 m apart and separated by levee roads or sloughs, and the three blocks were 1 100 to 8 500 m apart (Fig. 1b). Although salt marsh harvest mice are capable of moving distances similar to the distances between our study wetlands, we did not observe movement between these wetlands during our study. All tidal wetlands were subject to full natural tidal action. All diked wetlands were cut off from water access for the summer and were essentially dry.

Vegetation on the study sites included pickleweed (*Salicornia virginica*, also known as Virginia glasswort), Olney's threesquare bulrush (*Schoenoplectus americanus*, also known as chairmaker's bulrush), cattail (*Typha* spp.), saltgrass (*Distichlis spicata*), tule (*Sc. acutus*), common reed (*Phragmites australis*), Baltic rush (*Juncus balticus*) and other halophytic species. Plant community composition was broadly similar between natural tidal and diked areas, although tidal areas tended to be dominated by reeds and rush (mean height ± 1 SE $= 138.13 \pm 27.84$ cm), while the diked areas tended to be dominated by pickleweed (mean height ± 1

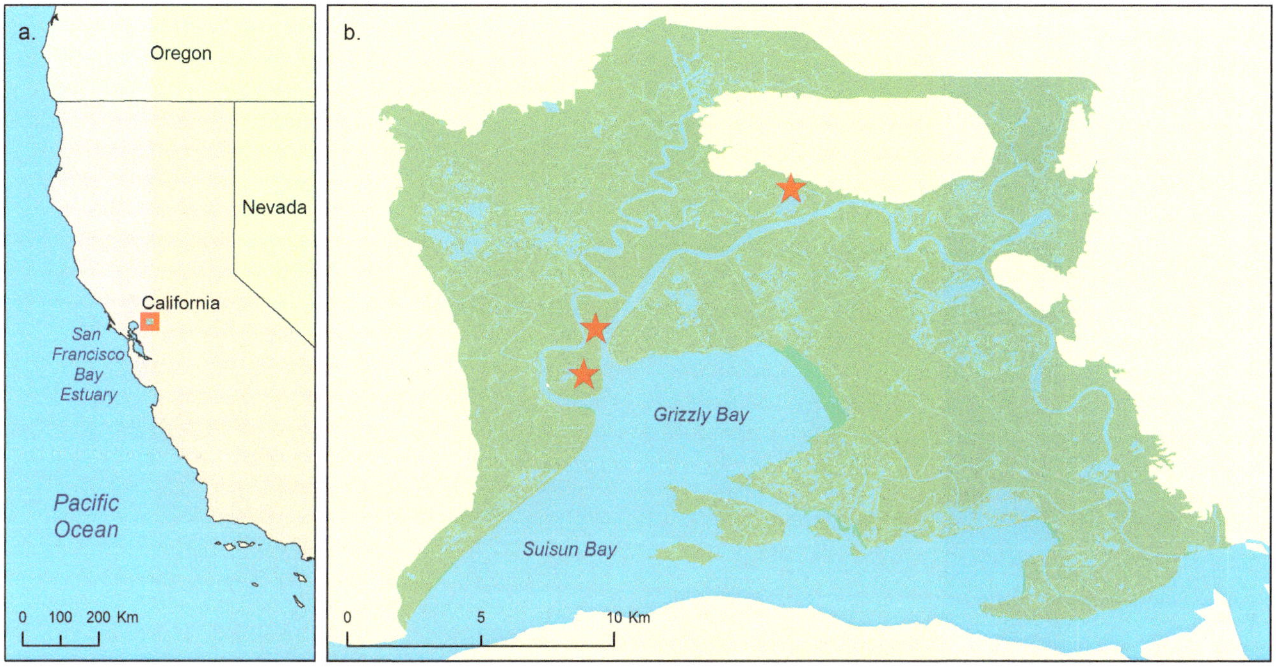

Figure 1. Maps of study location and areas. (a) The south-west coast of the United States with the location of the Suisun Marsh indicated by the boxed area. (b) The Suisun Marsh with 3 study blocks indicated by red stars. Data Source: California Department of Fish And Game: Vegetation - Suisun Marsh 2009 [ds711], Ocean Delta Water; California Digital Conservation Atlas. © Katherine Smith

SE = 42.40±6.00 cm). Although the vegetation was similar overall, diked wetlands had a higher proportion of bare ground, which lead to a lower calculated mean height.

Live Trapping

During more than 3 500 trap nights, we captured small mammals using collapsible Sherman live traps (SFA, 7.62×8.89×23.50 cm; H.B. Sherman Traps Inc., Tallahassee, FL). We placed trap grids in vegetation types known to support large salt marsh harvest mouse populations, including pickleweed and Olney's threesquare bulrush [33–35]. Traps were placed at 10 meter spacing within grids consisting of 50 or 100 traps, depending upon wetland configuration and mouse density. We placed traps flat on the ground in diked wetlands. In natural tidal wetlands, we placed traps above the water line, nested securely in the vegetation. We placed paperclips at the top of trap doors to leave a small space between the door and the body of the trap to prevent the tails of salt marsh harvest mice from getting caught when traps closed. We baited traps with ground walnut and mixed birdseed, and added cotton batting to the traps for warmth. Traps were opened at sunset, and checked and closed at sunrise. We trapped and tracked mice simultaneously in paired natural tidal and diked wetlands within each block during the same moon phases. At each block, trapping was initiated 4–5 days before the highest tide of the full and new moons. Tracking began 2–3 days before the highest high tide of the lunar phase, and typically continued until 2–3 days after the highest tide.

Radio Tracking

To maximize the potential effect of tidal inundation on movement, we tracked mice during several days surrounding the full moon, which is typically when the highest tides occur. To control for potential effects of bright lunar illumination during the full moon, we also tracked during the new moon when tides are also very high but lunar illumination is low. We tracked mice during sequential full and new moon phases for each block, with a different set of mice tracked during each lunar phase. This yielded a 2×2 factorial arrangement with four "treatments": tidal wetland during the full moon, tidal wetland during the new moon, diked wetland during the full moon, and diked wetland during the new moon.

To track movements of salt marsh harvest mice, we fitted individuals with BD-2NC radiotransmitters (Holohil, Inc., Carp, Ontario, Canada). Transmitters were equipped with the smallest battery possible, resulting in a total transmitter package mass of approximately 0.5 g. Because of the small size of salt marsh harvest mice (adult mass = 7.6–14.5 g, [20]), only the largest individuals could be collared in order to meet the recommendation that collars not exceed 5% of body mass [36]. Because females of sufficient size were commonly pregnant or lactating, we used only males weighing ≥10 g. We collared up to five of the largest individuals per wetland per lunar cycle (up to ten individuals per block per lunar cycle), yielding a total of 42 mice, 23 in the tidal wetlands and 19 in the diked wetlands. We collared mice in the field without using anesthesia and held them for observation for ten minutes following the collaring procedure; none showed any adverse reactions. Following the observation period, we released mice at their capture location and began tracking their movement during the second high tide following release (approximately 12 hours later). We tracked each individual for up to six days (mean; SD = 4.3, 0.99 days, min = 2 days, max = 6 days), collecting location data approximately once an hour for 2 to 3 hours on either side of the peak of both diurnal and nocturnal high tides. When tracking was complete, we re-trapped mice and removed their collars. Due to some dropped collars and unexpectedly short battery lives, not all collared individuals yielded sufficient data for analysis.

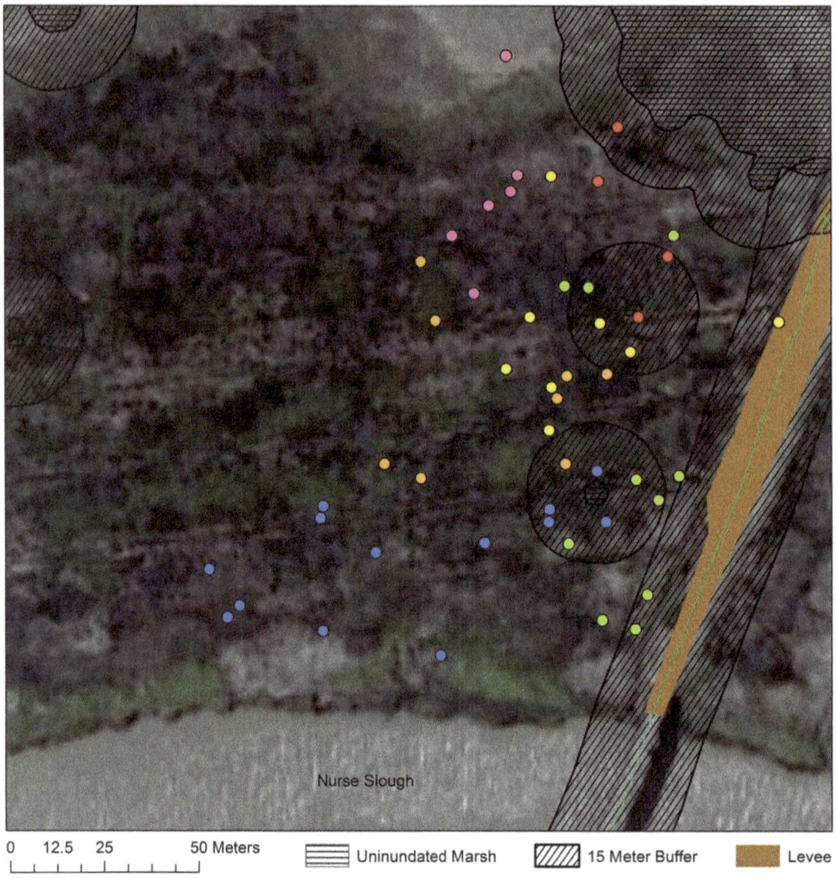

Figure 2. Individual locations at Joice Island Tidal. Individual salt marsh harvest mouse locations at high tide at Joice Island Tidal. Each individual is represented by a different color. Each point represents a mouse location within one hour of the high tide. Data Source: California Department of Fish And Game: NAIP 2010 Aerial Imagery. © Katherine Smith

We used standard radiotracking methods to estimate individual locations, using triangulation from known coordinates [37], with two bearings recorded per location by two observers (a total of four bearings from different points were used to triangulate each location). By allowing the location of animals from a distance, triangulation minimized the influence of observer presence on mouse movement and reduced damage to habitat. It also allowed us to track mice in flooded areas that were inaccessible to humans on foot. Location estimates were generated using Lenth's technique for maximum-likelihood estimates (MLE) in the computer program LOCATE III [38]. To ensure that mice retained their collars and were not caught in vegetation, we homed in on each mouse following each high tide. We then recorded location coordinates using a handheld GPS unit with submeter accuracy (Trimble GeoExplorer 3; Trimble Navigation Limited, Sunnyvale, CA, USA). Homing locations also allowed us to estimate the accuracy of triangulated locations. When a homing location was recorded within 15 minutes of a triangulated location, we calculated the distance between the homing location and the triangulated location to estimate triangulation error. The distance between triangulated and homed points was 15.51; 8.83 meters (mean; *SD*, N = 37). However, this error distance is probably an overestimate, because mice are capable of moving large distances in short periods of time, and because animals often move away from researchers during the homing process, which would increase the distance between triangulated and homed locations.

Determining Refuge Use

Our first goal was to determine whether salt marsh harvest mice moved vertically into emergent vegetation or horizontally into upland areas during the high tide. We examined mouse locations that were recorded within one hour of the high tide in relation to the tide height. We created maps of inundation for each tidal site using water surface data. To measure water surface level, we placed tide level markers throughout tidal areas within and around trap grids at 20 meter spacing, and measured the maximum height reached by water following each high tide. Tide markers were wooden dowels marked each centimeter and dusted with colored chalk. Tide height was estimated by examining dowels following high tide and noting the height to which chalk had been washed away. We recorded coordinates of all tide markers using a handheld GPS unit with submeter accuracy (Trimble GeoExplorer 3; Trimble Navigation Limited, Sunnyvale, CA, USA).

We input all tide measurements as points into ArcMap 10 and interpolated a raster surface using an Inverse Distance Weight method (Inverse Distance Weight, ArcGIS 10.0; Environmental Systems Research Institute, Redlands, CA, USA) which overlaid well with aerial imagery and was consistent with observations in the field. Over this raster surface, we overlaid all mouse locations recorded "during" the high tide. Points occurring within one hour before or after the time of the highest point of the tide were

Figure 3. Individual locations at Lower Joice Island North Tidal. Individual salt marsh harvest mouse locations at high tide at Lower Joice Island North Tidal. Each individual is represented by a different color. Each point represents a mouse location within one hour of the high tide. Data Source: California Department of Fish And Game: NAIP 2010 Aerial Imagery. © Katherine Smith

categorized as "during" the high tide. If a mouse location fell in an area where water level was ≤1 cm we considered it "upland". If a mouse location fell in an area where water level was >1 cm it was considered "over water". If a location was determined to be upland, we assumed the mouse moved horizontally to escape the tide. If a location was categorized as over water, we assumed that the mouse had remained in emergent vegetation to escape the tide. Refuge use was determined using the lowest high tide of both the full and new moon periods during which mice were tracked, thus represented the most conservative categorization of locations as over water. That is, locations were only considered to be over water if they would have been over water during the lowest high tide that we observed at a site. Locations that fell within 15 m of the boundary between inundated and unindated areas were excluded from analysis of refuge use. Because these "boundary" locations were within the average triangulation error distance, we could not definitively determine whether they were over water or not.

We also performed an analysis using minimum convex polygons (MCP) of salt marsh harvest mouse locations to test for differences in space use between mice in tidal and diked wetlands. Individual locations were grouped as "before", occurring more than one hour before high tide; "during", within one hour before or after the high tide; and "after", occurring more than one hour following the high tide. Using the Minimum Bounding Geometry-Convex Hull tool in ArcMap we created "before", "during", and "after" polygons for each mouse during each time period for which we recorded ≥6 locations. Then using the Intersect tool in ArcMap we calculated what percentage of the "before" and "after" MCPs overlapped the "during" MCP.

Estimating Movement Rates

Our second goal was to determine whether there were differences in movement distances between mice in natural tidal wetlands and mice in diked wetlands. However, because the time intervals between location estimates were not equal, it was necessary to first standardize movement data as rates. Movement rate is the Euclidean distance between two consecutive point locations divided by the time elapsed between those points. We calculated movement rates using all locations of an individual that were recorded less than three hours apart before, during and after high tide events (mean ± 1 *SE* $= 83.94 \pm 1.40$ minutes between successive points, min $= 21$ minutes, max $= 180$ minutes). Based on average times of sunrise and sunset during the study, we categorized movements occurring between 0601 and 2000 hours as "diurnal" and movements between 2001 and 0600 as "nocturnal", and calculated average movement rates separately for day and night for each mouse. We omitted from analysis 18 movement rates that spanned the diurnal/nocturnal time cut-offs. The number of movement rates per mouse per time of day (diurnal vs. nocturnal) ranged from 1 to 17, with an average of 7.28; 3.87 (mean; *SD*) rates per individual per time of day (total movement rates $= 411$). Mice with <5 movement rates were excluded from analysis, leaving a total of 29 mice for which we

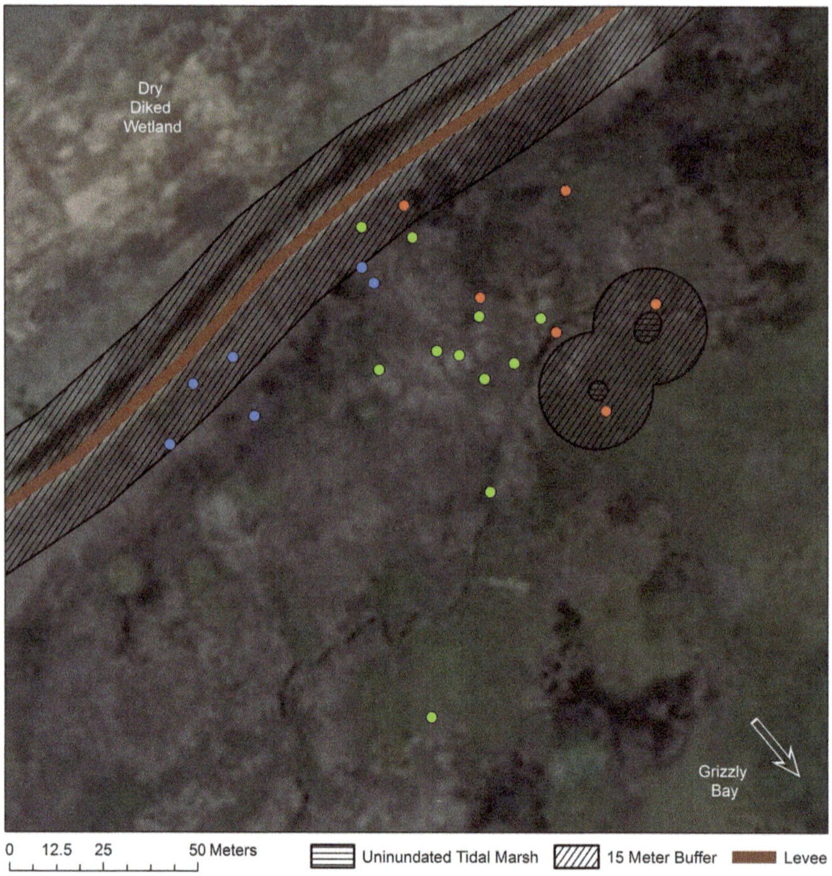

Figure 4. Individual locations at Lower Joice Island South Tidal. Individual salt marsh harvest mouse locations at high tide at Lower Joice Island South Tidal. Each individual is represented by a different color. Each point represents a mouse location within one hour of the high tide. Data Source: California Department of Fish And Game: NAIP 2010 Aerial Imagery. © Katherine Smith

analyzed movement rate data, 15 in tidal wetlands and 14 in diked wetlands.

Analysis

Statistical analysis of the MCP data was performed using paired t-tests to compare space use before and after the high tide to space use during the high tide. Statistical analysis of movement rates was conducted using PROC MIXED in SAS 9.3 (SAS Institute Inc., Cary, NC, USA). We used a replicated block design with blocks and the combination of wetland type and moon phase as fixed effects (four "treatments": tidal full, tidal new, diked full, and diked new), and mouse identification number as a nested random effect, with time of day (diurnal or nocturnal) as a repeated factor. Tukey post-hoc tests were used to detect differences among treatments using an alpha level of 0.05 for statistical significance. Only locations from mice in natural tidal wetlands were used for the refuge use analysis, while locations from all mice with sufficient data were used in the space use and movement rate analysis.

Results

Refuge Use

For mice in natural tidal wetlands, we recorded a total of 167 locations within 1 hour of high tide from 16 mice (mean; *SD* = 11.48; 4.02 points per mouse). Forty-two of these 167 locations (~25%) were within 15 m of the boundary between areas that we considered "upland" versus "over water" (Figs. 2–4). Because these locations were within our average triangulation error from the boundary, we could not classify them to either category with confidence, and they were eliminated from further analysis. Of the remaining 126 locations, 125 (>99%) were definitively in vegetation over water and 1 (<1%) was upland. It is worth noting that even if locations falling within 15 m of the boundary between inundated and unindated areas were included, only 6% (10/167) would have been categorized as "upland". Because we used the most conservative definition of "over water" (e.g., that the position would have been over water during the lowest high tide during which mice were tracked), 6% is likely an overestimate of the frequency with which mice used upland refuges.

When comparing the MCP of points for individuals before (t-test: $t_{15} = 1.327$, $P = 0.20$; diked: $n = 8$, mean ± SE $= 0.61 \pm 0.09$; tidal: $n = 9$, mean ± SE $= 0.44 \pm 0.09$) and after (t-test: $t_{18} = -0.55$, $P = 0.59$; diked: $n = 10$, mean ± SE $= 0.56 \pm 0.09$; tidal: $n = 10$, mean ± SE $= 0.49 \pm 0.08$) high tide to the MCP during high tides we saw no significant difference between tidal and diked wetlands (Figs. 5–6). We found no evidence that salt marsh harvest mice in tidal wetlands shift their space use more or less than mice in diked wetlands during the same high tide periods.

Movement rates

Because mice remained over water instead of moving large distances to use upland refuges to escape high tides in natural tidal wetlands, it was not surprising that we found no significant difference in movement rates among treatments (combinations of

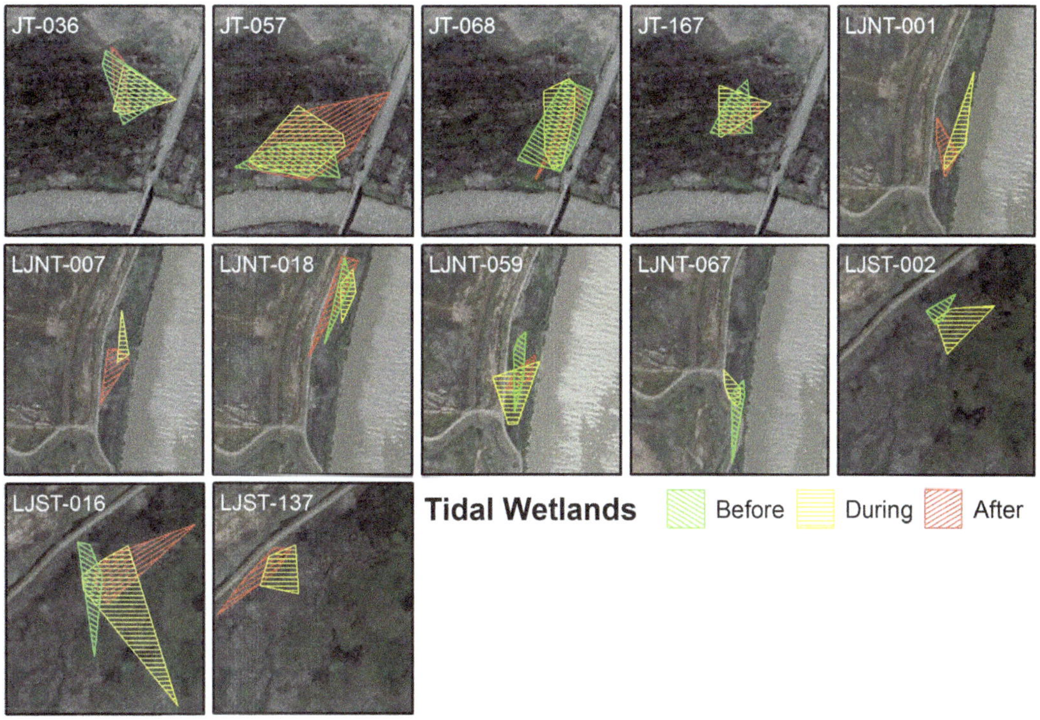

Figure 5. Minimum convex polygons for mice in tidal wetlands. Minimum convex polygons for salt marsh harvest mice in tidal wetlands. Green polygons represent points occurring more than one hour before high tide. Yellow polygons represent points falling within one hour before or after high tide. Red polygons represent points occurring more than one hour following high tide. Each block represents an individual mouse. Some blocks have all three polygons, while those that lack 6 or more individual points during the before or after time period, only have two polygons.

wetland type and moon phase; [repeated-measures ANOVA: $F_{3,24} = 0.92$, $P = 0.44$]). As expected for a primarily nocturnal species, we did find an effect of time: diurnal movement rates were lower than nocturnal movement rates (repeated-measures ANOVA: $F_{1,\ 23} = 7.46$, $P = 0.01$). Post-hoc comparisons of diurnal and nocturnal movement rates within treatments suggested that this difference was primarily due to diked habitats, in which diurnal movement rates during the new moon were considerably lower than nocturnal movement rates (Tukey post-hoc test: $P = 0.012$, Fig. 7). In contrast, in natural tidal wetlands, diurnal and nocturnal movement rates were very similar (Table 1). These

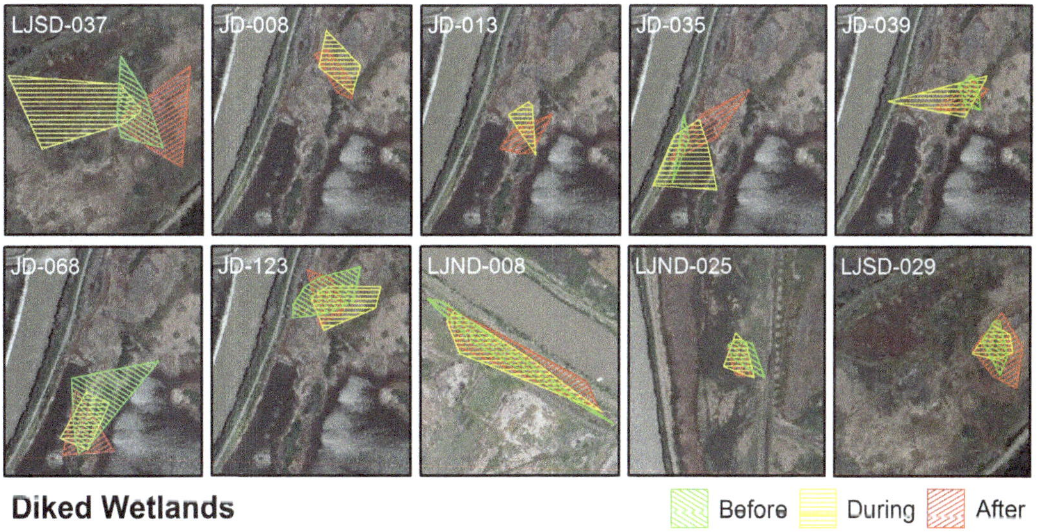

Figure 6. Minimum convex polygons for salt marsh harvest mice in diked wetlands. Green polygons represent points occurring more than one hour before high tide. Yellow polygons represent points falling within one hour before or after high tide. Red polygons represent points occurring more than one hour following high tide. Each block represents an individual mouse. Some blocks have all three polygons, while those that lack 6 or more individual points during the before or after time period, only have two polygons.

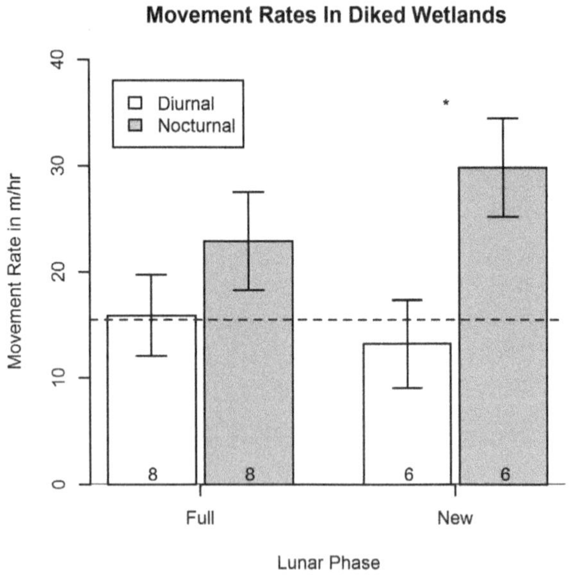

Figure 7. Movement rates for mice in diked wetlands. Movement rates for salt marsh harvest mice in diked wetlands. Dashed line represents 15 meters, corresponding to the level of error associated with the inundated/uninundated area boundary. Significant difference is marked with an asterisk; $P = 0.012$.

data indicate that in natural tidal wetlands, mice are moving almost as much during the day as at night (Fig. 8).

Discussion

Understanding how animals cope with both naturally-occurring heterogeneity and HIREC in their environment can be crucial to the conservation of threatened species such as the salt marsh harvest mouse [11], [39]. Behavioral flexibility represents a coping mechanism likely to be crucial for many animals in the face of anthropogenic change in their environment [11–13]. The purpose of this study was to determine how salt marsh harvest mice respond to both natural heterogeneity and HIREC. Following the framework of Berger-Tal et al. [7], we answered two questions. First, we considered behavior-based management and asked *do salt marsh harvest mice move vertically (into tall vegetation) or horizontally (into upland areas) to escape the high tide?* Determining high tide refuge use by the salt marsh harvest mouse is a high conservation priority for this endangered species [26]. We used radio telemetry to show that the vast majority of the time, male salt marsh harvest mice in the Suisun Marsh used emergent vegetation to escape tidal inundation, suggesting that maintenance of healthy intertidal vegetation is important in the conservation of small mammals and protection of biodiversity in this system. This conclusion is supported by extensive live-trapping data from a previous study that indicated that salt marsh harvest mice are only rarely trapped in upland areas [35], and is not surprising in the Suisun Marsh where vegetation is generally tall and thick providing extensive structure and cover. Our behavioral results will allow for improved habitat management, allowing managers to concentrate conservation efforts on maintaining intertidal habitat rather than expending efforts enhancing upland habitats that salt marsh harvest mice rarely use.

Secondly, we addressed whether and how HIREC causes changes in behavior [7]; specifically, do salt marsh harvest mice in anthropogenically altered diked wetlands behave differently than those subject to tidal influence? The lack of a significant difference in nocturnal movement rates between natural tidal and diked wetlands would suggest that this anthropogenic change has not affected the behavior of this species. However, diurnal movement rates were consistently lower than nocturnal movement rates in diked wetlands, a pattern not observed in tidal wetlands (Figs. 7–8). This result indicates that mice in human-influenced diked wetlands may have altered their diurnal movement behaviors in anthropogenically-modified diked wetlands, moving less during the day. Current telemetry efforts support these results and are revealing similar trends throughout the day, not just during high tides (K.R. Smith, unpublished data).

There are two potential explanations for the observed decrease in movement rate in diked wetlands during the day. First, anthropogenic habitat modification may have had an adverse effect on salt marsh harvest mice. For example, differences in habitat structure or a lack of inundation may either allow for increased access by diurnal predators or decreased cover for mice, potentially leading to reduced diurnal movement rates in diked wetlands (mice are *forced* to seek refuge during the day). A second potential explanation is that removal of tidal influence has instead had a positive effect on salt marsh harvest mice. For example, mice in diked wetlands may have increased availability and/or accessibility to resources, or an increased ability to cache resources, which could allow mice to restrict their foraging to night-time hours, remaining in refuges during the day when predation pressure is high (mice are *choosing* to seek refuge during the day) [40]. This possibility is supported by our observations of salt marsh harvest mice using underground refuges, and taking cover in voids below a hummocks of cattails during the day in three separate diked wetlands, behavior that would not be possible in a tidal wetland [41].

Habitat loss has been the primary threat to the salt marsh harvest mouse, and with the level of existing and continuing development in the San Francisco Bay Estuary [42–44], subsidence, and sea level rise [28], [45] the marshes utilized by

Table 1. Average movement rates of salt marsh harvest in meters per hour for treatments.

		Nocturnal		Diurnal	
Wetland Type	**Moon Phase**	**Mean**	***SE***	**Mean**	***SE***
Tidal	Full	24.89	4.22	24.11	4.22
Tidal	New	29.51	4.08	21.67	4.07
Diked	Full	22.91	4.14	15.91	3.82
Diked	New	29.82	4.63	13.23	4.63

All rates before, during, and after high tides are included in this analysis.

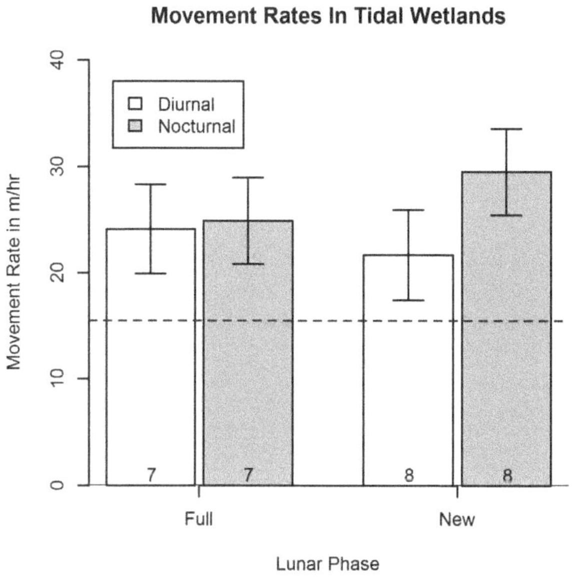

Figure 8. Movement rates for mice in tidal wetlands. Movement rates for salt marsh harvest mice in tidal wetlands. Dashed line represents 15 meters, corresponding to the level of error associated with the inundated/uninundated area boundary.

the salt marsh harvest mouse are vulnerable to additional habitat loss. Every remaining hectare of salt marsh harvest mouse habitat is crucial, and maximizing the value of this habitat through restoration and enhancement will be key to the persistence and recovery of the species [26], [45]. There are currently at least three plans that call for restoration of diked wetlands to tidal influence in the Suisun Marsh in the immediate future: Suisun Marsh Habitat Management, Preservation, and Restoration Plan [46], the Bay Delta Conservation Plan [17], and the Fish Restoration Program Agreement [48]. Refuge use by salt marsh harvest mice is a behavior that must be taken into account when planning and implementing these tidal restorations.

There is currently little research investigating the differences in value of various habitat types to the salt marsh harvest mouse. Our trapping was not designed to assess effects of wetland type on populations or densities of salt marsh harvest mice, but another study suggests that diked wetlands may be at least as valuable to conservation efforts as tidal wetlands, by supporting higher densities of mice [35]. Additionally, large populations of salt marsh harvest mice have recently been found in diked wetlands, even during flooded winter months (K.R. Smith, unpublished data). This suggests that this species is capable of thriving in flooded diked wetlands by remaining in tall vegetation over standing water. Given that much of the remaining salt marsh harvest mouse habitat in the Suisun Marsh exists as diked wetlands [45], the behavior and ecology of individuals living in these modified habitats warrants further exploration.

It is important to note that our findings may not necessarily apply to the entire range of the salt marsh harvest mouse. In the Suisun Marsh tidal vegetation is tall and marshes are wide, and there is an abundance of diked (21 044 hectares) and upland (11 210 hectares) habitat available, while only 21% of historic tidal habitat remains (2 550 hectares) [49]. The two other major strongholds for this species differ in marsh width and vegetation type: in San Pablo Bay many marshes are wide, but vegetation is short, and in the South San Francisco Bay marshes are narrow and vegetation is short [49]. In these areas, mice may employ alternate strategies when escaping the tide; for example, if vegetation is shorter than the water height at high tide, mice may have no choice but to move upland where they may be exposed to predation.

Finally, due to a number of limiting factors including minimum transmitter size and logistical constraints, we were unable to track female or young salt marsh harvest mice, so any speculation about their behavior should be made with care. Indeed, a recent study of another endangered wetland rodent, the New Mexico jumping mouse (*Zapus hudsonius luteus*), found different patterns of habitat use by males and females [50]. As technology advances and radio collars become smaller, it would be prudent to repeat this study using both male and female mice, as well as sub-adults and juveniles.

The results of this study provide insight to a question that researchers have been investigating for over 55 years: where do salt marsh harvest mice go during the high tide [29]? This study increases our understanding of the behavior of endangered salt marsh harvest mice in tidal and diked wetlands, suggesting that these small mammals may be capable of coping with the challenge of living in tidal wetlands with little to no emergent land as long as vegetation is sufficiently thick and tall. They also indicate that small mammals may exhibit different behaviors in diked wetlands than in tidal wetlands, highlighting the need to consider behavior when planning for conservation or management projects and shedding light on the need for comprehensive management strategies that account for potential behavioral differences in populations. Behavioral flexibility may be critical for threatened species in the face of HIREC [12]. As we cause further changes to the environment both directly, through processes such as urbanization, and indirectly, though processes such as climate change, the work of behaviorists will be critical in understanding behavioral responses of threatened species, like the salt marsh harvest mouse, to HIREC [6], [11], [51].

Acknowledgments

For invaluable field assistance we thank S. Estrella, D. Fidler, M. Riley, R. Pardee, L. Preteska, L. Patterson. For logistical support we thank S. Chappell and B. Wickland and the Suisun Resource Conservation District, and the California Department of Water Resources.

Author Contributions

Conceived and designed the experiments: KRS LBT KEM. Performed the experiments: KRS LBT. Analyzed the data: KRS WG KEM. Contributed reagents/materials/analysis tools: KRS LBT WG KEM. Wrote the paper: KRS LBT WG KEM.

References

1. Strier KB (1997) Behavioral Ecology and Conservation Biology of Primates and Other Animals. In: Slater PJB, Rosenblatt JS, Snowdon CT, Milinski M. Advances in the Study of Behavior: Academic Press. pp. 26, 101–158.
2. Shumway CA (1999) A neglected science. Applying behavior to aquatic conservation. Environ Biol Fishes 55: 183–201.
3. Buchholz R (2007) Behavioural biology: an effective and relevant conservation tool. Trends Ecol Evol 22: 401–407.
4. Angeloni L, Schlaepfer MA, Lawler JJ, Crooks KR (2008) A reassessment of the interface between conservation and behaviour. Anim Behav 75: 731–737.
5. Moore JA, Bell BD, Linklater WL (2008) The debate on behavior in conservation: New Zealand integrates theory with practice. BioScience 58: 454–459.
6. Caro T, Sherman PW (2013) Eighteen reasons animal behaviourists avoid involvement in conservation. Anim Behav 85: 305–312.

7. Berger-Tal O, Polak T, Oron A, Lubin Y, Kotler BP, et al. (2011) Integrating animal behavior and conservation biology: a conceptual framework. Behav Ecol 22: 236–239.
8. Wildermuth RP, Anadón JD, Gerber LR (2012) Monitoring behavior: assessing population status with rapid behavioral assessment. Cons Lett 6: 86–97.
9. Wong MYL (2013) Abiotic stressors and the conservation of social species. Biol Cons 155: 77–84.
10. West-Eberhard MJ (1989) Phenotypic plasticity and the origins of diversity. Annu Rev Ecol Evol Syst 20: 249–278.
11. Sih A, Ferrari MCO, Harris DJ (2011) Evolution and behavioural responses to human-induced rapid environmental change. Evol Appl 4: 367–387.
12. Sih A, Stamps J, Yang LH, McElreath R, Ramenofsky M (2010) Behavior as a key component of integrative biology in a human-altered world. Integr Comp Biol 50: 1–11.
13. Candolin U, Wong BB (2012) Behavioural responses to a changing world: mechanisms and consequences. New York: Oxford University Press. 256 p.
14. Ghalambor CK, Angeloni L, Carroll SP (2010) Behavior as phenotypic plasticity. In Fox C, Westneat D, editors. Evolutionary Behavioral Ecology. New York: Oxford University Press. pp. 90–107.
15. Wright T, Eberhard J, Hobson E, Avery M, Russello M (2010) Behavioral flexibility and species invasions: The adaptive flexibility hypothesis. Ethol Ecol Evol 22: 393–404.
16. Fraterrigo JM, Pearson SM, Turner MG (2009) Joint effects of habitat configuration and temporal stochasticity on population dynamics. Landsc Ecol 24: 863–877.
17. Oliver T, Roy DB, Hill JK, Brereton T, Thomas CD (2010) Heterogeneous landscapes promote population stability. Ecol Lett 13: 473–484.
18. Kuroe M, Yamaguchi N, Kadoya T, Miyashita T (2011) Matrix heterogeneity affects population size of the harvest mice: Bayesian estimation of matrix resistance and model validation. Oikos 120: 271–279.
19. Moyle PR, Lund JR, Bennett WA, Fleenor WE (2010) Habitat variability and complexity in the upper San Francisco estuary. SF Estuary Watershed Sci 8: 1–24.
20. Fisler GF (1965) Adaptations and speciation in harvest mice of the marshes of San Francisco Bay. PhD Dissertation, University of California Berkeley.
21. Coulombe HN (1970) The role of succulent halophytes in the water balance of salt marsh rodents. Oecologia. 4: 223–247.
22. Pearson OP (1960) The oxygen consumption and bioenergetics of harvest mice. Physiol Zool 33: 152–160.
23. Geissel W, Shellhammer H, Harvey HT (1988) The ecology of the salt-marsh harvest mouse (*Reithrodontomys raviventris*) in a diked salt marsh. J Mammal. 69: 696–703.
24. Federal Register 50 CFR 17.11; U.S. Fish and Wildlife Service. Conservation of Endangered Species and Other Fish or Wildlife; United States List of Endangered Native Fish and Wildlife, 35 Federal Register 199 (13 October 1970), pp. 16047–16047.
25. California Code of Regulations Title 14, Section 670.5[a][6][F]. Natural Resources; Fish and Game Commission-Department of Fish and Game; General Regulations; Miscellaneous; Animals of California Declared To Be Endangered or Threatened, 14 California Code of Regulation § 670.5 (28 May 71).
26. United States Fish Wildlife Service (2010) Draft recovery plan for tidal marsh ecosystems of northern and central California. Sacramento Fish and Wildlife Office of the US Fish and Wildlife Service, Sacramento, California, 408 pp.
27. Philip Williams & Associates, Faber PM (2004) Design guidelines for tidal wetland restoration in San Francisco Bay. The Bay Institute and California State Coastal Conservancy. 83 pp.
28. Greenberg R, Maldonado J, Droege S, McDonald M (2006) Tidal marshes: Home for the few and the highly selected. Stud Av Bio. 32: 675–685.
29. Johnston RF (1957) Adaptation of salt marsh mammals to high tides. J Mammal 38: 529–531.
30. Hadaway HC, Newman JR (1971) Differential responses of five species of salt marsh mammals to inundation. J Mammal 52: 818–820.
31. Hulst MD, Hall L, Morrison M, Bias M (200). Assessing salt marsh harvest mouse movements during high tides, San Pablo Bay, California. Trans West Sect Wildlife Society 37: 88–91.
32. Blumstein DT, Fernández-Juricic E (2004) The emergence of conservation behavior. Cons Bio 18: 1175–1177.
33. Shellhammer H, Duke HR, Orland MC (2010) Use of brackish marshes in the South San Francisco Bay by salt marsh harvest mice. Cal Fish Game 96: 256–259.
34. Shellhammer H, Jackson R, Davilla W, Gilroy A, Harvey H, et al. (1982) Habitat preferences of salt marsh harvest mice (*Reithrodontomys raviventris*). Wasmann Journ Biol 40: 102–114.
35. Sustaita D, Quickert PF, Patterson L, Barthman-Thompson L, Estrella S (2011) Salt marsh harvest mouse demography and habitat use in the Suisun Marsh, California. J Wild Manage 75: 1498–1507.
36. Sikes RS, Gannon WL (2011) Guidelines of the American Society of Mammalogists for the use of wild mammals in research. J Mammal 92: 235–253.
37. White GC, Garrott RA (1990) Analysis of wildlife radio-tracking data. San Diego: Academic Press. 383 pp.
38. Nams VO (2005) Using animal movement paths to measure response to spatial scale. Oecologia 143: 179–188.
39. Fahrig L, Merriam G (1994) Conservation of fragmented populations. Cons Bio 8: 50–59.
40. Johnston RF (1956) Predation by Short-eared Owls on a Salicornia salt marsh. Wilson Bulletin 91–102.
41. Smith KR (2012) Refuge use and movement of the salt marsh harvest mouse (*Reithrodontomys raviventris halicoetes*) in response to environmental heterogeneity. M. Sc. Thesis, New Mexico State University.
42. California Department of Water Resources, U.S. Department of the Interior, National Oceanic and Atmospheric Administration, U.S. Fish and Wildlife Service (2012) Bay Delta Conservation Plan. 1145 pp.
43. Shellhammer H, Duke R, Harvey H, Jennings V, Johnson V, et al. (1988) Salt marsh harvest mice in the diked marshes of southern San Francisco Bay. Wasmann Journ Biol 46: 89–103.
44. Takekawa JY, Woo I, Spautz H, Nur N, Grenier JL, et al. (2006) Environmental threats to tidal-marsh vertebrates of the San Francisco Bay Estuary. Stud Av Biol 32: 176–197.
45. Shellhammer HS (1989) Salt marsh harvest mice, urban development, and rising sea levels. Cons Biol 3: 59–65.
46. United States Department of the Interior, Bureau of Reclamation, U.S. Fish and Wildlife Service, California Department of Fish and Game, and IFC International (2011) Suisun Marsh habitat management, preservation, and restoration plan, Sacramento, California, 42 pp.
47. BDCP Steering Committee (2013) Bay Delta Conservation Plan Highlights. 61 pp.
48. California Department of Water Resources and California Department of Fish and Game (2010) DWR and DFG Fish Restoration Program Agreement. Sacramento, California, 26 pp.
49. Goals Project (1999) Baylands ecosystem habitat goals. A report of habitat recommendations prepared by the San Francisco Bay Area Ecosystems Goals Project. U.S. Environmental Protection Agency, San Francisco, California, and San Francisco Bay Regional Water Quality Control Board. 328 pp.
50. Wright G (2012) Multiple scale habitat selection by a small mammal habitat specialist (*Zapus hudsonius luteus*) in a managed floodplain landscape. M. Sc. Thesis, New Mexico State University.
51. Caro T Sherman PW (2011) Endangered species and a threatened discipline: behavioural ecology. Trends Ecol Evol 26: 111–118.

8

Quantifying Fish Assemblages in Large, Offshore Marine Protected Areas: An Australian Case Study

Nicole A. Hill[1]*, Neville Barrett[1], Emma Lawrence[2], Justin Hulls[1], Jeffrey M. Dambacher[3], Scott Nichol[4], Alan Williams[5], Keith R. Hayes[3]

1 Institute for Marine and Antarctic Studies, University of Tasmania, Hobart, Tasmania, Australia, **2** Digital Productivity Flagship, Commonwealth Scientific and industrial Research Organisation (CSIRO), Brisbane, QLD, Australia, **3** Digital Productivity Flagship, Commonwealth Scientific and industrial Research Organisation (CSIRO), Hobart, Tasmania, Australia, **4** Geoscience Australia, Canberra, ACT, Australia, **5** Oceans and Atmosphere Flagship, Commonwealth Scientific and industrial Research Organisation (CSIRO), Hobart, Tasmania, Australia

Abstract

As the number of marine protected areas (MPAs) increases globally, so does the need to assess if MPAs are meeting their management goals. Integral to this assessment is usually a long-term biological monitoring program, which can be difficult to develop for large and remote areas that have little available fine-scale habitat and biological data. This is the situation for many MPAs within the newly declared Australian Commonwealth Marine Reserve (CMR) network which covers approximately 3.1 million km^2 of continental shelf, slope, and abyssal habitat, much of which is remote and difficult to access. A detailed inventory of the species, types of assemblages present and their spatial distribution within individual MPAs is required prior to developing monitoring programs to measure the impact of management strategies. Here we use a spatially-balanced survey design and non-extractive baited video observations to quantitatively document the fish assemblages within the continental shelf area (a multiple use zone, IUCN VI) of the Flinders Marine Reserve, within the Southeast marine region. We identified distinct demersal fish assemblages, quantified assemblage relationships with environmental gradients (primarily depth and habitat type), and described their spatial distribution across a variety of reef and sediment habitats. Baited videos recorded a range of species from multiple trophic levels, including species of commercial and recreational interest. The majority of species, whilst found commonly along the southern or south-eastern coasts of Australia, are endemic to Australia, highlighting the global significance of this region. Species richness was greater on habitats containing some reef and declined with increasing depth. The trophic breath of species in assemblages was also greater in shallow waters. We discuss the utility of our approach for establishing inventories when little prior knowledge is available and how such an approach may inform future monitoring efforts within the CMR network.

Editor: David Mark Bailey, University of Glasgow, United Kingdom

Funding: This work was undertaken for the Marine Biodiversity Hub, a collaborative partnership supported through funding from the Australian Government's National Environmental Research Program (NERP). NERP Marine Biodiversity Hub partners include the Institute for Marine and Antarctic Studies, University of Tasmania; CSIRO, Geoscience Australia, Australian Institute of Marine Science, Museum Victoria, Charles Darwin University and the University of Western Australia. The funders had no role in study design, data collection and analysis, decision to publish, or preparation of the manuscript.

Competing Interests: The authors have declared that no competing interests exist.

* Email: Nicole.Hill@utas.edu.au

Introduction

Marine ecosystems face pressures from a range of sources, including pollution, fishing, habitat destruction, invasive species and a changing climate [1]. These pressures threaten the essential processes and resources that marine ecosystems provide, such as climate regulation, nutrient cycling and the provision of food, as well as threatening their intrinsic natural and cultural value [2]. As a consequence, management strategies have been adopted to safeguard marine ecosystems and their associated biodiversity. These strategies include the establishment of Marine Protected Areas (MPAs) which offer varying levels of protection according to their zoning or IUCN designation [3]. Countries such as the United States, Australia, South Africa, New Zealand, Kenya and the Philippines have a long history of establishing MPAs [4,5]. However, the absence of a truly global system of MPAs prompted the Convention of Biological Diversity [6] and the Jakarta Mandate [7] to provide a framework and renewed international impetus for protecting marine biodiversity. This framework has resulted in a significant increase in recent years in the declaration of MPAs worldwide, with many forming part of large networks or systems of MPAs [8].

The increasing global commitment to the sustainability of healthy marine ecosystems is encouraging, but the declaration alone of MPAs does not ensure the conservation of biodiversity or ecosystems [4]. Management authorities must: clearly articulate the objectives of individual MPAs and MPA networks, both of which may contain multiple IUCN zones; formulate and enforce management plans [9] and periodically revisit and evaluate their management plans, which in turn requires that they establish a long-term biological monitoring program [4]. Developing monitoring programs for individual MPAs within networks of MPAs is often a difficult task. There is generally a mismatch between the spatial scales of information used to design MPA networks and those required to manage individual MPAs. Inventories of the types and distribution of habitats, communities and species within individual MPAs are often required before monitoring programs can be designed. In addition, for offshore or difficult to access MPAs, there are significant logistical and statistical challenges to working in remote environments.

Australia, a signatory to the CBD, provides a good example of the challenges involved in ultimately developing long-term monitoring programs in large MPA networks. In 2012, the Australian government announced an expansion of its MPA network to include 33 new MPAs. This makes a total of 60 MPAs covering approximately 3.1 million square kilometres, divided between six planning bioregions [10]. Although called the "Commonwealth Marine Reserve" or "CMR" network, areas with IUCN zones ranging from la (Sanctuary Zone) to VI (General Use Zone) are distributed throughout the network. From this point forward we adopt the above terminology and refer to MPAs within this network as reserves regardless of their IUCN zoning. All of the reserves in the CMR are however, located in offshore waters, most are large and encompass a broad depth range, and many are remote.

The Australian CMR network aims to meet the principles of being Comprehensive, Adequate and Representative (the CAR principles) [11] and utilised the national Integrated Marine and Coastal Regionalisation of Australia (IMCRA; [12]). The IMCRA bio-regionalisation is based on breaks in the distribution of demersal fish communities (together with some physical datasets) and is the best available continental-scale regionalisation of Australian marine fauna. Hence it was the most appropriate mechanism for delineating management regions and informing the placement of reserves. When designing biological monitoring programs for individual reserves however, much finer scale information is needed on the spatial distribution of important habitats (such as reefs) and the biological communities they contain. For many reserves in the new network this level of information is simply unavailable. It is therefore necessary to provide a robust inventory of the biological communities and key species within the reserves, delineate their spatial distribution and quantify the relationship between these distributions and key environmental drivers to enable the development of monitoring programs that will ultimately evaluate the effectiveness of each reserve.

Inventory, and ultimately monitoring, of large, deep reserves is challenging and relatively expensive. It requires vessels large enough to conduct surveys safely in offshore waters, together with a suite of non-extractive survey methods that can be deployed at depth, and sampling designs that are efficient and flexible enough to accommodate multiple (and potentially changing) objectives. A suite of non-extractive survey methods are currently available including multibeam sonar (MBS) for characterising the seafloor and associated habitat types [13], and image-based methods, such as towed camera systems, cameras attached to autonomous or remotely operated vehicles and baited cameras, for observing benthic fauna. Here we trialled MBS and Baited Remote Underwater Video (BRUVs), deployed within a spatially balanced design, for inventorying demersal fish in the multiple use zone (IUCN VI) of one reserve in the Australian network, the Flinders CMR. We use BRUVs because of their demonstrated ability to survey fish in deep waters [14] and because they have been extensively used in Australian inshore environments for monitoring and other ecological applications [15–17]. We trialled a spatially-balanced (i.e. evenly spread out in space) design known as the Generalised Random- Tessellation Stratified (GRTS) design [18] because it is flexible and provides a good way to obtain a representative sample that respects the spatial distribution of habitats and communities in the target population [19].

In this study we: identify patterns in demersal fish assemblages across the continental shelf (a multiple-use zone, IUCN VI) of the Flinders CMR; quantify these patterns in relation to environmental gradients; examine the spatial distribution of assemblages across the reserve; and describe how our approach and the information gained may be useful to the development of a long-term biological monitoring program. We focus on fish communities because: broad-scale distributional records of demersal fish were instrumental in the marine bioregionalisation that underpins the design of the reserve network [20]; many species are of recreational and commercial value and are therefore likely to respond to current and future management actions within the reserve; and demersal fish are key components of benthic ecosystems.

Methods

Survey Area

The Flinders Commonwealth Marine Reserve (CMR) was established in 2007 and lies about 25 km offshore of northern Tasmania (Fig. 1). The reserve is 26,975 km^2 in size and extends as a west-east corridor from 35 m to > 3,000 m water depths. It consists of two zones: a multiple use zone (IUCN Category VI) that covers the majority of the continental shelf and slope; and a marine national park zone (IUCN Category II) that extends from the continental slope to the edge of the reserve at the 200 nautical mile limit of Australia's Exclusive Economic Zone (Fig. 1 inset). Our study area is the continental shelf, part of the multiple use zone (IUCN Category VI), where activities that impact on benthic habitats are prohibited, including demersal trawling, Danish seining and scallop dredging, or subject to permit requirements, for example mining activities [21]. Benthic habitats on the Flinders CMR shelf consist of sediment plains with patches of low profile and sand-inundated reefs, and steep rocky outcrops where canyon heads incise the shelf break (Lawrence, unpublished data). Within this environment, reefs and rocky outcrops have been identified as features likely to contain enhanced benthic diversity [21].

The Flinders CMR shelf straddles two provincial biogeographical regions: the Southeast Shelf Transition in the north and the Tasmanian Shelf Province in the South [12]. This region is characterised by variable but generally high exposure and strong tidal currents, especially in shallow areas between Flinders Island and the Tasmanian mainland [22,23]. The region is also influenced by southwards incursions of the East Australian Current (EAC) which brings warmer waters on to the shelf in summer [24]. The flora of the region is moderately rich and contains species common in cold temperate waters as well as low abundances of species common to warmer temperate waters [20].

Sampling design and methods

In contrast to the Flinders CMR continental slope, which has been comprehensively mapped using multibeam sonar (from which seafloor habitats have been inferred), very little spatially explicit habitat or biological information is available for the shelf. This represents a challenge for designing targeted sampling programs. Interpolated bathymetry for the Flinders shelf exists, but is at a gridded resolution of 250 m [25] (Fig. 1) which does not allow the identification of fine scale features of interest on the shelf, such as reefs. As a consequence of our limited *a priori* knowledge of the area, the inventory and description of fish assemblages

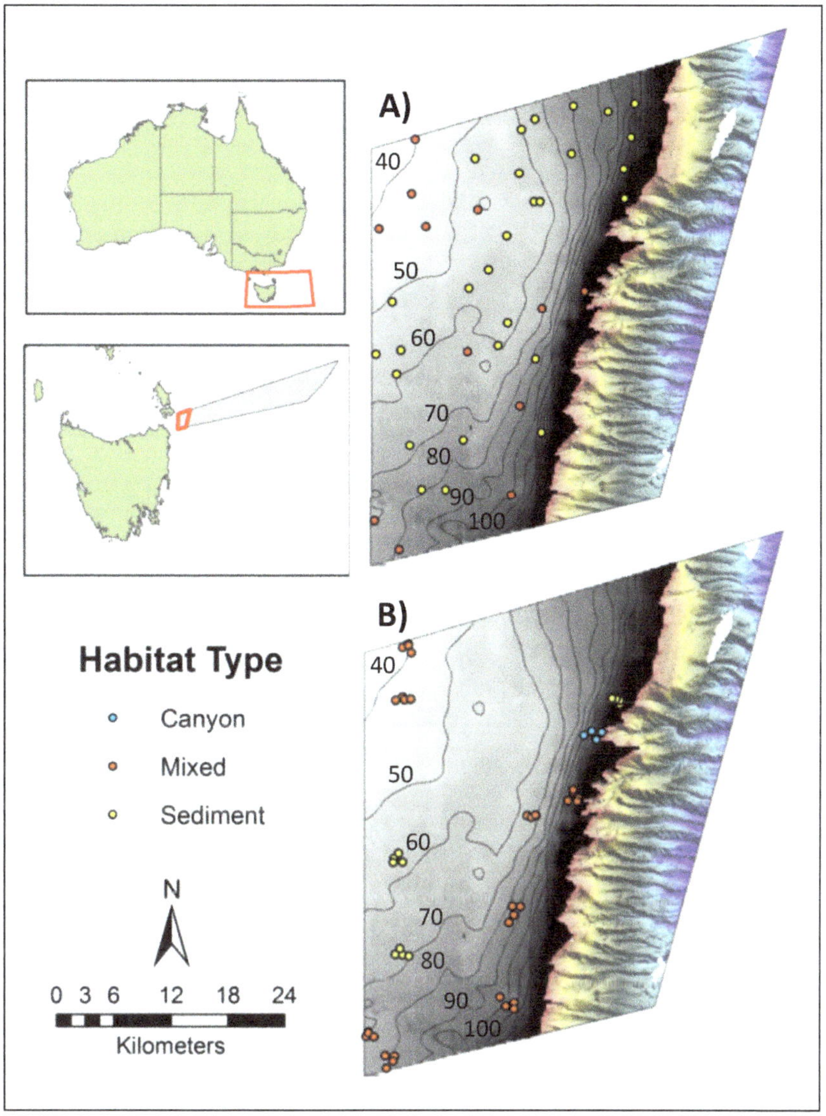

Figure 1. Location of the Flinders Commonwealth Marine Reserve (CMR) in Tasmania, Australia. Panels A and B show the CMR's Multiple Use Zone (IUCN VI) where survey work was conducted. The grey zone is the continental shelf (less than 200 m depth) where there was little pre-existing mapping data. Coarse bathymetry data (gridded at 250 m horizontal resolution) sourced from Geoscience Australia is shown with 10 m contour intervals overlain. The coloured area to the right shows the relatively steep and highly incised upper continental slope that had been mapped previously with multibeam sonar extending from 200 to 1500 m. A) Location of the 40 sites surveyed for habitat type in phase one of the sampling program. B) Location of the clustered sites where Baited Remote Underwater Videos (BRUVs) were deployed in phase two. Sites are coloured according to the broad habitat type: sediment (yellow); mixed, low-profile reef and sediments (red); canyon head (blue) recorded during phase one of sampling.

presented here forms one component of a multi-objective and multi-phase study within the Flinders CMR. Additional objectives of the broader study include examining the distribution and estimating the area of the different habitat types within the reserve, as well as quantifying benthic invertebrate communities using imagery captured by towed video and AUV, and these results will be reported elsewhere. To accommodate the objectives of the different components of the study, we used the probabilistic design, Generalised Random –Tessellation Stratified sampling technique (GRTS; [18]). GRTS ensures sampling sites are well spread out across the survey area (spatially-balanced), a desirable property for spatial sampling that enhances estimation efficiency and representativeness [18]. Other common sampling designs include simple or stratified random sampling or systematic sampling. However, simple random sampling often clumps sites [19] and stratified random sampling (based on habitat in our case) was not possible without more detailed knowledge of the study area. Systematic sampling prevents clumping but an accurate design-based variance estimator does not exist; a factor that was important for other aspects of the broader study [18,26]. provide a good discussion of the relative merits of various spatial sampling strategies. Another property of GRTS is that it produces an ordered list of sample sites and any set of consecutive sampling sites also maintain spatial balance, offering flexibility to adaptively change sample sizes in the field, if for example, sampling takes less time than expected. To date GRTS has primarily been used in natural resource assessment and monitoring in the United States e.g. [27].

Because of our lack of knowledge on the distribution of habitat types, we implemented a two-phase sampling program. In phase one, the distribution of habitat types across the reserve was quantified by characterising the first forty, 200 m square, sites from a master list of GRTS sites (essentially an ordered list of all possible GRTS sites in the study area. See [28] for a detailed discussion on the use of master lists) as either 'soft' sediments or 'mixed', low profile patchy reef using MBS and footage obtained from a drop camera (Fig. 1; Lawrence, unpublished data). In phase two of the sampling program, the subject of this paper, we sampled the fish assemblages using BRUVs. Ideally we would have revisited all forty of the phase one GRTS sites. The large size of the study area and the soak time required for each BRUV deployment however, meant that it was not possible to do so in the ship time available. Instead, we sampled clusters of sites surrounding a subset of the phase one GRTS- sites. In phase two, we sampled mixed reef habitats more intensively than sediment habitats because shelf reef systems within the reserve are recognised as an important biodiversity feature and a *priori* we expected that reefs would harbour a greater diversity of assemblage types than sediments. We utilised the fact that consecutive GRTS sites maintain spatial balance and selected (from the ordered list of phase one sites) the first eight mixed habitat sites and the first three sediment sites of the forty phase one sites as the basis of the phase two clusters (Fig. S1. shows the site numbers of the forty phase ones sites and lists those used as the basis of phase two clusters.) This amounts to selecting phase one sites for the BRUV sampling that are spatially balanced *within* each of the phase one habitat types, mixed reef and sediments, and presumably therefore also spread across the environmental gradients in the region. In addition, a site was added near the head of a canyon, to ensure representation of another identified feature of the reserve which occupies a relatively small proportion of space and is therefore less likely to be sampled in a probabilistic sampling design (Fig. 1B). Clusters of sites within 1 km of the subset of phase one sites were selected again using the GRTS master list, resulting in five sites per cluster (Fig. 1B).

At each site, one BRUV was deployed using systems and deployment conditions widely used in Australia [29]. Each stereo-BRUV consisted of a frame that houses a stereo camera pair, a bait bag attached to an arm within field of view of the cameras, a diode for synchronising imagery between camera pairs. In water deeper than approximately 70 m, a light with a blue filter was also attached [15]. Stereo- BRUVs were baited with 1 kg of crushed pilchards and deployed for 60 min (soak time). Appropriate ethics (University of Tasmania Animal Ethics Permit: A12514) and fieldwork (Australian Government Director of National Parks Approval of Research Activities in the Southeast Commonwealth Marine Reserve Network. Ref No 07/10622) approvals were obtained for this work.

Analysis

Imagery collected using stereo-BRUVs was scored using standard metrics including scoring the maximum number of fish occurring in any one frame for each species (MaxN) [15,30]. Scoring was completed with Event Measure software [31].

To determine the types of fish assemblages present at sites sampled on the Flinders CMR shelf, fuzzy clustering was performed on multivariate fish composition data using the 'cluster' package [32] in R [33]. Multivariate data were square root transformed to reduce the influence of the most abundant species and a Bray Curtis dissimilarity matrix generated. The number of fuzzy clusters was set to six, following preliminary analysis using several different clustering methods, and the membership exponent (which controls the 'fuzziness' of clusters) set to 1.3 after examining silhouette profiles of group membership.

A constrained ordination (Canonical Analysis of Principle Co-ordinates; CAP) [34] was used to test if the assemblage groups determined by fuzzy clustering were significantly different from each other (using a permutational test). In addition, jack-knife sampling was used to examine the overall and individual classification accuracy of the six groups. To enable the ecological interpretation of assemblages, the trophic level and habitat preference of species was defined. Species were assigned one of seven habitat preference categories: 1) pelagic; 2) wide-ranging demersal; 3) sediment associated; 4) reef and sediment associated; 5) reef and seagrass associated; 6) reef associated; and 7) demersal generalists (found in more than two habitats) based on species attributes described in [35,36]. The trophic level of species was extracted from FishBase [37]. Differences in the trophic level of species comprising the six groups was assessed with a one factor Analysis of Variance (ANOVA) after transforming data to satisfy the assumptions of ANOVA. To understand which species were the key contributors to each of the six assemblage groups, a SIMPER analysis was performed. SIMPER decomposes the Bray-Curtis dissimilarity matrix between all pairs within a group into the percentage contributions from each species [34].

Correlations between fish assemblages and their environment were examined against three easily quantifiable environmental variables that, from the literature, may influence benthic assemblages in our system; latitude, depth and the substratum type. At each cluster of five sites, only the central site had been surveyed with MBS and drop camera in phase one of the sampling program. Therefore, substratum type was derived from the BRUV field of view and characterised as either: 'sediment' if no hard substratum or no hard substratum-associated organisms (i.e. sponges etc.) were present; 'mixed' if some hard substratum was visible or if some hard substratum-associated organisms were present; and 'reef' if the majority of the field of view contained hard substratum or hard substratum-associated organisms. Latitude was taken from the recorded location of BRUV drops and depth was derived using the 250 m resolution bathymetry grid available from Geoscience Australia [25]. Relationships between assemblage groups and environmental variables were inferred by examining correlations with CAP axes and overlaying vectors of the environmental variables on CAP plots. Ordination and SIMPER analysis were performed in PRIMER + with PERMA-NOVA [34].

The influence of environmental variables in determining patterns in species richness (the number of species present) was examined using Generalised Linear Mixed Models (GLMMs). Observations were modelled using a Poisson distribution with a log link function and sampling cluster was included as a normally distributed random factor. Mixed models were fit in R [33] with penalised quasi-likelihood, using the glmmPQL function in the package MASS [38].

Results

Distribution of habitats across the CMR

The results of phase one sampling indicated that the majority (70%) of the 40 sites on the shelf were sedimentary habitats, and the remainder 'mixed' reef and sediment habitat. Mixed reef habitat occurred in the north-west, south-west, and near the edge of the shelf of the study area (Fig. 1A; Lawrence, unpublished data).

Table 1. List of species recorded in Baited Remote Underwater Video deployments across the continental shelf in the Flinders Commonwealth Reserve.

			Across CMR			Mean (SE) MaxN by group					
Family	Species name	Common name	Total MaxN	Max MaxN in any drop	Prevalence (% drops)	Group 1	Group 2	Group 3	Group 4	Group 5	Group 6
Blenniidae, Gobiidae, Tripterygiidae	*Blenniidae, Gobiidae, Tripterygiidae*	Blennies, Gobies, Triplefins	7	3	10	0.38 (0.18)	0.4 (0.31)	0	0	0	0
Callorhinchidae	*Callorhinchus milii*	Elephant shark	1	1	2	0	0.1 (0.1)	0	0	0	0
	Nemadactylus douglasii	Grey morwong	7	2	10	0	0.2 (0.13)	0.33 (0.24)	0.33 (0.33)	0	0
	Nemadactylus macropterus	Jackass morwong	379	43	59	0	10.7 (2.04)	23.22 (4.78)	0	5.25 (1.23)	0
Cyttidae	*Cyttus australis*	Silver dory	6	1	12	0	0.1 (0.1)	0	0	0.42 (0.15)	0
Diodontidae	*Diodon nicthemerus*	Globe fish	1	1	2	0	0	0	0	0.08 (0.08)	0
Gempylidae	*Thyrsites atun*	Barracouta	14	3	20	0.38 (0.37)	0.1 (0.1)	0.11 (0.11)	0	0.75 (0.22)	0
Gerreidae	*Parequula melbournensis*	Silverbelly	79	10	31	0	0.5 (0.27)	0	0	5.83 (0.6)	0.67 (0.67)
Heterodontidae	*Heterodontus portusjacksoni*	Port Jackson shark	4	1	8	0	0.3 (0.15)	0	0	0.08 (0.08)	0
Labridae	*Ophthalmolepis lineolatus*	Southern maori wrasse	1	1	2	0	0	0.11 (0.11)	0	0	0
	Pseudolabrus rubicundus	Rosy wrasse	72	13	25	0	1.4 (0.88)	0.67 (0.67)	0	4.33 (1.38)	0
Lamnidae	*Isurus oxyrinchus*	Shortfin Mako shark	1	1	2	0.12 (0.12)	0	0	0	0	0
Latridae	*Latris lineata*	Striped trumpeter	95	27	22	0	0.2 (0.13)	10.11 (3.47)	0.33 (0.33)	0	0
Loliginidae	*Sepioteuthis australis*	Southern calamari	21	3	29	0	0.2 (0.13)	0	0	1.5 (0.23)	0.17 (0.17)
Monacanthidae	*Acanthaluteres vittiger*	Toothbrush leatherjacket	15	3	18	0	0	0	0	1.25 (0.28)	0
	Meuschenia freycineti	Six-spine leatherjacket	3	1	6	0	0.2 (0.13)	0	0	0.08 (0.08)	0
	Meuschenia scaber	Velvet leatherjacket	97	11	53	0	3.3 (0.87)	0.78 (0.43)	0	4.25 (0.97)	1 (0.63)
	Meuschenia venusta	Stars and stripes leatherjacket	1	1	2	0	0.1 (0.1)	0	0	0	0
	Thamnaconus degeni	Degen's leatherjacket	471	89	41	0	1 (0.42)	0.11 (0.11)	0	36.92 (7.37)	2.83 (1.9)
Moridae	*Pseudophycis barbata*	Bearded cod	8	2	12	0	0.1 (0.1)	0.78 (0.28)	0	0	0
Mullidae	*Upeneichthys vlamingii*	Southern goatfish	16	8	16	0	0.3 (0.21)	0.11 (0.11)	0	1 (0.65)	0
Neosebastidae	*Neosebastes scorpaenoides*	Common gurnard perch	44	4	41	0	1.1 (0.31)	0	0	1.5 (0.42)	2.5 (0.5)
Ommastrephidae	*Ommastrephidae sp.*	Squid	9	2	12	0.5 (0.27)	0	0	0.83 (0.4)	0	0
Ostraciidae	*Aracana aurita*	Shaw's cowfish	7	2	12	0	0.1 (0.1)	0	0	0.5 (0.19)	0
Ostraciidae	*Aracana ornata*	Ornate cowfish	1	1	2	0	0.1 (0.1)	0	0	0	0
Palinuridae	*Jasus edwardsii*	Southern rocklobster	2	1	4	0	0	0.22 (0.15)	0	0	0
Paraulopidae	*Paraulopus nigripinnis*	Blacktip cucmberfish	47	12	22	2 (0.65)	0.1 (0.1)	0	5 (2.16)	0	0
Pinguipedidae	*Parapercis allporti*	Barred Grubfish	36	5	35	2.12 (0.52)	0.8 (0.25)	0.44 (0.44)	0.5 (0.34)	0	0.67 (0.42)
Platycephalidae	*Platycephalus bassensis*	Sand flathead	33	6	24	0.12 (0.12)	0	0	0.17 (0.17)	0.83 (0.42)	3.5 (0.96)
	Platycephalus richardsoni	Tiger flathead	21	4	25	0	0.2 (0.13)	0	1.67 (0.56)	0.17 (0.17)	1.17 (0.31)
Pristiophoridae	*Pristiophorus cirratus*	Longnose sawshark	2	1	4	0	0	0	0	0.08 (0.08)	0.17 (0.17)
Rajidae	*Dentiraja lemprieri*	Thornback skate	2	1	4	0	0	0	0	0.17 (0.11)	0

Table 1. Cont.

			Across CMR			Mean (SE) MaxN by group					
Family	Species name	Common name	Total MaxN	Max MaxN in any drop	Prevalence (% drops)	Group 1	Group 2	Group 3	Group 4	Group 5	Group 6
	Spiniraja whitleyi	Melbourne skate	9	1	18	0	0.3 (0.15)	0	0.33 (0.21)	0.33 (0.14)	0
Scyliorhinidae	*Cephaloscyllium laticeps*	Draughtboard shark	63	4	57	0.12 (0.12)	1.7 (0.5)	0.44 (0.34)	1.33 (0.49)	2.33 (0.31)	0.83 (0.48)
Sebastidae	*Helicolenus percoides*	Red gurnard perch	65	12	31	0.25 (0.25)	2.5 (0.81)	4.22 (1.28)	0	0	0
Serranidae	*Caesioperca lepidoptera/rasor*	Butterfly and Barber perch	140	60	14	0	6.1 (5.99)	2.11 (1.42)	0	5 (2.79)	0
Squalidae	*Squalus acanthias*	Spiny dogfish	1	1	2	0	0	0	0.17 (0.17)	0	0
	Squalus megalops	Shortnose spurdog	15	3	14	0.75 (0.49)	0	0	1.5 (0.43)	0	0
Triakidae	*Mustelus antarcticus*	Gummy shark	18	2	33	0.25 (0.16)	0	0	1.17 (0.17)	0.25 (0.13)	1
Triglidae	*Lepidoperca pulchella*	Tasmanian perch	11	5	10	0	0.1 (0.1)	1.11 (0.56)	0	0	0
	Lepidotrigla sp	Gurnards	3	1	6	0	0.1 (0.1)	0.11 (0.11)	0.17 (0.17)	0	0
Urolophidae	*Urolophus cruciatus*	Banded stingaree	2	1	4	0	0.1 (0.1)	0	0	0	0.17 (0.17)
	Urolophus paucimaculatus	Sparsely spotted stingaree	2	1	4	0	0.2 (0.13)	0	0	0	0
	Urolophidae sp.	Stingarees	4	1	8	0.12 (0.12)	0.2 (0.13)	0	0	0	0.17 (0.17)

The Total number of individuals recorded at Max N across all deployments, the greatest MaxN recorded in any single deployment and the prevalence (the percentage of deployments where a species was recorded) are presented as well as the mean (and standard error) MaxN for each assemblage group.

Table 2. Comparison between the number of sites assigned to each of the six groups using fuzzy clustering and Canonical Analysis of Principle coordinates (CAP).

		CAP Classified Group						Total	Correct
		1	**2**	**3**	**4**	**5**	**6**		**(%)**
Fuzzy Group	1	8	0	0	0	0	0	8	100
	2	0	9	1	0	0	0	10	90
	3	0	1	8	0	0	0	9	89
	4	0	0	0	5	0	1	6	83
	5	0	0	0	0	12	0	12	100
	6	0	0	0	0	0	6	6	100

Jack knife validation in the CAP procedure was used to assess overall and group-wise accuracy of the six groups determined by fuzzy clustering. Classification accuracy of all groups is good, indicating that the groups determined by the fuzzy clustering are robust.

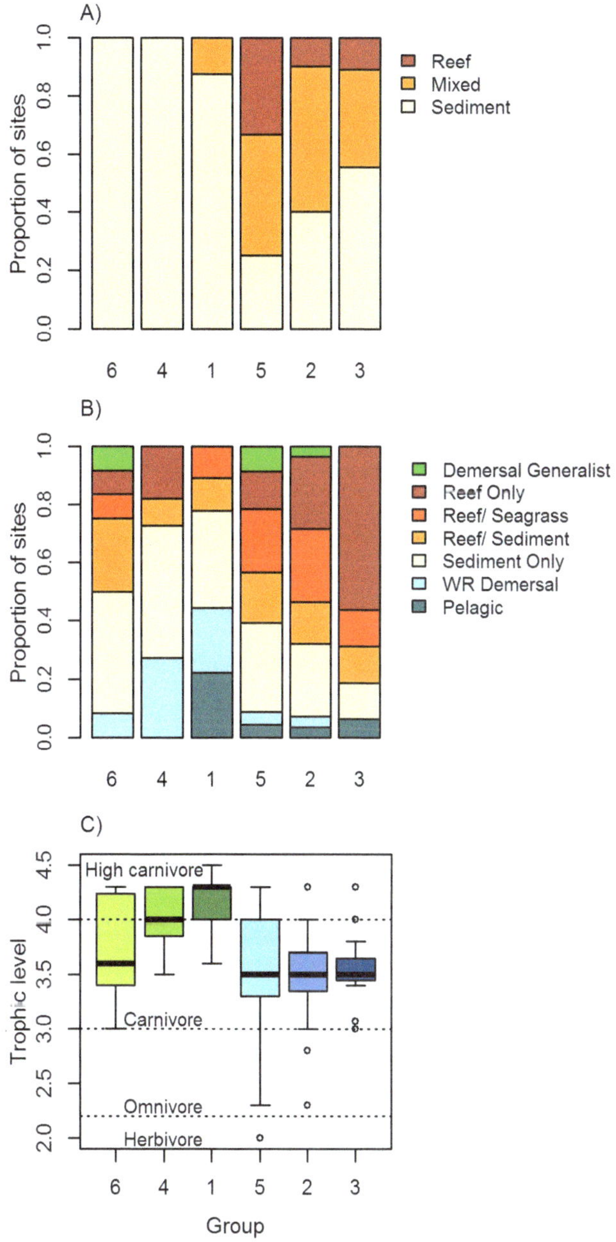

Figure 2. Characteristics of the six assemblage groups identified using fuzzy-clustering. A) Proportion of sites within each group classified as sediment, mixed or reef habitat based on the BRUVs footage; B) the habitat preference of species within each assemblage group; C) boxplot of the trophic level of species within each assemblage group; the box represents the 1st and 3rd quartiles, and circles denote potential outliers. In all panels, groups are ordered based on dominant habitat contained within groups (sediment versus mixed reef/reef) and from shallow to deep within each broad habitat type (based on spatial distribution of groups presented in Fig. 5) In plot C) sediment- associated groups are coloured green and reef-associated groups, blue.

Fish assemblages present in the CMR

Sixty BRUV deployments were completed in phase two, of which 51 yielded video footage of sufficient quality for scoring fish assemblage composition. Overall, 45 species were observed, with a total of 1,837 individuals recorded (at Max N) (Table 1). Of these species, 66% are endemic to Australia. Leatherjackets (Monacanthidae) were the most numerically abundant group, and

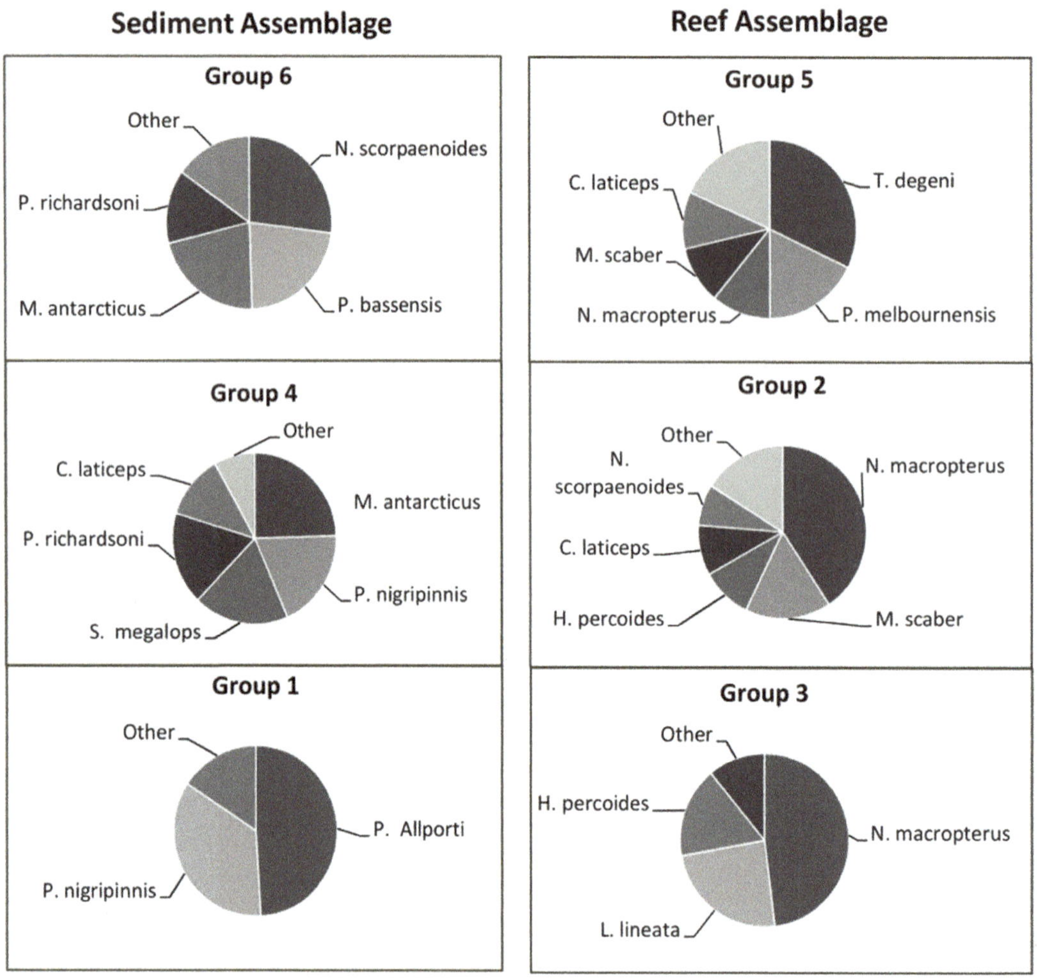

Figure 3. Species contributing up to 80% towards the similarity of the predominantly sediment- and predominantly reef-associated groups identified using fuzzy clustering.

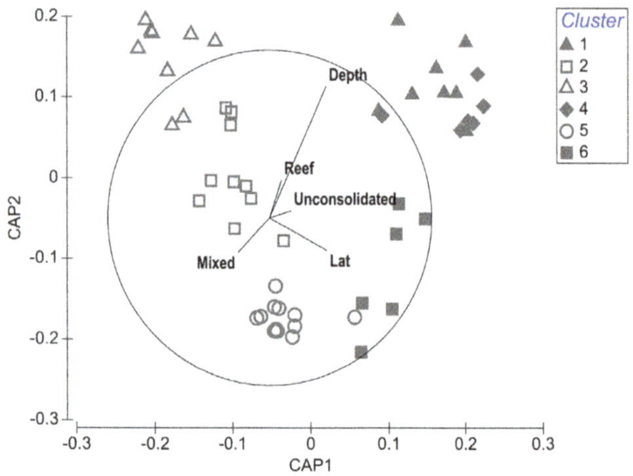

Figure 4. Canonical Analysis of Principle coordinates (CAP) plot discriminating sites based on fuzzy clustering. Each site was assigned to the cluster for which it had the highest probability of membership. Sites are shown as symbols: filled grey symbols represent sand-associated assemblages and open symbols represent reef-associated assemblages. Vectors of all environmental variables are overlaid and are proportional to their correlation with either CAP axis one or two.

comprised mostly *Thamnaconus degeni* (Degan's leatherjacket) and *Meuschenia scaber* (velvet leatherjacket) with, respectively, a total of 471 and 97 individuals recorded and MaxN of 89 and 11. *Nemadactylus macropterus* (jackass morwong), and *Caesioperca Lepidoptera and C. razor* (butterfly and barber perch) were the next most abundant species (379 and 140 individuals recorded and MaxN of 89 and 60). Other abundant species included *Latris lineata* (striped trumpeter), *Parequula melbournensis* (silverbelly), *Pseudolabrus rubicundus* (rosy wrasse), *Helicolenus percoides* (reef ocean perch), and *Cephaloscyllium laticeps* (draughtboard shark). The five most prevalent species were *N. macropertus* (59% of sites), *C. laticeps* (57% of sites), *M. scaber* (53% of sites), *Neosebastes scorpaenoides* (common gurnard perch; 41% of sites) and *T. degeni* (41% of sites).

The structure of fish assemblages surveyed was adequately represented by six groups. The overall classification accuracy of the six assemblage groups was high (94%) and a majority of sites (82%) were also assigned to groups with a high degree of confidence (greater than 0.85 probability of membership). Groups four and three were the least certain, but still had a high classification accuracy of 83% and 89%, respectively (Table 2). In addition, these six groups were statistically meaningful with significantly different multivariate community composition (P = 0.001).

Table 3. Results of Poisson GLMM for species richness recorded in BRUV deployments.

Factor	Co-efficient	Std Error	df	t-value	p-value
Depth	−0.277	0.067	36	−4.1489	0.0002
Mixed Habitat	0.330	0.092	36	3.575	0.0010
Reef Habitat	0.382	0.121	36	3.169	0.0031

Predictor variables have been scaled and centered and co-efficients are on the scale of the link function (log). Spatial clusters introduced as part of the sampling design were included as a random effect and the effects of habitat type (sediment, mixed or reef) are presented relative to sediment habitat.

The six assemblage groups were broadly aligned with habitats observed in the BRUV footage. Sites belonging to three groups (groups 1, 4, 6) contained exclusively or predominately sediment habitat (Fig. 2A) and contained greater numbers of species with a preference for sediment habitats. However, species with a strong affinity for reef were also found in the predominantly sediment-dominated habitat group, with groups 4 and 6 containing disproportionately more reef species than suggested by the available habitat (Fig. 2B). Sediment- associated assemblages (particularly groups 1 and 4) also contained a relatively high proportion of wide-ranging demersal species. Groups 2, 3, and 5 contained moderate to high proportions of mixed reef and reef habitat with the highest proportion of reef-associated species observed in group 3 (Fig. 2A, B).

Species representing a range of trophic levels were recorded in BRUV footage. Carnivores were the most numerous species, as expected given the use of baits. Even so, the average trophic level of species differed between the six assemblage groups ($df = 5$, $F = 5.46$, $P < 0.001$; Fig. 2C). Groups 1 and 4 (sediment groups) contained species with significantly higher trophic level (higher carnivores) than groups 2 and 3 (mixed-reef) which contained primarily carnivores. The remaining two groups (5 and 6) contained species from a relatively wider range of trophic levels (Fig. 2C) with group 5 represented by the largest range in values and the only herbivore observed.

The characteristic species of each assemblage group (making the highest contributions to within-group similarity) re-enforced the broad habitat-related patterns described above (Fig. 3). Sediment-associated species such as flathead (*Platycephalus bassensis*, *Neoplatycephalus richardsoni*) and sharks (*Mustelus antarcticus*, *Squalus megalops*) as well as the common gurnard perch (*Neosebastes scorpaenoides*) and cucumber fish (*Paraulopus nigripinnis*) were characteristic of groups 4 and 6. Group 1 also consisted of sediment-associated fish (eastern- barred grubfish - *Parapercis allporti*- as well as *P. nigripinnis*), but was differentiated from groups 4 and 6 by relatively low abundances and very few species. Of the predominantly mixed-reef associated groups, groups 2 and 3 were both dominated by jackass morwong (*Nemadactylus macropterus*). Group 2 was also composed of a variety of others species including the velvet leatherjacket (*Meuschenia scaber*), reef ocean perch (*Helicolenus percoides*) and draughtboard sharks (*Cephaloscyllium laticeps*). Group 3 was further characterised by relatively high abundances of striped trumpeter (*Latris lineata*) as well as reef ocean perch. Group 5 was differentiated from other groups by large numbers of leatherjackets (primarily *Thamnaconus degeni*), but was also characterised by silverbelly (*Parequula melbournensis*) and jackass morwong.

Relationships with environmental variables

Environmental factors correlated well with the fish assemblage groupings (Fig. 4). Depth delineated groups most strongly (a correlation with CAP axis 2 of 0.78) and groups 1, 4 and 3 were generally found at greater depth. Substratum type also influenced assemblages, with sediment-associated groups (1, 4 and 6) falling on the right hand-side of the CAP plot, confirming the patterns seen in the composition of species above. Latitude was also moderately correlated with axis one (0.35) and appears to separate the mixed/reef associated groups 2, 3 and 5.

Species richness was also correlated with environmental factors. Richness decreased with depth and mixed and reef substratum also supported more species than sediments (Table 3). Latitude however, was not correlated with species richness (Table 3). This resulted in spatial patterns as depicted in Fig. 5, where assemblages are generally richer towards the shallow, western side of the reserve.

Spatial distribution of fish assemblages across CMR

Fish assemblages showed distinct spatial patterns across the Flinders CMR (Fig. 5). When combined with depth, the assemblages can be broadly categorised as: shallow, reef-associated (group 5); intermediate reef-associated (group 2); deeper, reef-associated (group 3); shallow sediment-associated (group 6); and deeper, reef-associated (groups 1 and 4). While sites within the same spatial cluster often belonged to the same fish assemblage, this was not always the case. Clusters of sites close to the shelf edge and in the south of the reserve were often more heterogeneous, composed of two to three different assemblages within a 1 km radius.

Discussion

This study trialled the use of BRUVs and a spatially- balanced survey design to provide the first quantitative description of demersal fish species and assemblages occurring the multiple use zone of the Flinders CMR shelf. Non-extractive sampling with BRUVs identified characteristic species (including jackass morwong, leatherjackets, draughtboard sharks, reef ocean perch and common gurnard perch), several species of commercial interest (including gummy sharks, reef ocean perch, striped trumpeter and jackass morwong) and species from trophic groups ranging from omnivores through to higher predators. While many of the species present in the study area are widely distributed across southern and/or south-eastern Australia, the majority are endemic to Australian waters. This highlights the unique biodiversity represented in temperate Australia, primarily as a result of its relatively isolated evolutionary history [39,40], and reinforces the global importance of conserving representative fauna.

Spatial and environmental patterns in fish assemblages

Six assemblages with a distinct spatial pattern were observed at sites across the reserve. Many species contributed to more than one assemblage and differences in the relative abundance of

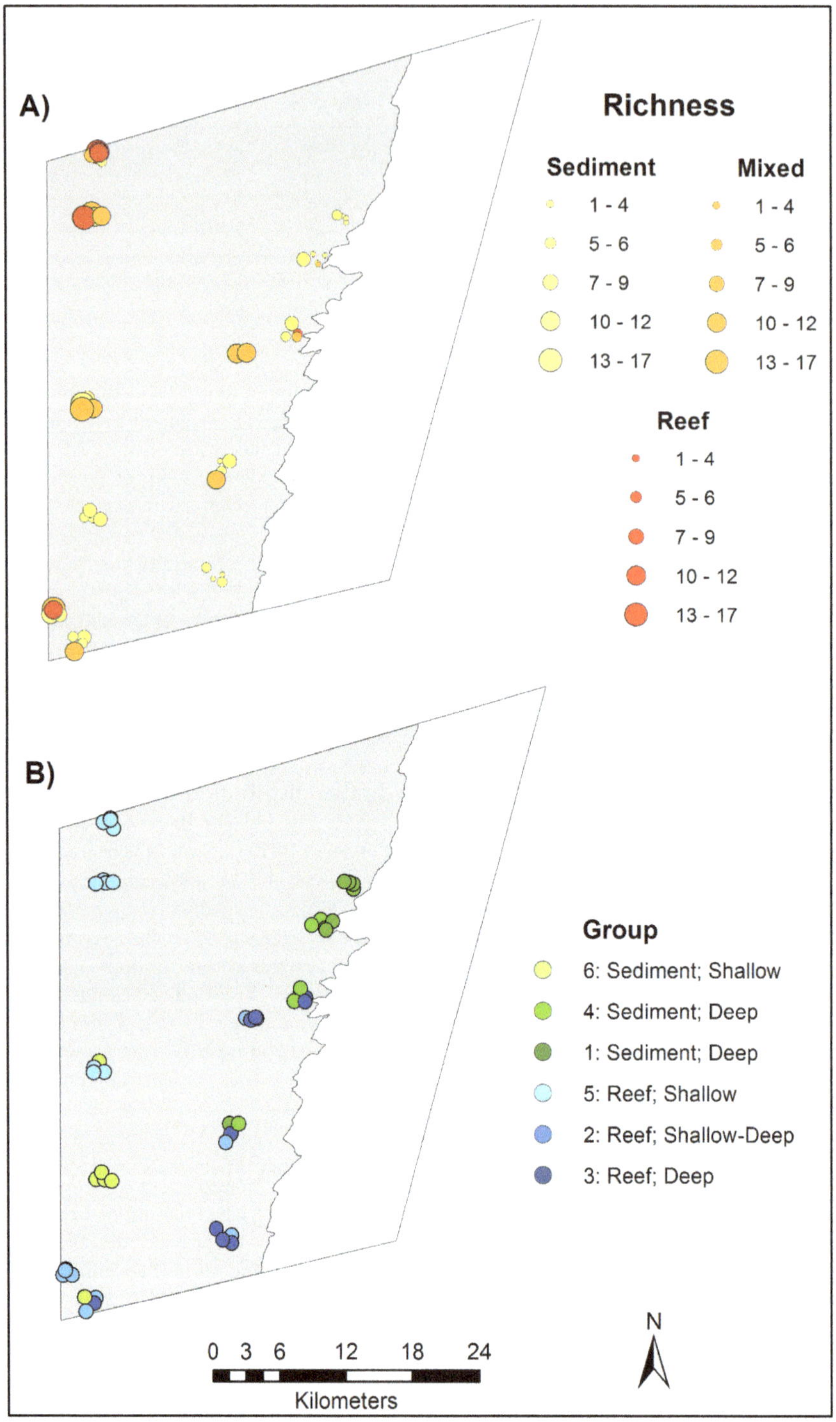

Figure 5. Spatial distribution of A) species richness B) fish assemblage groups on the Flinders CMR shelf (shaded grey). Increasing size of symbols in A) indicate increasing species richness. Symbols are colour coded according to the observed substratum type in BRUV footage: yellow = sediment; orange = mixed; red = reef. Assemblages in B) are coded by colour, with predominantly sediment-associated assemblages coloured green and predominantly reef-associated assemblages coloured blue.

species defined the differences between several assemblages. Assemblages were structured by depth and habitat type, and could be heterogeneous at relatively small scales (i.e. within a sampling cluster).

Depth-related assemblage transitions were most distinct and occurred in waters greater than approximately 80 m for sediment-associated assemblages and approximately 60 m for reef-associated assemblages, which is largely consistent with other studies on the southeast Australian shelf [41]. While some species were found across almost the full depth range of the study area (e.g. jackass morwong and the flathead, *Platycephalus richardsoni*) several species characterised particular substratum and depth zones. Leatherjackets (*T. degeni* in particular) for example, were characteristic of shallow reef sites; striped trumpeter were indicative of deeper, reef-associated assemblages; and cucumber fish were indicative of deeper, sediment-associated assemblages. These results accord with other studies where depth has a major influence on the distribution of fish assemblages [17,42].

As well as influencing assemblage composition, depth also influenced the diversity and ecology of assemblages. Species richness generally declined with depth consistent with observations elsewhere [39,43,44] and the trophic range of species occurring in shallow assemblages was generally greater than deeper assemblages. In reef-associated assemblages for example, larger trophic ranges of the shallow water assemblage were due to the co-occurrence of species such as leatherjackets (many of which are omnivorous) and draughtboard sharks (which are carnivores). Similarly, other studies have noted trophic-related changes with depth, such as increases in body size [42] and a transition towards higher –order predators [34].The combination of assemblage level patterns and changes in ecology and diversity with depth as seen here, have been used to suggest that the edge of the continental shelf forms an important faunal break. Consequently the shelf and slope environments should be considered separate management units [8,31] and not subsumed within the same management zone as currently occurs within the Flinders CMR.

Although a ubiquitous pattern, the processes involved in the depth structuring of benthic fish assemblages can be varied. In the shelf environments of the Flinders CMR, we suggest these are partly attributable to the strong tidal currents which affect the shallow, western section on the reserve [22]. In this region, low-profile reefs are dominated by tall biogenic structures with flexible forms such as sea whips and erect branching sponges (personal observation) that are indicative of high energy environments. This 3D biogenic structure is much reduced towards the deeper, eastern extent of the study area. These differences in reef structure [45], as well as physical limitations posed by increased current velocities [46], may have a large influence on the assemblages found in shallow regions of the reserve. Other physical and oceanographic conditions also co-vary with depth. For example, light availability decreases with depth, which affects primary productivity and therefore the range of food sources available to fish and appears to correspond with the observed changes in the trophic composition of assemblages. In addition, the oceanography in this region is complex, driven largely by inter-annual and inter-seasonal variability in the southward penetration of the warm-water, East Australian Current (EAC). The EAC can incur on the shelf or form fronts at the shelf break [24] and affects processes such as recruitment [47] and primary productivity [48] that are in turn likely influence the composition of fish assemblages, particularly on the outer shelf. Regardless of the proximal processes involved, the depth structuring of assemblages observed in the study area suggests that future sampling would benefit by breaking the region into shallow and deep strata around the 70 m depth mark.

Habitat or substratum type was also influential in discriminating between the assemblages observed in the Flinders CMR shelf. Despite preferentially sampling clusters of sites surrounding our initial mixed/reef sites, a large number of all sites surveyed contained only sediments, at least within the field of view of the BRUVs. This occurred because low profile shelf reefs in the CMR are small scale features (tens of meters) and their distribution is highly patchy. Sediment sites were generally less species rich than those found on mixed or reef substrata. However, the composition of these sites varied enough to form three distinct assemblage groups, primarily on the basis of their depth distribution. The lower diversity of sediment assemblages has been attributed to lower habitat complexity with fewer niches to support co-existing species [49]. Still, sediment-associated assemblages contained some species with affinity for reef habitats. This may be because reef habitat was present outside the BRUVs field of view but within the swimming distance of fishes resulting in a 'halo' effect where species with an affinity for reef and sediments co-occur [49]. In subtropical shallow reefs this halo can extend 200 m from the reef edge [49]. The importance of local substratum in structuring fish assemblages highlights the need to develop comprehensive, high resolution habitat maps for the reserve. Assembling these habitat maps for marine reserves will require a commitment to strategic mapping of the seabed using multibeam sonar, so that over time complete coverage is achieved. For the continental shelf, this mapping could take years to decades [50]. In the meantime, management will need to adopt spatially representative sampling such as presented in this study to provide habitat and biodiversity inventories and to underpin monitoring programs.

Geomorphic features of the seafloor, such as submarine canyons, can substantially alter water flows, enhance local productivity [51] [52] and influence fish assemblage patterns [53]. Canyons are a conservation feature of the Flinders CMR and several canyons intersect the shelf edge. Canyons were represented in our sampling by a cluster of sites near a canyon head in the north of the survey area. Assemblages adjacent to canyon heads were not however, unique to canyon heads and were also found in other deeper sections of the shelf. Neither were these assemblages particularly species rich. Despite our preliminary findings, a better understanding of the ecological significance of canyon heads in this system will require a more targeted and intensive sampling strategy within and around the canyon head systems, e.g. [51].

Implications for Monitoring and Management

An objective of the Flinders CMR – and the larger Southeast Marine Reserve Network of which it is part - is to protect the biodiversity, natural and cultural values of the region. These include "representative examples of the ecosystems, communities and habitats" associated with the Tasmanian Shelf and Southeast Shelf Transition Provinces [21]. While these values are defined in broad terms from the national bioregionalisation, these have yet to be translated into inventories of conservation values, with specific metrics for monitoring their status. This study is a first step towards developing both an inventory of biological values, and elements of an effective biological monitoring program, including relevant environmental influences, for fish communities represented in the shelf component of the Flinders CMR. We suggest that representative demersal fish communities include those found on dynamic, low- profile reefs in both shallow (<70 m) and deep (70–160 m) areas of the shelf as well as those found on sediments in the same depth strata, and here we have characterised each of these assemblages. Several prevalent commercial and recreational species, including striped trumpeter and jackass morwong, which are likely to be responsive to management actions may be good candidates as indicator species and this work will provide the basis for further research into deriving specific monitoring targets.

We used a non-standard sampling design, GRTS, to survey fish assemblages and to satisfy the objectives of the broader survey of the Flinders CMR. Due to the cost and logistics associated with sampling in offshore and/or remote environments, the use of survey designs that are efficient, representative and flexible enough to accommodate multiple objectives is likely to become increasingly important. This is particularly true when simultaneously considering the inventory and monitoring needs of multiple reserves within networks. In the present study, the GRTS sampling ensured that clusters of reef and sediment sites sampled with BRUVs are well spread out across the reserve (within the distribution of each habitat type) and consequently encompass broad spatial and environmental gradients. This in turn provides confidence in the patterns observed and provides a better basis for additional or repeated sampling in a monitoring context. The site-ordered sampling of GRTS has the potential disadvantage of using ship time inefficiently if there is a large cumulative distance between deployments. This would have been the case in our study, but was circumvented by clustering samples around a subset of original sites. Conversely, the site ordering of GRTS allows dynamic altering of samples according to situations in the field, while maintaining spatial balance within strata.

The GRTS methodology will also be useful when choosing samples for future sampling, typically as part of an adaptive monitoring program. The master list of samples, that encompasses all possible sampling locations [28], can be subset in different ways by varying the inclusion probabilities for factors such as habitat types or depth, to accommodate new research questions or changing management objectives. While such changes lead to higher or lower sampling rates for different habitats or strata, the set of sites selected always maintains spatial balance within the strata. Inclusion probabilities are taken into account when deriving representative estimates of the occurrence, proportions or abundance of metrics or indicators across the entire sampling area [18]. Once suitable indicator species and metrics have been selected, the GRTS master sample can provide the basis for rotating panel designs [28] for repeat sampling that aims to detect changes in the abundance of these species, and possibly also major shifts in the distribution of assemblage types. Another advantage of GRTS in this respect, is its use of a local neighbourhood variance estimator, which is generally unbiased and more precise than variance estimators from other types of sampling designs [26]. Future sampling within the context of monitoring however, will need to incorporate comparable areas outside of the multiple use management zone of the Flinders CMR. In addition sufficient sample sizes will be needed to provide enough power to distinguish between natural variability and biological trends in order to be able to assess whether current management controls are effective in achieving their stated objectives.

The use of baited systems to survey fish communities has sometimes been criticised because of the bias towards sampling scavenging and predatory species [54]. However, all fish sampling methods have selectivity biases, and alternatives for sampling deep continental shelf fishes (e.g. trawls, meshnets and traps [41]) are extractive (and some destructive to the habitats the reserve is intended to protect). We recognise that by using BRUVs we have preferentially observed a subset of species that is attracted to, or undeterred by, baits. However, BRUVs data have been shown to clearly discriminate between fish assemblages in a variety of environmental settings [43] and have better statistical power to detect spatial and temporal changes in assemblage structure and abundances of individuals than unbaited methods [54]. In the future, we may complement BRUVs with other non-extractive video techniques (e.g. forward looking cameras on remotely or autonomously operated vehicles) to capture other components of the community. Never-the-less, in our present study BRUVs successfully observed a broad range of species, including key ecological species and potential indicator species. Standardised methodologies, in combination with robust sampling designs, will be essential for ongoing monitoring and assessment across multiple reserves in Australia's reserve network. Finally, the challenges we outline for the inventory and long-term monitoring of large and remote marine reserves are not unique to the Australian reserve network, nor exclusively to marine reserves. Therefore the approach that we have demonstrated may be usefully applied in other systems and under other management scenarios.

Supporting Information

Figure S1 The location of the forty GRTS sites sampled across the Flinders CMR shelf in phase one. Sites are coloured according to the broad habitat type: sediment (yellow); mixed, low-profile reef and sediments (red); canyon head (blue) recorded during phase one of sampling. Sites are labelled with their GRTS site number which, when sampled in order, represents a spatially balanced sample. The table shows the subset of sites sampled as the basis of clusters in the phase two sampling with BRUVs. These sites are the first three GRTS sites classified as sediment and the first eight classified as mixed reef.

Data S1 Matrix of the species observed at each BRUV deployment. Environmental variables used in analyses are also included.

Data S2 Attributes of species used in analysis, including trophic level, habitat preference and endemicity. Summary statistics on the prevalence and abundance of species across the survey area are also given.

Acknowledgments

The authors would like to thank the crew of *Challenger*, Matt Francis and Tim Green, for their expertise and dedication during fieldwork, Richard Zavalas for assistance with fieldwork and video scoring, Johnathan Kool for assistance with fieldwork, and Vanessa Lucieer for assistance in the planning phase. SN publishes with permission of the Chief Executive Officer of Geoscience Australia.

Author Contributions

Conceived and designed the experiments: NAH NB EL JMD SN KRH. Performed the experiments: NAH NB JH SN. Analyzed the data: NAH NB EL JH KRH. Contributed reagents/materials/analysis tools: NAH NB EL JH SN AW KRH. Wrote the paper: NAH NB EL JMD SN AW KRH.

References

1. Halpern BS, Walbridge S, Selkoe KA, Kappel CV, Micheli F, et al. (2008) A global map of human impact on marine ecosystems. Science 319: 948–952.
2. Halpern BS, Longo C, Hardy D, McLeod KL, Samhouri JF, et al. (2012) An index to assess the health and benefits of the global ocean. Nature: 615–620.
3. Day J, Dudley N, Hockings M, Holmes G, Laffoley D, et al. (2012) Guidelines for applying the IUCN Protected Area Management Categories to Marine Protected Areas. Gland, Switzerland: IUCN. 36 p.
4. Kemp J, Gibson R, Atkinson R, Gordon J, Hughes R, et al. (2012) Measuring the performance of spatial management in marine protected areas. Oceanography and Marine Biology: An Annual Review 50: 287–314.
5. Gubbay S, editor (1995) Marine Protected Areas: principles and techniques for management. London: Chapman and Hall.
6. CBD (1992) Convention on Biological Diversity. 5 June 1992 Rio de Janerio (Brazil).
7. CBD (1995) The Jakarta Mandate on Marine and Coastal Biological Diversity. Decisions of the Second Meeting of the Conference of the Parties to the Convention on Biological Diversity. Jakarta, Indonesia, 6–17 November 1995: UNEP.
8. Spalding MD, Meliane IN, Milam A, Fitzgerald C, Hale LZ (2013) Protecting marine spaces: Global targets and changing approaches. Ocean Yearbook Online 27: 213–248.
9. Edgar GJ, Stuart-Smith RD, Willis TJ, Kininmonth S, Baker SC, et al. (2014) Global conservation outcomes depend on marine protected areas with five key features. Nature advance online publication.

10. Barr LM, Possingham HP (2013) Are outcomes matching policy commitments in Australian marine conservation planning? Marine Policy 42: 39–48.
11. ANZECC (1998) Guidelines for establishing the National Representative System of Marine Protected Areas. Canberra: Australian and New Zealand Environment and Conservation Council (ANZECC) Task Force on Marine Protected Areas.
12. Commonwealth of Australia (2006) A Guide to the Integrated Marine and Coastal Regionalisation of Australia Version 4.0. In: Heritage DotEa, editor. Canberra, Australia.
13. Lucieer V, Lamarche G (2011) Unsupervised fuzzy classification and object-based image analysis of multibeam data to map deep water substrates, Cook Strait, New Zealand. Continental Shelf Research 31: 1236–1247.
14. Bailey DM, King NJ, Priede IM (2007) Cameras and carcasses: historical and current methods for using artificial food falls to study deep-water animals. Marine Ecology Progress Series 350: 179–191.
15. Willis TJ, Babcock RC (2000) A baited underwater video system for the determination of relative density of carnivorous reef fish. Marine and Freshwater Research 51: 755–763.
16. Langlois TJ, Harvey ES, Meeuwig JJ (2012) Strong direct and inconsistent indirect effects of fishing found using stereo-video: Testing indicators from fisheries closures. Ecological Indicators 23: 524–534.
17. Harvey ES, Cappo M, Kendrick GA, McLean DL (2013) Coastal fish assemblages reflect geological and oceanographic gradients within an Australian zootone. Plos One 8.
18. Stevens DL, Olsen AR (2004) Spatially balanced sampling of natural resources. Journal of the American Statistical Association 99: 262–278.
19. Olsen AR, Kincaid TM, Payton Q (2012) Spatially balanced survey designs for natural resources. In: Gitzen RA, Millspaugh JJ, Cooper AB, Licht DS, editors. Design and Analysis of Long-term Ecological Monitoring Studies. Cambridge, England: Cambridge University Press. pp. 126–150.
20. Commonwealth of Australia (2006) A Guide to the Integrated Marine and Coastal Regionalisation of Australia Version 4.0. Canberra, Australia: Department of the Environment and Heritage. 16 p.
21. Director of National Parks (2013) South-east Commonwealth Marine Reserves Network management plan 2013–23. Director of National Parks, Canberra.
22. Fandry C (1983) Model for the three-dimensional structure of wind-driven and tidal circulation in Bass Strait. Marine and Freshwater Research 34: 121–141.
23. National Tidal Centre (2014) Tide predictions for Australia, South Pacific and Antarctica.
24. Harris G, Nilsson C, Clementson L, Thomas D (1987) The water masses of the East Coast of Tasmania: Seasonal and interannual variability and the influence on phytoplankton biomass and productivity. Australian Journal of Marine and Freshwater Research 38: 569–590.
25. Whiteway TG (2009) Australian Bathymetry and Topography Grid. Canberra, Australia: Geoscience Australia. 46 p.
26. Stevens DL, Olsen A (2003) Variance estimation for spatially balanced samples of environmental resources. Environmetrics 14: 593–610.
27. Dambacher JM, Jones KK, Larsen DP (2009) Landscape-level sampling for status review of great basin redband trout. North American Journal of Fisheries Management 29: 1091–1105.
28. Larsen D, Olsen A, Stevens D (2008) Using a Master Sample to Integrate Stream Monitoring Programs. Journal of Agricultural, Biological, and Environmental Statistics 13: 243–254.
29. Harvey ES, Newman SJ, McLean DL, Cappo M, Meeuwig JJ, et al. (2012) Comparison of the relative efficiencies of stereo-BRUVs and traps for sampling tropical continental shelf demersal fishes. Fisheries Research 125–126: 108–120.
30. Cappo M, De'ath G, Speare P (2007) Inter-reef vertebrate communities of the Great Barrier Reef Marine Park determined by baited remote underwater video stations. Marine Ecology Progress Series 350: 209–221.
31. Seager J (2014) Transect Measure. 2.30 ed: SeaGIS.
32. Maechler M, Rousseeuw P, Struyf A, Hubert M, Hornik K (2013) cluster: Cluster Analysis Basics and Extensions. R package. 1.14.4 ed.
33. R Development Core Team (2008) R: A language and environment for statistical computing. Vienna, Austria: R Foundation for Statistical Computing.
34. Anderson MJ, Gorley RN, Clarke KR (2008) PERMANOVA+ for PRIMER: Guide to software and statistical methods. Plymouth, UK: PRIMER-E. 214 p.
35. Edgar GJ (2008) Australian Marine Life: the plants and animals of temperate waters. New Holland, Sydney. Sydney: Reed New Holland.
36. Gomon MF, Bray DJ, Kuiter RH (2008) Fishes of Australia's Southern Coast. Chatswood, Australia: Reed New Holland.
37. (2014) FishBase. In: Froese R, Pauly D, editors. Available: www.fishbase.org: World Wide Web electronic publication.
38. Venables WN, Ripley BD (2002) Modern Applied Statistics with S. New York: Springer.
39. Last PR, White WT, Gledhill DC, Pogonoski JJ, Lyne V, et al. (2011) Biogeographical structure and affinities of the marine demersal ichthyofauna of Australia. Journal of Biogeography 38: 1484–1496.
40. Phillips J (2001) Marine macroalgal biodiversity hotspots: why is there high species richness and endemism in southern Australian marine benthic flora? Biodiversity and Conservation 10: 1555–1577.
41. Williams A, Bax N (2001) Delineating fish-habitat associations for spatially based management: an example from the south-eastern Australian continental shelf. Marine and Freshwater Research 52: 513–536.
42. Fitzpatrick BM, Harvey ES, Heyward AJ, Twiggs EJ, Colquhoun J (2012) Habitat specialization in tropical continental shelf demersal fish assemblages. PLoS ONE 7: e39634.
43. Zintzen V, Anderson MJ, Roberts CD, Harvey ES, Stewart AL, et al. (2012) Diversity and composition of demersal fishes along a depth gradient assessed by Baited Remote Underwater Stereo-Video. PLoS ONE 7: e48522.
44. Chatfield BS, Van Niel KP, Kendrick GA, Harvey ES (2010) Combining environmental gradients to explain and predict the structure of demersal fish distributions. Journal of Biogeography 37: 593–605.
45. Gratwicke B, Speight MR (2005) The relationship between fish species richness, abundance and habitat complexity in a range of shallow tropical marine habitats. Journal of Fish Biology 66: 650–667.
46. Fulton CJ, Bellwood DR (2004) Wave exposure, swimming performance, and the structure of tropical and temperate reef fish assemblages. Marine Biology 144: 429–437.
47. Bruce BD, Evans K, Sutton CA, Young JW, Furlani DM (2001) Influence of mesoscale oceanographic processes on larval distribution and stock structure in jackass morwaong (*Nemodactlus macropertus:* Cheilodactylidae). ICES Journal of Marine Science 58: 1072–1080.
48. Bax NJ, Burford M, Clementson L, Davenport S (2001) Phytoplankton blooms and production sources on the south-east Australian continental shelf. Marine and Freshwater Research 52: 451–462.
49. Schultz AL, Malcolm HA, Bucher DJ, Smith SDA (2012) Effects of reef proximity on the structure of fish assemblages of unconsolidated substrata. Plos One 7.
50. Kloser RJ, Williams A, Butler A (2006) Exploring surveys of seabed habitats in Australia's deep ocean using remote sensing – needs and realities. In: Todd BJ, Greene HG, editors. Mapping the Seafloor for Habitat Characterization. pp. 93–109.
51. Currie DR, McClatchie S, Middleton JF, Nayar S (2012) Biophysical factors affecting the distribution of demersal fish around the head of a submarine canyon off the Bonney Coast, South Australia. Plos One 7.
52. Vetter EW, Smith CR, De Leo FC (2010) Hawaiian hotspots: enhanced megafaunal abundance and diversity in submarine canyons on the oceanic islands of Hawaii. Marine Ecology 31: 183–199.
53. Leathwick JR, Elith J, Francis MP, Hastie T, Taylor P (2006) Variation in demersal fish species richness in the oceans surrounding New Zealand: an analysis using boosted regression trees. Marine Ecology Progress Series 321: 267–281.
54. Harvey E, Cappo M, Butle RJ, Hall N, Kendrick G (2007) Bait attraction affects the performance of remote underwater video stations in assessment of demersal fish community structure. Marine Ecology Progress Series 350: 245–254.

9

Predicting Vulnerabilities of North American Shorebirds to Climate Change

Hector Galbraith[1,4], David W. DesRochers[2], Stephen Brown[1], J. Michael Reed[3]*

1 Manomet Center for Conservation Sciences, Manomet, Massachusetts, United States of America, **2** Department of Natural Sciences, Dalton State College, Dalton, Georgia, United States of America, **3** Department of Biology, Tufts University, Medford, Massachusetts, United States of America, **4** National Wildlife Federation, Springfield, Massachusetts, United States of America

Abstract

Despite an increase in conservation efforts for shorebirds, there are widespread declines of many species of North American shorebirds. We wanted to know whether these declines would be exacerbated by climate change, and whether relatively secure species might become at-risk species. Virtually all of the shorebird species breeding in the USA and Canada are migratory, which means climate change could affect extinction risk via changes on the breeding, wintering, and/or migratory refueling grounds, and that ecological synchronicities could be disrupted at multiple sites. To predict the effects of climate change on shorebird extinction risks, we created a categorical risk model complementary to that used by Partners-in-Flight and the U.S. Shorebird Conservation Plan. The model is based on anticipated changes in breeding, migration, and wintering habitat, degree of dependence on ecological synchronicities, migration distance, and degree of specialization on breeding, migration, or wintering habitat. We evaluated 49 species, and for 3 species we evaluated 2 distinct populations each, and found that 47 (90%) taxa are predicted to experience an increase in risk of extinction. No species was reclassified into a lower-risk category, although 6 species had at least one risk factor decrease in association with climate change. The number of species that changed risk categories in our assessment is sensitive to how much of an effect of climate change is required to cause the shift, but even at its least sensitive, 20 species were at the highest risk category for extinction. Based on our results it appears that shorebirds are likely to be highly vulnerable to climate change. Finally, we discuss both how our approach can be integrated with existing risk assessments and potential future directions for predicting change in extinction risk due to climate change.

Editor: Grant Ballard, Point Blue Conservation Science, United States of America

Funding: This study was internally funded. Each of the authors were supported primarily by their employers; Galbraith's time was supported by Manomet Center for Conservation Sciences and by the National Wildlife Federation; DesRochers was supported by Dalton State College; Brown was supported by Manomet Center for Conservation Sciences; and Reed was supported by Tufts University and by Manomet Center for Conservation Sciences. The funders had no role in study design, data collection and analysis, decision to publish, or preparation of the manuscript.

Competing Interests: The authors have declared that no competing interests exist.

* Email: michael.reed@tufts.edu

Introduction

Shorebirds are important components of the ecosystems in which they live, they are valued by the general public, can exhibit extremely large and impressive aggregations during migration, and they can act as sentinels of global environmental change [1–3]. There also is a growing demand to move beyond evaluating climate change impacts on single species or habitats and to evaluate expected broad scale ecological impacts on communities and ecosystems [4]. Consequently, we are concerned about the current documented widespread declines of many species of North American shorebirds [1,5–8], particularly the recent steep declines in Atlantic populations of Red Knots (scientific names of North American shorebirds are given below) [9–10] and Semipalmated Sandpipers [11–12].

The U.S. Fish and Wildlife Service currently lists three North American shorebirds as Threatened or Endangered [13]. IUCN lists five shorebird species in North America as Near Threatened or at higher risk, and four additional species in these categories for the Western Hemisphere [14]. The causes of these declines are not well understood but most likely include loss of breeding, migration, and wintering habitats, and disturbance and exploitation [1,15–17]. It should be recognized, however, that the factors causing such changes could be global, since population reductions have been seen in virtually all shorebird flyways from North and South America, to East Africa, to Asia and Australia, e.g., [18–19].

Global climate change is an anthropogenic stressor that could adversely affect shorebird populations across species' ranges. Shorebirds that breed and/or winter at high latitudes may be among the most sensitive of bird species to this stressor because this is where climate change is expected to be most severe [20]. They also have several additional risk factors, including lengthy, energetically expensive migrations where they may be vulnerable to changes in wind patterns, dependence upon coastal migration stopover sites that are vulnerable to sea level rise, and dependence upon ecological synchronicities that may be disrupted by a changing climate [16,21–23]. Small-Lorenz *et al.* [24] point out that assessments of vulnerability to climate change often ignore problems associated with a migratory life–history, causing them to underestimate vulnerabilities. Shorebirds are already in a vulnerable condition and climate change may exacerbate this.

If we are to understand what may happen to shorebirds in the near future and initiate appropriate conservation measures it is essential that we be able to predict the likely vulnerabilities of shorebird species to various aspects of the changing climate, cf. [25]. To be useful for conservation, predictive frameworks should be based on the ecologies and life histories of the species, should incorporate what we know about how the planet's climate will alter, and should generate at least qualitative estimates of species vulnerabilities, e.g. [26–27].

Categorizing vulnerability to extinction based on a suite of characteristics, such as population size and rate of decline, is used widely e.g. [28–31]. The best known models are those of Partners in Flight and IUCN (also used by BirdLife International) [32]. Their categorization approach to vulnerability also can be used to evaluate species' changes in vulnerability as ecosystems change over time, e.g. [33]. Partners in Flight (PIF) uses a model to assess vulnerability based on population trend, relative abundance, threats during the breeding and non–breeding seasons, and breeding and non–breeding range sizes. For each category, each species receives a score of 1 to 5, with 5 associated with greatest risk. These scores are summed using several different formulas, each of which is used to determine species of conservation concern for particular reasons. A similar system was developed based on the same set of basic variables for the U.S. Shorebird Conservation Plan [1], although the resulting risk categories are defined somewhat differently. None of these systems includes risk due to climate change. In this paper, it was our overarching goal to determine the degree to which climate change will alter the extinction risk level assigned to shorebird species in the U.S. Shorebird Conservation Plan, and for this method to be compatible with the PIF ranking system.

We approached our reconsideration of risks under climate change by developing an assessment framework, and then used it to evaluate the vulnerabilities to climate change of North American (north of Mexico) shorebird species, whose life histories extend across wide ecological and behavioral spectra. Specifically, we (1) identified risk factors, (2) created a framework for quantifying the change in risk due to climate change for each of the factors, including the possibility of decreased extinction risk due to climate change, (3) identified the effects climate change would have on the risk factors, (4) reviewed the literature on each shorebird species we assessed to determine species–specific risk for each factor, and (5) assigned species to their new extinction–risk categories. We also (6) did a sensitivity analysis to determine how the results were affected by different decision rules for changing PIF risk categories.

Methods

We included 49 species in our assessment. For three species (Willet, Piping and Snowy Plover) we evaluated two distinct populations each, so in all 52 taxa were evaluated. We excluded Eskimo Curlew (*Numenius borealis*) from our analysis because it is likely extinct [34]. Our assessments are compatible with both the U.S. Shorebird Conservation Plan and PIF frameworks, although because of the increased risk to some species already at the highest risk categories, we needed to add a new risk category – critical – to distinguish species at greatly increased risk.

To achieve the goal of creating a framework that could be integrated with both PIF and the U.S. Shorebird Conservation Plan, we first evaluated other existing approaches. The State of the Birds [35] developed a framework to assess changes in risk due to climate change, with the goal of applying it to all bird species. They included migration distance and timing as bivariate factors (birds that migrate long distances and use daylight cues = 1; else 0); degree of breeding habitat obligation (high = 1 vs. not = 0); dispersal ability (1 vs. 0); niche specificity (1, 0); reproductive potential (lays one egg per year = 1, else 0); and habitat susceptibility (divided into 3 levels, 2, 1, 0, from highest to lowest susceptibility). Scores were summed to assess overall risk. This approach apparently ignores risks associated with migration and wintering habitat obligation, does not allow for extinction risk to decrease due to climate change, and there is heavy weighting of reproductive potential, which is evaluated on a narrow scale that distinguishes only between one–egg clutches and all other clutch sizes. Also, while reproductive potential may be important for population size recovery following sudden decline, it may be less important with respect to gradual climate change. This approach is applicable to other species included in the PIF prioritization system.

We included six factors in our risk framework, each of which had 3–5 risk levels. Factors included: expected losses or gains in (1) breeding, (2) migration, and (3) non–breeding habitat (4) degree of dependence on ecological synchronicities; (5) migration distance; and (6) degree of habitat specialization (on breeding, migration, and non–breeding grounds). All risk factors were given equal weight in the assessment, and each factor is described in detail below.

Expected Losses or Gains in Breeding, Migration, and Non–breeding Habitat (1–3)

We accepted that the atmospheric concentrations of greenhouse gases will approximately double (over pre–industrial levels) by the middle to the end of the century [36–37]. We then summarized, based on current understanding reported in the literature, the effects climatic change should have on habitats used by our focal shorebirds in the western hemisphere. What follows is our assessment of these changes (designated B1–B5), and brief statements about our confidence in these changes. These approximate confidence levels of >70%, 30–70%, and <30% are modified from the 5–category scale developed by [38] for the Intergovernmental Panel on Climate Change Third Assessment Report. We reduced the number of categories because we did not think the implied precision of 5 levels of confidence was defensible.

(B1) Northern hemispheric boreal and arctic areas. Tundra habitat will be reduced in extent as the tree line moves poleward; areas that persist as tundra will become less dominated by graminoids and other low–growth species and will become increasingly dominated by more shrubby species, reducing the habitat value for breeding shorebirds [39–47]. Also, the boreal forest will extend its range northward as it replaces tundra, but its southern distribution will contract northwards [37–44]. Although it is true that new areas of bare ground are likely to be created by ice cap and glacial recession in high tundra areas, we do not believe that this will result in more habitat for most breeding shorebirds since it will persist as gravel or bouldery moraine for a long period until vegetated and soil–forming processes can occur. Confidence = medium.

Changes in precipitation and evapotranspiration are also likely, but the aggregate effects on tundra hydrology are difficult to predict [48]. Drier overall conditions may be likely, and may reduce food availability during the breeding season [48]. It is unclear how climate change will affect the water balance on tundra breeding habitats due to the complex interaction of several factors, including amounts and timing of precipitation events, timing and extent of spring thaw, depth of the active layer, and erosion events [48–53]. While annual rainfall is predicted to increase throughout the breeding range, evapotranspiration is also expected to increase enough to more than offset the effect of increased precipitation [36]. The result may be a loss of some wetland breeding habitat to dryer conditions, but this is unclear. Confidence = medium.

(B2) North American Great Plains. Much of the climate modeling that has been performed indicates that these interior grassland regions will become hotter and drier [44,54–57]. This is likely to result in adverse impacts to shorebird species that depend on seasonally or permanently flooded wetlands for their migration stopovers. Confidence = medium

(B3) Coastal habitats. Based on IPCC [36] and more recent modeling [58–60] we assume that sea levels will rise globally by between 1 and 2 meters, resulting in the loss of coastal shorebird habitats. This applies to North, Central and South America [61], and will be worst in areas with, for example, high tidal amplitudes in shallow lagoons and broad estuaries [62–65]. Consequently, we anticipate major loss of coastal wintering habitat for shorebirds, particularly in areas where the land surface is subsiding or accretion rates of intertidal habitats are low (e.g., most Gulf Coast sites) [66–67]. If coastal habitats are able to move inland in response to sea level rise, it could offset losses, but at many sites this will be precluded by human infrastructure and interventions [21,65,68–69]. Confidence = high

(B4) Interior South America. Ecological modeling based on climate change models indicates that increased aridification in South America will have the following effects: first it is likely to result in the replacement of currently forested areas in the Amazon by savanna habitat and seasonal forests [70–71]. Experimental droughts in the eastern Brazilian Amazon resulted in increased tree mortality, which also supports the expectation of declining rainforest habitat [72]. This is unlikely to benefit shorebirds as few use the existing savanna habitats in central South America. Second, the existing grassland areas in central and southern South America will become drier [36,73], but the effect on the grassland habitats on which North American shorebirds currently winter is uncertain. Confidence = low.

(B5) Eastern North American forests. The only North American shorebird species that primarily uses temperate forest habitat for breeding is the American Woodcock. The species prefers young forest with openings, and the species tolerates a wide a range of tree species [74]. In much of the woodcock's range afforestation is occurring due to ecological succession resulting from abandonment of historical agricultural areas [75]. As a result, young forests adjacent to fields or containing areas of open habitat are declining, resulting in loss of required breeding habitat. Additionally, climate change likely will result in increased vegetation growth at higher latitudes in North America [38–41]. This will result in the establishment of more woody vegetation and a subsequent increase in young forest habitat in the north. It is unclear if northward expansion of the woodcock's range is occurring, so changes in forest landscape may outpace range expansion. Another potential concern for forest breeding habitat is climate change's impact on tree mortality. There is growing evidence that drought resulting from climate change leads to increased tree mortality [76–78]. This may open breeding areas for American Woodcock locally, but widespread forest loss could result in loss of breeding habitat. Confidence = Medium.

(B6) Ocean. One of the primary mechanisms through which climate change could impact oceanic habitats is through acidification [79–80]. This likely will reduce the quality of marine habitats, but the extent to which this might affect pelagic non–breeding shorebirds is uncertain [36,81–82]. One hypothesis is that ocean acidification could reduce the fitness of many plankton species by reducing calcification and other physiological processes [83–84]. If ocean acidification does negatively impact marine plankton food resources, the decrease could be offset, however, by increased ocean upwelling which could function to increase food resources [85]. Confidence = low.

Ecological Synchronicities (4)

We recognize two types of ecological synchronicities important to shorebirds that we think could be affected by climate change.

(ES1) Breeding season food resources. Arctic temperatures are rising and are projected to further increase in the future, resulting in earlier spring thaws and ice melts [36]. This likely will result in earlier invertebrate hatches because arctic invertebrate emergence is temperature dependent [86]. Long–term field observations and recent experimental warming studies of arctic plots support this hypothesis [87–88]. If birds are unable to alter migration timing, then arctic nesting shorebirds may have insufficient food resources for young.

(ES2) Migration food resources. Some migrants depend on highly seasonal food sources during migration [89]. For example, shorebirds such as Ruddy Turnstones, Red Knots, Sanderlings, and Semipalmated Sandpipers are highly reliant on American Horseshoe Crab (*Limulus polyphemus*) eggs for refueling during northward migration stopovers [90–91]. If climate change affects timing of horseshoe crab breeding, this would disrupt synchronicity between horseshoe crab egg laying and spring migration.

Migration Distance (5)

We treat migration distance as others have, as a surrogate for things that can go wrong that have not been captured by other factors [31,35,92]. The assumption is that the farther a species has to migrate, the more ecological disruption can occur [92–94]. In the context of climate change, for example, migratory connectivity interacts with habitat loss from sea level rise [95–96] and species may encounter more severe weather during migration [97–98]. Our separation of species into distance categories was done by looking for natural breaks in the migration distance data, resulting in distances being divided into 5 categories (Fig. 1). Migration distances were calculated from the approximate center of each species' breeding range to the approximate center of each species' wintering range using data from NatureServe [99]. The two exceptions were Bristle–thighed Curlew and Bar–tailed Godwit, which are not covered by this database. Known migration distances placed these species in the greatest–distance category.

Degree of Habitat Specialization (6)

This variable refers to degree of specialization to a certain habitat type, rather than the vulnerability of the habitat type. We assert that being specialized increases your extinction risk to climate change because of reduced response capability. If a species specializes on a habitat type at any time in its life cycle (breeding, migration, non–breeding), it was considered to be specialized. We divided this risk factor into three categories (Table 1).

Assessment Framework Development

Each risk factor was assessed for each species using information from the literature regarding the natural history of the species and anticipated changes due to climate change. A summary of each species' risk level associated with climate change for each risk factor narrative, as well as confidence scores can be found in Appendices S1 and S2. For each risk factor, for each species, we

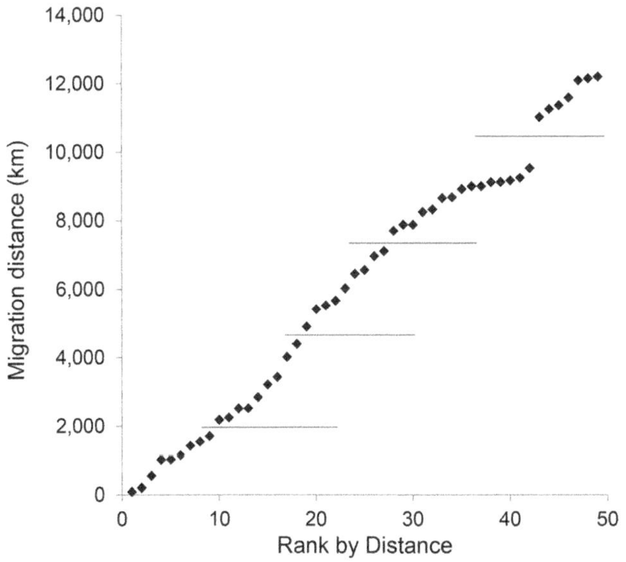

Figure 1. One-way migration distances calculated as mid-point to mid-point of their summer and winter geographic ranges. Ranges were downloaded from the NatureServe database. Horizontal lines separate dispersal distances as ranked in Table 1, with the shortest distances associated with rank 1 and the greatest distances with rank 5. The exceptions are the Bristle-thighed Curlew and Bar-tailed Godwit, which do not overwinter in the New World so they are not covered by the database. They fall into the greatest migration distance category, and are represented arbitrarily in the figure by the 2 points showing the greatest migration distances.

also included a subjective confidence score (1 = low to 5 = high confidence). We recognize that a species might have increased extinction risk due to climate change, but it might not increase enough to change risk categories.

We described changes in risk using two systems: a numeric scoring system that had maximum values for each factor of 5, and a graphical depiction of the change in risk using arrows because we thought they were more intuitive for rapid visual assessment of changes and patterns. Risk factors were scaled from 0 to 5 to match PIF scaling. For the three habitat factors (1–3), we allowed for the possibility of improved conditions due to climate change. Improvement resulted in negative scores (or down-arrows) to show reduced risk. The factors, and their subdivision and scoring, are shown in Table 1.

For our purposes, we decided that an increase in risk score of 10 (equivalent to 4 ↑ s; the arrows indicate the direction and degree of effect) was sufficient to increase by a single risk category because a score of 10 would mean that a species is at extreme risk in two of the six categories. This assignment is a first approximation based on best professional judgment cf. [32], but should be revisited as more information about shorebird ecology and vulnerability to habitat changes becomes available.

To investigate the importance of our decision for how much change in risk is sufficient to cause a change in risk category, we did a sensitivity analysis. Specifically, we assessed the sensitivity of our results – which species were placed into which risk category – to the amount of change in extinction risk that was required for a species to change risk categories. We did this by making the criterion for changing categories more sensitive, requiring the accumulation of only 3 arrows to make the transition between risk categories. We also evaluated the effect of making the criterion less sensitive, evaluating the effects of requiring 5, 6 and 7 arrows to allow a species to change risk categories. If our method is insensitive to this criterion, we would expect little change in categorization with changing criteria.

Results

Each species' account and changes in risk level are found in Appendices S1 and S2, but we briefly go through the account for the Semipalmated Sandpiper to demonstrate the procedure. (1) We anticipate moderate loss of breeding habitat (score 3; 1 ↑). Our reasoning is based largely on the expectation that tundra breeding habitat will be reduced over the longer term by the increase of woody vegetation, which will invade current areas of tundra [43]. Additional impacts may also occur from changes in precipitation, but it is unclear how climate change will affect the water balance on tundra breeding habitats due to the complex interaction of several factors, including amounts and timing of precipitation, timing of spring thaw, and depth of the active layer [48]. While annual rainfall is predicted to increase throughout the breeding range, evapotranspiration is also expected to increase enough to more than offset the effect of increased precipitation. The result may be a loss of some wetland breeding habitat to dryer conditions, but this is unclear. Our confidence in the assessment of the overall score for moderate loss of breeding habitat is low. (2) We anticipate major loss of wintering habitat (score 5; 2 ↑ s) because winter range includes almost exclusively coastal shoreline habitat, so sea level rise (SLR), storm surges, and changing fresh-salt water mixes pose a large threat. Since the species uses estuaries with large tidal amplitudes in Brazil, this may buffer against the SLR impacts, at least locally. Our confidence in this estimate is high. (3) We anticipate moderate loss of migration habitat (score 3; 1 ↑) because SLR likely will cause the loss of some coastal migratory areas. Expected decrease in rainfall in southern areas of North America will cause a decrease in spring migration habitat. In contrast, rainfall is expected to increase in northern portions of North America during spring migration, likely resulting in increased habitat in the interior. Our confidence in this estimate is high. (4) This species has a high degree of dependence on ecological synchronicities (score 5; 2 ↑ s). Arctic temperatures are expected to increase, resulting in earlier spring thaws and ice melts. This, in turn, will likely result in earlier invertebrate emergence. If birds are unable to alter migration timing, then arctic nesting shorebirds may have insufficient food resources to support reproduction. Our confidence in this estimate is high. (5) Migration distance is 7886 km (score 4; 2 ↑ s). (6) We categorize this species as being moderately specialized in its habitat use (score 4; 2 ↑ s). It has fairly specific wintering habitat requirements, including shorelines with wide intertidal mudflats, near shallow lagoons, and wide estuaries with large tidal amplitudes. Our confidence in this estimate is high. This assessment generates a total score of 24 (9 ↑ s), which is enough in our protocol to push the species up two risk categories from its place in the current U.S. Shorebird Conservation Plan, from a species of Moderate Concern to Highly Imperiled.

Of the 52 taxa we evaluated, 45 (87%) are predicted to qualitatively increase their risks of extinction as a result of climate change; 33 by one level in the U.S. Shorebird Conservation Plan, and 12 by 2 levels (Table 2, Fig. 2). Only three species had risk factors that we predict will lower a species' extinction risk due to climate change: Solitary Sandpiper, due to the creation of more breeding habitat; Bristle-thighed Curlew, due to the expansion of breeding and wintering habitat; and White-rumped Sandpiper, due to more wintering habitat. The U.S. Shorebird Conservation Plan currently lists 29 species at risk levels of High Concern or

Table 1. List of risk factors evaluated for species sensitivity to climate change.

1) Loss/gain in breeding habitat under climate change:	**Score**	**Arrow**
Major losses (>50%)	5	↑↑
Moderate losses (10–50%)	3	↑
Limited or no losses (−10–10%)	0	0
Moderate increase (10–50%)	−3	↓
Major increase (>50%)	−5	↓↓
2) Loss/gain in wintering habitat under climate change:		
Major losses (>50%)	5	↑↑
Moderate losses (10–50%)	3	↑
Limited or no losses (−10–10%)	0	0
Moderate increase (10–50%)	−3	↓
Major increase (>50%)	−5	↓↓
3) Loss/gain in migration habitat under climate change:		
Major losses (>50%)	5	↑↑
Moderate losses (10–50%)	3	↑
Limited or no losses (−10–10%)	0	0
Moderate increas (10–50%)	−3	↓
Major increase (>50%)	−5	↓↓
4) Degree of dependence on ecological synchronicities:		
High	5	↑↑
Moderate	3	↑
Low	0	0
5) Migration distance (= surrogate for a suite of issues):		
see figure 1 for distance categories	5	↑↑
	4	↑
	3	↑
	2	0
	1	0
6) Degree breeding, wintering or migration habitat specialization		
Highly specialized	5	↑↑
Specialized	4	↑↑
Somewhat specialized	3	↑
Not specialized	0	0

Values are given by scores (similar to the PIF approach) and by arrows. Note that negative scores/down arrows indicate a decreased extinction risk due to climate change. Our current assessment is based on 4 arrows in the same direction (up or down) being sufficient to shift a species to the next risk category.

higher, and no species are considered Not at Risk. Based on our assessments, we categorize 43 taxa (species+races, hereafter 'species' or 'taxa') as High Concern or higher due to increased risks resulting from climate change, with 15 of these being in the newly created Critical category (Table 2).

Of the 52 taxa assessed, 38 (73%) showed increased vulnerabilities due to effects of climate change on breeding habitat, 36 (69%) due to effects on wintering habitat, and 34 (65%) due to migration habitat (Table 2). More taxa also exhibited maximal negative responses (criteria in Table 1) to climate change on the breeding grounds than to winter or migration habitat (24 taxa vs. 19 and 16, respectively). That is, more taxa exhibited increased risk due to climate change on the breeding grounds than for the wintering and migration grounds, and the risks were higher. The number of taxa predicted to have no response or a positive response to climate change was similar across breeding, winter, and migration habitat (13, 15, and 18 taxa respectively). Ecological synchronicity and migration distance, by comparison, had less of an effect on extinction risk due to climate change, with 17 (33%) and 14 (27%) species, respectively, showing no negative effect due to climate change. The greatest risk factor of those assessed, however, was degree of habitat specialization, with 47 (90%) of the taxa showing a negative response to climate change (Table 2).

A natural potential comparison of our results is with those of the State of the Birds [35]. This is a somewhat difficult comparison to make, however, because we used different scales for our risk categories. However, there appears to be general, qualitative concordance for many species. For example, of the 12 species where they predict no (0 score out of 5) or a low (1 score) increase of extinction risk due to climate change, we predict no or low effects on all of them; i.e., our results leave the species in the same risk category or increase by one category (Table 2). However, we predict an increase of only a single risk category on an additional 19 species where State of the Birds predicts greater impacts of

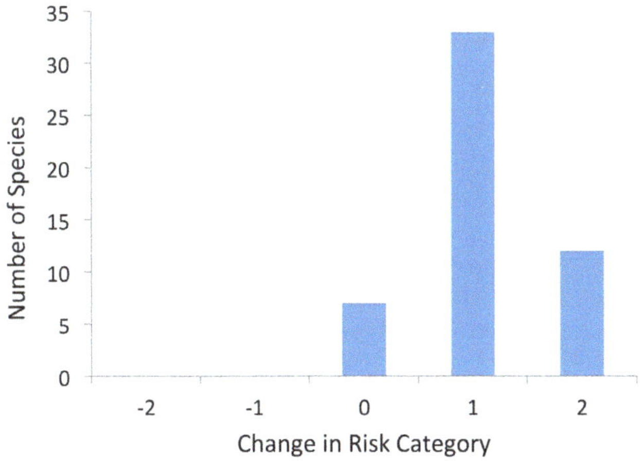

Figure 2. Number of species that we predict will not change U.S. Shorebird Conservation Plan Risk Categories due to climate change (0), and the number that will have increased risk of extinction (positive values); we predicted no species to have reduced risk (negative values). Data are summarized from Table 2 (differences between last two columns).

climate change (scores of ≥2). Although there is a lot of variability, our results are generally, but not closely, consistent with those of the State of the Birds ($r^2 = 0.27$). Our biggest difference occurs for the Purple Sandpiper, where we predict no change in risk category due to climate change, while State of the Birds predicts a strong response (score = 4). Although not to the same degree, we also predict substantially lower increases in extinction risk due to climate change for Black Oystercatcher, Wandering Tattler, Bristle–thighed Curlew, Hudsonian Godwit, Surfbird, Western Sandpiper, and Rock Sandpiper (Table 2).

The number of species that change risk categories in our assessment was sensitive to how much of an effect of climate change is required to cause the shift (Table 3; Appendices S3 and S4). When we make it easier to shift categories (3 arrows to change), we are left with only five species in the moderate or lower concern categories and 22 species in the highest (newly created 'critical') risk category, compared to 9 and 15, respectively, when 4 arrows are required to change categories. There is less sensitivity in the other direction. Even when we require 7 arrows to change risk category, we still have 20 species in the highly imperiled or critical risk categories, compared to only 6 when climate change is not considered (Table 3; Appendices S3 and S4). Consequently, one might argue about the most appropriate degree of increased risk required to change risk categories; however, regardless of the threshold used, we conclude that there is an important shift in the numbers of North American shorebirds species at risk of extinction due to climate change.

Discussion

Many species of shorebirds are the focus of conservation efforts aimed at reversing population declines e.g. [9,100], so there is a need to prioritize conservation actions that can have the largest impact on the species most in need. The system currently in use for prioritizing shorebird conservation efforts in the United States was developed in 1999–2001 [1], and did not explicitly include vulnerability to the impacts of a changing climate, e.g. [101]. Many studies have shown that climate change poses risks to populations of plants and animals and that impacts to vulnerable species are already occurring, e.g. [102–104]. It is expected that such adverse impacts will become more severe and widespread in the future as the climate continues to change. One major application of the system developed in the present study would be to revise the priority scores given to shorebird species by updating the threat scores with the information presented here regarding vulnerability to climate change. We recommend that the U.S. Fish and Wildlife Service revise shorebird priority scores as suggested here, so that the impacts of a changing climate can be more fully integrated into efforts to conserve shorebirds. In addition to applying this information to shorebird species, the same approach could also be applied to other birds. The Partners–in–Flight prioritization system also could be updated to include the approach presented here, if the information on relative risks were collected for other species. This would allow a similar update to reflect vulnerability to climate change across a wide range of bird taxa. We do note that the species assessments and criteria assigned in this manuscript should be considered as first approximations, and will undoubtedly be revised with further discussion by a wider audience. Our primary goal was to establish a system for evaluating the increased risk to species from climate change with respect to existing threat assessments, and to start a discussion about the appropriate values for various species.

Shorebird populations and flyways across the planet are currently being affected by other stressors, many of them unknown, in addition to climate change, e.g., [5]. These impacts are resulting in severe population reductions [1,6–8]. Based on our analyses, adding the stresses and risks imposed by a changing climate to this already threatened baseline renders shorebirds even more vulnerable to extinction. If we are correctly to understand the risks to which shorebirds are exposed, and to identify and implement effective conservation strategies and actions, it is important that we understand these vulnerabilities, particularly those that will occur due to climate change. The purpose of this study was to assess the climate change risks to shorebirds and incorporate these into existing vulnerability evaluations so that we gain a better understanding of the entire panoply of risk factors to which these species are exposed, and their resulting overall vulnerabilities.

Based on our results it appears that shorebirds, as a group, are likely to be highly vulnerable to the changing climate. These vulnerabilities are due to a number of factors. First, many species breed, migrate through, or winter in areas that are likely to be severely impacted by climate change (particularly arctic tundra, coastal breeding, and wintering, and migration stopover sites). Second, the extensive migrations that many of them undertake expose them to risks of changing weather patterns (increased frequencies and intensities of hurricanes, for example) [98]. Shorebirds that require particular staging areas might be more vulnerable to climate change than are those species using stopover sites [95,105]. Lastly, the ecological synchronicities that many shorebirds depend on (e.g., the complementary timing of the arctic snowmelt and invertebrate prey availability) might suffer disruptions [16,21–24]. Our results reflect these vulnerabilities.

Of the 52 shorebird taxa (49 species, 3 split into 2 populations) that breed in North America and that we evaluated, 45 (87%) were predicted to exhibit an increased extinction risk when the risks posed by climate change were added to their current vulnerabilities as estimated in the U.S. Shorebird Conservation Plan [1]. No species was reclassified into a lower–risk category, although prior to the analysis it had been a possibility. The factors responsible for these increased vulnerabilities were risks of: loss of breeding habitat (particularly for arctic– and coastal breeders); loss of coastal and inland migration stopover habitats due to sea level rise and drought; and loss of coastal wintering habitat due to sea level

Table 2. Results of predicted change in extinction risk to shorebird species based on climate change.

Common name	Scientific name	Breeding habitat	Wintering habitat	Migration habitat	Ecological synchronicity	Migration distance	Degree of habitat specialization	Sum of change (arrows)	State of the Birds 2010[1]	Current risk category	Revised risk category[2]
Black-necked Stilt	*Himantopus mexicanus*	↑↑	↑	↑	0	0	↑↑	6	0	2	3
American Avocet	*Recurvirostra americana*	↑↑	↑↑	↑↑	0	0	↑	7	1	3	4
American Oystercatcher	*Haematopus palliatus*	↑↑	↑↑	↑↑	0	0	↑↑	8	4	4	6
Black Oystercatcher	*Haematopus bachmani*	↑	↑	↑	0	0	↑↑	5	4	4	5
Black-bellied Plover	*Pluvialis squatarola*	↑↑	↑	↑	↑	↑	↑↑	8	3	3	5
American Golden-Plover	*Pluvialis dominica*	↑↑	↑	↑	↑	↑↑	↑	8	2	4	6
Pacific Golden-Plover	*Pluvialis fulva*	↑↑	0	0	↑	0	↑↑	5	3	4	5
Snowy Plover - coastal	*Charadrius nivosus*	↑↑	↑↑	↑↑	0	0	↑↑	8	2	5	6
Snowy Plover - inland		↑	↑	↑↑	0	0	↑↑	6	2	5	6
Wilson's Plover	*Charadrius wilsonia*	↑↑	↑↑	↑↑	0	0	↑↑	8	3	4	6
Semipalmated Plover	*Charadrius semipalmatus*	↑	↑	↑	↑	↑	↑↑	7	1	2	3
Piping Plover - coastal	*Charadrius melodus*	↑↑	↑↑	↑↑	0	0	↑↑	8	3	5	6
Piping Plover - inland		↑↑	↑↑	↑	0	0	↑↑	7	3	5	6
Killdeer	*Charadrius vociferous*	0	0	0	0	0	0	0	0	3	3
Mountain Plover	*Charadrius montanus*	↓	↑	↑	0	0	↑↑	3	1	5	5
Spotted Sandpiper	*Actitis macularius*	0	0	0	0	↑	0	1	1	2	2
Solitary Sandpiper	*Tringa solitaria*	↓	↑↑	↑↑	↑	↑	↑↑	7	2	4	5
Wandering Tattler	*Tringa incana*	0	↑	0	↑	↑	↑	4	4	3	4
Greater Yellowlegs	*Tringa melanoleuca*	0	0	0	↑	↑	↑↑	4	1	3	4
Willet - eastern	*Tringa semipalmata*	↑↑	↑↑	↑↑	0	0	↑	7	2	3	4
Willet - western		↑	↑	↑↑	0	0	↑	5	2	3	4
Lesser Yellowlegs	*Tringa flavipes*	↓	↑	↑	↑	↑	↑	4	1	3	4
Upland Sandpiper	*Bartramia longicauda*	0	0	0	0	↑	↑↑	3	2	4	4
Whimbrel	*Numenius phaeopus*	↑↑	↑↑	↑↑	↑	↑	↑	9	3	4	6
Bristle-thighed Curlew	*Numenius tahitiensis*	↓	↑↑	↑	↑	↑↑	↑	6	4	4	5
Long-billed Curlew	*Numenius americanus*	↑↑	↑	0	0	0	↑↑	5	2	5	6
Hudsonian Godwit	*Limosa haemastica*	0	↑	0	↑	↑↑	0	4	4	4	5
Bar-tailed Godwit	*Limosa lapponica*	↑↑	↑↑	↑↑	↑	↑↑	↑↑	11	4	4	6
Marbled Godwit	*Limosa fedoa*	↑	↑	↑	0	0	↑↑	5	1	4	5
Ruddy Turnstone	*Arenaria interpres*	↑	↑↑	↑↑	↑↑	↑	↑↑	10	3	4	6
Black Turnstone	*Arenaria melanocephala*	↑	↑	↑	↑	0	↑↑	6	4	4	5
Red Knot	*Calidris canutus*	↑	↑↑	↑↑	↑↑	↑	↑↑	10	4	4	6
Surfbird	*Calidris virgata*	↑	↑↑	↑↑	↑	↑	↑↑	9	5	4	6
Stilt Sandpiper	*Calidris himantopus*	↑↑	0	0	↑	↑	↑↑	6	3	3	4

Table 2. Cont.

Common name	Scientific name	Breeding habitat	Wintering habitat	Migration habitat	Ecological synchronicity	Migration distance	Degree of habitat specialization	Sum of change (arrows)	State of the Birds 2010[1]	Current risk category	Revised risk category[2]
Sanderling	*Calidris alba*	↑↑	↑↑	↑↑	↑↑	↑	↑↑	11	3	4	6
Dunlin	*Calidris alpina*	↑↑	↑↑	↑	↑	0	↑↑	8	3	3	5
Rock Sandpiper	*Calidris ptilocnemis*	↑	0	0	↑	0	↑↑	4	4	3	4
Purple Sandpiper	*Calidris maritima*	↑	0	0	0	0	↑↑	3	4	2	2
Baird's Sandpiper	*Calidris bairdii*	0	0	0	↑	↑↑	↑↑	5	3	2	3
Least Sandpiper	*Calidris minutilla*	0	↑	↑	↑	↑	↑	5	3	3	4
White-rumped Sandpiper	*Calidris fuscicollis*	↑↑	↓	↑	↑	↑↑	↑	6	3	2	3
Buff-breasted Sandpiper	*Calidris subruficollis*	↑↑	0	0	↑	↑↑	↑↑	7	3	4	5
Pectoral Sandpiper	*Calidris melanotos*	↑↑	0	0	↑	↑	↑↑	6	3	2	3
Semipalmated Sandpiper	*Calidris pusilla*	↑↑	↑↑	↑	↑↑	↑	↑↑	10	4	3	5
Western Sandpiper	*Calidris mauri*	↑	↑↑	↑	↑	↑	↑	7	4	4	5
Short-billed Dowitcher	*Limnodromus griseus*	0	↑↑	↑↑	↑	↑	↑	7	4	4	5
Long-billed Dowitcher	*Limnodromus scolopaceus*	↑↑	↑	↑	↑	↑	↑	7	3	4	5
Wilson's Snipe	*Gallinago delicata*	0	0	0	0	↑	↑	2	0	3	3
American Woodcock	*Scolopax minor*	↑	↓	↓	0	0	↑↑	1	1	4	4
Wilson's Phalarope	*Phalaropus tricolor*	↑	↑	↑	0	↑	0	4	3	4	5
Red-necked Phalarope	*Phalaropus lobatus*	↑↑	0	0	↑	↑	↑↑	6	1	3	4
Red Phalarope	*Phalaropus fulicarius*	↑↑	0	0	↑	↑	↑↑	6	3	3	4

Arrows depict extent and direction of change in risk associated with climate change. See Table 1 for description of risk factors and scoring, and see Appendices S1 and S2 for species specific discussion. Also included in current U.S. Shorebird Conservation Plan (USSCP) Risk Categories and State of the Birds vulnerability score and our proposed revised USSCP risk categories based on the added effects of climate change. Our Assessment of change in risk due to climate change is assessed by adding arrows across rows (1 up arrow and 1 down arrow result in a net of 0 arrows), and using the decision rule of 4 arrows (net, up or down) to shift risk categories.

[1]The lower scores, which were not published as part of the State of the Birds, were provided by the U.S. NABCI Committee.

[2]Based on our risk analysis, this would be the new U.S. Shorebird Conservation Plan risk category; we added a new risk category – 6 (Critical, Table 3) – to account for qualitatively greater risk than the current U.S. Shorebird Conservation Plan assessment allows.

Table 3. Results of sensitivity analysis of risk categorization for shorebird species.

		Projected under with climate change, from most (7) to least (3) conservative transition criterion				
Risk Category	**Current**	**7↑**	**6↑**	**5↑**	**4↑**	**3↑**
Not at risk	0	0	0	0	0	0
Low concern	7	6	3	2	2	1
Moderate concern	16	12	12	11	7	4
High concern	23	14	13	13	13	11
Highly imperiled	6	17	18	17	15	14
Critical	–[1]	3	6	9	15	22

What is shown in the first column of results is the current distribution of taxa across risk categories by the U.S. Shorebird Conservation Plan (USSCP). The columns that follow are the predicted distributions under different criteria for changing risk category. In Table 2 we assume that the accumulation of 4 arrows across risk factors is sufficient for a species to change risk category; this table shows the sensitivity of this result using more liberal (3 arrows) and more conservative (5, 6, and 7 arrows) criteria for changing risk category. We added a new risk category to those used by Partners-in-Flight (PIF) and the USSCP, Critical, to account for species being at categorically greater risk than previously considered. (See Table 2 and Appendices S3 and S4 for species-specific assessments and summaries.)
[1]Category does not exist in current PIF framework.

rise. Of particular note, for high-Arctic breeders, there is minimal latitude and land for northward range expansion. Extreme weather events were also projected to increase vulnerabilities due to negative effects on habitat, migration mortality, and disruption of ecological synchronicities, e.g. [94].

The increased vulnerabilities of 10 species could not be accommodated using the existing PIF scoring system and we had to create an even higher level of risk than is currently available. These Critical species (including coastal Snowy, Piping, and Wilson's plovers, and inland species such as Mountain Plover and Long–billed Curlew) are already at a high risk level due to other stressors (particularly anthropogenic habitat destruction) and their populations are already declining and jeopardized [32]. The addition of climate change to their risk factors raises them to an even higher level of vulnerability, which may pose even higher threats to their continued existence.

Also of concern is that the addition of climate change to the vulnerability calculations elevates another 18 species to the highest U.S. Shorebird Conservation Plan risk category. Thus, a total of 28 of 49 species are now at the highest risk category under the U.S. Shorebird Conservation Plan, or they exceeded this risk level and had to have an additional category created. The degree to which species changed risk categories was sensitive to our rules of category change. To some degree, as with population viability analyses using stochastic simulation models, which rule we use for category change is a value judgment [106]. Regardless of what rules are used, however, our analysis suggests that shorebirds will have increased vulnerability under climate change, perhaps to a large extent.

Our assessment of extinction risk might be criticized because it does not allow for adaptive capacity in shorebird populations. That is, shorebirds might modify their breeding, migratory, and/or wintering habitat use, foraging, and/or timing to accommodate the changing climate. We already know that some shorebirds in Western Europe have apparently truncated their fall migrations to winter in the Baltic, rather than in oceanic Atlantic countries, such as the UK [104]. Previously, the winter conditions in the Baltic were so harsh that birds had to move farther to exploit the milder conditions of the UK, Holland, etc. Thus, the ameliorating winter conditions in the Baltic have encouraged changes in migration distance [104]. Similarly, In North America, some migratory populations of Hudsonian Godwits have advanced their timing of migration during warm periods, which allows their breeding to synchronize with peak food abundance, while other populations have not [107]. There also is some evidence that Semipalmated and Pectoral sandpipers and Red–necked and Red phalaropes have been observed breeding earlier during warm years [108]. As another example of adapting to changing conditions, Dunlin nestlings can exhibit accelerated growth during periods of low food availability during warm conditions [109]. However, it would be unwise of us to assume that such adaptive capacities were likely to apply across all shorebird species because there is evidence that high Arctic shorebird species may have little capacity for adaptation due to low genetic variability resulting from bottleneck events from previous climate shifts [110]. Time constraints can also cause conflicts among competing life–history requirements, as has been reported in Pied Flycatchers *Ficedula hypoleuca* [111]. Clearly more research needs to be done to determine the degree to which climate adaptation might occur in shorebirds.

What would it take to accurately and precisely predict change in extinction risk due to climate change for migratory shorebirds, or for any species, rather than taking the relatively coarse approach we did in this paper? Certainly there have been detailed assessments of expected regional changes in shorebird populations in response to climate change [112–113], and one could create models to link species to landscapes via simulation. But what would be required for accurate, reliable predictions? Strictly speaking, to build a convincing case for an accurate prediction, the first thing we would need is accurate models of climate change. Although there are many models of climate change, and they agree in general with climate trends, there is still a great deal of uncertainty in the exact amount of changes in expected temperature and precipitation, e.g. [114–115], particularly at the fine geographic scale that would be needed to understand biotic responses, including the effects of changes in wind patterns [116]. Because hydrological models are complex (i.e., non–linear, with feedback and chaotic dynamics), more accurate data are unlikely to improve model predictions [117]. In addition, accurate regional and local downscaling of global climate models might not be possible [118–120].

The next requirement is accurate models linking climate change to hydrologic responses, so we could accurately determine changes in hydrology, amount of sea level rise, the degree to which plant communities will change in response to climate change, in both inland and coastal regions,. Accurate models that allow these

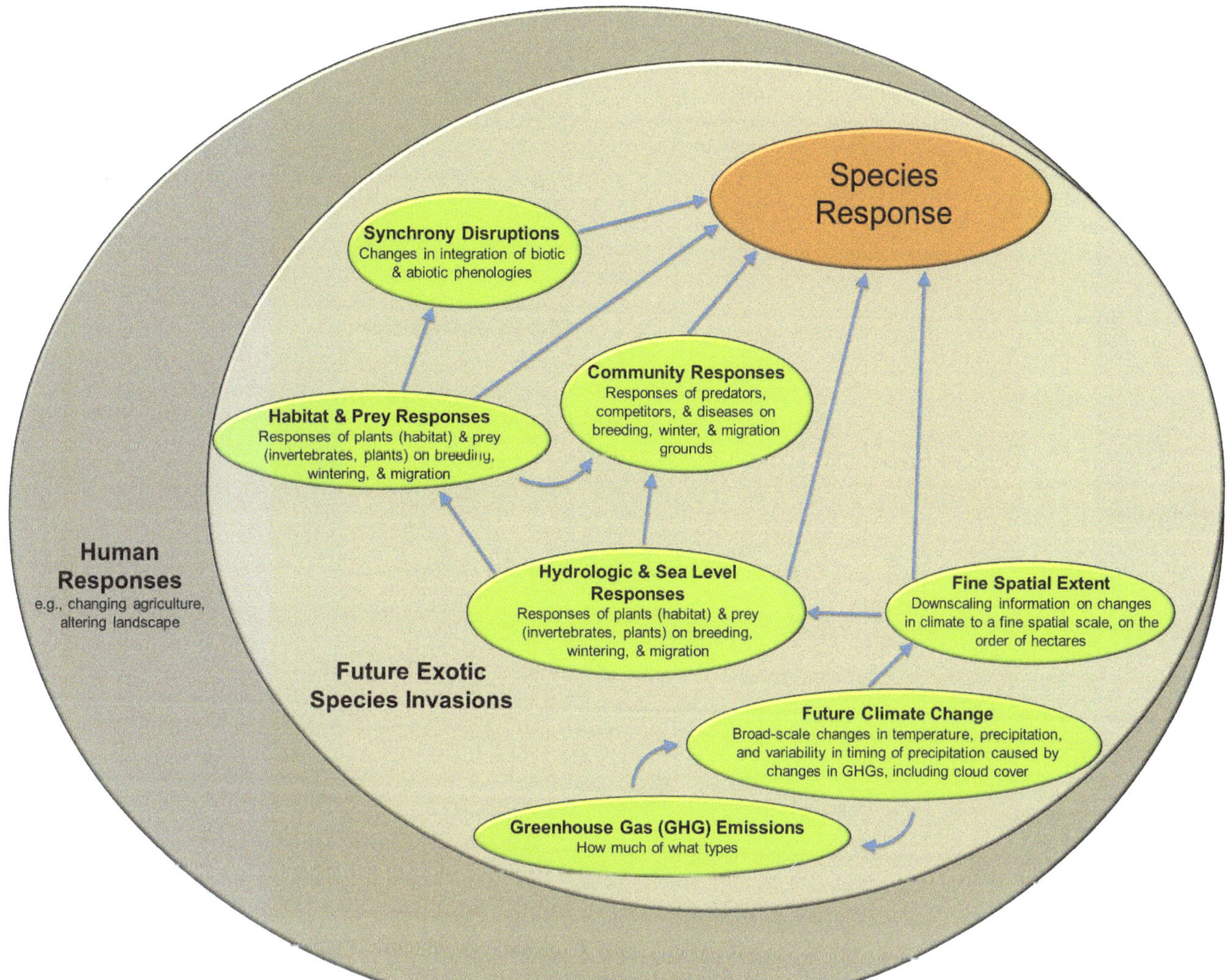

Figure 3. Digraph showing relationships (arrows) for which we need accurate information in order to accurately predict species-specific shorebird responses to climate change. By accurate, we mean variation explained between nodes is >90% or near that, not merely determining statistically significant relationships. Subheadings specify the relationships, and 'species response' includes adaptive responses as well as non-adaptive responses. 'Fine spatial extent' refers to downscaling climate change estimates to the spatial scale at which species respond; factors at this scale affect species' responses directly and indirectly. The digraph is nested within the contexts of future introductions of exotic, invasive species, and human responses to climate change to indicate that all of the relationships from the digraph can be affected by these particular occurrences or responses.

predictions do not exist [117,121–123]. Even if we had accurately developed models, we also would need accurate assessments of species' ranges as well as niche–based models for each species we want to evaluate that accurately predicts, with a very high level of variability in distribution explained, the distribution of species, cf. [106]. We do not yet have these, e.g. [124–127], and it is not clear to what extent or rate different bird might respond behaviorally to climate change [128]. Finally, we need models that accurately depict community–wide biotic responses to climate change, including accurate anticipation of inter–specific interactions, how local species invasions and extinctions will affect resource availability, how they might change as niches shift [129–133]. We do not have these either, and we might be unlikely to accurately anticipate shifting realized niches for a variety of practical reasons [134–136]. These challenges are exacerbated by migration because the relationships must be known in breeding, non–breeding, and migration habitats [137]. These relationships we just described are depicted in Fig. 3. Even the highly restrictive requirements we just presented might ultimately be insufficient, because they do not take into account human responses to climate change. For example, what will be the human responses in changes to agricultural practices, relocation away from coastal areas, and so–called adaptive response measures, e.g. [138–140], and how will they affect the capacity for ecosystems and shorebirds to respond?

Consequently, we suspect that detailed regional and local biological forecasting of the effects of climate change, even if the correct (but currently unknown) IPCC scenario is selected, is likely to be only generally accurate. Therefore, we think that the relatively coarse assessment of changes in extinction risk that we present here is a useful level of assessment for species at a continental scale; see [31] for another example of a categorical risk assessment at a smaller geographic scale. We stress that the somewhat bleak picture we paint regarding prediction accuracy at small spatial scales should not be used as an excuse to not make models or predictions, or to avoid planning for climate change. Rather, we encourage model development and testing, followed by model revision as more data become available. As with all models, we suggest treating the structure, parameter values, and predictions as hypotheses to test. We also support alternative modeling approaches that might be effective at accommodating model uncertainty, such as robust decision-making [141].

Acknowledgments

We thank B. Harrington and N. Warnock for many discussions about shorebird ecology and for commenting on an earlier draft of this manuscript; B. Tavernia for his ArcGIS work to determine migration distances; A. Madden for presentation advice on figure 3; B. Andres for sharing U.S. NABCI Committee unpublished data; and three anonymous reviewers provided useful input on an earlier draft of this manuscript. Data provided by NatureServe in collaboration with Robert R., J. Zook, The Nature Conservancy – Migratory Bird Program, Conservation International – CABS, World Wildlife Fund – US, and Environment Canada – WILDSPACE.

Author Contributions

Conceived and designed the experiments: HG DWD SB JMR. Performed the experiments: HG DWD SB JMR. Analyzed the data: HG DWD SB JMR. Contributed reagents/materials/analysis tools: HG DWD SB JMR. Wrote the paper: HG DWD SB JMR.

References

1. Brown S, Hickey C, Harrington B, Gill R (2001) United States shorebird conservation plan, 2nd ed. Manomet: Manomet Center for Conservation Sciences. 61 p.
2. Warnock N, Elphick C, Rubega MA (2002). Shorebirds in the marine environment. In: Schreiber EA, Burger J, editors. Biology of marine birds. Boca Raton: CRC Press. pp. 581–615.
3. Piersma T, Lindström Å (2004) Migrating shorebirds as integrative sentinels of global environmental change. Ibis 146(S), 61–69.
4. Russel BD, Harley CDG, Wernberg T, Mieszkowska N, Widdicombe S, et al. (2012) Predicting ecosystem approaches that integrate the effects of climate change across entire systems. Biology Letters 8: 164–166.
5. International Wader Study Group (2003) Waders are declining worldwide. Conclusions from the 2003 International Wader Study Group Conference, Cadiz, Spain. Wader Study Group Bulletin 101/102: 8–12.
6. Morrison RIG, McCaffery BJ, Gill RE, Skagen SK, Jones SL, et al. (2006) Population estimates of North American Shorebirds, 2006. Wader Study Group Bulletin 111: 67–85.
7. Thomas GH, Lanctot RB, Székely T (2006) Can intrinsic factors explain population declines in North American breeding shorebirds? A comparative analysis. Animal Conservation 9: 252–258.
8. Bart J, Brown S, Harrington B, Morrison RIG (2007) Survey trends of North American shorebirds: population declines or shifting distributions? Journal of Avian Biology 38: 73–82.
9. Morrison RIG, Ross PK, Niles LJ (2005) Declines in wintering populations of Red Knots in southern America. Condor 106: 60–70.
10. Federal Register (2006) Vol. 71, No. 176. Tuesday, September 12, Proposed Rules.
11. Jehl J (2007) Disappearance of breeding Semipalmated Sandpipers from Churchill, Manitoba: More than a local phenomenon. Condor 109: 351–360.
12. Morrison RIG, Mizrahi DS, Ross RK, Ottema OH, De Pracontal N, et al. (2012) Dramatic declines of Semipalmated Sandpipers on their major wintering areas in the Guianas, northern South America. Waterbirds 35: 120–134.
13. US Fish and Wildlife Service (2011) Available: http://ecos.fws.gov/tess_public/SpeciesReport.do?groups=B&listingType=L&mapstatus=1, Accessed 2011 Jun 22.
14. International Union for the Conservation of Nature (2013) IUCN Red List of Threatened Species. Version 2012.2. Available: http://www.iucnredlist.org. Accessed 2013 Jan 13.
15. Howe MA, Geissler PH, Harrington BA (1989) Population trends of North American shorebirds based on the International Shorebird Survey. Biological Conservation 49: 185–199.
16. Gratto-Trevor CL (1997) Climate change: proposed effects on shorebird habitat, prey, and numbers in the Outer Mackenzie Delta. Mackenzie Basin Impact Study Final Report, pp. 205–210.
17. Austin GE, Rehfisch MM (2003) The likely impact of sea level rise on waders (Charadrii) wintering on estuaries. Journal of Nature Conservation 11: 43–58.
18. Milton D (2003) Threatened shorebird species of the East Asian–Australasian Flyway: significance for Australian wader study groups. Wader Study Group Bulletin 100: 105–110.
19. Stroud DA, Davidson NC, West R, Scott DA, Haanstra L, et al, compilers(2004) Status of migratory wader populations in Africa and Western Eurasia in the 1990s. International Wader Studies 15: 1–259.
20. Parry ML, Canziani OF, Palutikof JP, van der Linden PJ, Hanson CE, editors(2007) Contribution of Working Group II to the Fourth Assessment Report of the Intergovernmental Panel on Climate Change, 2007. Cambridge University Press, Cambridge, United Kingdom and New York, NY, US.
21. Galbraith H, Jones R, Park R, Clough J, Herrod-Julius S, et al. (2002) Global climate change and sea level rise: potential losses of intertidal habitat for shorebirds. Waterbirds 25: 173–183.
22. Hedenström A, Barta Z, Helm B, Houston AI, McNamara JM, et al. (2007) Migration speed and scheduling of annual events by migrating birds in relation to climate change. Climate Res 35: 79–91.
23. Colwell MA (2010) Shorebird Ecology, Conservation, and Management. Berkeley: Univ. of California Press. 344 p.
24. Small-Lorenz SL, Culp LA, Ryder TB, Will TC, Marra PP (2013) A blind spot in climate change vulnerability assessments. Nature Climate Change 3: 91–93.
25. Lee TM, Jetz W (2008) Future battlegrounds for conservation under global change. Proceedings of the Royal Society, B 275: 1261–1270.
26. Galbraith H, Price J (2009) A Framework for Categorizing the Relative Vulnerability of Threatened and Endangered Species to Climate Change. U.S. Environmental Protection Agency. EPA/600/R-09/011. Available: http://cfpub.epa.gov/ncea/cfm/recordisplay.cfm?deid=203743# Accessed 2011 Jun 20.
27. Lee TM, Jetz W (2011) Unravelling the structure of species extinction risk for predictive conservation science. Proceedings of the Royal Society, B 278: 1329–1338.
28. Millsap BA, Gore JA, Runde DR, Cerulean CI (1990) Setting priorities for the conservation of fish and wildlife species in Florida. Wildlife Monographs 111.
29. Herman TB, Scott FW (1994) Protected areas and global climate change: assessing the regional or local vulnerability of vertebrate species. In: Pernetta JC, Leemans R, Elder D, Humphrey S, editors Impacts of climate change on ecosystems and species: implications for protected areas. Gland, Switzerland: IUCN. pp. 13–27.
30. Reed JM (1995) Relative vulnerability to extirpation of montane breeding birds in the Great Basin. Great Basin Naturalist 55: 342–351.
31. Gardali T, Seavy NE, DiGaudio RT, Comrack LA (2012) A climate change vulnerability assessment of California's at-risk birds. PLoS ONE 7(3): e29507. doi:10.1371/journal.pone.0029507
32. Panjabi AO, Dunn EH, Blancher PJ, Hunter WC, Altman B, et al. (2005) The Partners in Flight handbook on species assessment. Version 2005. Partners in Flight Technical Series No. 3. Rocky Mountain Bird Observatory website, Available: http://www.rmbo.org/pubs/downloads/Handbook2005.pdf
33. Higdon JW, MacLean DA, Hagan JM, Reed JM (2006) Risk of extirpation for vertebrate species on an industrial forest in New Brunswick, Canada: 1945, 2002, and 2027. Canadian Journal of Forest Research 36: 467–481.
34. Elphick CS, Roberts DL, Reed JM (2010) Estimated dates of recent extinctions for North American and Hawaiian birds. Biological Conservation 143: 617–624.

35. North American Bird Conservation Initiative, U.S. Committee (2010) The State of the Birds 2010 Report on Climate Change, United States of America. Washington, DC: U.S. Department of the Interior.
36. Intergovernmental Panel on Climate Change (2007) Climate Change 2007: Mitigation. Contribution of Working Group III to the Fourth Assessment Report of the Intergovernmental Panel on Climate Change. Metz B, Davidson OR, Bosch PR, Dave R, Meyer, LA, editors. Cambridge, UK: Cambridge University Press.
37. Chapin FS III, Shaver GR, Giblin AE, Nadelhoffer KG, Laundre JA (1995) Response of arctic tundra to experimental and observed changes in climate. Ecology 76: 694–711.
38. Moss RH, Schneider SH (2001) Towards Consistent Assessment and Reporting of Uncertainties in the IPCC TAR. In: Pachauri R, Taniguchi T, editors. Cross–Cutting Issues in the IPCC Third Assessment Report. Global Industrial and Social Progress Research Institute (for IPCC). Tokyo.
39. Keeling CD, Chin JFS, Whorf TP (1996) Increased activity of northern vegetation inferred from atmospheric CO_2 measurements. Nature 382: 146–149.
40. Zhou L, Tucker CJ, Kaufmann RK, Slayback D, Shabanov NV, et al. (2001) Variation in northern vegetation activity inferred from satellite data of vegetation index during 1981–1999. Journal Geophysical Research–Atmospheres 106(D17), 20069–20083.
41. Lucht W, Prentice IC, Myneni RB, Sitch S, Friedlingstein P, et al. (2002) Climatic control of the high–latitude vegetation greening trend and Pinatubo effect. Science 296: 1687–1689.
42. Arctic Climate Impact Assessment (2004) Impacts of a Warming Arctic. Arctic Climate Impact Assessment. Cambridge: Cambridge University Press. 144 p.
43. Sturm M, Schimel J, Michaelson G, Welker JM, Oberbauer SF, et al. (2005) Winter biological processes could help convert arctic tundra to shrubland. Bioscience 55: 17–26.
44. Karl T, Melillo M, Peterson T (2009) Global Climate Change Impacts in the United States. Cambridge University Press. 192 p.
45. Bhatt US, Walker DA, Raynolds MK, Comiso JC, Epstein HE, et al. (2010) Circumpolar Arctic tundra vegetation change is linked to sea ice decline. Earth Interactions 14: 1–20.
46. Elmendorf SC, Henry GHR, Hollister RD, Björk RG, Boulanger–Lapointe N, et al. (2012) Plot–scale evidence of tundra vegetation change and links to recent summer warming. Nature Climate Change 2: 453–457.
47. Raynolds MK, Walker DA, Epstein HE, Pinzon JE, Tucker CJ (2012) A new estimate of tundra–biome phytomass from trans–Arctic field data and AVHRR NDVI. Remote Sensing Letters 3: 403–411.
48. Martin PD, Jenkins JL, Adams FJ, Jorgenson MT, Matz AC, et al. (2009) Wildlife response to environmental arctic change: predicting future habitats of arctic Alaska. Report of the Wildlife Response to Environmental Arctic Change (WildREACH): Predicting Future Habitats of Arctic Alaska Workshop, 17–18 November 2008. Fairbanks, Alaska: U.S. Fish and Wildlife Service. 138 p.
49. Karlsson JM, Bring A, Peterson GD, Gordon LJ, Destouni G (2011) Opportunities and limitations to detect climate–related regime shifts in inland Arctic ecosystems through eco–hydrological monitoring. Environmental Research Letters 6: 1–9.
50. MacDonald LA, Turner KW, Balasubramaniam AM, Wolfe BB, Hall RI, et al. (2012) Tracking hydrological responses of a thermokarst lake in the Old Crow Flats (Yukon Territory, Canada) to recent climate variability using aerial photographs and paleolimnological methods. Hydrological Processes 26: 117–129.
51. Assini J, Young KL (2012) Snow cover and snowmelt of an extensive High Arctic wetland: spatial and temporal seasonal patterns. Hydrological Sciences Journal 57: 738–755.
52. Prowse T (2012) Lake and river ice in Canada. In: French H, Slaymaker O, editors. Changing cold environments: a Canadian perspective. Sussex: Wiley–Blackwell. pp. 163–181
53. Woo M–K, Pomeroy J (2012) Snow and runoff: Processes, sensitivity and vulnerability. In: French H, Slaymaker O, editors. Changing cold environments: a Canadian perspective. Sussex: Wiley–Blackwell. pp. 105–125
54. Sheffield J, Wood WF (2008) Projected changes in drought occurrence under future global warming from multi–model, multi–scenario, IPCC AR4 simulations. Climate Dynamics 31: 79–105.
55. Johnson WC, Werner B, Guntenspergen GR, Voldseth RA, Millett B, et al. (2010) Prairie wetland complexes as landscape functional units in a changing climate. BioScience 60: 128–140.
56. Strzepek K, Yohe G, Neumann J, Boehlert B (2010) Characterizing changes in drought risk for the United States from climate change. Environmental Research Letters 5: 1–9.
57. Dai A (2011) Drought under global warming: a review. Wiley Interdisciplinary Reviews: Climate Change 2: 45–65.
58. Rahmstorf S (2007) A Semi–Empirical Approach to Projecting Future Sea–Level Rise. Science 315: 368–370.
59. Pfeffer WT, Harper JT, O'Neel SO (2008) Kinematic Constraints on Glacier Contributions to 21st–Century Sea–Level Rise. Science 321: 1340–1343.
60. Mcleod E, Poulter B, Hinkel J, Reyes E, Salm R (2010) Sea–level rise impact models and environmental conservation: A review of models and their applications. Ocean & Coastal Management 53: 507–517.
61. Menon S, Soberón J, Li X, Peterson AT (2010) Preliminary global assessment of biodiversity consequences of sea level rise mediated by climate change. Biodiversity and Conservation 19: 1599–1609.
62. Shaw J, Taylor RB, Forbes DL, Ruz MH, Solomon S (1998). Sensitivity of the coasts of Canada to sea–level rise. Ottawa: Geological Survey of Canada, Report No. 505. 79 p.
63. Boruff BJ, Emrich C, Cutter SL (2005) Erosion hazard vulnerability of U.S. coastal counties. Journal of Coastal Research 21: 932–942.
64. Nicholls RJ, Wong PP, Burkett VR, Codignotto JO, Hay JE, et al. (2007) Coastal systems and low–lying areas. In: Parry ML, Canziani OF, Palutikof JP, van der Linden PJ, Hanson CE, editors. Climate change 2007: impacts, adaptation and vulnerability. Contribution of Working Group II to the Fourth Assessment Report of the Intergovernmental Panel on Climate Change. Cambridge: Cambridge University Press. pp. 315–356.
65. Strauss BJ, Ziemlinski R, Weiss JL, Overpeck JT (2012) Tidally adjusted estimates of topographic vulnerability to sea level rise and flooding for the contiguous United States. Environmental Research Letters 7: 014033.
66. Chu–Agor ML, Muñoz–Carpena R, Kiker G, Emanuelsson A, Linkov I (2011) Exploring vulnerability of coastal habitats to sea level rise through global sensitivity and uncertainty analyses. Environmental Modelling & Software 26: 593–604.
67. Stralberg D, Brennan M, Callaway JC, Wood JK, Schile LM, et al. (2011) Evaluating tidal marsh sustainability in the face of sea–level rise: A hybrid modeling approach applied to San Francisco Bay. PLoS One 6(11): e27388. doi:10.1371/journal.pone.0027388.
68. Titus JG, Hudgens DE, Trescott DL, Craghan M, Nuckols WH, et al. (2009) State and local governments plan for development of most land vulnerable to rising sea level along the US Atlantic coast. Environmental Research Letters 4: 044008.
69. Seavey JR, Gilmer B, McGarigal KM (2011) Effect of sea–level rise on piping plover (*Charadrius melodus*) breeding habitat. Biological Conservation 144: 393–401.
70. Magrin G, Gay García C, Cruz Choque D, Giménez JC, Moreno AR, et al. (2007) Latin America. In: Parry ML, Canziani OF, Palutikof JP, van der Linden PJ, Hanson CE, editors. Climate change 2007: impacts, adaptation and vulnerability. Contribution of Working Group II to the Fourth Assessment Report of the Intergovernmental Panel on Climate Change. Cambridge: Cambridge University Press. pp. 581–615.
71. Malhi Y, Aragão LEOC, Galbraith D, Huntingford C, Fisher R, et al. (2009) Exploring the likelihood and mechanism of a climate–change–induced dieback of the Amazon rainforest. Proceedings of the National Academy of Sciences USA 106: 20610–20615.
72. Lola da Costa AC, Galbraith D, Almeida S, Portela BTT, da Costa M, et al. (2010) Effect of 7 yr of experimental drought on vegetation dynamics and biomass storage of an eastern Amazonian rainforest. New Phytologist 187: 579–591.
73. Hirota M, Nobre C, Oyama MD, Bustamante MMC (2010) The climatic sensitivity of the forest, savanna and forest–savanna transition in tropical South America. New Phytologist 187: 707–719.
74. Keppie D, Whiting JRM (1994) American Woodcock (*Scolopax minor*). The Birds of North America Online, Poole A, editor. Ithaca: Cornell Lab of Ornithology; Retrieved from the Birds of North America Online: http://bna.birds.cornell.edu/bna/species/100.
75. Flinn K, Vellend M (2005) Recovery of forest plant communities in post–agricultural landscapes. Frontiers in Ecology and the Environment 3: 243–250.
76. Adams HD, Guardiola–Claramonte M, Barron–Gafford GA, Villegas JC, Breshears DD, et al. (2009) Temperature sensitivity of drought–induced tree mortality: implications for regional die–off under global–change type drought. Proceedings of the National Academy of Sciences USA 106: 7063–7066.
77. Allen CD, Macalady AK, Chenchouni H, Bachelet D, McDowell N, et al. (2010) A global overview of drought and heat–induced tree mortality reveals emerging climate change risks for forests. Forest Ecology and Management 259: 660–684.
78. Jactel H, Petit J, Desprez–Loustau M–L, Delzon S, Piou D, et al. (2012) Drought effects on damage by forest insects and pathogens: a meta–analysis. Global Change Biology 18: 267–276.
79. Raven J, Caldeira K, Elderfield H, Hoegh–Guldberg O, Liss P, et al. (2005) Ocean Acidification due to Increasing Atmospheric Carbon dioxide. Policy Document 12/05. London: Royal Society. 60 p.
80. Flynn KJ, Blackford JC, Baird ME, Raven JA, Clark DR, et al. (2012) Changes in pH at the exterior surface of plankton with ocean acidification. Nature Climate Change 2: 510–513.
81. Blackford JC (2010) Predicting the impacts of ocean acidification: Challenges from an ecosystem perspective. Journal of Marine Systems 81: 12–18.
82. Milligan AJ (2012) Plankton in an acidified ocean. Nature Climate Change 2: 489–490.
83. Fabry VJ, Seibel BA, Feely RA, Orr JC (2008) Impacts of ocean acidification on marine fauna and ecosystem processes. ICES Journal of Marine Science 65: 414–432.
84. Taylor AR, Chrachri A, Wheeler G, Goddard H, Brownlee C (2011) A voltage–gated H^+ channel underlying pH homeostasis in calcifying coccolithophores. PLoS Biology 9e1001085. doi:10.1371/journal.pbio.1001085.

85. Snyder MA, Sloan LC, Diffenbaugh NS, Bell JL (2003) Future climate change and upwelling in the California Current. Geophysical Research Letters 30: 8–1–8–4.
86. Sweeney BW, Jackson JK, Newbold JD, Funk DH (1992) Climate change and the life histories and biogeography of aquatic insects in eastern North America. In: Firth P, Fisher SG, editors. Global change and freshwater ecosystems. Springer. pp. 143–176
87. Hodkinson ID, Webb NR, Bale JS, Block W, Coulson SJ, et al. (1998) Global change and Arctic ecosystems: Conclusions and predictions from experiments with terrestrial invertebrates on Spitsbergen. Arctic and Alpine Research 30: 306–313.
88. Tulp I, Schekkerman H (2008) Has prey availability for Arctic birds advanced with climate change? Hindcasting the abundance of tundra arthropods using weather and seasonal variation. Arctic 61: 48–60.
89. Castro G, Myers JP (1993) Shorebird predation on eggs of horseshoe crabs during spring stopover on Delaware Bay. The Auk 110: 927–930.
90. Baker AJ, Gonzalez PM, Piersma T, Niles LJ, De Serrano Do Nascimento I (2004) Rapid population decline in Red Knots: fitness consequences of decreased refuelling rates and late arrival in Delaware Bay. Proceedings of the Royal Society of London B 271: 875–882.
91. Mizrahi DS, Peters KA (2009) Relationships between sandpipers and horseshoe crab in Delaware Bay: A synthesis. In: Tanacredi JT, et al. editors. Biology and Conservation of Horseshoe Crabs, DOI 10.1007/978-0-387-89959-6_4, Springer.
92. Jones T, Cresswell W (2010) The phenology mismatch hypothesis: are declines of migrant birds linked to uneven global climate change? Journal of Animal Ecology 79: 98–108.
93. Both C, Van Turnhout CAM, Bijlsma RG, Siepel H, Van Strien AJ, et al. (2010) Avian population consequences of climate change are most severe for long-distance migrants in seasonal habitats. Proceedings of the Royal Society B 1685: 1259–1266.
94. Saino N, Ambrosini R, Rubolini D, von Hardenberg J, Provenzale A, et al. (2011) Climate warming, ecological mismatch at arrival and population decline in migratory birds. Proceedings of the Royal Society B 278: 835–842.
95. Iwamura T, Possingham HP, Chadès I, Minton C, Murray NJ, et al. (2013) Migratory connectivity magnifies the consequences of habitat loss from sea–level rise for shorebird populations. Proceedings of the Royal Society B 280: 20130325.
96. Convertino M, Bockelie A, Kiker GA, Muñoz-Carpena R, Linkov I (2012) Shorebird patches as fingerprints of fractal coastline fluctuations due to climate change. Ecological Processes 1:9.
97. Robinson RA, Crick HQP, Learmoth JA, Maclean IMD, Thomas CD (2009) Travelling through a warming world: climate change and migratory species. Endangered Species Research 7: 87–99.
98. Klaassen M, Hoye BJ, Nolet BA, Buttemer WA (2012) Ecophysiology of avian migration in the face of current global hazards. Philosophical Transactions of the Royal Society B 367: 1719–1732.
99. Ridgely RS, Allnutt TF, Brooks T, McNicol DK, Mehlman DW, et al. (2007) Digital distribution maps of the birds of the western hemisphere, version 3.0. Arlington: NatureServe.
100. Nebel S, Porter JL, Kingsford RT (2008) Long–term trends of shorebird populations in eastern Australia and impacts of freshwater extraction. Biological Conservation 141: 971–980.
101. Godet L, Jaffré M, Devictor V (2011) Waders in winter: long–term changes of migratory bird assemblages facing climate change. Biology Letters 7: 714–717.
102. Parmesan C, Galbraith H (2004) Observed ecological impacts of climate change in North America. Arlington: Pew Center for Global Climate Change. Available: http://www.c2es.org/publications/observed-impacts-climate-change-united-states.
103. Parmesan C (2006) Ecological and evolutionary responses to recent climate change. Annual Review of Ecology and Systematics 37: 637–669.
104. MacLean IMD, Austin GE, Rehfisch MM, Blew J, Crowe O, et al. (2008) Climate change causes rapid changes in the distribution and site abundance of birds in winter. Global Change Biology 14: 2489–2500.
105. Warnock N (2010) Stopover vs. staging: the difference between a hop and a jump. Journal of Avian Biology 41: 621–626.
106. Reed JM, Akçakaya HR, Burgman M, Bender D, Beissinger S., et al. (2006) Critical habitat. In: Scott JM, Goble DD, Davis FW, editors. The Endangered Species Act at thirty: conserving biodiversity in human–dominated landscapes. Washington D.C.: Island Press. pp. 164–177
107. Senner NR (2012) One species but two patterns: Populations of the Hudsonian Godwit (*Limosa haemastica*) differ in spring migration timing. The Auk 129: 670–682.
108. Liebezeit JR, Gurney KEB, Budde M, Zack S, Ward D (2014) Phenological advancement in arctic bird species: relative importance of snow melt and ecological factors. Polar Biology in press, 10.1007/s00300–014–1522–x.
109. McKinnon L, Nol E, Juillet C (2013) Arctic–nesting birds find physiological relief in the face of trophic constraints. Scientific Reports 3:1816 DOI: 10.1038/srep01816
110. Meltofte H, Piersma T, Boyd H, McCaffery B, Ganter B, et al. (2007) Effects of climate variation on the breeding ecology of Arctic shore birds. Danish Polar Center, Meddelelser om Grønland Bioscience 59, Copenhagen, 1–48.
111. Both C, Visser ME (2001) Adjustment to climate change is constrained by arrival date in a long–distance migrant bird. Nature 411:296–298.
112. Rehfisch MM, Austin GE, Freeman SN, Armitage MJS, Burton NHK (2004) The possible impact of climate change on the future distributions and numbers of waders on Britain's non–estuarine coast. Ibis 146 (Suppl. 1): 70–81.
113. Aiello–Lammens ME, Chu–Agor ML, Convertino M, Fischer RA, Linkov I, et al. (2011) The impact of sea–level rise on Snowy Plovers in Florida: integrating geomorphological, habitat, and metapopulation models. Global Change Biology 17: 3644–3654.
114. Araújo MB, Whittaker RJ, Ladle RJ, Erhard M (2005) Reducing uncertainty in projections of extinction risk from climate change. Global Ecology and Biogeography 14: 529–538.
115. Davidson EA, Janssens IA (2006) Temperature sensitivity of soil carbon decomposition and feedbacks to climate change. Nature 440: 165–173.
116. Gill R Jr, Tibbitts T, Douglas D, Handel C, Mulcahy D, et al. (2009) Extreme endurance flights by landbirds crossing the Pacific: ecological corridor rather than barrier? Proceeding of the Royal Society B 276: 447–457.
117. Vano JA, Udall B, Cayan DR, Overpeck JT, Brekke LD, et al. (2014) Understanding uncertainties in future Colorado River streamflow. Bulletin of the American Meteorological Society January, 59–78.
118. Anagnostopolos GG, Koutsoyiannis D, Christofides A, Efstratiadis A, Mamassis N (2010) A comparison of local and aggregated climate model outputs with observed data. Hydrological Sciences Journal 55:1094–1110.
119. Kundzewicz ZW, Stakhiv EZ (2010) Are climate models "ready for prime time" in water resources management applications, or is more research needed? Hydrological Sciences Journal 55:1085–1089.
120. Wilby RL (2010) Evaluating climate model outputs for hydrological applications. Hydrological Sciences Journal 55:1090–1093.
121. He Y, Wetterhall F, Bao H, Cloke H, Li Z, et al. (2010) Ensemble forecasting using TIGGE for the July–September 2008 floods in the Upper Huai catchment: a case study. Atmospheric Science Letters 11: 132–138.
122. Convertino M, Welle P, Muñoz–Carpena R, Kiker GA, Chu–Agor MaL., et al. (2012) Epistemic uncertainty in predicting shorebird biogeography affected by sea–level rise. Ecological Modelling 240: 1–15.
123. O'Gorman PA (2012) Sensitivity of tropical precipitation extremes to climate change. Nature Geoscience 5: 697–700.
124. Pearson RG, Dawson TP (2003) Predicting the impacts of climate change on the distribution of species: are bioclimate envelope models useful? Global Ecology & Biogeography 12: 361–371.
125. Araújo MB, Pearson RG, Thuillers W, Erhard M (2005) Validation of species–climate impact models under climate change. Global Change Biology 11: 1504–1513.
126. Rotenberry JT, Wiens JA (2009) Habitat relations of shrubsteppe birds: A 20–year retrospective. Condor 111: 401–413.
127. Essl F, Rabitsch W, Dullinge S, Moser D, Milasowszky N (2013) How well do we know species richness in a well–known continent? Temporal patterns of endemic and widespread species descriptions in the European fauna. Global Ecology and Biogeography 22: 29–39.
128. Devictor V, Julliard R, Couvet D, Jiguet F (2008) Birds are tracking climate warming, but not fast enough. Proceedings of the Royal Society B 275: 2743–2748.
129. Smith PA, Gilchrist G, Forbes MR, Martin J–L, Allard K (2010) Inter–annual variation in the breeding chronology of arctic shorebirds: effects of weather, snow melt and predators. Journal of Avian Biology 41: 292–304.
130. Pe'er G, Henle K, Dislich C, Frank K (2011) Breaking functional connectivity into components: a novel approach using an individual–based model, and first outcomes. PLoS ONE 6. doi:10.1371/journal.pone.0022355
131. Schmidt NM, Ims RA, Høye TT, Gilg O, Hansen LH, et al. (2012) Response of an arctic predator guild to collapsing lemming cycles. Proceedings of the Royal Society, B 279: 4417–4422.
132. Cahill AE, Aiello–Lammens ME, Fisher–Reid MC, Hua X, Karanewsky CJ, et al. (2013) How does climate change cause extinction? Proceedings of the Royal Society, B 280, doi: 10.1098/rspb.2012.1890.
133. Brantley S, Ford CR, Vose JM (2013) Future species composition will affect forest water use after loss if eastern hemlock from southern Appalachian forests. Ecological Applications 23: 777–790.
134. Williams JW, Jackson ST (2007) Novel climates, no–analog communities, and ecological surprises. Frontiers in Ecology and the Environment 5: 475–482.
135. Correa–Metrio A, Bush MB, Cabrera KR, Sully S, Brenner M, et al. (2012) Rapid climate change and no–analog vegetation in lowland Central America during the last 86,000 years. Quaternary Science Reviews 38: 63–75.
136. Veloz SD, Williams JW, Blois JL, He F, Otto–Bliesner B, et al. (2012) No–analog climates and shifting realized niches during the late quaternary: implications for 21st–century predictions by species distribution models. Global Change Biology 18: 1698–1713.
137. Knudsen E, Lindén A, Both C, Jonzén N, Pulido F, et al. (2011) Challenging claims in the study of migratory birds and climate change. Biological Reviews 86: 928–946.
138. Olesen JE, Trnka M, Kersebaum KC, Skjelvag AO, Seguin B, et al. (2011) Impacts and adaptation of European crop production systems to climate change. European Journal of Agronomy 34: 96–112.
139. Khailani DK, Perera R (2013) Mainstreaming disaster resilience attributes in local development plans for the adaptation to climate change induced flooding: A study based on the local plan of Shah Alam City, Malaysia. Land Use Policy 30: 615–627.

140. Lloyd MG, Peel D, Duck RW (2013) Towards a social–ecological resilience framework for coastal planning. Land Use Policy 30: 925–933.
141. Veloz SD, Nur N, Salas L, Jongsoomijt D, Wood J, et al. (2013) Modeling climate change impacts on tidal marsh birds: restoration and conservation planning in the face of uncertainty. Ecosphere 4(4):49. http://dx.doi.org/10.1890/ES12-00341.1.

10

Residency Patterns and Migration Dynamics of Adult Bull Sharks (*Carcharhinus leucas*) on the East Coast of Southern Africa

Ryan Daly[1,2]*, Malcolm J. Smale[2,4], Paul D. Cowley[3], Pierre W. Froneman[1]

1 Department of Zoology & Entomology, Rhodes University, Grahamstown, South Africa, **2** Port Elizabeth Museum at Bayworld, Humewood, Port Elizabeth, South Africa, **3** South African Institute for Aquatic Biodiversity (SAIAB), Grahamstown, South Africa, **4** Department of Zoology, Nelson Mandela Metropolitan University, Port Elizabeth, South Africa

Abstract

Bull sharks (*Carcharhinus leucas*) are globally distributed top predators that play an important ecological role within coastal marine communities. However, little is known about the spatial and temporal scales of their habitat use and associated ecological role. In this study, we employed passive acoustic telemetry to investigate the residency patterns and migration dynamics of 18 adult bull sharks (195–283 cm total length) tagged in southern Mozambique for a period of between 10 and 22 months. The majority of sharks (n = 16) exhibited temporally and spatially variable residency patterns interspersed with migration events. Ten individuals undertook coastal migrations that ranged between 433 and 709 km (mean = 533 km) with eight of these sharks returning to the study site. During migration, individuals exhibited rates of movement between 2 and 59 km.d^{-1} (mean = 17.58 km.d^{-1}) and were recorded travelling annual distances of between 450 and 3760 km (mean = 1163 km). Migration towards lower latitudes primarily took place in austral spring and winter and there was a significant negative correlation between residency and mean monthly sea temperature at the study site. This suggested that seasonal change is the primary driver behind migration events but further investigation is required to assess how foraging and reproductive activity may influence residency patterns and migration. Results from this study highlight the need for further understanding of bull shark migration dynamics and suggest that effective conservation strategies for this vulnerable species necessitate the incorporation of congruent trans-boundary policies over large spatial scales.

Editor: David Hyrenbach, Hawaii Pacific University, United States of America

Funding: Funding and support were provided by the Professional Development Programme of the South African National Research Foundation (www.nrf.ac.za), Port Elizabeth Museum at Bayworld (www.bayworld.co.za), Rhodes University (www.ru.ac.za), and Bayworld Centre for Research and Education who administered the grant, 3 Fathoms Research, the Acoustic Tracking Array Platform maintained by the South African Institute for Aquatic Biodiversity (www.saiab.co.za) with support from the Ocean Tracking Network (www.oceantrackingnetwork.org) funded in part by the Save Our Seas Foundation (www.saveourseas.com). The funders had no role in study design, data collection and analysis, decision to publish, or preparation of the manuscript. One of the contributing funders of this study (the National Research Foundation of South Africa) only provided funding for one of the authors' salaries (RD). This funding was awarded to the Port Elizabeth Museum at Bayworld and administered by the Bayworld Centre for Research and Education. The Port Elizabeth Museum at Bayworld and Bayworld Centre for Research and Education are not commercial businesses. The Port Elizabeth Museum at Bayworld had no role in the study design, data collection and analysis, decision to publish, or preparation of the manuscript.

Competing Interests: RD and MJS have an affiliation with the institute (Port Elizabeth Museum at Bayworld) that received funding from the National Research Foundation of South Africa for this study. This does not alter the authors' adherence to PLOS ONE policies on sharing data and materials.

* Email: ryandaly.mail@gmail.com

Introduction

The spatial ecology of apex predatory shark species is important to understand when considering their role in structuring marine communities and proposing effective conservation strategies [1,2,3,4]. Large predatory shark species may exert broad scale influences on multiple levels of community structure through predation and associated risk affects [5,6,7]. Quantifying their movement patterns is important for understanding the associated spatial and temporal scales over which these sharks influence their marine communities [8]. Elucidating the role of top predatory sharks in marine communities is important as top marine predatory populations are increasingly under threat from over-exploitation [6,9,10].

Bull sharks (*Carcharhinus leucas*) are large (>3 m), globally distributed sharks that typically occur in tropical and warm temperate coastal marine environments [11]. Adult bull sharks are known to consume a variety of prey from multiple trophic levels [12,13] and may influence marine communities through predation over a broad geographical range [13]. However, little is known about the long-term (>1 yr) temporal and spatial scales over which adult bull sharks typically move. Previous studies suggest that adult bull sharks exhibit high levels of residency interspersed with some degree of short to medium scale geographical migrations linked with seasonal change [11,14,15]. On the east coast of South Africa, bull sharks are known to undertake seasonal migrations into more temperate latitudes during summer [11,16] but have been thought to be largely resident with limited home ranges [12,17].

For a slow growing and late maturing species such as the bull shark [18], exhibiting periods of residency and undertaking large-scale migrations are important life history characteristics that may make the species especially vulnerable to both fishing pressure and habitat loss [19,20]. Evidence suggests that bather protection nets have already depleted bull shark populations on the east coast of South Africa [12,21] and understanding the residency patterns and migration dynamics of this species will help to mitigate future threats and improve conservation strategies [22,23]. In this study, we employed passive acoustic telemetry to investigate the habitat use of adult bull sharks at a remote natural aggregation site in southern Mozambique that is minimally influenced by human activity. Our aim was to investigate the temporal and spatial dynamics of adult bull shark residency patterns at the study site and quantify the timing, frequency and distance of migration events.

Methods

Ethics statement

All research in this investigation was conducted under the permit number 0002/2010 issued by The Mozambican Directorate of National Conservation Areas. The Animal Ethics Committee of the Department of Zoology and Entomology at Rhodes University approved the research protocol used in this study (ethical clearance number ZOOL-14-2012).

Study site

This investigation took place in southern Mozambique within the Ponta do Ouro Partial Marine Reserve (Fig. 1a). This region of coastline lies in a biogeographical transition zone between the subtropical east coast of South Africa and the tropical coast of central Mozambique [24] referred to as the Delagoa bioregion [25]. The focal study site was a reef complex called the Pinnacle reef, located approximately 3.7 km offshore (S26° 44.900′ E32° 56.000′). The reef is a fossilized coastal dune [26] running parallel to the coast and consisting of a ridge approximately 1.2 km long with a series of shallower pinnacles along its spine (Fig. 1b). The reef sandstone substrate is interspersed with sand patches and has scattered alcyonarian coral growth with limited scleractinian coral growth [27].

Shark capture and tagging

Between January 2012 and February 2013, 18 bull sharks were captured and tagged at the Pinnacle reef. Sharks were captured from a 6 m vessel using baited Mustad 18/o circle hooks connected to 450 pound carbon coated steel trace attached to 20 m of nylon rope and two 15-liter surface buoys. Once a shark was hooked, it was left to tire for 20–30 minutes before being brought alongside the vessel where it was inverted to induce a state of tonic immobility and then secured alongside the vessel while remaining partially submerged. Captured sharks were first measured (TL) and then surgically equipped with a VEMCO V16 coded acoustic transmitter (VEMCO Ltd., Halifax, Canada). The tags were implanted into the peritoneal cavity via a 2–3 cm incision in the abdominal wall that was subsequently sutured closed using a surgical needle and thread [28]. Tags were deployed in four main batches and all transmitted at 69 kHz at low power with a nominal delay of 120 seconds (range: 80–160 seconds). Before release, a uniquely coded plastic dart tag (ORI, South Africa) was externally attached to the base of the first dorsal fin to identify internally tagged individuals. To release the shark, the hook was removed from the mouth before turning the shark into an upright position and releasing it. The time from securing the shark alongside the vessel to its release ranged between 10 to 15 minutes. Tagged sharks ranged between 195 and 283 cm TL (mean TL = 254.4 cm) with a M:F ratio of 11:7 (Table 1). Based on the measured length, all sharks were considered sexually mature adults [12,18] and all males (with the exception of ID # 13) had calcified claspers.

Receiver deployment

An array of three VEMCO VR2 acoustic receivers was deployed at the study site (the Pinnacle reef) between November 2011 and January 2012. All receivers were attached directly to the reef substrate using a 4 mm stainless steel anchor chain connected to a 3 m long 14 mm nylon rope suspended by a 200 mm trawl float with the receiver attached vertically to the rope 2.5 m from the substrate. The deployment of three receivers on the Pinnacle reef occurred in locations where reef profile was low and within safe diving limits (Fig. 1b). These three receivers collected data continuously from November 2011 to November 2013. Two additional receivers were deployed inshore of the Pinnacle reef approximately 700 m from the shore and spaced 2.4 km apart. The inshore receivers were maintained from January 2012 to November 2013 although data were retrieved from only one inshore receiver due to equipment failure.

In addition to the local receiver array, data were also obtained from VEMCO VR2W acoustic receiver arrays along the coast and within estuaries in South Africa maintained by the Acoustic Tracking Array Platform (Hout Bay, False Bay, Gansbaai, Mossel Bay, Algoa Bay, Port Alfred, Dwesa MPA, Port St Johns, South Africa/Mozambique border) and the Marine Megafauna Foundation (Tofinho, Pomene and Bazaruto) as well as an array at Aliwal Shoal maintained by the Department of Environmental Affairs: Oceans and Coasts (Fig. 1a). All receivers at the above mentioned locations were operational over the same period as data were collected for this study.

Two temperature loggers (Hobo Water Temp Pro v2), set to record temperature every hour, were deployed at the Pinnacle reef from November 2011 until November 2013 (ongoing). One logger was attached to receiver number 1 (Fig. 1b) at 30 m depth approximately 2 m from the bottom. The second logger was located 3 km inshore of the Pinnacle reef at a depth of 16 m, approximately 2 m from the bottom.

Range testing

Receiver range testing was conducted at the Pinnacle reef to help interpret the detection data at the study site [29,30]. Results from the range test suggested that detection range at the study site was relatively poor (<100 m) but within the same range as previous studies using the same transmitters (VEMCO V16-6L) in similar environments [31,32]. A full description of the range test experiment methods and results are presented in the supporting information of the manuscript (Text S1, Table S1, Figure S1). Based on the results from the range test it can be assumed that receivers 1–3 do not have overlapping detection ranges.

Data analysis

Detection data from the Pinnacle reef study site was downloaded and initially analyzed using the software package VUE version 2.0.6 (VEMCO 2013). Data from all receivers were combined to investigate overall presence and absence at the study site. Daily detections were plotted on a time line. A shark was considered present on a certain day when at least one detection was recorded within 24 hours. Potentially false detection data were examined by

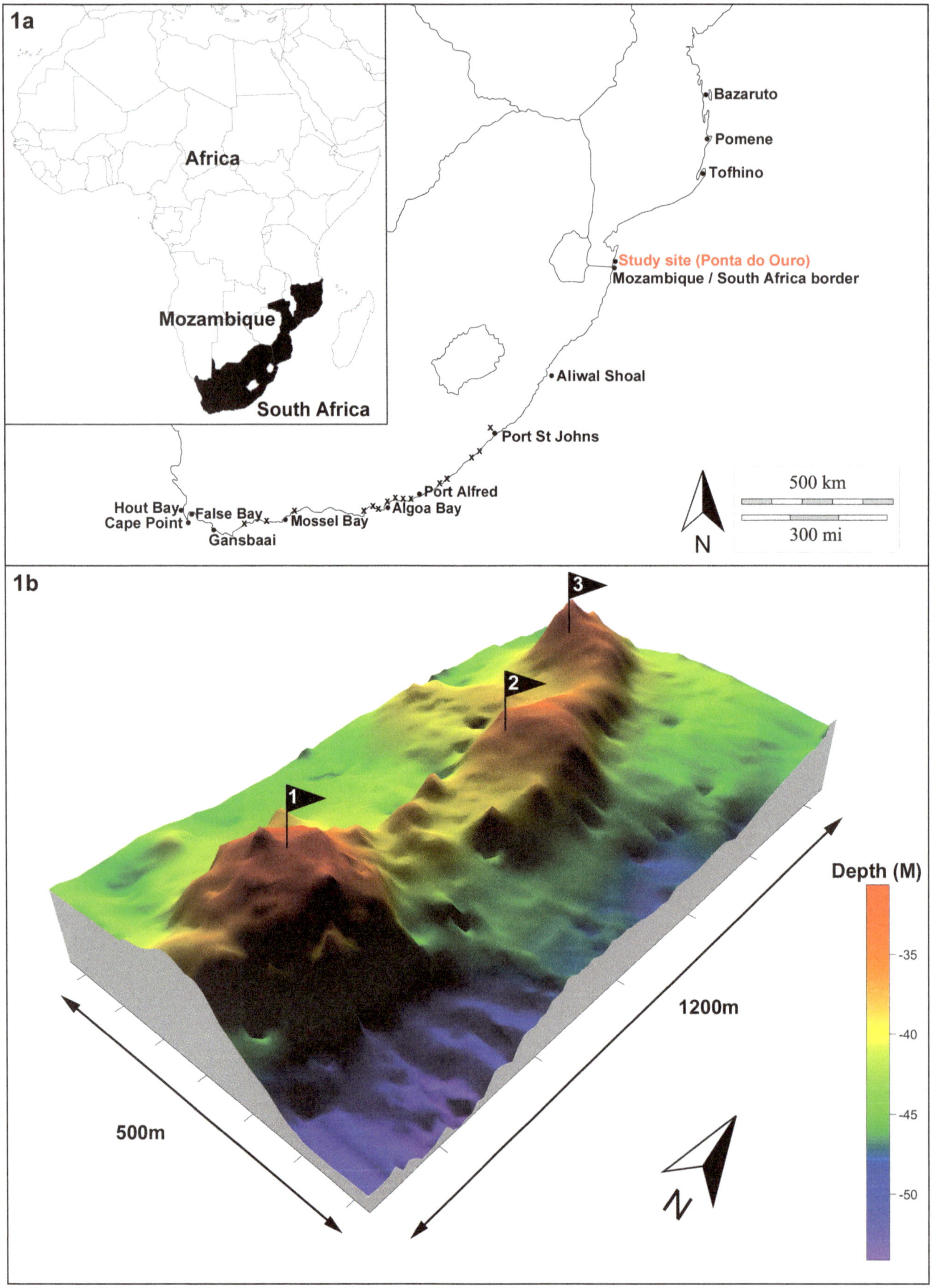
1a
Africa
Mozambique
South Africa
Bazaruto
Pomene
Tofhino
Study site (Ponta do Ouro)
Mozambique / South Africa border
Aliwal Shoal
Port St Johns
Port Alfred
Algoa Bay
Mossel Bay
Hout Bay
False Bay
Cape Point
Gansbaai
500 km
300 mi
N
1b
1
2
3
Depth (M)
-35
-40
-45
-50
1200m
500m
N

Figure 1. Study site Map and Pinnacle reef bathymetry. Figure 1a shows a map of southern Africa with the location of the study site at Ponta do Ouro in southern Mozambique. All other named locations represent acoustic receiver sites on the coast (represented by a circle) or within an estuary (represented by an x) from which additional bull shark detection data was obtained. Figure 1b shows the bathymetry of the Pinnacle reef complex with the colour scale indicating the corresponding depth of the reef showing the shallow spine surrounded by reef drop offs. Numbered flags represent the approximate location of three acoustic receivers on the reef.

considering realistic detection time frames and potential tag collisions and any erroneous data were removed. Combined daily detection data from all Pinnacle receivers for males and females were analyzed for differences using a Students t-test.

Data from all three receivers at the Pinnacle reef were combined to investigate the temporal habitat use of individual bull sharks during periods of continuous presence (Fig 2). Detection data from individual sharks were binned into hourly periods and the total number of detections recorded within each hour of a 24-hour period was represented in rose diagrams using the statistical software package Oriana (version 4, Kovach Computing Services). Rao's Spacing Test [33] was employed to investigate if the temporal presence (represented by the number of detections recorded for each hour of a 24-hour period) of individual bull sharks at the Pinnacle reef was homogenous. To test for uniformity in the temporal detection data, Rao's Spacing Test investigates the null hypothesis that the recorded temporal detection data are evenly distributed throughout each hour of a 24-hour period. The level of statistical significance is determined from a table of simulated critical points [34] with the statistical significance set to $p<0.05$. The time of sunrise ranged between 04:51 h and 06:39 h and the time of sunset ranged between 17:07 h and 18:44 h during the length of this investigation.

Overall residency indices (RI) at the Pinnacle reef were calculated for each shark by dividing the number of days each shark was detected by the total number of days each shark was monitored [35,36]. This measure standardizes the detection data for each shark regardless of the monitoring period by providing a figure of between 0 (shark detected on zero days that it was monitored) and 1 (shark detected on every day that it was monitored) that allows us to compare residency indices between sharks that were monitored over different time periods. An overall residency index was calculated for the entire study period, as well as the mean monthly indices. To investigate any potential bias between the monitoring periods of individual sharks and the calculated residency indices a regression analysis and one-way ANOVA were employed. The overall residency indices of males and females were compared using a Student's t-test. A single linkage hierarchical cluster analysis was also employed to investigate groupings within the sample population based on overall residency indices. Finally, a regression analysis and one-way ANOVA were employed to investigate the relationship between monthly residency indices and mean monthly sea temperatures.

Migration events were defined as directed persistent one-way movements between two receivers over large spatial scales [37]. The overall migration distances for sharks were measured using Garmin Homeport (version 2.2.1) and calculated as the minimum distance along the coast between receiver locations [38]. The rate of travel ($km.day^{-1}$ and $km.h^{-1}$) by a shark was calculated from the minimum distance moved between two receivers (km) divided by the time at liberty (days) during a migration event. Regression analyses with a one-way ANOVA were employed to investigate

Table 1. Summary of all bull sharks tagged with acoustic transmitters at the Pinnacle reef.

Shark ID	Date tagged	TL (cm)	Sex	Total detections	Total number of days detected	Residency Index (RI)
1	21.01.2012	250	M	4135	172	0.26
2	23.01.2012	220	F	4649	86	0.13
3	30.01.2012	260	M	19	5	0.01
4	02.02.2012	280	F	117	15	0.02
5	23.11.2012	282	M	884	76	0.21
6	24.11.2012	270	M	4	1	<0.01
7	07.12.2012	253	F	24	7	<0.01
8	09.12.2012	244	M	1142	36	0.10
9	18.12.2012	263	M	1134	96	0.28
10	20.12.2012	264	M	1667	45	0.13
11	24.01.2013	250	F	1790	40	0.12
12	26.01.2013	262	M	128	8	0.03
13	29.01.2013	195	M	126	12	0.04
14	30.01.2013	251	F	145	21	0.04
15	31.01.2013	283	M	1677	76	0.26
16	03.02.2013	261	F	1546	44	0.15
17	18.02.2013	241	F	996	23	0.08
18	19.02.2013	251	M	10960	169	0.61

TL represents total length. Total detections refer to all detections on all receivers at the study site over the length of the study.

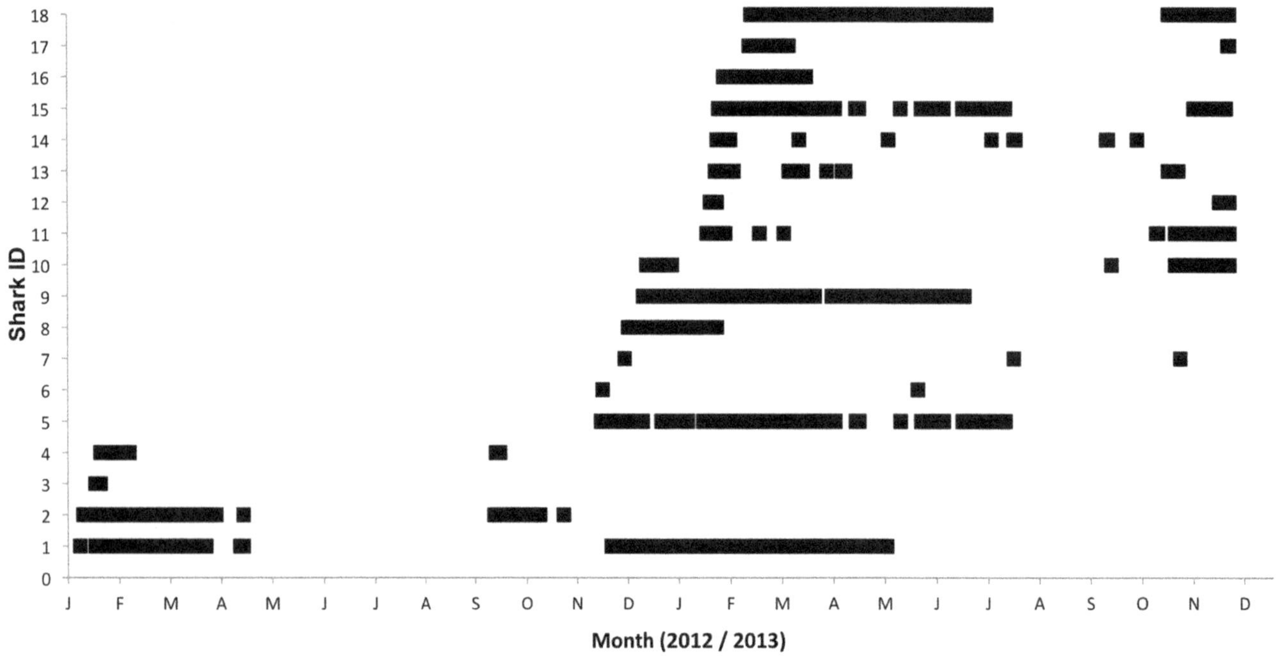

Figure 2. Daily bull shark detection data from all receivers at the study site. Detection data is represented by black squares and presented on a timeline from January 2012 to December 2014 for all 18 tagged sharks. The first detection data point for an individual shark corresponds to the day the shark was tagged at the study site.

any relationships between shark length and migration distance and speed.

Unless otherwise stated, all data analyses were conducted using the statistical package R (CRAN 2009).

Results

Presence/absence at the study site

Bull sharks were monitored at the study site for 664 days between January 2012 and November 2013 (Fig. 2). All sharks were detected on the receiver array at the Pinnacle reef for at least one day after tagging. Individual tag detection totals ranged between 3 and 10958 detections (1724.1±2664.6, mean ± SD) over a period of 1 to 172 days (50.3±52.9, mean ± SD) (Table 1). Shark presence was variable with some sharks (ID # 1,2,5,8,9,15,16,18) exhibiting continuous periods of presence for over one month that ranged between 46 and 158 days (Fig. 2). Although there was not a significant difference (t-test, DF = 6, $p > 0.05$) in the total number of days detected between males and females, males (62.8±62.0, mean ± SD), on average were detected for more days than females (mean = 30.6±27.9, mean ± SD). Most sharks (72%) were detected at the inshore receiver for an average of 5.8 (SD = 9.0) days but nearly all of these individuals (97.1%) were also detected at the Pinnacle reef on the same day.

The majority (15 out of 18) of tagged sharks were also detected at the array of five receivers located at the border of South Africa and Mozambique, approximately 12 km south of the Pinnacle reef. Detections at the border receiver array only occurred over periods when sharks were also detected at the Pinnacle reef without exceptions. The total number of days that sharks were detected on the border receiver array ranged between 1 and 102 days (16.8±29.3, mean ± SD) and the proportion of days that sharks were detected at both the border receiver array and the Pinnacle reef array over the same monitoring period ranged widely between 0.01% and 100% (49±41.6, mean ± SD).

Temporal residency patterns

Figure 3 represents the temporal detection patterns at the Pinnacle reef of all sharks with sufficient data (n = 10). Hourly detection frequencies for all sharks (except ID # 15 and 18) exhibited non-homogeneous distributions (Rao's spacing test, $p < 0.01$). There was some overlap in temporal use of the Pinnacle reef between individuals (ID # 1, 2,5,8,9,16) that appeared to be present primarily from morning (~7am) until early evening (~8pm). Other individuals, however, were primarily present during the evening (ID # 10 and 11) whilst others were present throughout the day and night (ID # 15 and 18). Most sharks (except ID # 15 and 18) exhibited increased habitat use at the Pinnacle reef over a 12-hour period followed by a 12-hour period of absence.

Residency

Bull sharks monitored at the Pinnacle reef exhibited residency indices (RI) ranging between 0.001 and 0.67 (0.14±0.15, mean ± SD) over the duration of the study (Table 1). Results from a hierarchical cluster analysis (Fig. 4) suggest that there were three main groupings of sharks based on their overall residency indices. Five of 18 tagged sharks (ID # 18,5,1,9,15) exhibited relatively high overall residency indices (RI>0.2) with periods of prolonged (>3 consecutive months) high monthly mean residency indices (> 0.3) interspersed with relatively short periods of absence (Fig. 2) and appeared to be seasonal residents. One individual (ID # 18) appeared as an outlier within the seasonal resident group with an overall residency index (RI = 0.6) substantially higher than other individuals within the group. The second group of seven individuals (ID # 3,6,7,13,14,4,12) exhibited very low resident indices (RI<0.1) with only a few detections over the entire study period (Fig. 2). The third group of sharks (ID # 17,16,8,11,2,10) exhibited intermediate residency indices (RI = 0.08–0.2) with

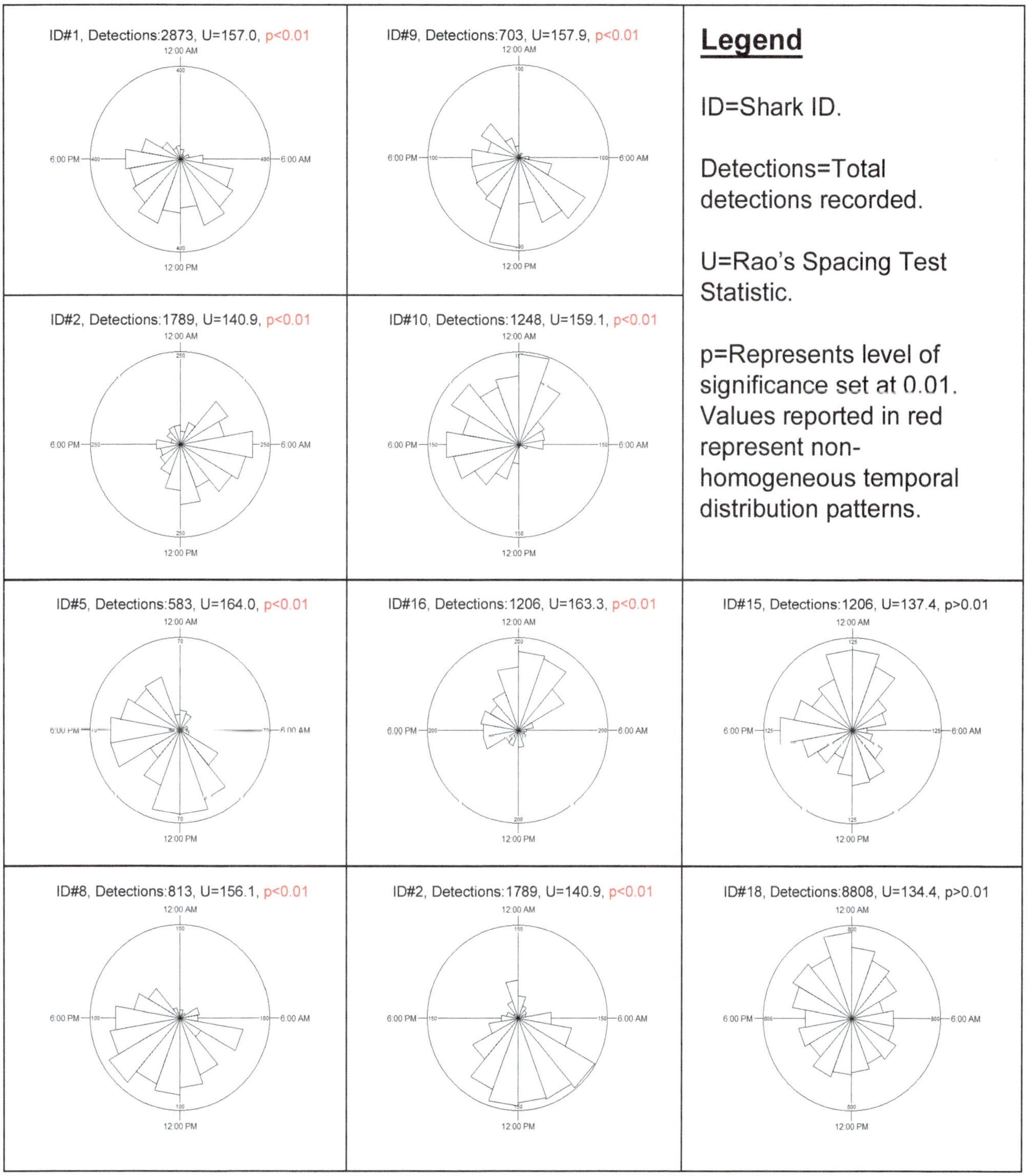

Figure 3. Circular plots showing the temporal detection distribution of bull sharks at the Pinnacle reef. Each plot represents the detection frequency within each hour of a 24-hour period. The temporal distribution of most sharks (except 15 and 18) was significantly non-homogenous (Rao's spacing test, $p<0.01$). Individual sharks not represented had insufficient data for the analysis.

periods (1–3 months) of high monthly mean residency indices (RI>0.3) interspersed with long periods of absence (Fig. 2).

There were no significant relationships between the total monitoring periods of individual sharks and the total number of detections or total days detected or calculated residency indices (ANOVA, DF = 3, $p>0.05$). There was no significant (t-test, DF = 6, $p>0.05$) difference in the overall residency indices exhibited by female or male sharks, however, the mean male residency index (0.18 ± 0.18, mean $\pm$ SD) was higher than the mean female residency index (0.08 ± 0.06, mean $\pm$ SD) and only male sharks exhibited residency indices over 0.2 ($n = 5$). Monthly mean residency indices were highest during summer (Dec-Feb)

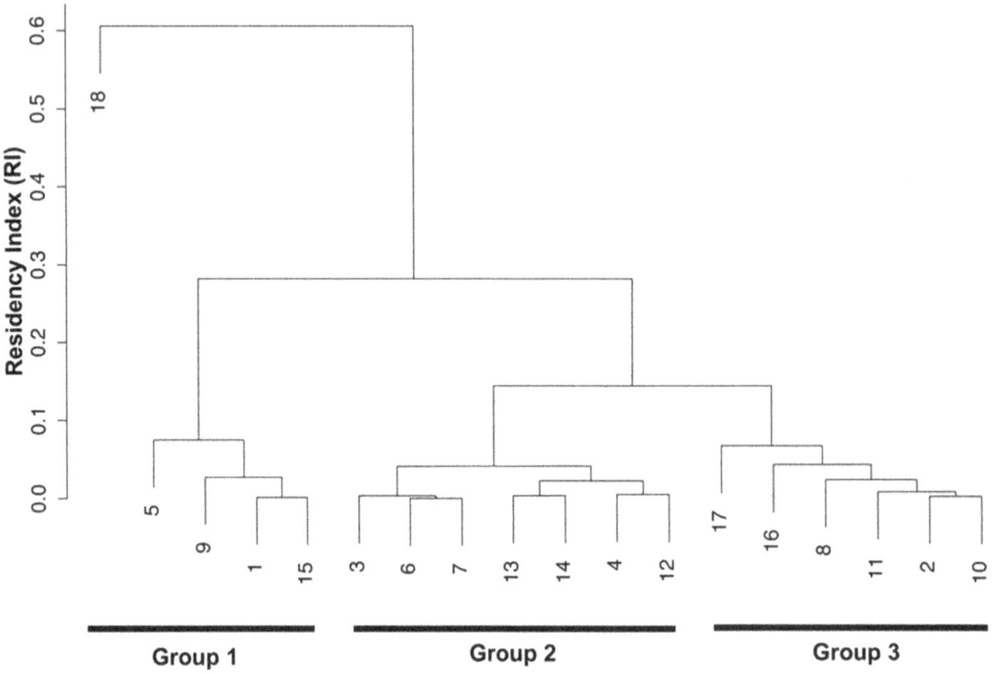

Figure 4. Cluster analysis results. Dendrogram showing the groupings of individual sharks based on their calculated residency indices at the study site.

(mean = 0.33) and autumn (Mar-May) (mean = 0.46), compared with winter (Jun-Aug) (mean = 0.08) and spring (Sep-Nov) (mean = 0.24), and showed a positive correlation ($R^2 = 0.84$, ANOVA, DF = 1, $p<0.05$) with increased water temperature at the study site (Fig. 5).

Migration

Ten out of 18 tagged sharks undertook large-scale migrations (> 400 km) primarily in northerly direction (Fig. 6) with only 1 shark (ID # 7) undertaking a southward migration of 450 km to Aliwal Shoal, South Africa (Fig. 1a). No other detections were recorded at any of the receivers located in South African coastal waters (Fig. 1a). Eight out of ten sharks returned to the study site after

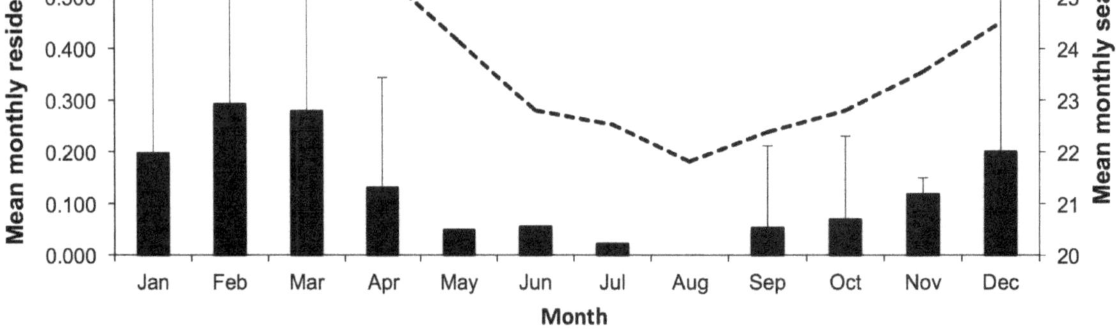

Figure 5. Mean monthly residency indices and mean monthly sea temperature. Bar charts represent the mean monthly residency indices from all tagged bull sharks detected at the Pinnacle reef. Error bars representing standard deviation. The dashed line represents the mean monthly sea temperatures (°C) recorded at the study site.

migration events (ID # 2,4,7,10,11,14,17,18) and one shark undertook multiple return migrations (ID # 14). Individuals that undertook migrations ranged in size between 220 and 280 cm TL (mean = 258.8 cm TL) and 70% were females. The minimum distance that sharks travelled during migration events (persistent and directed one-way movements between receivers) ranged between 433 km and 709 km (533±122.8, mean ± SD) with daily rates of movement ranging between 2.0 km.day^{-1} and 59.1 km.day^{-1} (17.6±15.5, mean ± SD) and with corresponding speed ranging between 0.08 and 2.5 km.h^{-1} (0.75±0.65, mean ± SD). Total annual minimum distances travelled by sharks ranged between 450 and 3760 km (mean = 1163 km). There was no significant relationship (ANOVA, DF = 2, p>0.05) between total length and distance travelled or speed. Nor were there any significant relationships (ANOVA, DF = 2, p>0.05) between gender and distance travelled or speed, although females did exhibit a mean travelling distance (991.7 km±327.3 km, mean ± SD) which was greater than that of males (770 km±266.7 km, mean ± SD). Northward migration events took place between January and September with the majority (75%) during austral autumn and winter (March-August).

Discussion

Evidence provided in this study suggests that the Pinnacle reef in southern Mozambique is an important aggregation site for adult bull sharks and that they exhibit prolonged periods of residency, primarily during the austral summer months. Importantly, this study highlights the geographical scale of movement that adult bull sharks are capable of undertaking along the coast of southern Africa, and suggests that they are not as sedentary as previously thought [12,17]. This study also documents the first return migration events for the species along this coast and emphasizes the need for greater understanding of bull shark migration dynamics.

Presence/absence

Data from this study suggests that the Pinnacle reef is an important habitat for adult bull sharks with tagged individuals exhibiting periods of prolonged presence (Fig. 2). Detection data from the nearby receiver array at the south African/Mozambican border also suggests that bull sharks at the study site do undertake medium-scale movements (<12 km) but the frequency, direction and range of these movements may vary widely between individuals. The majority of tagged sharks (72%) also undertook inshore forays at the study site but further investigation is required to determine the importance of inshore habitat use.

Temporal residency patterns

The majority of tagged sharks appeared to exhibited diel patterns in habitat use during this investigation, which is consistent with previous observations of multiple shark species across a range of habitats [39,40,41]. Previous studies investigating diel patterns in bull shark habitat use found that sharks exhibited increased site fidelity during daylight hours followed by absence or broader habitat use at night [31]. Findings from this investigation showed that the majority of individuals appeared to demonstrate similar diel patterns, although some individuals exhibited higher site fidelity at night while others exhibited equal habitat use throughout a 24-hour cycle. The behavioral drivers behind diel habitat use may vary and include foraging [8], predator avoidance [42] and thermal regulation [43]. In the absence predator constraints, it is likely that foraging behavior is an important driver of temporal habitat use. The significantly biased and varied temporal habitat use patterns of sharks at the study site suggest that individuals may employ a variety of temporally segregated foraging strategies.

Residency

The overall range of residency indices recorded for bull sharks during this study were lower than a previous investigation in Fiji [31], although in this study there was no food provisioning, the receiver array was smaller and the study site was located in an environment with greater seasonal change. There appeared to be individual variation in the levels of residency behavior that could be categorized into three general groupings (Fig. 4) but overall austral summer and autumn residency indices were substantially higher for individuals and mean monthly sea temperatures appeared to be a major driver of habitat use at the study site. Bull shark seasonal habitat use is well documented and may be more pronounced in sub-tropical environments such as the study site [12,16,44].

Although the seasonal change in water temperature appeared to be an important driver of residency at the study site, the wide range of residency indices suggests other drivers, such as reproductive activity and foraging strategies, may also be important factors determining habitat use by bull sharks [20,45,46,47]. The onset of increasing levels of residency exhibited by bull sharks at the study site appeared to be associated with large aggregations of Carangid fish species (mainly *Caranx ignobilis* but also *Carangoides fulvoguttatus, Carangoides gymnostethus* and *Caranx sexfasciatus*), known to be bull shark prey items (Video S1) [12]. However, it was not clear if bull sharks were responding to increased prey abundance or if both the sharks and the fish aggregations were corresponding to environmental cues. Prey availability is a known driver of shark habitat use [48,49], thus it is likely that the increase in seasonal potential prey abundance at the study site influences the foraging behavior and associated residency patterns of bull sharks at the study site.

Migration

Evidence from this study confirmed that bull sharks are capable of substantial migrations [15] and provided the first description of the distance and speed bull sharks travel on return migration events. Migration distances recorded for individuals in this investigation were within range reported for bull sharks [15] but present the longest travel distances recorded for bull sharks in southern Africa. Previous studies in South Africa using conventional tagging [17] and catch data [12] showed that bull sharks were a sedentary species occupying small home ranges. In contrast, this investigation shows that it is more likely that bull sharks along the southern African coast exhibit periods of residency interspersed with substantial geographical migration events.

Most migration events recorded in this investigation were linked with seasonal change at the study site with a northward migration to warmer latitudes during austral winter and spring and a southward movement into more temperate latitudes during summer. This is consistent with previous literature that has shown bull sharks may be more abundant at higher latitudes along the east coast of South Africa during warmer austral summer months [12]. However, one shark (a mature female, ID # 14) undertook multiple return migration events over multiple seasons, suggesting that seasonal change may not be the only driver of migration events. Other large shark species may undertake migrations driven by reproductive activity [38] and prey availability [49]. As female sharks in this investigation were recorded migrating further and undertaking the majority of migration events (70%), it is possible

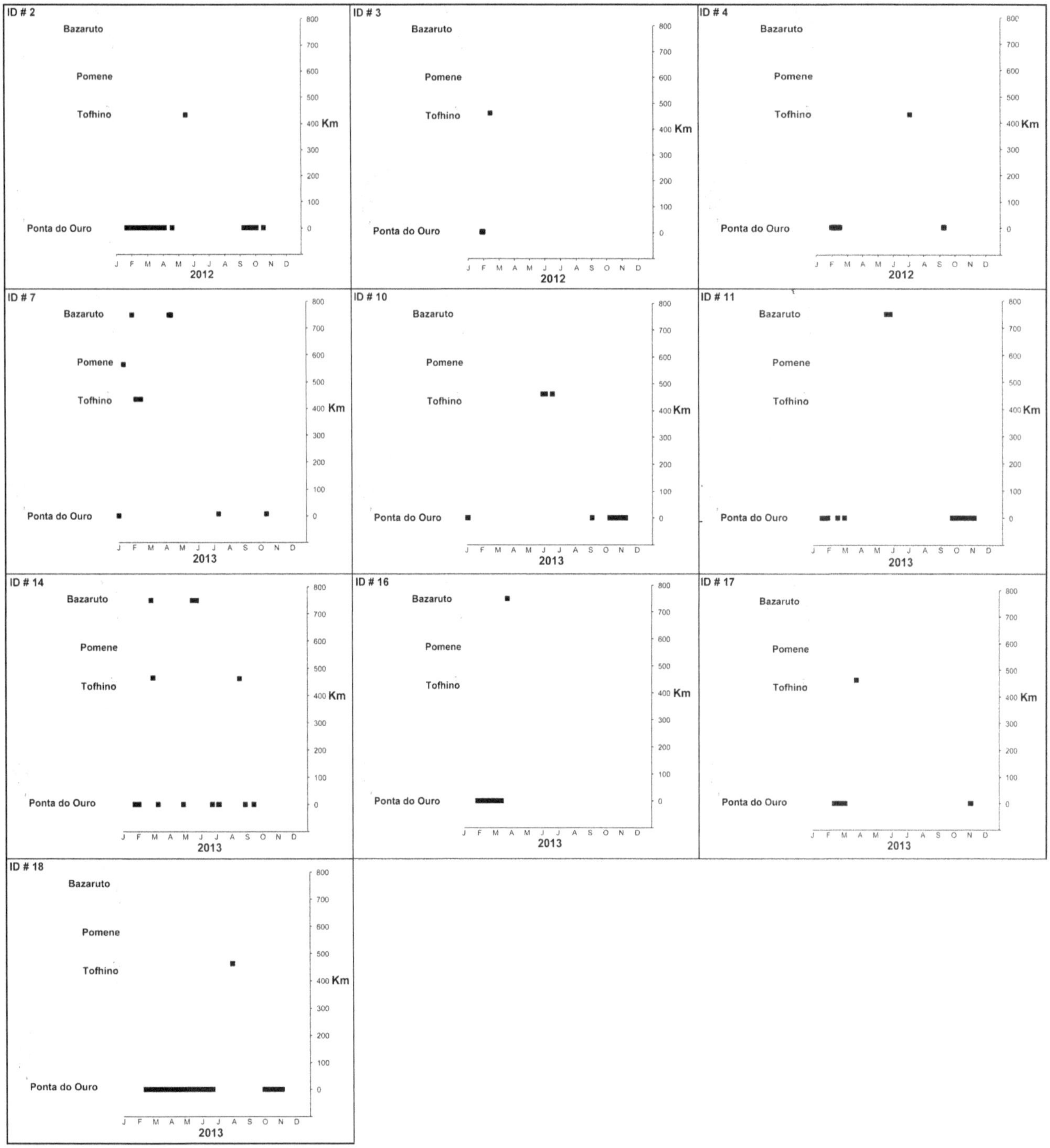

Figure 6. Bull shark migration events along the coast of Mozambique. The x-axis represents a time line showing months and year. The y-axis shows distance along the coast in kilometers (starting at Ponta do Ouro) corresponding with the coastal map of southern Mozambique. The solid square markings show annual bull shark detection events at the study site and other locations within Mozambique.

that gender specific migration linked to reproductive cues were a factor that determined the distance, direction and frequency of migration events.

Evidence from this study also indicate that bull sharks are capable of covering substantial distances over short periods of time with sustained mean (0.74 $km.h^{-1}$) and maximum (2.46 $km.h^{-1}$) rates of movement faster than those previously recorded for bull sharks [15]. Recorded rates of movement were still, however, substantially slower than tiger (3–4 $km.h^{-1}$) and white sharks (4.7 $km.h^{-1}$) undertaking sustained movements [50,51,52] but showed that within a year, adult bull sharks are capable of covering a broad geographical range that may include coastal waters of multiple countries. This supports evidence from a previous investigation that showed adult bull sharks have a geographically broad foraging range along the east coast of southern Africa [13]. Although in this investigation bull sharks were only detected as far north as Bazaruto, it is possible that some sharks travel substantially further north and additional investigation is required to establish the extent of bull sharks northward range.

Summary

Previous investigations using telemetry to monitor adult bull shark habitat use have shown that bull sharks have an affinity for coastal environments, exhibit periods of residency and are capable of substantial geographical migrations [14,15,31,53]. Results of the current study generally supported these findings and provided further understanding of the residency patterns and migration dynamics of bull sharks in southern Africa with the first description of return migration events for the species. Our results suggest that coastal migrations are a frequent and important component of bull sharks habitat use and are primarily driven by seasonal change. Seasonal change is often the driver of animal migration but further investigation is required to understand how spatiotemporally fluctuating resources and reproductive activity influence bull shark residency patterns and migration dynamics [37]. Improving our understanding of bull shark migration dynamics is important to elucidate the spatial scales of their ecological role [54] and plan effective conservation management strategies for this species [23]. A broad range of management approaches encompassing trans-boundary cooperation may be required to effectively conserve critical bull shark habitat.

Supporting Information

Figure S1 Range test results. Proportion of transmissions received vs the expected detections received represented as a percentage (0–100%) from tags deployed at the study site at varying distances, depths and times of day. Error bars represent standard deviation.

Table S1 Range test results represented as mean detection rates expressed as percentages. Distance corresponds to the distance of the receiver from the transmitter (5 m, 100 m, 200 m, 300 m, 400 m), Day or Night corresponds to the time of day detections were recorded and Surface (10 m) or Bottom (25 m) corresponds with the depth at which the transmitter was positioned in the water column. Standard deviations are presented in parenthesis.

Text S1 Range test experiment. Methods, results and discussion of the range test experiment.

Video S1 Carangid fish aggregation video. Video showing adult bull sharks associated with a Carangid fish (*Caranx ignobilis*) aggregation at the study site.

Acknowledgments

We wish to thank the Marine Megafauna Foundation and DEA: Oceans and Coasts for providing data from their respective acoustic receivers. Specific thanks are extended to Daan van Duinkerken from the Marine Megafauna Foundation and their funders (Shark Foundation and the Wildlife Conservation Society) who support their receiver array. Thanks are also due to the Mozambican Department of Conservation Areas and the Ponta do Ouro Partial Marine Reserve and manager Miguel Gonçalves for their support. Thank you also to J.D. Filmalter for his assistance with data analyses and, finally, we wish to thank all those who provided invaluable fieldwork assistance including C.A. Daly, B.G. Skinstad, J.B. Blake and the residents of Ponta do Ouro.

Author Contributions

Conceived and designed the experiments: RD MJS PDC. Performed the experiments: RD. Analyzed the data: RD. Contributed reagents/materials/analysis tools: MJS PDC PWF. Wrote the paper: RD MJS PDC PWF. Obtained permission to conduct the study: RD.

References

1. Speed CW, Field IC, Meekan MG, Bradshaw CJA (2010) Complexities of coastal shark movements and their implications for management. Marine Ecology Progress Series 408: 275–293.
2. Abrantes KG, Barnett A (2011) Intrapopulation variations in diet and habitat use in a marine apex predator, the broadnose sevengill shark *Notorynchus cepedianus*. Marine Ecology Progress Series 442: 133–148.
3. Simpfendorfer CA, Heupel MR, White WT, Dulvy NK (2011) The importance of research and public opinion to conservation management of sharks and rays: a synthesis. Marine and Freshwater Research 62: 518–527.
4. Speed C, Meekan M, Field I, McMahon C, Abrantes K, et al. (2012) Trophic ecology of reef sharks determined using stable isotopes and telemetry. Coral Reefs 31: 357–367.
5. Myers RA, Baum JK, Shepherd TD, Powers SP, Peterson CH (2007) Cascading Effects of the Loss of Apex Predatory Sharks from a Coastal Ocean. Science 315: 1846–1850.
6. Ferretti F, Worm B, Britten GL, Heithaus MR, Lotze HK (2010) Patterns and ecosystem consequences of shark declines in the ocean. Ecology Letters 13: 1055–1071.
7. Heithaus MR, Wirsing AJ, Dill LM (2012) The ecological importance of intact top-predator populations: a synthesis of 15 years of research in a seagrass ecosystem. Marine and Freshwater Research 63: 1039–1050.
8. Speed CW, Meekan MG, Field IC, McMahon CR, Stevens JD, et al. (2011) Spatial and temporal movement patterns of a multi-species coastal reef shark aggregation. Marine Ecology Progress Series 429: 261–275.
9. Worm B, Davis B, Kettemer L, Ward-Paige CA, Chapman D, et al. (2013) Global catches, exploitation rates, and rebuilding options for sharks. Marine Policy 40: 194–204.
10. Dulvy NK, Fowler SL, Musick JA, Cavanagh RD, Kyne PM, et al. (2014) Extinction risk and conservation of the world's sharks and rays. eLife 3: http://dx.doi.org/10.7554/eLife.00590
11. Compagno LJV (1984) Sharks of the world. An annotated and illustrated catalogue of shark species to date. Part II (*Carcharhiniformes*). FAO Fisheries Synopsis: 250–655.
12. Cliff G, Dudley SFJ (1991) Sharks caught in the protective gill nets off Natal, South Africa. 4. The bull shark *Carcharhinus leucas* Valenciennes. South African Journal of Marine Science 10: 253–270.
13. Daly R, Froneman PW, Smale MJ (2013) Comparative Feeding Ecology of Bull Sharks (*Carcharhinus leucas*) in the Coastal Waters of the Southwest Indian Ocean Inferred from Stable Isotope Analysis. PLoS ONE 8: e78229.
14. Brunnschweiler JM, Queiroz N, Sims DW (2010) Oceans apart? Short-term movements and behaviour of adult bull sharks *Carcharhinus leucas* in Atlantic and Pacific Oceans determined from pop-off satellite archival tagging. Journal of Fish Biology 77: 1343–1358.
15. Carlson JK, Ribera MM, Conrath CL, Heupel MR, Burgess GH (2010) Habitat use and movement patterns of bull sharks *Carcharhinus leucas* determined using pop-up satellite archival tags. Journal of Fish Biology 77: 661–675.
16. Compagno LJV, Smale MJ (1986) Recent records of four warm-water elasmobranchs from the Eastern Cape province, South Africa. South African Journal of Marine Science 4: 11–15.
17. Mann BQ (2000) Southern African Marine Linefish Status Reports. Oceanographic Research Institute Special Report: 265p.
18. Wintner SP, Dudley SFJ, Kistnasamy N, Everett B (2002) Age and growth estimates for the Zambezi shark, *Carcharhinus leucas,* from the east coast of South Africa. Marine and Freshwater Research 53: 557–567.
19. Smith SE, Au DW, Show C (1998) Intrinsic rebound potentials of 26 species of Pacific sharks. Marine and Freshwater Research 49: 663–678.
20. Barnett A, Abrantes KG, Stevens JD, Semmens JM (2011) Site fidelity and sex-specific migration in a mobile apex predator: implications for conservation and ecosystem dynamics. Animal Behaviour 81: 1039–1048.
21. Dudley SFJ, Simpfendorfer CA (2006) Population status of 14 shark species caught in the protective gillnets off KwaZulu-Natal beaches, South Africa, 1978–2003. Marine and Freshwater Research 57: 225–240.
22. Warkentin IG, Hernández D (1996) The conservation implications of site fidelity: A case study involving nearctic-neotropical migrant songbirds wintering in a Costa Rican mangrove. Biological Conservation 77: 143–150.
23. Shuter JL, Broderick AC, Agnew DJ, Jonzen N, Godley BJ, et al. (2011) Conservation and management of migratory species. In: Milner-Gulland EJ,

Fryxell JM, Sinclair ARE, editors. Animal migration: a synthesis: Oxford University Press. pp. 172–206.
24. Turpie JK, Beckley LE, Katua SM (2000) Biogeography and the selection of priority areas for conservation of South African coastal fishes. Biological Conservation 92: 59–72.
25. Sink K, Harris J, Lombard A (2004) South African National Spatial Biodiversity Assessment 2004: Technical Report Vol. 4 Marine Component 11.
26. Ramsay PJ (1994) Marine geology of the Sodwana Bay shelf, southeast Africa. Marine Geology 120: 225–247.
27. Riegl B, Schleyer MH, Cook PJ, Branch GM (1995) Structure of Africa's Southernmost Coral Communities. Bulletin of Marine Science 56: 676–691.
28. Stevens JD (1999) Shark tagging: a brief history of methods. Fish Movement and Migration 65–68.
29. Kessel ST, Cooke SJ, Heupel MR, Hussey NE, Simpfendorfer CA, et al. (2013) A review of detection range testing in aquatic passive acoustic telemetry studies. Reviews in Fish Biology and Fisheries: 1–20.
30. Welsh JQ, Fox RJ, Webber DM, Bellwood DR (2012) Performance of remote acoustic receivers within a coral reef habitat: implications for array design. Coral Reefs 31: 693–702.
31. Brunnschweiler JM, Barnett A (2013) Opportunistic Visitors: Long-Term Behavioural Response of Bull Sharks to Food Provisioning in Fiji. PLoS ONE 8: e58522.
32. How JR, de Lestang S (2012) Acoustic tracking: issues affecting design, analysis and interpretation of data from movement studies. Marine and Freshwater Research 63: 312–324.
33. Batschelet E (1981) Circular statistics for biology. London Academic Press.
34. Russell GS, Levitin DJ (1996) An expanded table of probability values for Rao's Spacing Test. Communications in Statistics: Simulation and Computation 24: 879–888.
35. Bond ME, Babcock EA, Pikitch EK, Abercrombie DL, Lamb NF, et al. (2012) Reef sharks exhibit site-fidelity and higher relative abundance in marine reserves on the Mesoamerican barrier reef. PLoS ONE 7: e32983.
36. Welsh JQ, Bellwood DR (2012) How far do schools of roving herbivores rove? A case study using *Scarus rivulatus*. Coral Reefs 31:991–1003.
37. Dingle H, Drake VA (2007) What is migration? Bioscience 57: 113–121.
38. Bansemer CS, Bennett MB (2011) Sex- and maturity-based differences in movement and migration patterns of grey nurse shark, *Carcharias taurus*, along the eastern coast of Australia. Marine and Freshwater Research 62: 596–606.
39. Heupel MR, Yeiser BG, Collins AB, Ortega L, Simpfendorfer CA (2010) Long-term presence and movement patterns of juvenile bull sharks, *Carcharhinus leucas*, in an estuarine river system. Marine and Freshwater Research 61: 1–10.
40. Barnett A, Abrantes KG, Seymour J, Fitzpatrick R (2012) Residency and Spatial Use by Reef Sharks of an Isolated Seamount and Its Implications for Conservation. PLoS ONE 7: e36574.
41. Filmalter JD, Dagorn L, Cowley PD (2013) Spatial behaviour and site fidelity of the sicklefin lemon shark *Negaprion acutidens* in a remote Indian Ocean atoll. Marine Biology 160: 2425–2436.
42. Heupel MR, Simpfendorfer C (2005) Quantitative analysis of aggregation behavior in juvenile blacktip sharks. Marine Biology 147: 1239–1249.
43. DiGirolamo AL, Gruber SH, Pomory C, Bennett WA (2012) Diel temperature patterns of juvenile lemon sharks *Negaprion brevirostris*, in a shallow-water nursery. Journal of Fish Biology 80: 1436–1448.
44. Brunnschweiler JM, Baensch H (2011) Seasonal and Long-Term Changes in Relative Abundance of Bull Sharks from a Tourist Shark Feeding Site in Fiji. PLoS ONE 6: e16597.
45. Weng KC, Castilho PC, Morrissette JM, Landeira-Fernandez AM, Holts DB, et al. (2005) Satellite Tagging and Cardiac Physiology Reveal Niche Expansion in Salmon Sharks. Science 310: 104–106.
46. Papastamatiou YP, Meyer CG, Carvalho F, Dale J, Hutchinson M, et al. (2013) Telemetry and random walk models reveal complex patterns of partial migration in a large marine predator. Ecology 94: 2595–2606.
47. Werry JM, Planes S, Berumen ML, Lee KA, Braun CD, et al. (2014) Reef-Fidelity and Migration of Tiger Sharks, *Galeocerdo cuvier*, across the Coral Sea. PLoS ONE 9: e83249.
48. Fitzpatrick R, Thums M, Bell I, Meekan MG, Stevens JD, et al. (2012) A comparison of the seasonal movements of tiger sharks and green turtles provides insight into their predator-prey relationship. PLoS ONE 7: e51927.
49. Kock A, O'Riain MJ, Mauff K, Meÿer M, Kotze D, et al. (2013) Residency, Habitat Use and Sexual Segregation of White Sharks, *Carcharodon carcharias*, in False Bay, South Africa. PLoS ONE 8: e55048.
50. Holland KN, Wetherbee BM, Lowe CG, Meyer CG (1999) Movements of tiger sharks (*Galeocerdo cuvier*) in coastal Hawaiian waters. Marine Biology 134: 665–673.
51. Bonfil R, Meyer M, Scholl MC, Johnson R, O'Brien S, et al. (2005) Transoceanic migration, spatial dynamics and population linkages of White Sharks. Science 310:100–103.
52. Heithaus MR, Wirsing AJ, Dill LM, Heithaus LI (2007) Long-term movements of tiger sharks satellite-tagged in Shark Bay, Western Australia. Marine Biology 151: 1455–1461.
53. Hammerschlag N, Luo J, Irschick DJ, Ault JS (2012) A Comparison of Spatial and Movement Patterns between Sympatric Predators: Bull Sharks (*Carcharhinus leucas*) and Atlantic Tarpon (*Megalops atlanticus*). PLoS ONE 7: e45958.
54. Holdo RM, Holt RD, Sinclair ARE, Godley BJ, Thirgood S (2011) Migration impacts on communities and ecosystems: empirical evidence and theoretical insights. In: Animal migration, a synthesis. Eds Milner-Gulland EJ, Fryxell JM and Sinclair ARE. Oxford University Press 2011. pp. 131–143.

11

Effect of Site Level Environmental Variables, Spatial Autocorrelation and Sampling Intensity on Arthropod Communities in an Ancient Temperate Lowland Woodland Area

Jakub Horak*

Department of Forest Protection and Entomology, Faculty of Forestry and Wood Sciences, Czech University of Life Sciences Prague, Prague, Czech Republic

Abstract

The interaction of arthropods with the environment and the management of their populations is a focus of the ecological agenda. Spatial autocorrelation and under-sampling may generate bias and, when they are ignored, it is hard to determine if results can in any way be trusted. Arthropod communities were studied during two seasons and using two methods: window and panel traps, in an area of ancient temperate lowland woodland of Zebracka (Czech Republic). The composition of arthropod communities was studied focusing on four site level variables (canopy openness, diameter in the breast height and height of tree, and water distance) and finally analysed using two approaches: with and without effects of spatial autocorrelation. I found that the proportion of variance explained by space cannot be ignored (≈20% in both years). Potential bias in analyses of the response of arthropods to site level variables without including spatial co-variables is well illustrated by redundancy analyses. Inclusion of space led to more accurate results, as water distance and tree diameter were significant, showing approximately the same ratio of explained variance and direction in both seasons. Results without spatial co-variables were much more disordered and were difficult to explain. This study showed that neglecting the effects of spatial autocorrelation could lead to wrong conclusions in site level studies and, furthermore, that inclusion of space may lead to more accurate and unambiguous outcomes. Rarefactions showed that lower sampling intensity, when appropriately designed, can produce sufficient results without exploitation of the environment.

Editor: Wolfgang Arthofer, University of Innsbruck, Austria

Funding: This study was supported by a grant of the Ministry of Agriculture of the Czech Republic (no. QH91097). The funders had no role in study design, data collection and analysis, decision to publish, or preparation of the manuscript.

Competing Interests: The author has declared that no competing interests exist.

* E-mail: jakub.sruby@gmail.com

Introduction

Research on diversity of arthropods, management of populations and their interaction with the environment is one of the main topics of the present ecological agenda, especially with respect to potential biotical and abiotic threats [1]. However, recent research on biota suffers from many pitfalls [2], which may lead to biased conclusions [3].

Arthropods are, mostly quickly, responding to environmental changes [1] and knowledge of their response to habitat parameters continues to increase [4,5]. Many arthropods are influenced by the fact that they are often dispersal-limited and thus not able to reach more distant habitats [6,7]. Response of arthropods to the environmental variables may be biased by spatial structure of their distribution [8].

With respect to data analyses, spatial autocorrelation is one of the statistical problems encountered when modelling species-environmental relationships [9,10]. Spatial heterogeneity is defined as either the variation in space in the distribution of a point pattern, or as variation in the qualitative or quantitative value of a surface pattern [11,12], which can be caused also by site level factors [13]. Spatial dependency within geographic space leads to the spatial autocorrelation [14]. It is known that all of nature is autocorrelated that everything is related to everything else and that objects close to each other are more related than those that are further apart [15] – thus, spatial autocorrelation generates possible bias [16]. Although the existence of spatial autocorrelation does not in itself constitute real bias, it does in terms of what ecologists want to understand [9]. Spatial autocorrelation can be fundamental to building a spatial component into statistical models [17]. If spatial autocorrelation is ignored it is always hard to determine if results can in any way be trusted [3].

Present studies indicate that potential bias of spatial autocorrelation is not absolutely dependent on scale of trapping designs [8,9,17,18]. Thus, mass trapping is not necessary for site level type of studies. Lower trapping intensity in appropriately designed studies gain useful results and may lead to a higher significance without exploitation of the environment [19].

The main goals of this study were to determine the response of arthropod communities to environmental variables at the site level in an area of continuous lowland woodland and to determine how their response could be influenced by sampling intensity and spatial autocorrelation, while employing commonly used trapping methods at the same trees during two seasons.

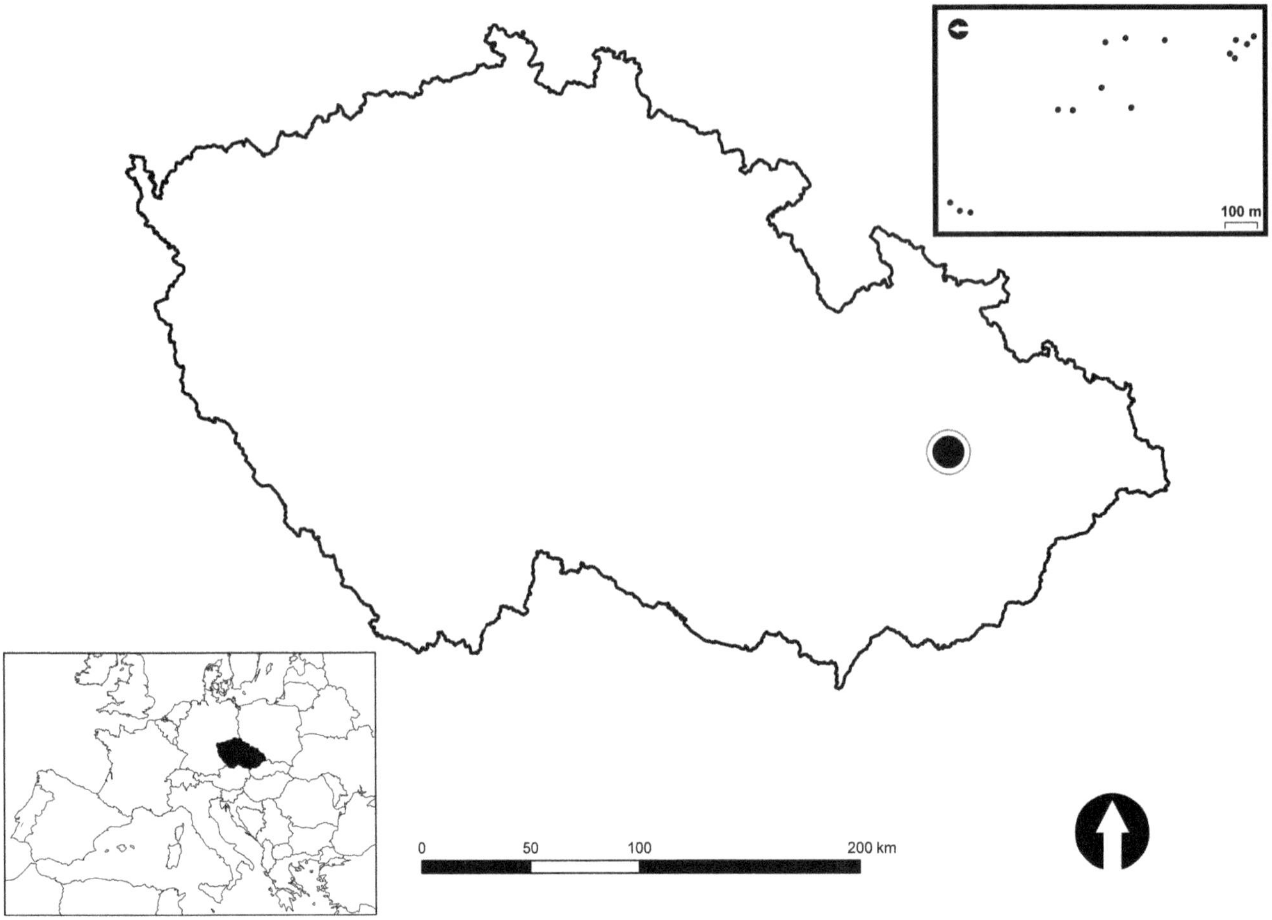

Figure 1. Map of the study sites in Zebracka (Czech Republic).

Materials and Methods

Ethics Statement

All necessary permits were obtained for the described study. As the study area was a part of the protected area, this study was undertaken with the permission of the Czech Government (no. 1473/09).

Study Area

All sample occasions were situated within approximately 50 ha situated in the national nature reserve of Zebracka (Prerov,

Table 1. Description and comparison of trapping methods used in deciduous woodland area of Zebracka.

Character	Window trap	Panel trap
Season	2009	2010
Number	15	15
Intercept area (m^2)	0.800	0.063
Colour	Blank	Yellow
Activity	Passive	Active
Medium	Water+NaCl+detergent [22]	Soveurode® Witasek [23]
Selection	Not known	Larger (>1.5 cm) individuals
Trapping activity	20.V.–12.IX.	9.IV.–11.IX.
Trapping days per trap	115	155
Placement	Trunk	Trunk
Height of the centre (m)	1.3	1.3
Irradiance	South	South

Olomouc Region, Central Moravia, Czech Republic; Fig. 1), one of the most continuous (i.e. ancient) deciduous woodlands near the River Becva [20], known from at least the 1750s (i.e. the time of the first Austro-Hungarian Military Mapping). Zebracka is known to be one of the species rich woodlands in Moravia [21].

Sampling Methods

During the 2009 season, fifteen crossed-panel window traps [22] were used, replaced in 2010 by panel traps [23] – two commonly used methods. All traps were placed on the stems of standing veteran poplars (*Populus*) at 1.3 m. The sampling methods are described in Table 1. All trapped arthropods were identified to the taxon level listed in Table 2.

Study Variables

I focused on four site level variables that potentially most influenced the occurrence of the studied taxa of arthropods (Table 3): (1) Canopy openness (an expression of light conditions) was measured under full foliage using a Sigma 4.5 mm F2.8 EX DC Circular FISHEYE HSM on July 15, 2010. Photographs were evaluated using a Gap Light Analyser 2.0 [24]. (2) Diameter at breast height of tree (an expression of tree diameter) was calculated from the circumference of a tree at 1.3 m from ground level. (3) Height of a tree (an expression of vertical biological frontier) was estimated and rounded in metres in the field. (4) Water distance (an expression of humidity) was calculated from the Euclidean distance of a tree to the River Becva.

Table 2. Results of trapped taxa (sorted in alphabetical order) and abundances during the study seasons 2009 and 2010 in a deciduous woodland area of Zebracka.

Taxa	2009	2010	Total
Araneae	383	184	567
Coleoptera	1,756	2,256	4,012
Collembola	15	–	15
Dermaptera	1,090	1	1,091
Diptera	546	9,485	10,031
Ephemeroptera	9	2	11
Glomerida	104	–	104
Hemiptera: Heteroptera	95	–	95
Hemiptera: Sternorrhyncha	106	19,019	19,125
Hymenoptera	2,413	1,269	3,682
Chilopoda	3	–	3
Isopoda	41	–	41
Ixodida	20	–	20
Julida	182	6	188
Lepidoptera	284	74	358
Mecoptera	10	12	22
Neuroptera	–	11	11
Opilionida	84	4	88
Orthoptera	37	–	37
Prostigmata	5	–	5
Pseudoscorpionida	6	2	8
Pulmonata	76	–	76
Raphidioptera	14	–	14
Trichoptera	39	1	40

Table 3. Results of descriptive statistics of study site level variables in a deciduous woodland area of Zebracka (sorted in alphabetical order).

Variable	Mean	S.E.	Minimum	Maximum
Canopy openness (%)	28.64	3.97	11.29	57.95
Diameter (cm)	82.29	6.35	49.68	135.35
Height (m)	13.73	0.50	10.00	15.00
Water distance (m)	104.87	18.47	3.00	230.00

Statistical Analyses

For identification of sufficient number of trapping occasions, sample-based rarefaction with 95% confidence intervals computed using a Mao Tau function [2] and the Chao estimation function [25] were used. Analyses were computed in EstimateS 8.2 [26]. The number of randomisations was set at 1,000, with strong hash encryption and randomisation of samples without replacement. The upper abundance limit for rare or infrequent species was set at 10. I used the classic formulae for Chao for bias correction [26].

Correlation of taxa between seasons was evaluated using Spearman correlation coefficient in R.

Spatial autocorrelation was preliminary tested by randomized Geary's C test using packages spdep and RANN in R [27].

For the final analyses with and without space, I used multivariate statistical methods provided by CANOCO for Windows version 4.5 [28]. All species data were square-root transformed, as is recommended for trapping designs [29]. The length of gradient in detrended correspondence analysis (DCA) in each season was at <2, which demonstrated that the data did not reveal high heterogeneity [28]. I therefore used redundancy analysis (RDA), a constrained linear ordination method [29]. I used focused scaling on inter-species correlations, and species

Table 4. Correlation of taxa distribution in traps between the study seasons 2009 and 2010 in a deciduous woodland area of Zebracka (sorted in alphabetical order).

Taxa	*r*	*p*
Araneae	−0.05	0.87
Coleoptera	0.10	0.72
Dermaptera	−0.17	0.55
Diptera	0.21	0.46
Ephemeroptera	−0.14	0.64
Hemiptera	0.12	0.67
Hymenoptera	0.27	0.34
Julida	0.89	<0.0001
Lepidoptera	0.41	0.13
Mecoptera	0.38	0.16
Opilionida	−0.23	0.25
Pseudoscorpionida	−0.15	0.58
Trichoptera	−0.34	0.21

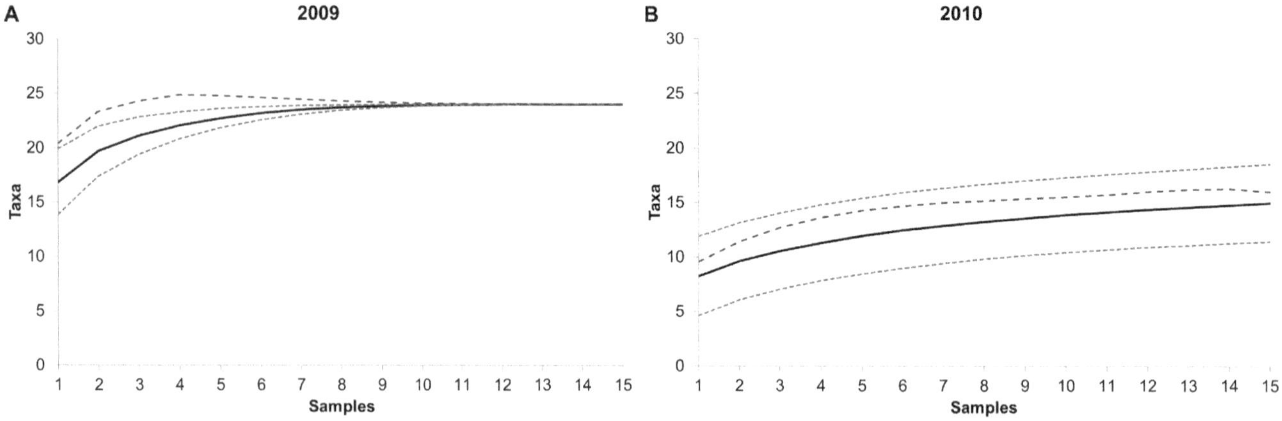

Figure 2. Taxa rarefactions and estimates of total richness of the trapped arthropods in the deciduous woodland. Complete data for taxa from all samples are included from a) 2009 and b) 2010. The solid black line shows a sample-based rarefaction of assemblages; the two surrounding light-grey dashed lines are Mao Tau estimates with 95% confidence intervals and the dark-grey dashed line is the Chao 1 estimate of the total number of taxa.

scores divided by S.D., Monte-Carlo permutation tests with significance of canonical axes together (9,999 permutations) under the full model were used [29].

During the process of variance partitioning [30], two sets of environmental categories were used. Each fraction was measured based on three multivariate analyses using combinations of taxa, variables and co-variables [9,31]. The first set of explanatory variables was composed of five most commonly spatial variables [32], namely geographical coordinates of the sampling points (x, y) and their squares and cross-product terms (x^2, y^2, xy) [8]. The cubic terms [30] were not included, keeping a comparable number of factors for the three explanatory data sets [8]. Thus, I used firstly space (x, y, xy, x^2 and y^2) as variables and habitat (Canopy openness, Diameter, Height and Water distance) as co-variables, then the same in reverse order, and finally with all variables included [9]. I visualised the results of variance partitioning using a two-circle Venn diagram.

The shared explained variance and p values of habitat variables were employed with a Monte Carlo permutation test (9,999 permutations) under the full model [29]. For the resulting ordination diagrams, I used RDA environmental ordination plots created in CanoDraw 4.14 [28].

Results

The number of trapped taxonomical groups (Table 2) was 23 (7,318 individuals) in 2009 and 14 (32,326 individuals) in 2010. Total number of taxa was 24 (39,644 individuals), while 13 taxa overlapped in both seasons (Table 2; 4).

Taxa Accumulations and Correlations in Distribution between Seasons

Sample-based rarefactions were made separately for each season (Fig. 2) and indicated that trapping success was higher for window traps than for panel traps. The curve reached its asymptote in 2009 (Fig. 2a), but not in 2010. However, the Chao estimate seemed to approach the total number of taxa (Fig. 2b), suggesting that the majority of the taxa in the study area were represented in the forthcoming analysis and that the number of samples was sufficient. Except of millipedes from order Julida, there was no correlation structure between 13 taxa overlapped in both seasons (Table 4).

Preliminary Tests on Spatial Autocorrelation of Taxa Richness and Habitat Variables

Richness of study taxa was not spatially autocorrelated in either year (Geary's $C_{2009} = 0.78$, $p = 0.19$; Geary's $C_{2010} = 0.94$, $p = 0.41$), with the same being the case for Diameter (Geary's

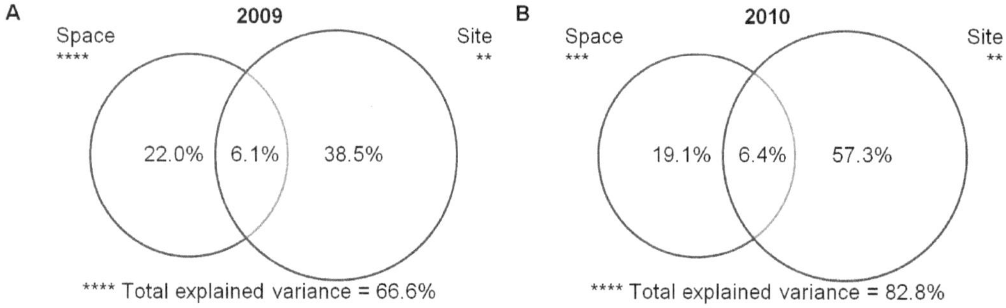

Figure 3. Venn diagrams. Figures are demonstrating percentage values of total, shared (values in circle overlaps) and independent explained variance of space and site using variance partitioning in a) 2009 and b) 2010 in a deciduous woodland area of Zebracka (**$p<0.01$; ***$p<0.001$; ****$p=0.0001$).

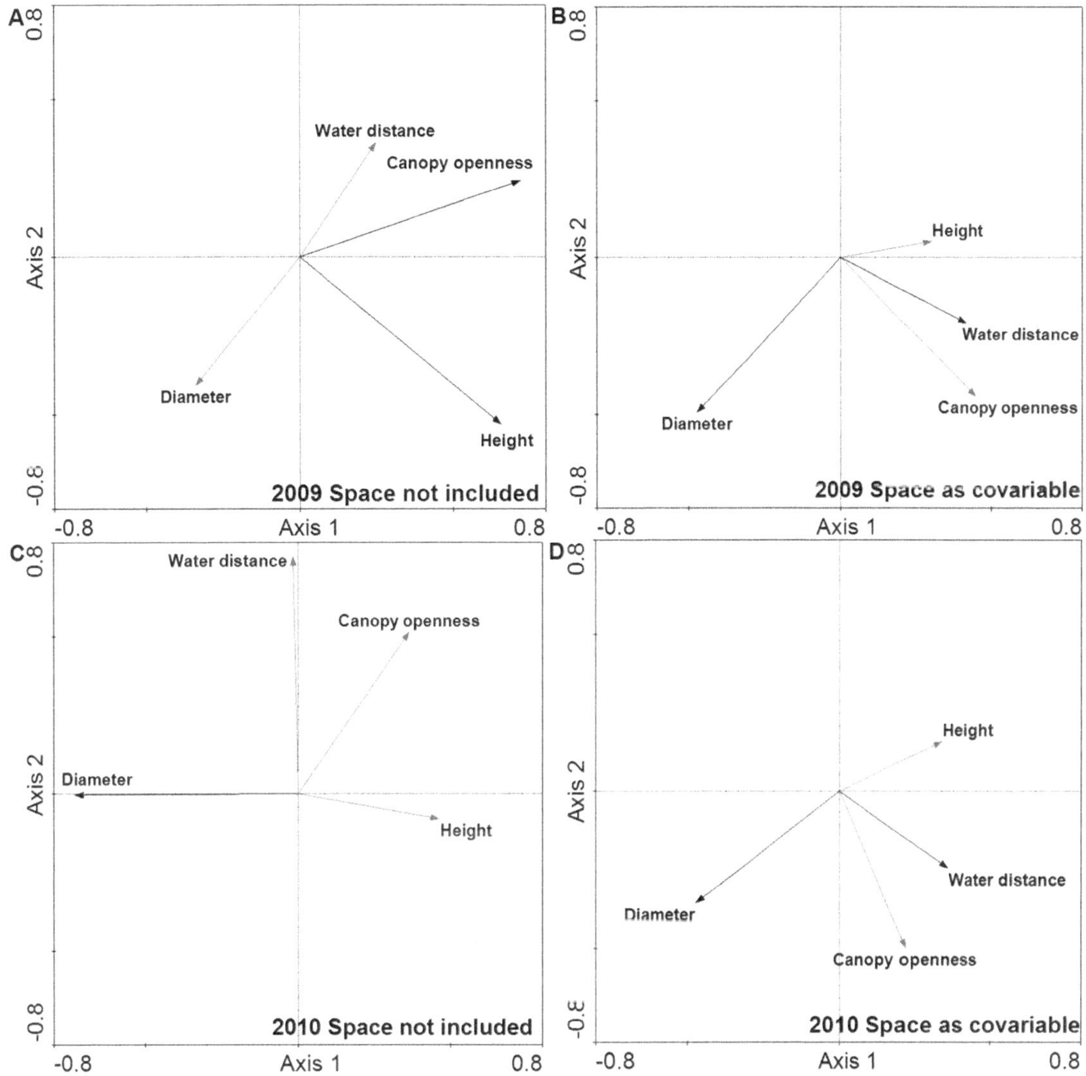

Figure 4. Relative position of environmental site level variables in RDA environmental ordination plots. Figures are with respect to taxa composition in a) 2009 without spatial covariables, b) 2009 with spatial covariables, c) 2010 without spatial covariables, and d) 2010 with spatial covariables, all in a deciduous woodland area of Zebracka (habitat variables $p<0.05$, black; p = n.s., grey).

C = 0.67, p = 0.10) and Height of tree (Geary's C = 0.63, p = 0.08). On the other hand, two site level variables were spatially autocorrelated: Canopy openness (Geary's C = 0.34, p = 0.0060) and Water distance (Geary's C = 0.13, p = 0.0005).

Response of Arthropod Communities using Multivariate Statistics

Venn diagrams (Fig. 3) demonstrate that the total explained variance was high in both seasons. Site level variables explained a high level of variance (especially in 2010) and the percentage of shared variance did not much differ between study seasons. In contrast, the significance of space was high in both seasons. This indicated potential bias in the response of arthropods to site level variables in analyses that do not include space.

The main difference was that environmental site level variables were in all cases differently rotated along the first axis. It can be seen that Diameter is separated from other variables on the second axis, and that Height has a different influence than Canopy openness and Water distance, both of which have nearly the same effect on taxa composition (Fig. 4).

The analyses showed that the level of total explained variance was lower when the effect of space was included in the analyses. On the other hand, the significance of the results at the site level increased (Table 5). Furthermore, the values of shared variance and significance differed strikingly between analyses with and without spatial variables (Table 5) and the relative position of habitat variables in RDA was highly variable (Fig. 4). Including space led to more accurate results, because Water distance and Diameter were significant, showing approximately the same ratio of variance and direction in both seasons (Table 5). The results without spatial co-variables were much more disordered (e.g. the effect of Diameter) and were also relatively difficult to explain (Table 5; Fig. 4).

Discussion

As arthropods dominate the landscape with respect to their abundance and richness, local trapping success may be highly

Table 5. Results of taxa composition response to study site level variables in a deciduous woodland area of Zebracka, with space included as co-variable (left) and not included (right).

Variable	Trace	F	*p*	Variable	Trace	F	*p*
Space as co-variable				Space not included			
2009							
Total	0.385	2.72	0.0024	Total	0.446	2.01	0.0118
Water distance	0.234	5.88	0.0001	Canopy openness	0.180	2.85	0.0265
Diameter	0.135	2.72	0.0287	Height	0.169	2.65	0.0331
Canopy openness	0.100	1.87	0.12	Water distance	0.061	0.84	0.46
Height	0.073	1.30	0.25	Diameter	0.059	0.82	0.50
2010							
Total	0.573	6.66	0.0013	Total	0.636	4.38	0.0033
Water distance	0.310	7.49	0.0027	Diameter	0.323	6.21	0.0100
Diameter	0.217	4.27	0.0358	Height	0.138	2.08	0.15
Height	0.128	2.15	0.14	Canopy openness	0.089	1.28	0.28
Canopy openness	0.045	0.66	0.50	Water distance	0.018	0.24	0.76

Variables are sorted by value of trace (from highest to lowest).

dependent on habitat and landscape heterogeneity [33]. Much of the current research uses mass trapping designs, not only for eradication of potential pests, but also for estimating biodiversity [34,35,36]. It is known, but often overlooked, that a large number of samples does not necessarily imply more significant results [37,38,39]. From this point of view, appropriate statistical design [9,29,31,40] may lead to a higher significance without major exploitation of the environment, which is illustrated on rarefaction curves therein.

The trapping methods used in this study are mainly used for flying arthropods [41,42]. However, when traps are suitably designed, they are known also to trap flightless fauna [18]. This is illustrated by sedentary taxa such as snails (Gastropoda) [43] or pillbugs (Isopoda) in this study.

Økland [44] writes that the total explained variance in ordination methods using variance partitioning, even using sets of carefully selected explanatory variables, is normally 20–50%, and occasionally up to 65%. On the other hand, inclusion of the spatial autocorrelation [9,45] into analyses often leads to lowering of the explained variance of habitat variables [46], as in this study.

Previous studies, similar to this case, have used multivariate analysis with variation partitioning to separate the confounding effects of space and environment [8,47,48]. The results show that including spatial autocorrelation in analyses may lead to a more accurate outcome while its exclusion may lead to wrong conclusions [10,17]. Even if there were differences in trapped arthropods between seasons, the response to habitat variables was, surprisingly, nearly the same in both seasons.

The results showed that it is necessary, for the study of organismal distribution spatial patterns using site level variables, to take spatial autocorrelation into account [46,47,48]. This also makes it possible to distinguish when the spatial structure is mainly due to biotic interactions with an underlying unmeasured environmental factor, or a common spatial gradient shared by data on taxa and environmental variables [8,30]. Results of variance partitioning showed that spatial terms gave a lesser explanation of variance than did site, although significance of space and shared variation led to a much better explanation than when spatial autocorrelation was not included [49].

In the context of this article, there is a strong need for spatial autocorrelation to be included in the analysis, as it is also in similar relatively small-scale studies.

Distance to the River Becva and Diameter of tree were significant habitat variables, when spatial autocorrelation was included. Water distance reflected the humidity of habitats, which is known to influence arthropod communities [50,51], Diameter is a traditional habitat variable in studies, used to explain diameter and age of the tree, especially for saproxylic communities [4,52] and thus the response of communities to Diameter was not surprising.

Conclusions

This study showed that neglecting spatial autocorrelation could possibly lead to wrong conclusions in small site level studies and, furthermore, that inclusion of spatial terms may lead to more accurate and less ambiguous outcomes. Lower sampling intensity, when appropriately designed, is able to gain sufficient results and may lead to a higher significance without major exploitation of the environment.

Acknowledgments

I would like to thank I. Neradilová for her help in the field, sorting and determinations, J. Klápště with Geary's C and Z.F. Fric for his many good suggestions on the early version of the manuscript, two anonymous referees provided constructive comments. Thanks is also owed to P. Stloukal for his support, to A. Kohutka for help with FITs logistics, L. Hort for fish eye, Š. Vodka for fish eye and construction of FITs, the non-governmental organisation Lesák (www.lesak.eu) for logistics, and A. Scott for English correction.

Author Contributions

Conceived and designed the experiments: JH. Performed the experiments: JH. Analyzed the data: JH. Wrote the paper: JH.

References

1. Samways M (2007) Insect conservation: a synthetic management approach. Annual Review of Entomology 52: 465–487.
2. Gotelli NJ, Colwell RK (2001) Quantifying biodiversity: procedures and pitfalls in the measurement and comparison of species richness. Ecology Letters 4: 379–391.
3. Diniz-Filho JAF, Bini LM, Hawkins B (2002) Spatial autocorrelation and red herrings in geographical ecology. Global Ecology and Biogeography 12: 53–64.
4. Buse J, Schröder B, Assmann T (2007) Modelling habitat and spatial distribution of an endangered longhorn beetle – a case study for saproxylic insect conservation. Biological Conservation 137: 372–381.
5. Horak J, Chumanova E, Hilszczanski J (2012) Saproxylic beetle thrives on the openness in management: a case study on the ecological requirements of *Cucujus cinnaberinus* from Central Europe. Insect Conservation and Diversity 5: 403–413.
6. Götmark F, Von Proschwitz T, Franc N (2008) Are small sedentary species affected by habitat fragmentation? Local vs. landscape factors predicting species richness and composition of land molluscs in Swedish conservation forests. Journal of Biogeography 35: 1062–1076.
7. WallisDeVries MF (2004) A quantitative conservation approach for the endangered butterfly *Maculinea alcon*. Conservation Biology 18: 489–499.

8. Barbaro L, Rossi JP, Vetillard F, Nezan J, Jactel H (2007) The spatial distribution of birds and carabid beetles in pine plantation forests: the role of landscape composition and structure. Journal of Biogeography 34: 652–664.
9. Legendre P (1993) Spatial autocorrelation: trouble or new paradigm? Ecology 74: 1659–1673.
10. Mauricio Bini L, Diniz-Filho JAF, Rangel TF, Akre TS, Albaladejo RG, et al. (2009) Coefficient shifts in geographical ecology: an empirical evaluation of spatial and non-spatial regression. Ecography 32: 193–204.
11. Dutilleul P, Legendre P (1993) Spatial heterogeneity against heteroscedasticity: an ecological paradigm versus a statistical concept. Oikos 66: 152–171.
12. Vinatier F, Tixier P, Duyck P-F, Lescourret F (2011) Factors and mechanisms explaining spatial heterogeneity: a review of methods for insect populations. Methods in Ecology and Evolution 2: 11–22.
13. Tscharntke T, Steffan-Dewenter I, Kruess A, Thies C (2002) Characteristics of insect populations on habitat fragments: a mini review. Ecological Research 17: 229–239.
14. Dormann CF, McPherson JM, Araujo MB, Bivand R, Bolliger J, et al. (2007) Methods to account for spatial autocorrelation in the analysis of species distributional data: a review. Ecography 30: 609–628.
15. Tobler WR (1970) A computer movie simulating urban growth in the Detroit region. Economical Geography 46: 234–240.
16. Hawkins BA (2012) Eight (and a half) deadly sins of spatial analysis. Journal of Biogeography 39: 1–9.
17. Kühn I (2007) Incorporating spatial autocorrelation may invert observed patterns. Diversity and Distributions 13: 66–69.
18. Horak J, Vodka S, Pavlicek J, Boza P (2013) Unexpected visitors: flightless beetles in window traps. Journal of Insect Conservation 17: 441–449.
19. Horak J, Rebl K (2013) The species richness of click beetles in ancient pasture woodland benefits from a high level of sun exposure. Journal of Insect Conservation 17: 307–318.
20. Safar J (2003) Olomoucko. Chráněná území ČR VI. AOPK ČR Praha, EkoCentrum Brno.
21. Hradilek Z, Duchoslav M (2007) Flóra a vegetace Národní přírodní rezervace Žebračka u Přerova. Přírodopisný Časopis Slezského Muzea Opava (A) 56: 193–226.
22. Horak J (2011) Response of saproxylic beetles to tree species composition in a secondary urban forest area. Urban Forestry & Urban Greening 10: 213–222.
23. Francese JA, Oliver JB, Fraser I, Lance DR, Youssef N, et al. (2008) Influence of trap placement and design on capture of the emerald ash borer (Coleoptera: Buprestidae). Journal of Economic Entomology 101: 1831–1837.
24. Frazer GW, Canham CD, Lertzman KP (1999) Gap Light Analyzer (GLA): imaging software to extract canopy structure and gap light transmission indices from true-colour fisheye photographs, users manual and program documentation. Simon Fraser University, Burnaby, British Columbia, and the Institute of Ecosystem Studies, Millbrook, New York.
25. Chao A (1984) Non-parametric estimation of the number of classes in a population. Scandinavian Journal of Statistics 11: 265–270.
26. Colwell RK (2006) EstimateS: Statistical estimation of species richness and shared species from samples. Version 8. Available: http://viceroy.eeb.uconn.edu/estimates. Accessed 2010 May 5.
27. R Development Core Team (2008) R: A language and environment for statistical computing. R Foundation for Statistical Computing, Vienna, Austria.
28. ter Braak CJF, Smilauer P (2002) CANOCO reference manual and CanoDraw for Windows user's guide: Software for Canonical Community Ordination (version 4.5). Microcomputer Power, Ithaca, NY, USA.
29. Leps J, Smilauer P (2003) Multivariate analysis of ecological data using CANOCO. Cambridge University Press, UK.
30. Borcard D, Legendre P, Drapeau P (1992) Partialling out the spatial component of ecological variation. Ecology 73: 1045–1055.
31. Peres-Neto PR, Legendre P, Dray S, Borcard D (2006) Variation partitioning of species data matrices: estimation and comparison of fractions. Ecology 87: 2614–2625.
32. Borcard D, Legendre P, Avois-Jacquet C, Tuomisto H (2004) Dissecting the spatial structure of ecological data at multiple scales. Ecology 85: 1826–1832.
33. Weibull AC, Ostman O, Granqvist A (2003) Species richness in agroecosystems: the effect of landscape, habitat and farm management. Biodiversity and Conservation 12: 1335–1355.
34. El-Sayed AM, Suckling DM, Wearing CH, Byers JA (2006) Potential of mass trapping for long-term pest management and eradication of invasive species. Journal of Economical Entomology 99: 1550–1564.
35. Liebhold AM, Tobin PC (2008) Population ecology of insect invasions and their management. Annual Review of Entomology 53: 387–408.
36. Økland B (1996) A comparison of three methods of trapping saproxylic beetles. European Journal of Entomology 93: 195–209.
37. Basset Y (1988) A composite interception trap for sampling arthropods in tree canopies. Journal of Australian Entomological Society 27: 213–219.
38. Erwin TL (1982) Tropical forests: their richness in Coleoptera and other arthropod species. Coleopterists Bulletin 36: 74.
39. Fric Z, Konvicka M (2007) Dispersal kernels of butterflies: Power-law functions are invariant to marking frequency. Basic and Applied Ecology 8: 377–386.
40. Collins MD, Simberloff D (2009) Rarefaction and non random spatial dispersion patterns. Environmental and Ecological Statistics 16: 89–103.
41. Hosking GP (1979) Trap comparison in the capture of flying coleopteran. New Zealand Entomologist 7: 87–92.
42. Lassau SA, Hochuli DF (2005) Wasp community responses to habitat complexity in Sidney sandstone forests. Australian Ecology 30: 179–187.
43. Peltanova A, Dvorak L, Horak J (2012) Land snail community of fruit trees or what can be found in tree trunk window traps. Acta Musei richnovensis 19: 12–16.
44. Økland RH (1999) On the variation explained by ordination and constrained ordination axes. Journal of Vegetation Science 10: 131–136.
45. Koenig WD (1999) Spatial autocorrelation of ecological phenomena. Trends in Ecology and Evolution 14: 22–26.
46. Frutos A, Olea PP, Vera R (2007) Analyzing and modelling spatial distribution of summering lesser kestrel: the role of spatial autocorrelation. Ecological Modelling 200: 33–.
47. Heikkinen RK, Luoto M, Virkkala R, Rainio K (2004) Effects of habitat cover, landscape structure and spatial variables on the abundance of birds in an agricultural-forest mosaic. Journal of Applied Ecology 41: 824–835.
48. Titeux N, Dufrene M, Jacob JP, Paquay M, Defourny P (2004) Multivariate analysis of a fine-scale breeding bird atlas using a geographical information system and partial canonical correspondence analysis: environmental and spatial effects. Journal of Biogeography 31: 1841–1856.
49. Legendre P, Legendre L (1998) Numerical ecology. Elsevier, Amsterdam.
50. Landis DA, Wratten SD, Gurr GM (2000) Habitat management to conserve natural enemies of arthropod pests in agriculture. Annual Review of Entomology 45: 175–201.
51. Southwood TRE (1962) Migration of terrestrial arthropods in relation to habitat. Biological Reviews of Cambridge Philosophical Society 37: 171–214.
52. Grove SJ (2002) Saproxylic insect ecology and the sustainable management of forests. Annual Review of Ecology and Systematics 33: 1–23.

12

Seagrass Canopy Photosynthetic Response Is a Function of Canopy Density and Light Environment: A Model for *Amphibolis griffithii*

John D. Hedley[1]*, Kathryn McMahon[2], Peter Fearns[3]

1 Environmental Computer Science Ltd., Tiverton, Devon, United Kingdom, **2** School of Natural Sciences and Centre for Marine Ecosystems Research, Edith Cowan University, Joondalup, Western Australia, **3** Department of Imaging and Applied Physics, Curtin University of Technology, Perth, Western Australia

Abstract

A three-dimensional computer model of canopies of the seagrass *Amphibolis griffithii* was used to investigate the consequences of variations in canopy structure and benthic light environment on leaf-level photosynthetic saturation state. The model was constructed using empirical data of plant morphometrics from a previously conducted shading experiment and validated well to *in-situ* data on light attenuation in canopies of different densities. Using published values of the leaf-level saturating irradiance for photosynthesis, results show that the interaction of canopy density and canopy-scale photosynthetic response is complex and non-linear, due to the combination of self-shading and the non-linearity of photosynthesis versus irradiance (P-I) curves near saturating irradiance. Therefore studies of light limitation in seagrasses should consider variation in canopy structure and density. Based on empirical work, we propose a number of possible measures for canopy scale photosynthetic response that can be plotted to yield isoclines in the space of canopy density and light environment. These plots can be used to interpret the significance of canopy changes induced as a response to decreases in the benthic light environment: in some cases canopy thinning can lead to an equivalent leaf level light environment, in others physiological changes may also be required but these alone may be inadequate for canopy survival. By providing insight to these processes the methods developed here could be a valuable management tool for seagrass conservation during dredging or other coastal developments.

Editor: Kay C. Vopel, Auckland University of Technology, New Zealand

Funding: This research was funded by an Edith Cowan University (ECU) Industry Collaboration Scheme with BMT Oceanica Pty. Ltd., the industry partner, awarded to KM, JH, and PF. ECU Faculty Visiting Fellow Scheme supported JH for travel to ECU. KM was supported by the ECU Collaborative Research Network. Environmental Computer Science Ltd. provided support in the form of salary for author JH. The funders had no role in the study design, data collection and analysis, decision to publish, or preparation of the manuscript. The specific roles of these authors are articulated in the "author contributions" section.

Competing Interests: This work was partly funded by BMT Oceanica. The author J Hedley is an employee of Environmental Computer Science Ltd. This does not alter the authors' adherence to PLOS ONE policies on sharing data and materials.

* Email: j.d.hedley@envirocs.com

Introduction

Seagrass meadows are a dominant habitat of most coastal environments and provide important ecosystem services such as primary production, nutrient cycling, sediment stabilization, food and habitat for other organisms and trophic transfers to adjacent habitats [1]. Globally, these ecosystem services have been valued at an approximated US$ 19000 ha^{-1} yr^{-1} [2] but emerging understanding of the carbon storage capability of seagrass meadows implies this may be an underestimate [3]. Despite these recognized values, the area of seagrass is reducing world-wide at an increasing rate. Waycott et al. [4] estimated 29% of the known areal extent has disappeared since seagrass areas were initially recorded in 1879, and the rate of decline has accelerated in the last two decades.

The key anthropogenic pressures impacting seagrass meadows at local scales are urban, industrial and agricultural runoff, infrastructure development and dredging [5]. These pressures impact seagrasses directly via physical removal or indirectly through the introduction of pollutants such as nutrients, or suspended sediments that result in a reduction of light reaching seagrass meadows. Seagrasses are sensitive to light reduction as they are typically adapted to high light environments [1].

Increasing research is being undertaken to improve the management and conservation of seagrass meadows through improved understanding of the risks they face (e.g. [6]), developing bioindicators of the pressures they are exposed to [7] and thresholds of stressors such as light reduction which may differentiate sub-lethal effects from permanent loss of seagrass [8,9]. In general, leaf-level photosynthetic activity in response to irradiance follows a 'photosynthesis versus irradiance curve', which is linear for subsaturating irradiances but becomes non-linear, as progressively increasing irradiance causes saturation of the photosynthetic electron transport chain, and finally attains a plateau phase, which is defined as maximal photosynthesis rate (P_{max}) [10,11]. A key physiological parameter that represents a species response to a given light level is E_k, defined as the intersection between the initial linear slope and P_{max} on a P-I curve. E_k is frequently referred to as the 'saturating irradiance' [12,13] although technically it is slightly below the irradiance at which full photosynthetic saturation occurs, and above the irradiance at which saturation starts to cause deviation from linearity. E_k can be empirically determined and for each species may vary over a restricted range due to physiological acclimatization or factors such as temperature [14].

Various light threshold analyses have been proposed as having predictive capability for seagrass mortality. Dependent on available data, light levels can be assessed with respect to different factors or components of the environment, including the water column light attenuation coefficients [15] or Secchi disk depths [16]; light at the top of the seagrass canopy expressed as percentage of surface irradiance [17,18]; instantaneous or mean daily irradiance [8,19] or the number of hours of irradiance above E_k per day, H_{sat}, [8,20]. These thresholds can also be integrated over time, which is relevant to management when pressures persist over particular durations, e.g., dredging or flood plumes. The percentage of days below a particular mean daily irradiance [8] or the sum of the hours of irradiance below E_k compared to reference conditions [9] are two examples for which thresholds have been proposed to predict the onset of seagrass mortality.

One important component that all of these thresholds do not consider is the interaction of the seagrass canopy itself with the benthic light field, since it is the amount of light reaching individual leaves of a seagrass that governs the plants photosynthetic response [21]. The photosynthetic activity in turn influences how the seagrass meadow responds to the changes in light [12] and overall plant productivity [22]. Canopy structure of seagrass meadows can also vary markedly due to natural variations in light [23] or in response to light perturbations [9]. Due to canopy self-shading, light levels at the top of the canopy may be very different to light levels within the canopy, and will vary throughout the canopy in a manner dependent on the incident benthic light field, canopy structure and bending angle of the leaves, which vary under water motion [24,25,26]. Therefore, a mechanistic explanation of how light levels affect canopy sustainability must include the interaction of the canopy structure with the incident light field.

In this study we developed a 3D model of a complex seagrass canopy (*Amphibolis griffithii*) of varying structure, from low to high leaf area index (LAI), by adapting the model described in [25] and [27]. We modeled the exposure of these virtual canopies to a number of environmentally relevant levels of light reduction to assess the amount of light reaching each leaf surface and how this varies under different canopy densities and positions due to movement associated with water motion. Finally, we assessed the canopy saturation state by relating the light each leaf receives to values of leaf-level E_k for *A. griffithii* found in the literature. The modeling scenarios were based on empirically quantified canopy structures from specific plant morphologies, and were designed to be comparable to a shading experiment that was conducted on *A. griffithii* in 2005 [9].

In summary, the objectives were:

- To develop a 3D canopy model for a seagrass species with a complex canopy, hence demonstrating an advance in technical capability with respect to the simple *Thalassia* morphology model of Hedley and Enríquez [25].
- To understand the consequences on within-canopy light capture and canopy saturation state of 1) canopy position: upright vs. moving under high wave action, 2) canopy structure: low to high LAI (1.27 to 7.65), and 3) light reduction: 0–95% shading
- To identify potential descriptors of canopy light levels which could have use for the management of seagrass beds under light reduction events such as dredging or coastal pollution.

Methods

Canopy structures

The modelling experiment was designed to mirror aspects of a previously published empirical shading manipulation experiment [9,28]. The empirical study utilised an extensive (>6 ha) meadow of *Amphibolis griffithii* in 4.5 m water depth at Jurien Bay, Western Australia (30° 18′ 34″ S, 115° 00′ 26″ E; WGS84 datum). A control plus two-treatment shading experiment was conducted, the first phase of which ran from 10th March to 14th June 2005. Before and during the experiment individual *A. griffithii* plants were sampled and characterised in terms of stem and branch lengths, internodal distances, and number and dimensions of terminal leaves (Fig. 1a).

In the computer model, ten sets of individual plant data from the initial control sampling were replicated as vector mesh structures (Fig. 1f). The model plants were assembled into five canopies of leaf area index (LAI) from 1.27 to 7.65, by varying the choice and number of plants in a 20 cm×20 cm segment of substrate (Table 1, Fig. 2). The leaves and stems of the vector mesh structures were modelled as a point-mass and force system according to methods typically used for modelling cloth in the computer graphics industry [29]. A dynamic numerical integrator modelled the plant structures flexing naturalistically under a simple wave-action force model. Two wave actions, 'high' and 'low', were employed. In the dynamic model the low wave energy treatment plants were allowed to assume a typical upright position with no wave induced movement (Fig. 2). Under high wave energy plants underwent a vigorous cycle of forward and backward motion (Figs. 2f–i). From these dynamic models canopy structure treatments were extracted as instantaneous snapshots for each of the five LAI treatments: 1) a single snapshot for low wave action, 2) 14 snapshots through a cycle of movement for high wave action (Fig. 2). The 14 snapshots for high wave action were individually passed to the optical model (see below) and the results were averaged, thereby assuming the canopies undergo this movement continuously and photosynthetic response is the mean of the responses at any instant in time.

Water column optical model and shading

The canopy structures were input to the optical model for estimating diurnal leaf-incident irradiance. The model framework, previously described in [27] and [25], propagates sky radiance distributions through the canopy to give leaf incident irradiance in 17 wavebands of 20 nm width from 400–740 nm. Spectral irradiance can then be reduced to photosynthetically available radiation (PAR) at leaf level, and related to leaf tissue photosynthetic saturating irradiance, approximated by E_k (Fig. 1). To parameterise the model, hourly clear sky radiance distributions were produced using libRadtran and a directional radiance model [30,31] corresponding to the Jurien Bay site on 27th April; the middle of the post summer 3 month trial phase in Lavery et al. [9] (Fig. 1c).

The sky radiance distributions were input to PlanarRad (http://www.planarrad.com), a plane parallel water column model functionally similar to HydroLight [32,33] to estimate the hourly top of canopy radiance distribution (Fig. 1e). The model provides directional radiance tabulated over a hemisphere of zenith and azimuth angles, but to remove any dependency on sun azimuth and canopy orientation downwelling irradiances were azimuthally averaged to have only a zenith angle dependency (Fig. 1c, e). The water column utilised a library set of spectral inherent optical properties (IOPs, for details see [34]) which when input to the model produced a diffuse attenuation of planar PAR irradiance,

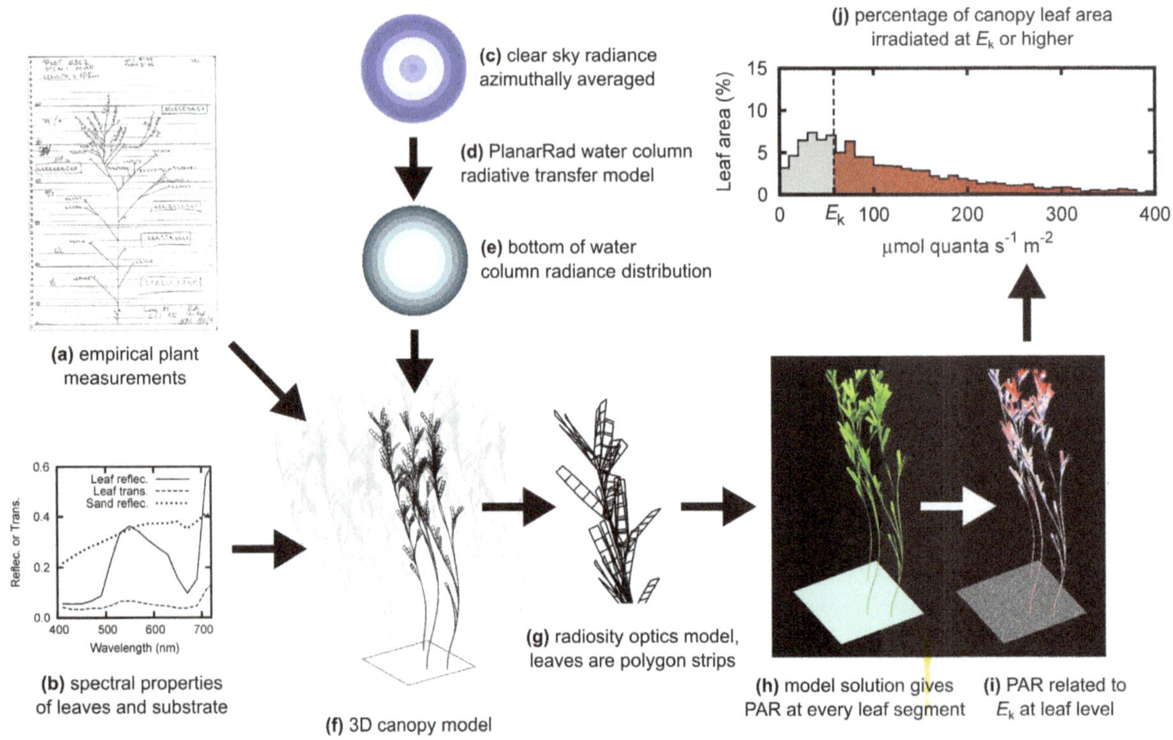

Figure 1. Overview of modelling system. (a, b) empirical data informs construction of 3D canopy model (f), (c, d, e) A plane-parallel model estimates directional radiance incident on the top of the canopy, (f, g, h) a geometric optical model handles radiative transfer to and between leaf segments, (i, j) PAR distribution over leaf area is reduced to the percentage of the canopy irradiated above leaf-level photosynthetic saturation, E_k.

k_d, of approximately 0.2 m^{-1}. In comparison, k_d values measured at the time of the empirical shading experiment ranged from 0.07 to 0.19 over a four month period but were 0.19 in April (Department of Parks and Wildlife, unpublished data). A set of nine modelled shading treatments were implemented by taking the top of canopy radiance distribution and reducing the values by 10%, 20%, etc. up to 95% (Table 1). Hence shading was spectrally neutral as was the shade cloth used in Lavery et al. [9], where the shading treatments were equivalent to 81–87% and 89–95% in our notation. The empirical study therefore represented quite a strong shading effect with respect to the modelled range. The water column optical model was additionally evaluated by comparing modelled top of canopy daily PAR irradiance against *in-situ* measurements from the associated study [28].

Canopy structure optical model

The top of canopy irradiance was propagated through a geometrical optics model [25,27] that accounts for inter-reflection and transmission between leaf segments. The spectral reflectance and transmittance of *A. griffithii* leaves was taken from the paper of Durako [35]. In this study we did not attempt to capture inter or intra-plant variability in leaf absorptance. This can be done [25], but the data collection requirements are onerous. All surfaces were considered Lambertian reflectors and transmitters. The underlying substrate reflectance was set from a library sand spectral reflectance that had a mean value of 0.33. The 20×20 cm modelled canopy segment was repeated periodically horizontally so the modelled canopy was of uniform LAI and has no edge (Fig. 1f).

Empirical measurements of PAR irradiance close to midday at both the canopy top and on the substrate underneath canopies of *A. griffithii* of differing LAIs were available for validation of the canopy optical model from the study of McMahon and Lavery [28]. Canopy transmission was measured in control and treatment plots of varying but known LAI through measuring the instantaneous photosynthetic photon flux density (PPFD, µmol m^{-2} s^{-1}) at the top and base of the canopy. The light sensor (Odyssey PAR sensor) was calibrated against a standard calibration light source (Quartz Tungsten Halogen Reference Lamp operated at 3150°K from a LI-1800-02 Optical Radiation Calibrator). The low wave energy structure model treatments at the hour closest to midday were used to perform this validation, and a number of additional runs with different LAIs to those in Table 1 were added to further populate the validation data. An additional quality assurance protocol for the canopy optical model is to set the within-canopy water absorptance to zero and then verify energy conservation between the top of canopy incident and exitant irradiances and energy absorbed by all surfaces in the model [25]. This was performed for a subset of the runs in Table 1.

Relation to photosynthetic properties

The model solution provided incident PAR at every point on every leaf at a resolution of approximately 0.5 cm^2. This was then related to the leaf level saturation irradiance for *A. griffithii*, approximated by E_k. Masini and Manning [12] evaluated E_k in *A. griffithii* as ranging from 25 to 55 µmol quanta m^{-2} s^{-1} for temperatures of 13°C to 23°C, of which the upper value is closer to the conditions of the empirical data from the associated study here. While Masini and Manning [12] did not assess physiological variation in E_k in *A. griffithii*, in the same study *Posidonia sinuosa* was shown to have E_k that varied from 50.4 to 39.1 µmol quanta m^{-2} s^{-1} in depths of 4 m and 12–15 m respectively. Therefore to accommodate a realistic variation in E_k in the absence of data, we

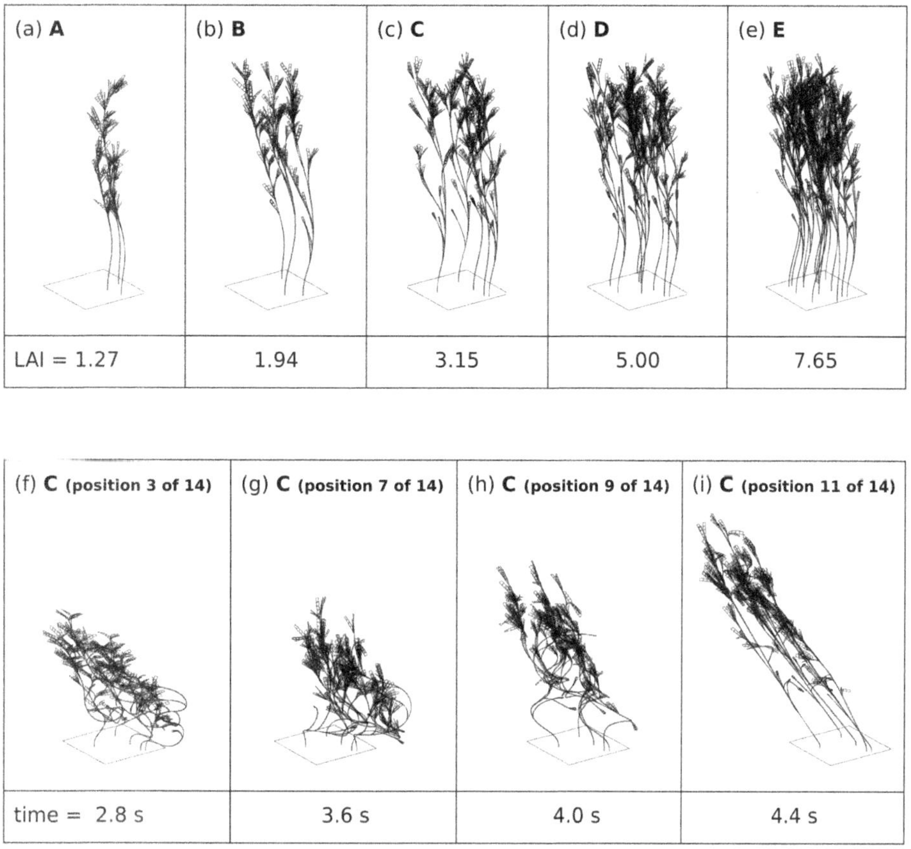

Figure 2. Example canopy structures and positions used in the model treatments. (a)–(e) low wave action canopy position for the five LAI treatments A to E. (f)–(i) subset of time sequence positions under high wave action for canopy C, all 14 positions were used in the optical model. In all cases the canopy structure is notionally repeated in all horizontal directions such that the square substrate section tessellates.

have used the comparable range of 45–55 μmol quanta m^{-2} s^{-1} to put all of our results into the context of potential physiological variation in E_k. To produce plots including the range and mid-point specific values of 45, 50 and 55 were used. Based on the value of E_k of 45 or 55 μmol quanta m^{-2} s^{-1}, the model can report the instantaneous proportion of the leaf area of the canopy that is irradiated at or above the saturation irradiance. To interpret these results, for each canopy structure and shading treatment, a value termed H^{A}_{sat} was calculated as the time integral of the percentage leaf area above saturation in a 24 hour period, with units % leaf area×hour. This measure is discussed later, but was intended to be analogous to H_{sat}, the daily top of canopy irradiated hours above saturation [9,20] but also factoring in the canopy self-shading.

Table 1. Modelling experiment design.

LAI	Structure	Shading (%)	Hour
A −1.27	low wave energy (1 position)	0	×12
B −1.94	high wave energy (14 positions)	10	
C −3.15		25	
D −5.00		40	
E −7.65		50	
		60	
		70	
		85	
		95	

A fully-factored set of model runs were performed for each of five LAI treatments, 15 canopy structures and nine shading treatments over 12 hourly diurnal intervals, a total of 8100 runs.

Results

Optical model validation

The sky radiance and water column model produced a daily top of canopy PAR dose of 11.0 mol quanta m^{-2}, whereas the comparable *in-situ* measured average daily PAR irradiance over 3 months was 19.0 mol quanta m^{-2} [9]. Since our study was primarily concerned with the relative effect of the shading treatments and LAI this discrepancy is not of great importance, but could be due to: 1) the accuracy of the libRadtran sky radiance model (no validation data available); 2) the accuracy of the Odyssey PAR sensors, which can have issues in long-term stability (Slivkoff, pers. comm.), or; 3) the model water column k_d(PAR), which was at the upper range compared to measurements taken during the empirical study (0.2 vs. 0.070.19). This deviation in k_d(PAR) does provide an almost exact explanation for the discrepancy, but since k_d(PAR) is a wavelength-integrated output of the model parameterised on spectral IOPs for absorption and backscatter it is not trivial to set an arbitrary value of k_d(PAR). In the scope of this study, using the closest IOP set from actual measured data [34] was considered adequate. In reality the daily measured PAR was sometimes above and sometimes below the model value, so all things considered the modelled canopy PAR dose was reasonable and the discrepancy is inconsequential to the subsequent interpretation of the results.

The percentage of the incident top of canopy PAR irradiance transmitted to the substrate, as a function of leaf area index, validated well against empirical data (Fig. 3). The empirical data showed wide variation, but the modelled transmitted irradiances corresponded very closely to the upper bound of the empirical data. This is to be expected, since some of the real canopies contained free standing and epiphytic macroalgae which would have reduced the transmission beyond that described by the *A. griffithii* LAI alone. The upper bound points most likely represent the most monospecific *A. griffithii* canopies and correspond best to the model. An exponential function fit to the model data ($n = 28$) gave r^2 of 0.96, the fit of all 27 empirical data points to that same function gives an r^2 of 0.73. However, if only four outliers are removed (Fig. 3) the empirical data r^2 rises to 0.90. The model and empirical data therefore compare well, especially given the practical difficulty in making accurate within-canopy light measurements.

The performance of the model in terms of energy conservation was demonstrated in the subset of runs for which water absorption was set to zero. For the majority of runs energy losses were less than 2% and for all runs they were less than 3%. In practice, when water absorption is non-zero, energy conservation performance would be better than these figures suggest. The current model implementation requires water absorption to be set to zero for energy accounting, but this in itself removes a damping effect on the multiple scattering and increases energy losses through numerical errors. Therefore the model solutions for leaf incident irradiance can be considered, at worst, slight underestimates by around 2%.

Effect of canopy structure and position on leaf level PAR

As expected, the distribution of leaf level PAR irradiance became increasingly skewed to lower values as LAI increased (Figs. 4a, e, i, m, q). In low LAI canopies the distribution of PAR over leaf area was almost flat: leaf tissue received a wide range of PAR with almost equal probability, and much of it was above saturating irradiance at mid-day under the model conditions of clear sky and moderately clear water. In denser canopies the leaf level light distribution had a long high-end tail: many leaves

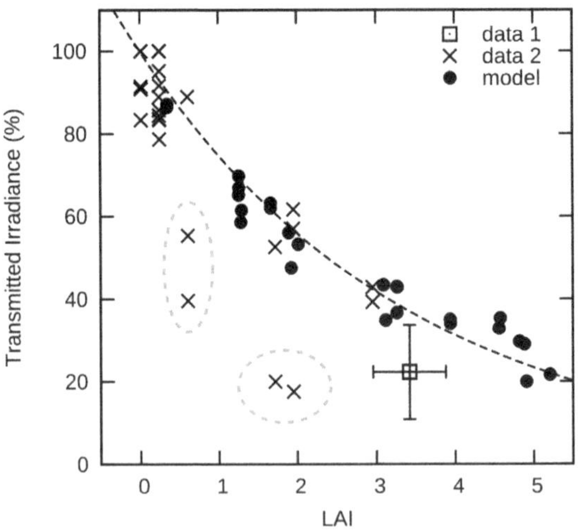

Figure 3. Percentage of downwelling top of canopy PAR irradiance reaching the substrate, as a function of LAI. Results for empirical in-situ measurements (data 1 and 2) and modelled estimates (model) are shown. Curve fit to model data points is $y = 100 * \exp(-0.29 \times \text{LAI})$.×(data 2) - are 26 individual coincident measurements of LAI and irradiance above and below the canopy. ⊡ (data 1) is a single point based on a site mean of 13 LAI determinations from 2.84 to 4.09 and a set of associated but not spatially co-incident light measurements, error bars are one S.D. The four encircled points are the outliers referred to in the text. Transmittance data collection is described in McMahon and Lavery 2014 and corresponding LAI is unpublished data from Lavery et al. 2009. ● are 25 model runs including both those described in detail and some additional runs. In the models runs solar zenith angle was approximately 28°.

received light below E_k, but a few leaves received very high light (Fig. 4i). Overall the pattern was clearly linked to the relative openness or self-shading within the canopies. The range of E_k of 45 to 55 μmol quanta m^{-2} s^{-1} was generally small compared to range of irradiances the leaves experienced, but this was more true for the lower LAI and unshaded treatments (Fig. 4).

The treatments of canopy position of upright or moving under wave action appeared to have little effect on leaf level PAR irradiance (e.g. Fig. 4a vs. 4b). Numerically the canopy movement slightly reduced the daily integrated percentage of saturated leaf area for all but the lowest LAI (Fig. 5). However overall there was not a statistically significant difference at either E_k of 45 or 55 μmol quanta m^{-2} s^{-1} (paired value t-test, $p>0.05$). Therefore there is no evidence from our data that canopy movement affects time-integrated light capture. However the instantaneous light capture has a high variation under movement. While the standard deviation was 10–20% of the mean (Fig. 5) at some individual time points the saturated leaf area was up to 50% more or less than the mean. As expected, shading scaled the x-axis position of leaf-level irradiance distribution plots by the corresponding factor. That is, halving the top of canopy irradiance halved the leaf level incident irradiance at every leaf (Figs. 4b, d, f, h, etc.).

Diurnally accumulated saturated leaf area

The accumulated percentage of leaf area above saturation over a twelve-hour day showed a complex relationship between both shading and LAI (Fig. 6). While, as expected, increasing either shading or LAI monotonically decreased the accumulated percentage of leaf area above saturation, the shape of the function was non-linear and there was an interaction between shading and LAI (Fig. 6a). The contour lines in Figure 6a make clear the trade-

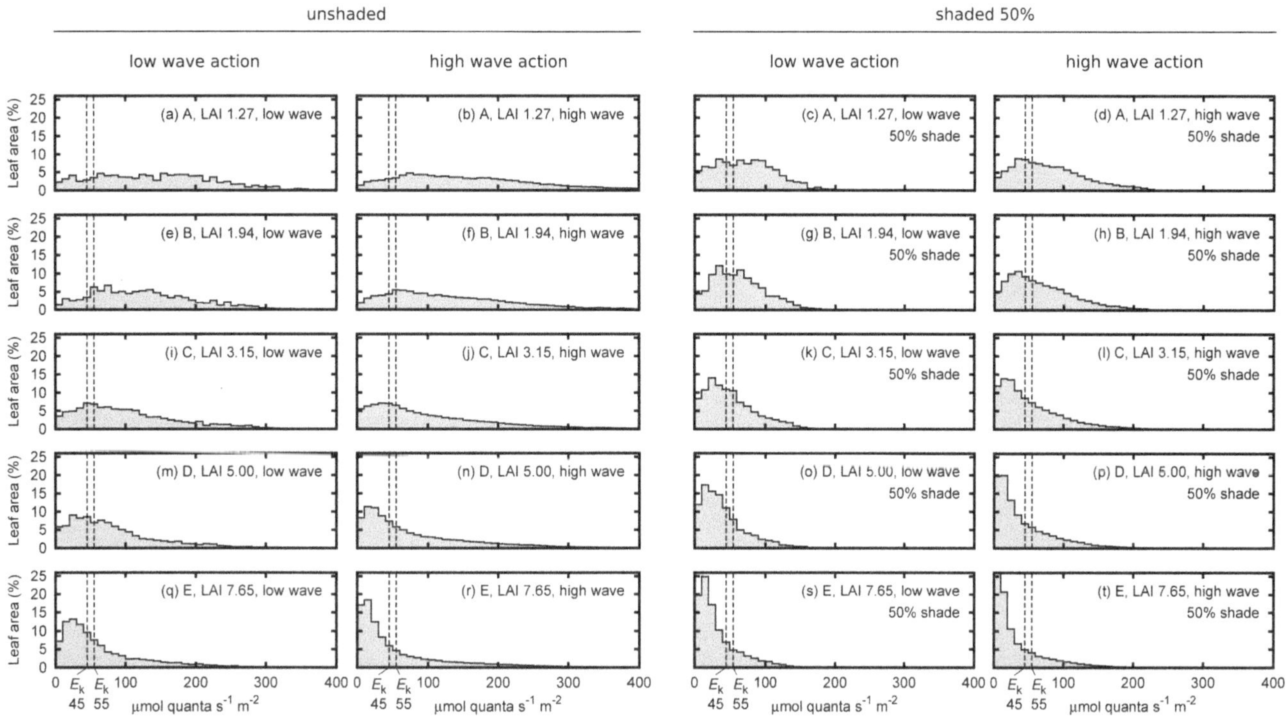

Figure 4. Distribution of leaf saturation state in the canopy in terms of percentage of leaf area at midday. Treatments are upright low wave action canopy positions and the average over the high wave action movement positions (left columns), and the same for 50% shading (right columns), for the five LAI treatments. The estimated photosynthetic saturation irradiances, E_k, of 45 and 55 μmol quanta m^{-2} s^{-1} inferred from Masini and Manning (1997) are shown as vertical dotted lines.

off between leaf area index and light with respect to the saturation state of the canopy. These lines show equal points in the LAI–shading function space, so for example a canopy of LAI 5.5 with no shading was equivalent to LAI 2.0 with 50% shading, with respect to the diurnally accumulated saturation of relative leaf area. The potential acclimation range of E_k from 45 to 55 μmol quanta m^{-2} s^{-1} (assuming water temperature at approx. 23°C) added a degree of freedom to the LAI-shading relationship approximately equivalent to 1 unit of LAI at low shading (e.g. along the *x*-axis of Fig. 6a), but this increased as shading increased to 60% or more (Fig. 6a). Therefore at low shading modifying E_k over the suggested range is equivalent to changing LAI by plus or minus one half.

For the high wave action treatment there is a small qualitative difference in the position of the contour lines in low LAI and low shading region as compared to the upright low wave action treatment (Fig. 6b vs. 6a). However a sensitivity analysis of the data tables underlying Figures 6a and 6b showed that these differences are equivalent to an error in the shading percentage of only 6 points or less. In other words, if in a practical application shading were quantified at discrete levels of 0, 5, 10, 15% etc. then the difference between upright and moving canopies would be negligible.

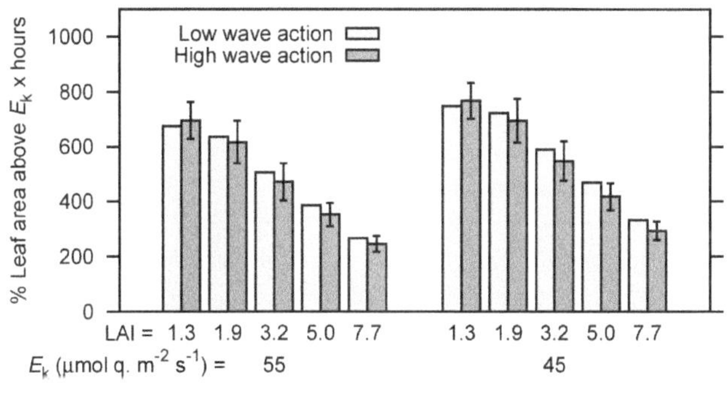

Figure 5. Time accumulated percentage of leaf area irradiated above photosynthetic saturation irradiance for low wave action and high wave action treatments, for E_k of 45 and 55 μmol quanta m^{-2} s^{-1}. Error bars on high wave action treatment are the standard deviation over all 14 movement positions and hence indicate the range in the instantaneous canopy saturation state.

Discussion

Geometric optical modelling of seagrass canopies and validation

In terms of the geometrical optical modelling of seagrass canopies, the results presented here corroborate those of Hedley and Enríquez [25], showing that it is possible to construct a physical three dimensional model of a seagrass canopy and obtain acceptable validation against *in-situ* light measurements. Through-canopy transmission was estimated accurately for pure *Amphibolis* canopies, but the importance of considering epiphytes or other canopy constituents was underlined by the high variability of the empirical data, which in some cases had lower light penetration than the model predicted based on *Amphibolis* LAI alone.

With respect to morphological complexity, *A. griffithii* is on the more complex end of the spectrum in comparison to strap leaf morphologies of *Thalassia* and many other seagrass species [36], to which this modelling framework was previously confined.

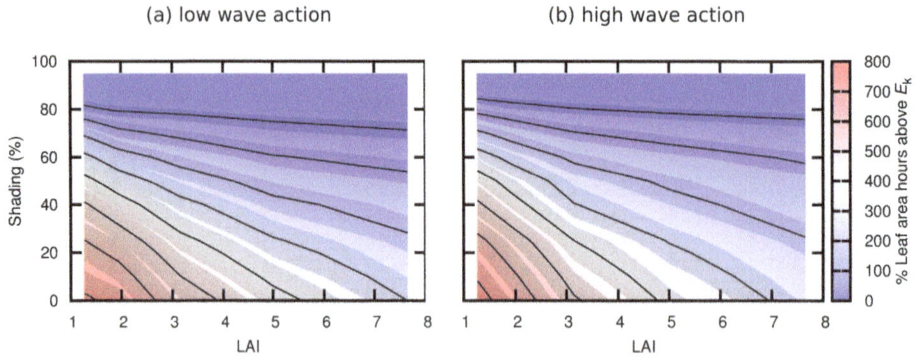

Figure 6. Time accumulated percentage of leaf area irradiated above photosynthetic saturation irradiance. The colour scale shows $H^A{}_{sat}$, the "percentage leaf area hours" above E_k, as a function of LAI and shading. (a) is for upright low wave action canopy structures, (b) is for the average over the high wave energy canopy positions. Contour lines are isoclines based on the mid value of E_k equal to 50 μmol quanta m^{-2} s^{-1} while the surrounding greyed region shows the limits for E_k of 45 to 55 μmol quanta m^{-2} s^{-1}. The isoclines are located at $H^A{}_{sat}$ of 50, 100, 200, and then steps of 100 up to 800% leaf area hours.

Therefore the potential for future models of other seagrass species is good as these two examples capture the range in canopy complexity. Hedley and Enríquez [25] used profiles of light through the canopy to derive a diffuse attenuation profile, k_d, for validation. In this study only light measurements at the top and bottom of the canopy were available. However, this simpler validation may be preferable and adequate. In practice, empirical measurements of light profiles within canopies are difficult to make, and rarely fit well to exponential attenuation with depth. The measurements at the top and bottom of the canopy are the strongest "signal" for within-canopy attenuation and can be used to derive k_d if canopy height is known. So in future empirical work to which such modelling may be subsequently applied, we recommend measuring downwelling irradiance at the top and bottom of canopies, together with canopy height and LAI.

Influence of shading, LAI and position on diurnal leaf photosynthetic saturation

The previous empirical shading study on *A. griffithii* [9] quantified the change in H_{sat} induced by shading, i.e. the total number of hours of top-of-canopy irradiance that was above photosynthetic saturation, as compared to the unshaded treatments. This quantity, summed over time, was demonstrated as a good indicator of changes in canopy biomass and capacity for subsequent recovery. However H_{sat} relates the top of canopy irradiance to leaf-level photosynthetic saturation irradiance and so ignores canopy self-shading and other structural factors, which therefore introduce an additional degree of freedom. Here, we factor in the canopy structure by considering the percentage of the canopy leaf area above saturating irradiance accumulated over time, $H^A{}_{sat}$, with units of % leaf area×hour. This descriptor extends H_{sat} by reducing to a single number the interaction of the duration of saturating irradiance and the canopy self-shading. It can be roughly interpreted as the daily ratio of saturated photosynthesis to leaf area at canopy scale, and ranges from 0 to 1200 for plants completely saturated for 12 hours of daylight.

Considering the variation in $H^A{}_{sat}$ with leaf area index and shading, as expected LAI has a strong effect on diurnal leaf saturation state (Fig. 6.). A change in LAI from 1 to 7 has as much effect as 60% shading (Fig. 6a), so the ambient light field cannot be treated independently from the canopy structure when photosynthetic processes at leaf level are of interest. Furthermore, the relationship between LAI and both shading and leaf saturation state is a non-linear interaction; Figure 6 represents a curved surface in both axes of shading and LAI. This occurs because while leaf level irradiance is a linear function of canopy level irradiance, the leaf level photosynthetic response is not a linear function of irradiance when the irradiance approaches or exceeds E_k. As leaf level irradiance approaches and exceeds E_k the photosynthetic response levels off at P_{max}. In general, since photosynthesis versus irradiance curves are non-linear in the region of the saturating irradiance any derived measure of leaf level photosynthetic activity will have a complex relationship with LAI unless all leaves are well below saturating irradiance.

Within this study there is no statistically significant evidence to support the statement that canopy movement effects light capture and photosynthetic response. Qualitatively it is interesting that under movement the lowest LAI canopy experienced an increase in daily saturation whereas the higher LAI canopies were systematically lower (Fig. 5). To test the statistical significance of this observation would require substantial further modelling effort and was outside the scope of this study, however at low LAIs sideways movement may serve to enhance light collection by spreading leaves out horizontally and making them insensitive to the directionality of incident light. Under wave action such flattening is intermittent and as we have shown here is not a factor of great photosynthetic significance. Additionally, under these conditions the optical consequences of surface waves and sediment resuspension should also be considered and may be more significant [37]. However, in other systems and species, canopy flattening can be a result of shallow water depth or tidal or estuarine flow [38]. In this case the semi-constant flattening of the canopy may be of optical significance.

Potential for LAI modification as an acclimation response

Figure 6 indicates that modification of LAI is a possible response to maintain the saturation state of the canopy under reduced or enhanced light conditions. This role of morphological plasticity has been demonstrated in a number of experimental studies (e.g. [39]) and hypothesized as a regulatory mechanism in *Thalassia* [13]. From our model data (Fig. 6), if a canopy of LAI 7 is observed to reduce to LAI 4 after a period of 40% reduction in light, this loss of biomass might be interpreted as a trajectory of canopy decline but alternative interpretation is that of an acclimation response to maintain the leaf level photosynthetic state. This interpretation is independent of the mechanism by which it occurs. Leaf mortality might be considered just a by-product of inadequate light to maintain respiration, but if the net effect is a return to an unstressed leaf-level light regime then the distinction between a compensatory morphological adjustment and a decline is at best ambiguous. This argument is of course dependent on the definition of 'decline'; it does not apply if net productivity per unit area rather than biomass is the criteria. Here, by 'decline' we mean an implied trajectory toward canopy eradication.

Under the interpretation of potential acclimation a key question is whether the reduction in LAI remains on the isocline for canopy saturation state (i.e. the contour lines in Fig. 6). A canopy that moves on a trajectory through LAI–shading 'space' such that it stays on a contour line is experiencing the same time integrated percentage of its leaf area at saturating irradiance (see A to E in Fig. 7). That is, it experiences the same daily photosynthetic saturation in relation to its leaf biomass. We might therefore hypothesise that if a canopy can sustainably exist at one point on an H^{A}_{sat} isocline, canopies can also sustainably exist at other points on that line, all other things being equal. Movement along an isocline can occur purely by modification of the LAI, alternative acclimation responses such as modification of E_k at the leaf-level will enable movement perpendicular to the isoclines, illustrated by the range around the isoclines in Figures 6 and 7 delimited by E_k of 45 to 55 µmol quanta m^{-2} s^{-1}. Ignoring the latter possibility, and focussing on LAI modification alone, the prospects for long term survival of a canopy under a change in light environment can be estimated by following the isocline from its current location in LAI-shading space to the new light environment (*y*-axis) location. If at this location the LAI is greater than zero (judged by extrapolation), then the canopy could survive by thinning out to this LAI, at which point it will have the same relative photosynthetic saturation of its leaf area. In the following section we use this concept to interpret the results from the previous empirical shading experiment [9].

Canopy trajectories in LAI and shading space

In the post-summer treatment of the shading experiment of Lavery et al. [9] canopies with an LAI of ~4 had reduced to an LAI of ~2 after three months of 84% shading. This change in LAI and shading can be represented as a trajectory on the H^{A}_{sat} map: Figure 7, point A to B. At 6 months of shading the LAI had reduced to one and by 9 months the canopy was almost eradicated

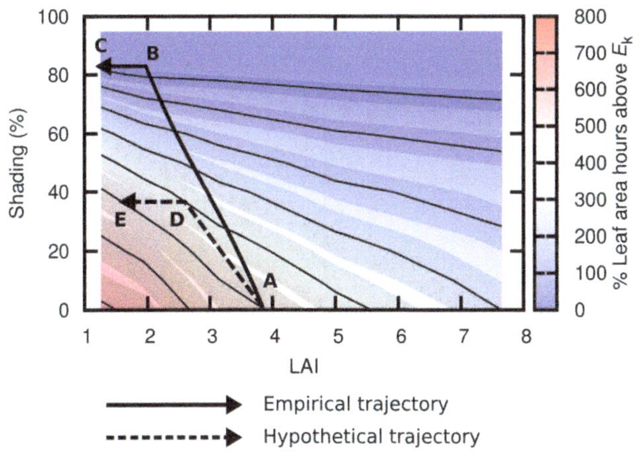

Figure 7. Trajectories of canopies from empirical shading experiment and hypothetical example. Underlying plot is as described in the caption of Figure 6a.

and did not subsequently recover (Fig. 7, point C). Assuming for the moment that H^{A}_{sat}, the accumulated percentage area of leaf saturation above E_k, is a measure relevant to canopy sustainability then Figure 7 indicates that such a measure could have predictive power for canopy survival. In the previous section we postulated that canopies can move along the isoclines by modification of LAI alone. The initial reduction of LAI in the empirical data (Fig. 7, point A to B) occurred in the first three months in response to 84% shading. The trajectory cuts across the isoclines because there is a time lag as the canopy cannot become thinner instantaneously. At three months (Fig. 7, point B) the LAI has reduced to 2 but the shading is extreme so at the leaf level the light environment is still very much reduced. There can be two response pathways, either physiological changes may allow the canopy to exist on the new isocline, such as the mobilisation of stored reserves [40] or reductions in the saturating irradiance [23], or if such processes cannot bring about sufficient change then further LAI reduction is required in an attempt to return to an isocline closer to the original. In this case following the isocline to the right and extrapolating to the intercept with the *x*-axis it is clear that the light environment is equivalent to a canopy with huge LAI of at least 20+ in the original un-shaded situation. For the canopy to survive would require physiological changes that would permit canopies of these high LAIs to exist normally in this environment. Such canopies did not exist, hence physiological changes are insufficient (it is clear from Fig. 7 that variation in E_k is inadequate), hence the LAI continued to decrease and eventually the canopy was eradicated (Fig. 6, Point C).

The previous example is a straightforward case of severe light limitation, but with moderate shading (for example 40%, Fig. 7) the situation is more complicated. The empirical data of Lavery et al. [9] only contained shading at a minimum of 81% so the example of 40% in Figure 7 is hypothetical. If a situation of 40% shading is introduced, assuming the validity of H^{A}_{sat}, it is clear that the canopy could survive by reducing LAI from 4 to 1 (moving along the isocline from point A to point E, Fig. 7). However, because of the time lag in reducing LAI, an initial trajectory in which LAI partially reduces (A to D, Fig. 7) is realistic and is likely to include physiological responses in tandem. For example E_k could decrease, but at LAI around 3 the range of E_k from 45 to 55 only allows accommodation of up to 20% shading at the most (Fig. 7). The existence of a time lag is supported by studies on different seagrass species that have incorporated less extreme shading treatments over short time-scales; physiological changes such as increases in chlorophyll and reduction in LAI occur after longer durations of reduced light [39]. At point D in Figure 7 the canopy lies on an isocline that represents a canopy of LAI 5.5 in the unshaded environment. If such canopies can sustainably exist, at the same depth, water clarity etc. then the canopy may survive at LAI of 2 to 3 (point D), rather than reducing LAI to 1 (point E). Either way, Figure 7 has predictive power for the canopy response in that if it is anticipated a 40% shading event may occur, e.g., from dredging activities, then it is clear that a canopy that is sustainable at LAI of 4 could reduce to LAI of 1, or, could induce physiological changes to maintain an LAI higher than 1, but in the latter case only if canopies greater than LAI of 4 currently exist in that environment.

The possible trajectories in LAI-shading space of Figure 7 are dependent on the capability and time constants of other physiological acclimation mechanisms. These mechanisms could include adjustments to photosystem kinetics to increase the efficiency (i.e. lower E_k), increases in chlorophyll content and a:b ratio to enhance light capture, or mobilisation of stored carbohydrates for maintenance and growth of the existing leaf biomass [7]. To our knowledge there is no published data on photo-acclimation in *A. griffithii* under changing light conditions. However, unpublished data by co-author McMahon shows that under high levels of shading there are reductions in the saturating irradiance and other photo-acclimation responses, which maintain electron transport rates at unshaded values, but there is a time-limit over which this photo-acclimation is maintained of around 21 days. Therefore, the model as we have developed here is very relevant for predicting impacts associated with longer term reductions in light of over three weeks or more.

Other measures of canopy scale photosynthetic response to light

In the previous discussion we have assumed that the concept of isoclines of equal light environment' with respect to H^{A}_{sat} is valid. Alternative measures may be more appropriate but this does not affect the primary concept that canopy self-shading can be equivalent to environmental shading, and that there are two mechanisms of photosynthetic acclimation: physiological and via canopy structure. Any alternative measure of the photosynthetically relevant light environment would likely have a similar form to that of Figures 6 and 7. The interaction of self-shading and the non-linearity of leaf-level photosynthesis must inevitably result in a complex canopy scale response to LAI and the light environment for all canopies that are subjected to irradiances above photosynthetic saturation. Another candidate measure would be the integration of photosynthesis over time, i.e. to propagate the leaf-level light through a photosynthesis versus irradiance (P-I) curve to give an integrated photosynthesis measure equivalent to µmol O_2 evolution. In addition, plots of actual top of canopy PAR light levels may have greater descriptive power than percentage shading (Figures 6 and 7). Lavery et al. [9] observed different canopy changes at similar shading levels and interpreted these as being due to differences in the absolute light levels. In this study we suggested H^{A}_{sat} as a simple extension of the top-of canopy H_{sat}, since that measure has been demonstrated to have predictive power for canopy sustainability [8,9] and has been used in management contexts, and percentage shading was employed as a mirror of the empirical treatments. Clearly, there are many opportunities for further experiments and modelling to determine the most relevant measure of canopy photosynthetic response, the key point being that that measure needs to include within-canopy light propagation.

Conclusions

Three dimensional canopy modelling of *Amphibolis griffithii* has revealed that the interaction of light levels and canopy density on canopy-scale photosynthetic activity is complex and non-linear, in particular due to the non-linearity of leaf-level photosynthesis at saturating irradiance. The accumulated percentage area of leaf saturation above saturating irradiance, H^{A}_{sat}, was proposed as a measure relevant to canopy sustainability, based on extension of the equivalent top-of-canopy measure H_{sat} that has previously proved useful. The available empirical data were not sufficient to evaluate the efficacy of H^{A}_{sat} due to lack of lower shading treatments. Evaluating this measure and other candidates such as integrated leaf-level photosynthesis requires further experimental work. Nevertheless the principle has been demonstrated that plots of equal light environment' (Fig. 6) produced for different seagrass species, water depths, and water column optical properties could have practical management applications for predicting and interpreting canopy changes under light reduction events. Reduction in seagrass density in response to shading must be interpreted in terms of the leaf-level light environment. While physiological responses are also important, existing canopies in the same environment can provide information of the limits of physiological acclimation, and indicate if change in light levels will induce a trajectory to steady state sustainability, or to eradication. An important future step is to understand the time constants in change and recovery trajectories, to determine how long shading events can be tolerated and the required recovery periods. This information will be invaluable to coastal management.

Acknowledgments

J Hedley wishes to acknowledge Susana Enríquez, whose ideas and insight have informed the work described here. K McMahon acknowledges the collaboration with P S Lavery, which provided background for this research, L Twomey and M Westera from BMT Oceanica Pty Ltd for insights contributing to this work and N Dunham for canopy measurements.

Author Contributions

Conceived and designed the experiments: JH KM PF. Performed the experiments: JH KM. Analyzed the data: JH KM PF. Contributed reagents/materials/analysis tools: JH KM. Contributed to the writing of the manuscript: JH KM PF.

References

1. Orth RJ, Carruthers TJB, Dennison WC, Duarte CM, Forqurean JW, et al. (2006) A global crisis for seagrass ecosystems. Bioscience 56: 987–996.
2. Costanza R, d'Arge R, de Groot R, Farber S, Grasso M, et al. (1997) The value of the world's ecosystem services and natural capital. Nature 387: 253–260.
3. Fourqurean JW, Duarte CM, Kennedy N, Marbà N, Holmer M, et al. (2012) Seagrass ecosystems as a globally significant carbon stock. Nature Geoscience 5: 505–509.
4. Waycott M, Duarte CM, Carruthers TJB, Orth RJ, Dennison WC, et al. (2009) Accelerating loss of seagrasses across the globe threatens coastal ecosystems. Proceedings of the National Academy of Sciences 106: 12377–12381.
5. Grech A, Chartrand-Miller K, Erftemeijer P, Fonseca M, McKenzie L, et al. (2012) A comparison of threats, vulnerabilities and management approaches in global seagrass bioregions. Environmental Research Letters 7: 024006.
6. Grech A, Coles R, Marsh H (2011) A broad-scale assessment of the risk to coastal seagrasses from cumulative threats. Marine Policy 35: 560–567.
7. McMahon K, Collier C, Lavery PS (2013) Identifying robust bioindicators of light stress in seagrasses: A meta-analysis. Ecological Indicators 30: 7–15.
8. Collier CJ, Waycott M, McKenzie LJ (2012) Light thresholds derived from seagrass loss in the coastal zone of the northern Great Barrier Reef, Australia. Ecological Indicators 23: 211–219.
9. Lavery PS, McMahon K, Mulligan M, Tennyson A (2009) Interactive effects of timing, intensity and duration of experimental shading on *Amphibolis griffithii*. Marine Ecology Progress Series 394: 21–33.
10. Lambers H, Chapin FS, Pons TL (1998) Plant physiological ecology. New York: Springer. 540 pp.
11. Kirk JTO (1994). Light and photosynthesis in aquatic ecosystems. Cambridge: Cambridge Press. 528 pp.
12. Ralph PJ, Durako MJ, Enríquez S, Collier CJ, Doblin MA (2007) Impact of light limitation on seagrasses. Journal of Experimental Marine Biology and Ecology 350: 176–193.
13. Cayabyab NM, Enríquez S (2007) Leaf photoacclimatory responses of the tropical seagrass *Thalassia testudinum* under mesocosm conditions: a mechanistic scaling-up study. New Phytol 176: 108–123.
14. Masini RJ, Manning CR (1997) The photosynthetic responses to irradiance and temperature of four meadow-forming seagrasses. Aquatic Botany 58: 21–36.
15. Duarte CM, Marba N, Krause-Jensen D, Sanchez-Camacho M (2007) Testing the predictive power of seagrass depth limit models. Estuaries and Coasts 30: 652–656.
16. O'Brien KR, Grinham A, Roelfsema CM, Saunders MI, Dennison WC (2011) Viability criteria for the presence of the seagrass *Zostera muelleri* in Moreton Bay, based on benthic light dose. In: MODSIM 2011: International Congress on Modelling and Simulation, Proceedings. Modelling and Simulation Society of Australia and New Zealand (MODSIM 2011), Perth, Australia, 12–16 December 2011. 4127–4133.
17. Kemp WM, Batiuk R, Bartleson R, Bergstrom P, Carter V, et al. (2004) Habitat requirements for submerged aquatic vegetation in Chesapeake Bay: Water quality, light regime, and physical-chemical factors. Estuaries 27: 363–377.

18. Dennison WC, Orth RJ, Moore KA, Court Stevenson J, Carter V, et al. (1993) Assessing water quality with submersed aquatic vegetation: habitat requirements as barometers of Chesapeake Bay health. BioScience 43: 86–94.
19. Gacia E, Marba N, Cebrian J, Vaquer-Sunyer R, Garcias-Bonet N, et al. (2012) Thresholds of irradiance for seagrass *Posidonia oceanica* meadow metabolism. Mar Ecol Prog Ser 466: 69–79.
20. Dennison WC, Alberte RS (1985) Role of daily light period in the depth distribution of *Zostera marina* (eelgrass). Mar Ecol Prog Ser 25: 51–61.
21. Enríquez S, Merino M, Iglesias-Prieto R (2002) Variations in the photosynthetic performance along the leaves of the tropical seagrass *Thalassia testudinum*. Marine Biology 140: 891–900.
22. Fourqurean JW, Zieman JC (1991) Photosynthesis, respiration and whole plant carbon budget of the seagrass *Thalassia testudinum*. Mar Ecol Prog Ser 69: 161–170.
23. Collier CJ, Lavery PS, Masini RJ, Ralph PJ (2007) Morphological, growth and meadow characteristics of the seagrass *Posidonia sinuosa* along a depth-related gradient of light availability. Mar Ecol Prog Ser 337: 103–115.
24. Carruthers TJB, Walker DI (1997) Light climate and energy flow in the seagrass canopy of *Amphibolis griffithii* (J.M.Black) den Hartog. Oecologia 109: 335–341.
25. Hedley JD, Enríquez S (2010) Optical properties of canopies of the tropical seagrass *Thalassia testudinum* estimated by a three-dimensional radiative transfer model. Limnology and Oceanography 55: 1537–1550.
26. Zimmerman R (2006) Light and photosynthesis in seagrass meadows. In: Larkum AWD, Orth RJ, Duarte C, editors. Seagrasses: Biology, ecology and conservation. Springer. 303–321.
27. Hedley JD (2008) A three-dimensional radiative transfer model for shallow water environments. Optics Express 16: 21887–21902.
28. McMahon K, Lavery PS (2014) Canopy-scale modifications of the seagrass *Amphibolis griffithii* in response to and recovery from light reduction. Journal of Experimental Marine Biology and Ecology 455: 38–44.
29. House DH, Breen DE (2000) Cloth modeling and animation. Massachusetts: A. K. Peters. 344 p.
30. Mayer B, Kylling A (2005) The libRadtran software package for radiative transfer calculations – description and examples of use. Atmospheric Chemistry and Physics 5: 1855–1877.
31. Grant RH, Heisler GM, Gao W (1996) Photosynthetically-active radiation: Sky radiance distributions under clear and overcast conditions. Agricultural and Forest Meteorology 82: 267–292.
32. Mobley CD (1994) Light and Water. San Diego: Academic Press. 608 p.
33. Mobley CD, Sundman L (2000) HydroLight 4.1 user's guide. Sequoia Scientific. Available: http://www.sequoiasci.com/products/Hydrolight.aspx.
34. Hedley JD, Roelfsema CM, Phinn SR, Mumby PJ (2012). Environmental and sensor limitations in optical remote sensing of coral reefs: Implications for monitoring and sensor design. Remote Sensing 4: 271–302.
35. Durako MJ (2007) Leaf optical properties and photosynthetic leaf absorptances in several Australian seagrasses. Aquatic Botany 87: 83–89.
36. Green EP, Short F (2004) World atlas of seagrasses. Berkeley: University of California Press. 320 p.
37. Pedersen TM, Gallegos CL, Nielsen SL (2012) Influence of near-bottom re-suspended sediment on benthic light availability. Estuarine, Coastal and Shelf Science 106: 93–101.
38. Koch E, Gust G (1999) Water flow in tide and wave dominated beds of the seagrass *Thalassia testudinum*. Mar Ecol Prog Ser 184: 63–72.
39. Collier CJ, Waycott M, Ospina AG (2012b) Responses of four Indo-West Pacific seagrass species to shading. Marine Pollution Bulletin 65: 342–354.
40. Brun FG, Vergara JJ, Hernádez I, Pérez-Lloréns JL (2003) Growth, carbon allocation and proteolytic activity in the seagrass *Zostera noltii* shaded by *Ulva* canopies. Funct Plant Biol 30: 551–560.

13

Invasibility of Mediterranean-Climate Rivers by Non-Native Fish: The Importance of Environmental Drivers and Human Pressures

Maria Ilhéu[1,2]*, Paula Matono[1,2], João Manuel Bernardo[1]

1 Departamento de Paisagem Ambiente e Ordenamento, Escola de Ciências e Tecnologia, Universidade de Évora, Évora, Portugal, **2** Instituto de Ciências Agrárias e Ambientais Mediterrânicas, Universidade de Évora, Évora, Portugal

Abstract

Invasive species are regarded as a biological pressure to natural aquatic communities. Understanding the factors promoting successful invasions is of great conceptual and practical importance. From a practical point of view, it should help to prevent future invasions and to mitigate the effects of recent invaders through early detection and prioritization of management measures. This study aims to identify the environmental determinants of fish invasions in Mediterranean-climate rivers and evaluate the relative importance of natural and human drivers. Fish communities were sampled in 182 undisturbed and 198 disturbed sites by human activities, belonging to 12 river types defined for continental Portugal within the implementation of the European Union's Water Framework Directive. Pumpkinseed sunfish, *Lepomis gibbosus* (L.), and mosquitofish, *Gambusia holbrooki* (Girard), were the most abundant non-native species (NNS) in the southern river types whereas the Iberian gudgeon, *Gobio lozanoi* Doadrio and Madeira, was the dominant NNS in the north/centre. Small northern mountain streams showed null or low frequency of occurrence and abundance of NNS, while southern lowland river types with medium and large drainage areas presented the highest values. The occurrence of NNS was significantly lower in undisturbed sites and the highest density of NNS was associated with high human pressure. Results from variance partitioning showed that natural environmental factors determine the distribution of the most abundant NNS while the increase in their abundance and success is explained mainly by human-induced disturbance factors. This study stresses the high vulnerability of the warm water lowland river types to non-native fish invasions, which is amplified by human-induced degradation.

Editor: Maura (Gee) Geraldine Chapman, University of Sydney, Australia

Funding: This study was partially funded by INAG (Instituto da Água) as part of a larger programme on the implementation of the Water Framework Directive and by the Foundation for Science and Technology (FCT) through the research project (PTDC/AAC-AMB/102541/2008). Additional funding was provided by FEDER under the programme COMPETE and by National Funds through FCT under the Strategic Project PEst-OE/AGR/UI0115/2014. P. Matono was supported by a PhD grant from FCT (SFRH/BD/23435/2005). The funders had no role in study design, data collection and analysis, decision to publish, or preparation of the manuscript.

Competing Interests: The authors have declared that no competing interests exist.

* Email: milheu@uevora.pt

Introduction

The rate and extent of invasions in freshwater ecosystems are particularly alarming in the Mediterranean region, which is among the most heavily invaded ecosystems in the world [1–3]. The Iberian freshwater habitats are one of the most paradigmatic examples where there are constant reports of new invading fish species and colonization of new areas (see [4–10] as examples).

The success of biological invasions can be explained by several factors, with environmental drivers being among the most important ones [11–12]. One of the most frequently stated hypotheses in the biological invasion literature is that species should have a greater chance of success if they are introduced to an area with a climate that closely matches that of their original range [13–15]. Other environmental drivers, such as spatial heterogeneity and environmental variability, may also be important [16–18]. The success of an invading fish may be predicted with reference to environmental conditions at different scales, ranging from habitat to river types or ecoregions. Some of the features of the recipient ecosystem and the success of an introduced species may be determined by human-induced pressure. The "human activity" hypothesis argues that human activities facilitate the establishment of non-native species (NNS) by disturbing natural landscapes and by increasing propagule pressure [2,8]. In Mediterranean-climate rivers, both landscape and human disturbance factors are expected to play a major role in the biological invasions, as these systems are largely governed by stochastic processes and have suffered a long history of human-induced pressure [19].

Although non-native species are not mentioned specifically in the European Union's Water Framework Directive (WFD) [20], in the context of the Directive's objectives NNS represent an important pressure (listed in the WFD annexes), since they can modify the structure of native biota and the ecological functioning of aquatic systems. NNS affect the ecological quality of natural environments in multiple ways and may represent a serious threat to native communities. Potential effects include genetic alterations within populations, spreading of pathogens and parasites, competition with and replacement of native species, and habitat deterioration or modification (see [21] for Iberian freshwater ichthyofauna). All the combined effects may result in changes in ecosystem function and global homogenization [1,3,22–23], with profound impact at ecological, evolutionary, genetic, and economic levels [24–25].

The apparently strong relationship between introduced fish species and degraded river conditions and their potential impact on native species and ecosystem suggest that NNS may be useful indicators of biological integrity and river health [26–28] and they are therefore incorporated in river monitoring schemes under the WFD and elsewhere. The incorporation of NNS into the ecological assessment will require knowledge about the density, distribution and potential risk of each species for each water body. According to the WFD, with the exception of particular cases, the ecological assessment involves the previous derivation of typologies for the aquatic systems. The ecological assessment is performed within groups of similar ecosystems and is expressed as a deviation from the reference conditions, i.e. conditions found in sites without human pressures. In fact, differences in climate, hydrology, geomorphology, geology, soil and vegetation make comparison of river communities difficult if not impossible. The use of river typology, by stratifying the spatial variability, makes the ecological assessment more practical, transparent and understandable by decision-takers, managers, technicians and public, and for these reasons it was the chosen approach on the WFD implementation. Moreover, approaching invasibility at a regional framework is useful for management purposes because it allows the identification of water bodies particularly vulnerable to NNS invasion and helps to regionalize monitoring schemes [29–31].

Understanding the factors promoting successful invasions is of great conceptual and practical importance. From a practical point of view, it should help to prevent future invasions and to mitigate the effects of recent invaders through early detection and prioritization of management measures. Thus, the identification of the environmental determinants of fish invasions is relevant in forecasting the overall impact of invasions on a global scale and prerequisite management authorities to adopt sound conservation policies.

The objectives of the present study were to determine: i) the patterns of non-native fish richness and abundance in Mediterranean-climate river types of Portugal; ii) the environmental drivers that favour the invasiveness of non-native fish within the morpho-climatic gradients; iii) the relationship between human-induced disturbance and non-native fish abundance and iv) the relative importance of environmental variables and human pressures for the occurrence and abundance of non-native fish species.

Methods

Study Area

The study considered 12 river types out of 15 defined for continental Portugal under the implementation of the WFD [32] (Fig. 1), as the 3 very large and highly modified river types were not considered.

The climate of Portugal presents high intra- and inter-annual precipitation and discharge variation, with severe and unpredictable floods between autumn and spring and persistent summer droughts [33], although the influence of factors such as topography and proximity to the Atlantic Ocean causes significant climatic contrasts. The general conditions of the atmospheric circulation cause a decrease in precipitation from north to south and from west (coast) to east (inland), enhanced by orographic asymmetry. Indeed, the mountain barrier in the north and center causes less rainfall in the interior regions. The temperature shows an opposite pattern, increasing from north to south. In a general way, the altitude causes a decrease in temperature and an increase in rainfall.

Portuguese river types reflect two main geographical and climatic gradients: north–south and west–east. The north–south gradient is associated with a decrease in altitude, precipitation, annual discharge, and increasing temperature. The west–east gradient is related to a continental effect, with a precipitation decrease and an increase in temperature extremes. Owing to these features, most rivers are permanent in the north and intermittent in the south. The most relevant climatic and morphological characteristics of the basins of the river-types are presented in Table 1 (see also [32]). The river types may be described as:

M - Mountain rivers of the North: trout rivers with small drainage area located in the North-West and North-Center regions with low mean temperature and high precipitation; N1< 100 km^2 - North rivers of small drainage area: low mean temperature, altitude mainly in the 200–600 m range; N1> 100 km^2 - North rivers of medium and large drainage area: characteristics similar to the previous type; N2 - Medium-large Alto Douro rivers: drainage area larger than 100 km^2, high temperature (similar to the south types), lower rainfall; N3 - Small Alto Douro rivers: drainage area smaller than 100 km^2 and characteristics similar to the previous type; N4 - North-South transition type: located at the center region with intermediate characteristics; L - Littoral type: coastal streams of small-medium dimensions located at the littoral center region, low altitude; S1< 100 km^2 - Small South rivers: a large group of low altitude temporary rivers, in a region of high temperature and low precipitation; S1>100 km^2 - South rivers of medium-large dimension: larger drainage temporary rivers with similar characteristics to the previous type; S2 - South mountain rivers: small streams in the mountain regions of the south with lower temperature and higher precipitation than in the other south river types; S3 - Rivers of Tagus and Sado sedimentary basins: small and medium dimension, low altitude and high mineralization, smaller rivers with temporary regime; S4 - Small chalk rivers of Algarve: low altitude, low precipitation and high temperature.

Native freshwater fish fauna of Portuguese rivers presents relatively low species richness per site and most fish species are endemic cyprinids with high conservation value, particularly in the south. Many of those species are threatened with extinction [34–35].

Sampling and data collection

Sampling was carried out between 2004 and 2006 during spring at 380 sites in the main Portuguese river basins (Fig. 1). The number of sampled sites in each river type was proportional to its basin area: M = 28 sites; N1<100 km^2 = 60 sites; N1> 100 km^2 = 68 sites; N2 = 20 sites; N3 = 28 sites; N4 = 16 sites; L = 33 sites; S1<100 km^2 = 26 sites; S1>100 km^2 = 33 sites; S2 = 22 sites; S3 = 37 sites; S4 = 9 sites. In each river type, sampling sites included undisturbed or least disturbed sites and disturbed ones. For the site selection, a preliminary human-induced pressure screening using GIS and information on pollution loads was followed. The final selection was based on the human disturbance level regarding ten variables [36–37]: land use, urban area, riparian zone, connectivity of the river segment, sediment load, flow regime, morphological condition, presence of artificial lentic water bodies, toxic and acidification levels, and nutrient/organic load. The spatial scale of the assessment depended on the character of each pressure, some were made at local scale, others at the fluvial segment, and some others at the basin scale. Each variable was scored from 1 (minimum

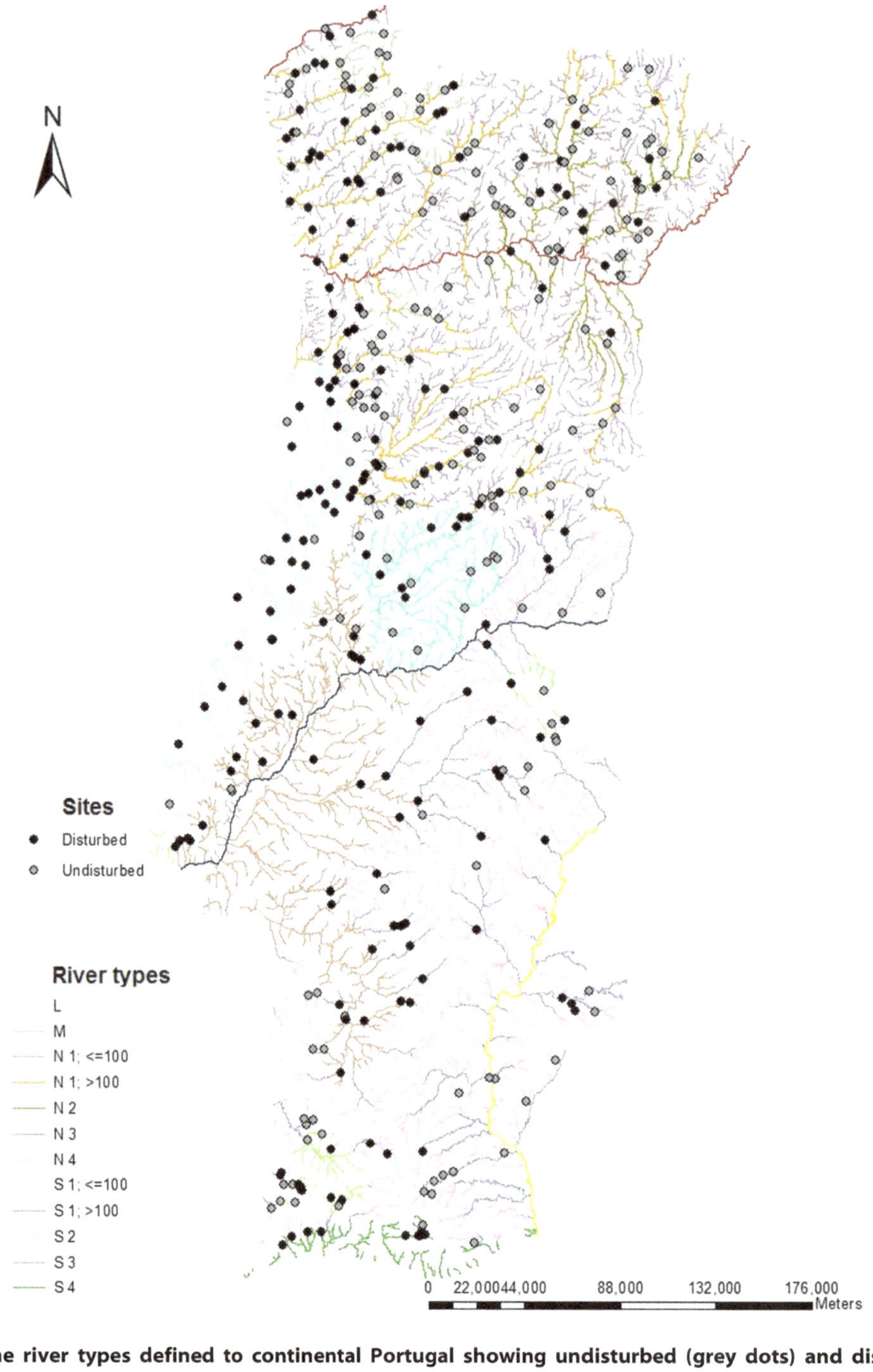

Figure 1. Map of the river types defined to continental Portugal showing undisturbed (grey dots) and disturbed (black dots) sampling sites.

disturbance) to 5 (maximum disturbance) (Tables 2 and 3) and only sites with scores of 1 and/or 2 and only one variable scored with a 3 were considered as undisturbed or least disturbed (i.e., reference sites). The sum of these scores represents the total human pressure. Several water quality variables complemented the evaluation of human pressure in each site, following Standard Methods for the Examination of Water and Wastewater [38]: Biochemical Oxygen Demand (BOD_5, mg L^{-1}), Chemical Oxygen Demand (COD, mg L^{-1}), Total Suspended Solids (TSS, mg L^{-1}), Soluble Reactive Phosphorous (SR-P, mg L^{-1}), nitrite (NO_2^-, mg L^{-1}), nitrate (NO_3^-, mg L^{-1}), ammonium (NH_4^+, mg L^{-1}), and total inorganic dissolved nitrogen (N, mg L^{-1}).

Environmental characterization of sites was based on regional and local variables. Regional variables were obtained from digital cartography with free Internet access and included the drainage area of the basin upstream the site (km^2), distance from source (km), altitude (m), slope (%), mean annual discharge (mm), mean annual air temperature (°C), and mean annual rainfall (mm). Rainfall, temperature, and flow variables were described from 30-year data series. Topographical variables were derived from a Digital Elevation Model (DEM), with a 90-m grid cell resolution

Table 1. Main morpho-climatic characteristics of river types in continental Portugal (mean and SD).

River-Types	Mean Annual Temperature (°C)	Mean Annual Precipitation (mm)	Altitude (m)	Drainage Area (km²)
M	11.0	1944	506	24.8
northern mountain streams	(1.5)	(379)	(300)	(17)
N1<100 km²	12.4	1190	413	33
north streams with small drainage area	(1.3)	(358)	(242)	(23)
N1>100 km²	12.6	1196	274	549
north streams with large drainage area	(1.2)	(374)	(205)	(65)
N2	13.1	596	300	960
streams from Alto Douro with large drainage area	(1.0)	(81)	(141)	(1115)
N3	13.0	671	432	32
streams from Alto Douro with small drainage area	(0.8)	(134)	(160)	(23)
N4	14.1	1065	280	151
transition streams between north and south	(0.7)	(168)	(122)	(361)
L	14.8	941	44	180
littoral streams from west/centre region	(0.3)	(118)	(44)	(671)
S1<100 km²	15.7	628	183	30
south streams with small drainage area	(0.9)	(86)	(75)	(21)
S1>100 km²	15.8	587	137	439
south streams with large drainage area	(0.9)	(84)	(68)	(579)
S2	15.4	743	175	60
southern mountain streams	(0.3)	(85)	(147)	(87)
S3	15.6	730	54	388
streams of sedimentary deposits in Tagus and Sado basins	(0.4)	(118)	(46)	(1081)
S4	16.9	632	54	67
southern carsick streams of Algarve	(0.5)	(60)	(57)	(89)

(CGIAR-CSI 2005), using ArcMap 9.1. Local variables were assessed *in situ* at each sampling occasion: water temperature (°C), conductivity (μS cm^{-1}), pH, dissolved oxygen (mg L^{-1}), mean stream wetted width (m), maximum and mean water depth (m), mean current velocity (m s^{-1}), dominant substrate class, adapted from the Wentworth scale [39]: 1 – mud and sand; 2 – gravel; 3 – pebble; 4 – cobble; 5 – boulders; 6 – boulders larger than 0.50 m, riparian vegetation (%), shadow (%), and proportion of different habitat types (pool, run, riffle).

Human disturbance level was evaluated *in situ* regarding the same ten variables described above for site selection. A total of 182 undisturbed and 198 disturbed sites were sampled covering the full human pressure gradient in each river type. The proportion of sites along the degradation gradient was sampled according to their availability within each river type.

Fish were collected by electrofishing following the WFD-compliant and CEN sampling protocol [40–41]. A backpack battery-powered electrofishing equipment was used, wading in shallow reaches (<1.2 m) or from a boat in deeper areas. The power and pulse frequency of the equipment were adjusted to ensure capture efficiency but also that fish were only stunned for a short period of time. Captured specimens were carefully handled, ensuring the minimum possible stress, and were quickly returned to the river. No specimens of endangered or protected species were injured or sacrificed.

The present study was part of a large programme coordinated by the National Water Institute (INAG, Instituto da Água, now part of APA, Agência Portuguesa do Ambiente). The National Institute for the Nature Conservation and Forestry (ICNF, Instituto de Conservação da Natureza e das Florestas (http://www.proforbiomed.eu/project/partners/institute-conservation-nature-and-forest-icnf) provided the necessary fishing permits. All sampling took place in national public rivers, which were under the jurisdiction of the National Water Institute and ICNF. As all the fish manipulation was restricted to the sampling procedures and animal care was taken into account, it was not necessary to obtain permission from the National *Animal Care* and Use *Committee* (Comissão de Ética, Bioética e Bem-estar Animal).

Captures were quantified as density (individuals/100 m^2) and biomass (grams/100 m^2). The degree of invasibility was measured using NNS richness and abundance.

A nonparametric ANCOVA performed using the rank transformation analysis of covariance evaluated the existence of significant differences in the density and number of NNS between river types [42,43]. The total human pressure was included as a covariable, in order to exclude its effects from the analysis.

Mann-Whitney test was used to identify significant differences in the number and density of NNS between undisturbed and disturbed sites. The Z test of proportions [44] evaluated the existence of significant differences in the frequency of occurrence of NNS between undisturbed and disturbed sites.

Table 2. Description, assessment scale and methods, and scoring criteria of the variables; land use, urban area, riparian vegetation, morphological condition and sediment load - used to evaluate the level of anthropogenic disturbance in sampled sites.

Variables	Description	Assessment scale	Score	Criteria	Methods
Land use	Impact of farming/ forestry practices	River segment	5	>40% Agricultural use (intensive agriculture), very severe impact (rice field)	Local expert assessment complemented with Corine Land Cover (2000, 2006)*
			4	>40% Strong impact (area with strong forestry, including clearcuts)	
			3	<40% Moderate impact (subsistence gardens, pastures)	
			2	<40% Small impact (cork and holm oaks, high-growth forest)	
			1	<10% No significant impacts (natural forest and bush)	
	Land cover and bankface characterization	Local	5	Irrigated crops and/or high stocking	Local expert assessment complemented with Corine Land Cover (2000, 2006)*
			4	Horticultural crops, semi-intensive grazing	
			3	Extensive cultures (e.g. pastures, cereal crops, pine, eucalyptus), extensive grazing	
			2	Cork and holm oaks	
			1	Natural	
Urban area	Impact of urban areas	River segment	5	Very severe (location near a city with basic sanitation needs)	Local expert assessment complemented with Corine Land Cover (2000, 2006) *
			4	Town	
			3	Village	
			2	Hamlet	
			1	Negligible (isolated dwellings)	
Riparian vegetation	Deviation from the natural state of the riparian zone	River segment	5	Lack of riparian shrubs and trees (only the presence of annual plants)	Local expert assessment
			4	Fragmented vegetation with bushes and/or the presence of reed	
			3	Second replacement step (dominance of dense brushwood)	
			2	First replacement step (presence of shrub or tree strata with some level of preservation).	
			1	Potential vegetation (presence of shrub and tree strata according to the geo-series)	
Morphological condition	Deviation from the natural state of the stream bed and banks	Local	5	Transverse and longitudinal profile of the channel completely changed, with very few habitats	Local expert assessment
			4	Channelized sector, missing most of the natural habitats	
			3	Channelized sector, missing some types of natural habitats, but maintaining much of the shape of the natural channel	
			2	Poorly changed sector, close to the natural mosaic of habitats.	
			1	Morphological changes absent or negligible	

Table 2. Cont.

Variables	Description	Assessment scale	Score	Criteria	Methods
Sediment load	Deviation from the natural sediment load (both carried in the water column and deposited on the riverbed)	River segment and local	5	>75% of coarse particles of the stream bed are covered with fine sediments (sand, silt, clay)	Local expert assessment
			4	50–75% of coarse particles of the stream bed are covered with fine sediments (sand, silt, clay)	
			3	25–50% of coarse particles of the stream bed are covered with fine sediments (sand, silt, clay)	
			2	5–25% of coarse particles of the bed are covered with fine sediments (sand, silt, clay)	
			1	<5% of coarse particles of the stream bed are covered with fine sediments (sand, silt, clay)	

*information available from http://sniamb.apambiente.pt/clc/frm/.

The response pattern of the most abundant NNS to the human-induced pressure gradient was tested with a nonparametric ANCOVA (detailed above) [42,43] of NNS density between five human pressure classes. The establishment of these five quality classes (High, Good, Moderate, Poor and Bad) for the stressor gradient (total human pressure) to which sites were assigned, followed the approach used in the REFCOND Guidance Document [37]. To account for the possible influence of the environmental variables, a Principal Components Analysis (outputs not shown) was performed in order to extract the most relevant environmental gradient (PCA1), which was then included as a covariable in this analysis. For each species, only data from river types with occurrences were considered.

Univariate Generalized Linear Models (GLM) were used for partitioning the variance [45] among three sets of explanatory variables - environmental, human-induced pressure and spatial trends on the occurrence and density of the most abundant NNS. For each species, only data from river types with NNS occurrences were considered. The set of environmental variables included altitude, mean annual discharge, drainage area of the basin, slope, mean annual temperature, mean water depth, mean stream width, mean current velocity, dominant substrate class, proportion of each habitat type, pH, dissolved oxygen and conductivity. The set of human pressure variables included BOD_5, COD, TSS, N, SRP, and the 10 human pressure variables evaluated at each site (Table 2). The spatial structure of data was incorporated into the analyses to prevent misinterpretation of relations between and within the spatially arranged datasets [46]. Owing to spatial autocorrelation, values of particular variables in neighbouring sites are more or less similar than they would be in a random set of observations [47]. Autocorrelation is a frequently observed feature in spatially sampled biological data that may make the identification of plausible relationships between biota and the environment difficult [48]. The spatial structure of data was explored, including geographical coordinates of sites and their higher and cross product terms, in the modelling procedure (x, y, xy, x^2, y^2, x^2y, xy^2, x^3 and y^3) [47]. The x and y coordinates were centred to zero mean before computing the matrix of spatial descriptors, to reduce collinearity between successive terms when fitting the polynomial. GLMs were performed using a forward selection procedure of the explanatory variables. The best models (minimal adequate) were estimated according to the lowest Akaike Information Criterion (AIC). Plots of residuals (not shown) were examined to complement AIC values and to confirm goodness-of-fit (see [49]). For the occurrence of species (presence–absence data), GLMs were performed using the binomial distribution and logit link function. In order to overcome the problems of distribution fitting resulting from the high number of absences in the density matrix, this continuous response variable was standardized to the maximum value for the species and converted into four classes: 0 (0% of individuals), 1 (0% - 10% of individuals), 2 (10% - 50%) and 3 (>50%). For this transformed density response variable (count data), GLMs were performed using the Poisson distribution and log link function. The existence of over-dispersion in data (variance higher than the mean; dispersion parameter>1) was checked during the analysis. If the value observed was higher than the threshold limit then quasi-binomial and quasi-Poisson distributions should be used, respectively. The total variation within the response variables was decomposed among the three sets of explanatory variables and the percentage of explained deviance calculated for eight different components [50]: (i) pure effect of environmental drivers, (ii) pure effect of anthropogenic disturbance, (iii) pure effect of spatial trends, (iv) combined variation due to the joint effect of environmental and anthropogenic components, (v) combined variation due to the joint effect of environmental and spatial components, (vi) combined variation due to the joint effect of anthropogenic and spatial components, (vii) combined variation due to the joint effect of the three components and (viii) variation not explained by the independent variables included in the analysis.

Multicollinearity among explanatory variables may result in the exclusion of ecologically meaningful variables from the models if another intercorrelated variable or variables happen to explain the variation better in statistical terms. Therefore, some of the most clearly intercorrelated variables (Spearman's rank correlation $|r|>0.75$; $P<0.05$) were initially excluded. The exclusion decision took into account the potential relevance of the variable in the occurrence and distribution of NNS. Furthermore variables were maintained in the models only if their addition did not cause any Variation Inflation Factor (VIF) to exceed 3.0, therefore ensuring that no covariation exists between the selected variables in the models.

Table 3. Description, assessment scale and methods, and scoring criteria of the variables: hydrological regime, toxic and acidification levels, organic and nutrient loads, artificial lentic waters - used to evaluate the level of anthropogenic disturbance in sampled sites.

Variables	Description	Assessment scale	Score	Criteria	Methods
Hydrological regime	Deviation from the natural hydrological regime (flow pattern and/or quantity). Includes all sources of hydrologic alteration, such as significant water abstraction	Local	5	<50% and strong deviation from the natural variability of the flow regime	Local expert assessment complemented with SNIRH
			4	<50% and moderate deviation from the natural variability of the flow regime	
			3	>50% and duration of flood periods close to the natural	
			2	>75% and duration of flood periods close to the natural	
			1	>90% and normal duration of natural flood periods	
		Local	5	<10% of mean annual discharge	Local expert assessment complemented with SNIRH
			4	<15% of mean annual discharge	
			3	>15% of mean annual discharge	
			2	>30% of mean annual discharge	
			1	>90% of mean annual discharge	
Toxic and acidification levels	Deviation from the natural state of toxicity conditions, including acidification and oxygen levels	Local	5	Constant for long periods (months) or frequent occurrence of strong deviations from natural conditions (e.g. pH <5.0, DO <30%)	Local expert assessment complemented with SNIRH
			4	Constant for long periods (months) or frequent occurrence of strong deviations from natural conditions (e.g. pH <5.5, DO <30–50%)	
			3	Occasional deviations (single measurements or episodic) in relation to natural conditions (e.g. pH <5.5, DO <30–50%)	
			2	Occasional deviations (single measurements or episodic) in relation to natural conditions (e.g. pH <6.0)	
			1	Conditions within the normal range of variation	
Organic and nutrient loads	Deviation from the normal values of BOD, COD, ammonium, nitrate and phosphate concentrations	Local	5	>20% of values in classes D or E	SNIRH (classification of water quality for multiple uses, according to the guidelines from the National Water Institute*), complemented with local expert assessment
			4	>10% of values in classes D or E	
			3	>10% of values in class C	
			2	No obvious or too small signs of eutrophication and organic loading	
			1	No signs of eutrophication and organic loading	
Artificial lentic water bodies	Impact related to the presence of artificial lentic water bodies upstream and/or downstream of the site (upstream change in thermal and flow regimes; downstream invasion by exotic species of lentic character)	Local	5	Local immediately downstream of a large reservoir or within the influence area of its backwater	SNIRH and available cartography

Table 3. Cont.

Variables	Description	Assessment scale	Score	Criteria	Methods
			4	Local immediately downstream of a mini-hydro or within the influence area of its backwater	
			3	Local downstream of a reservoir or within the influence area of the reservoir	
			2	Local downstream of a mini-hydro or within the influence area of its backwater	
			1	No influence of reservoirs	
Connectivity	Impact of artificial barriers to fish migration	River basin and segment	5	Permanent artificial barrier	SNIRH, available cartography, documental data and local expert assessment
			4	Occasional passage of some species	
			3	Passage of certain species or only in certain years	
			2	Passage of most species in most years	
			1	No barriers or existence of an effective pass-through device	

*information available from http://snirh.pt/snirh/_dadossintese/qualidadeanuario/boletim/tabela_classes.php.

Prior to the analyses, all data were either log (x+1) (linear measurements) or arcsin [sqrt(x)] (percentages) transformed to improve normality. Statistical analyses were performed using CANOCO 4.5, STATISTICA 6, PRIMER 6, SPSS 21 and BRODGAR 2.6 software applications.

Results

Patterns of richness, distribution and abundance of non-native species

A total of 41 fish species were captured, including 6 diadromous species and 10 NNS: pumpkinseed sunfish *Lepomis gibbosus* (L.), Iberian gudgeon *Gobio lozanoi* Doadrio and Madeira, mosquitofish *Gambusia holbrooki* (Girard), carp *Cyprinus carpio* L, goldfish *Carassius auratus* (L.), largemouth bass *Micropterus salmoides* (Lacépède), chameleon cichlid *Herichthys facetum* (Jenyns), black bullhead *Ameiurus melas* (Rafinesque), pike-perch *Sander lucioperca* (L.), and bleak *Alburnus alburnus* (L.) (Table 4).

NNS occurred in 33% of the sites, representing nearly 11% of both the total mean density (3.58 ind./100 m^2, SD = 12.54) and number of species (0.54, SD = 0.94) per site. Although the absolute values may seem low, they represent an important percentage of the total density and number of species, as Mediterranean streams usually show low species richness and density per site. Biomass values were not shown or included in the analysis, as they followed the density pattern and were therefore redundant.

There were significant differences in the density ($F_{(11,380)} = 8.04$; $P<0.001$) and proportion of non-native species ($F_{(11,380)} = 9.20$; $P<0.001$) among river types considering the total human pressure as a covariable in the analysis. NNS did not occur in the mountain river type or in the southern chalk river type. The small northern stream types, N1<100 km^2, N3 and N4, registered very low densities, frequency of occurrence, and percentage of NNS (Table 4). N2 and S2 also showed low densities of NNS, but these species represented more than 15% of the fish assemblages in each site and registered high values of frequency of occurrence (f. oc.). N1>100 km^2 and S1<100 km^2 presented high values of occurrence and density of NNS, also representing more than 10% of total density and species richness. Littoral and southern river types S1> 100 km^2 and S3 showed the highest occurrence and abundance of NNS; in Littoral type, NNS represented almost 12% of the total density and 15% of the total species richness, S1>100 km^2 and S3 registered percentages between 20% and 30%.

The most frequent and abundant NNS were *L. gibbosus* (mean density = 0.89 ind./100 m^2, SD = 3.3; f. oc. = 0.21), *G. lozanoi* (mean density = 1.79 ind./100 m^2, SD = 10.9; f. oc. = 0.14), and *G. holbrooki* (mean density = 0.54 ind./100 m^2, SD = 3.1; f. oc. = 0.11) (Table 4). The remaining species registered low occurrences and abundances. *L. gibbosus* showed a wide distribution in both northern and southern river types - N1> 100 km^2, N2, L, and mostly southern river types (Table 4), which are associated with high annual temperature and conductivity, large drainage area, and low altitude (Table 1). *G. holbrooki* were the most abundant species in the southern river types, showing a distribution almost limited to these river types, while *G. lozanoi* was the dominant species in north/central river types, exhibiting high occurrence in river types with large drainage area and low altitude, especially in the north/centre region - N1>100 km^2, N2, and L (Table 4). *C. carpio* and *M. salmoides* occurred in several river types, showing higher values in the south. *H. facetum*, *A. melas*, and *A. alburnus*, only occurred in the southern river types S1<100 km^2, S1>100 km^2, and S3 and *S. lucioperca* only occurred in N1>100 km^2, all presenting very low frequencies of occurrence and abundance. *C. auratus* registered nearly vestigial occurrence in N1>100 km^2, N2, and S3 (Table 4).

Overall, disturbed sites showed significantly higher occurrence (f. oc. = 0.46, $P<0.001$) and abundance (mean = 6.41, SD = 16.8, $U_{380} = 12155.00$, $Z = -6.56$, $P<0.001$) of NNS than undisturbed ones (f. oc. = 0.18; mean = 0.5, SD = 1.7). *M. salmoides*, *A. melas*, *S. lucioperca*, and *A. alburnus* only occurred in disturbed sites. All NNS but *H. facetum*, exhibited significantly higher occurrence ($P<0.01$) in disturbed sites. Fish density was significantly higher in

Table 4. Frequency of occurrence (f.oc.), number of species (%) and density (mean and SD) (ind/100 m^2) of non-native fish species (NNS) for total data (a), undisturbed (b) and disturbed (c) sites in each river-type (see Table 1) with NNS occurrence.

		NNS f.oc.	Total	N1<100 km^2	N1>100 km^2	N2	N3	N4	L	S1<100 km^2	S1>100 km^2	S2	S3
NNS f.oc.	a)	0.30		0.1	0.4	0.7	0.07	0.2	0.5	0.4	0.7	0.3	0.5
	b)	0.18		0	0.23	0.67	0.12	0.11	0	0.13	0.50	0.25	0.17
	c)	0.46		0.25	0.58	0.80	0	0.29	0.58	0.70	0.79	0.40	0.52
Nbr. NNS	a)		12.0	3.0	13.1	19.3	2.9	4.6	14.8	13.1	31.1	16.3	19.6
(%)			(21.0)	(9.2)	(19.3)	(14.2)	(11.2)	(10.0)	(19.6)	(23.2)	(30.1)	(31.1)	(25.9)
	b)		5.3	0	5.6	19.0	4.9	2.2	0	2.6	17.3	7.4	4.2
			(12.0)		(10.9)	(15.2)	(14.2)	(6.7)		(7.2)	(19.5)	(15.6)	(10.2)
	c)		18.2	6.4	19.0	20.0	0	7.7	18.8	29.8	41.2	27.0	22.6
			(25.2)	(12.7)	(22.4)	(12.6)		(13.1)	(20.3)	(30.1)	(32.8)	(41.7)	(27.0)
NNS Mean density	a)		3.6	2.3	4.8	1.2	0.07	0.3	9.3	3.6	6.4	1.1	6.3
(ind/100 m^2)			(12.5)	(9.1)	(12.8)	(1.6)	(0.3)	(0.8)	(30.1)	(8.6)	(11.5)	(2.1)	(12.5)
	b)		0.5	0	0.4	1.2	0.1	0.3	0	1.6	1.4	0.7	0.2
			(1.7)		(1.2)	(1.8)	(0.5)	(1.0)		(4.4)	(1.8)	(1.6)	(0.4)
	c)		6.4	4.8	8.2	1.2	0	0.2	11.8	6.9	10.1	1.5	7.5
			(16.8)	(13.0)	(16.3)	(0.9)		(0.4)	(33.6)	(12.4)	(14.1)	(2.7)	(13.3)
Lepomis gibbosus		0.21	0.9	0.1	1.1	0.6	0.01	0.3	0.8	2.3	2.1	0.06	2.3
			(3.3)	(0.9)	(3.7)	(0.9)	(0.02)	(0.8)	(2.1)	(6.7)	(4.4)	(0.3)	(5.4)
Cyprinus carpio		0.03	0.1	0	0	0.01	0	0	0.09	0.01	0.9	0	0.08
			(1.6)			(0.01)			(0.4)	(0.06)	(5.5)		(0.4)
Carassius auratus		0.008	0.01	0	0.01	0.01	0	0	0	0	0	0	0.01
			(0.03)		(0.06)	(0.01)							(0.02)
Gobio lozanoi		0.14	1.8	2.2	3.4	0.6	0.07	0	7.5	0	0	0	1.6
			(10.9)	(9.1)	(10.6)	(1.2)	(0.3)		(29.9)				(6.7)
Micropterus salmoides		0.02	0.07	0	0.2	0	0	0	0.2	0	0.1	0.07	0.08
			(0.7)		(1.5)				(0.6)		(0.6)	(0.3)	(0.5)
Herichtys facetum		0.01	0.03	0	0	0	0	0	0	0.3	0.1	0	0
			(0.4)							(1.3)	(0.3)		
Ameiurus melas		0.003	0.1	0	0	0	0	0	0	0	0	0	1.2
			(2.2)										(7.1)
Gambusia holbrooki		0.11	0.5	0	0.03	0	0	0.01	0.7	1.1	2.7	0.9	1.1
			(3.1)		(0.2)			(0.06)	(2.0)	(2.5)	(9.0)	(2.1)	(3.2)
Sander lucioperca		0.003	0.01	0	0.04	0	0	0	0	0	0	0	0
			(0.1)		(0.3)								
Alburnus alburnus		0.005	0.03	0	0	0	0	0	0	0	0.4	0	0
			(0.5)								(1.6)		

disturbed sites only for the most abundant species, *L. gibbosus* ($U_{380} = 15029.50$, $Z = 3.94$, $P<0.001$), *G. lozanoi* ($U_{380} = 15594.50$, $Z = 3.73$, $P<0.001$), *G. holbrooki* ($U_{380} = 14676.50$, $Z = 5.68$, $P<0.001$), and *M. salmoides* ($U_{380} = 17199.00$, $Z = 2.91$, $P<0.01$). In all river types but N3, the occurrence of NNS was significantly higher in disturbed sites ($P<0.01$). The abundance and percentage of NNS were higher in human-disturbed sites in L, N1<100 km^2, N1>100 km^2, S1<100 km^2, and S1>100 km^2 (Table 4). The majority of river types that did not revealed significant differences between undisturbed and disturbed sites presented null or very low abundance of non-native fishes.

No significant correlations (Spearman's rank correlations |r| < 0.5) were observed between the density of the most abundant NNS and total human pressure, as this relationship was not linear. A marked increase in fish density along the human disturbance gradient considering the environmental gradient (PCA1) as a covariable was observed, but there was a decrease in fish abundance in the extreme of the gradient, when the pressure was maximum (Fig. 2). This was particularly evident for *G. lozanoi* ($F_{(4,246)} = 5.34$, $P<0.001$) and *G. holbrooki* ($F_{(4,235)} = 7.07$, $P<0.001$), as mean fish densities reached the highest values in the poor quality class. The increase in fish density from moderate to poor quality classes was particularly marked for *G. holbrooki*. This species showed a comparatively higher mean density in the bad quality class and a smooth decrease between the poor and bad quality classes. *L. gibbosus* densities showed a considerable dispersion in each quality class. Nevertheless, the highest mean density was observed in moderate human pressure conditions, decreasing in the poor quality class and presenting very low values in the extreme of the pressure gradient (Fig. 2). Overall, the response of this species along the disturbance gradient was less marked ($F_{(4,343)} = 0.30$, $P>0.05$) than that observed for the other species.

Relative importance of pure environmental and human pressure variables

The results obtained above showed that both environmental and human pressure variables play a determinant role in the occurrence and abundance of non-native species. Moreover, the performed analyses revealed the existence of significant covariation ($P<0.001$) between these two groups of variables. Therefore, the analysis of the relationship between the occurrence and abundance of the most abundant NNS and the environmental and human pressure gradients was complemented with a GLM approach in order to simultaneously evaluate the relative influence of environmental variables, human-induced pressure, and spatial trends. All GLMs showed over-dispersion parameters under the threshold limit of 1, and were therefore performed using binomial and Poisson distributions as previously described. For all the response variables, most of the explained variability was accounted for by unique, or "pure", components. Shared effects of explanatory variables were responsible for a comparatively small proportion of data variability (Fig. 3). Together, environmental variables and human pressure accounted for the majority of the explained variation in the occurrence and abundance of the three considered NNS (31.7% to 40.2%) (Fig. 3).

The partition of the variance showed some differences in the relative influence of each variable set among the three NNS. The occurrence of NNS was mainly determined by environmental variables, whereas abundance showed a proportionally higher influence of human pressure variables. *L. gibbosus* occurrence and abundance revealed a high dependence on environmental variables. *G. lozanoi* presented a balanced influence of environmental and pressure variables. *G. holbrooki* showed a strong influence of human pressure on both occurrence and abundance (Fig. 3). The results suggest that these three species have different levels of tolerance to human pressure. In all models, spatial variables (autocorrelation) represented the smaller fraction of the explained variation in both occurrence (0.8% to 4.8%) and abundance of the three NNS (1.4% to 2.5%) (Fig. 3). A higher influence of these variables on the occurrence of *G. lozanoi* and *G. holbrooki* was observed, which is in accordance with a more restricted and concentrated distribution of these two species.

Overall, both regional and local explanatory variables were present in the best ecological models in environmental and human pressure GLMs (Table 5). Amongst environmental GLMs, drainage area, in particular, showed a significant reduction in deviance, and positively shaped all the response variables. Slope (positive coefficient) and percentage of riffles (negative coefficient) were ranked as the second most important predictors respectively in G. lozanoi and L. gibbosus occurrence and abundance. For G. holbrooki only mean annual temperature (positive coefficient) revealed a higher positive influence than drainage area. Concerning human pressure variables, toxicity and acidification levels were the most important predictors and were negatively related to the occurrence and abundance of the three NNS. For L. gibbosus,

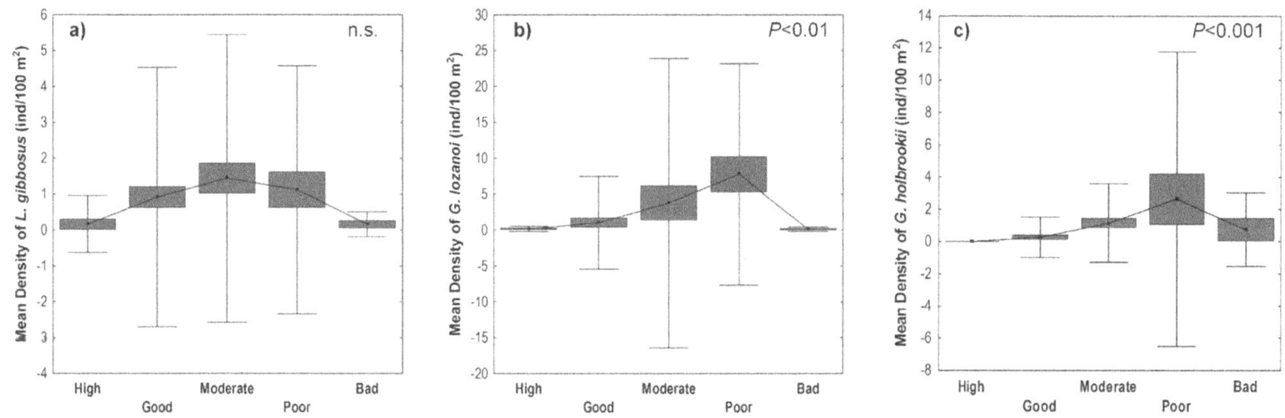

Figure 2. Response pattern of the most abundant NNS to human-induced pressure gradient, established according to 5 environmental quality classes: a) Mean density of *L. gibbosus*; b) Mean density of *G. lozanoi*; c) Mean density of *G. holbrooki*. (■): Mean; box:±SE; whisker:±SD. Significance of ANCOVA results with total human pressure as a covariable is shown.

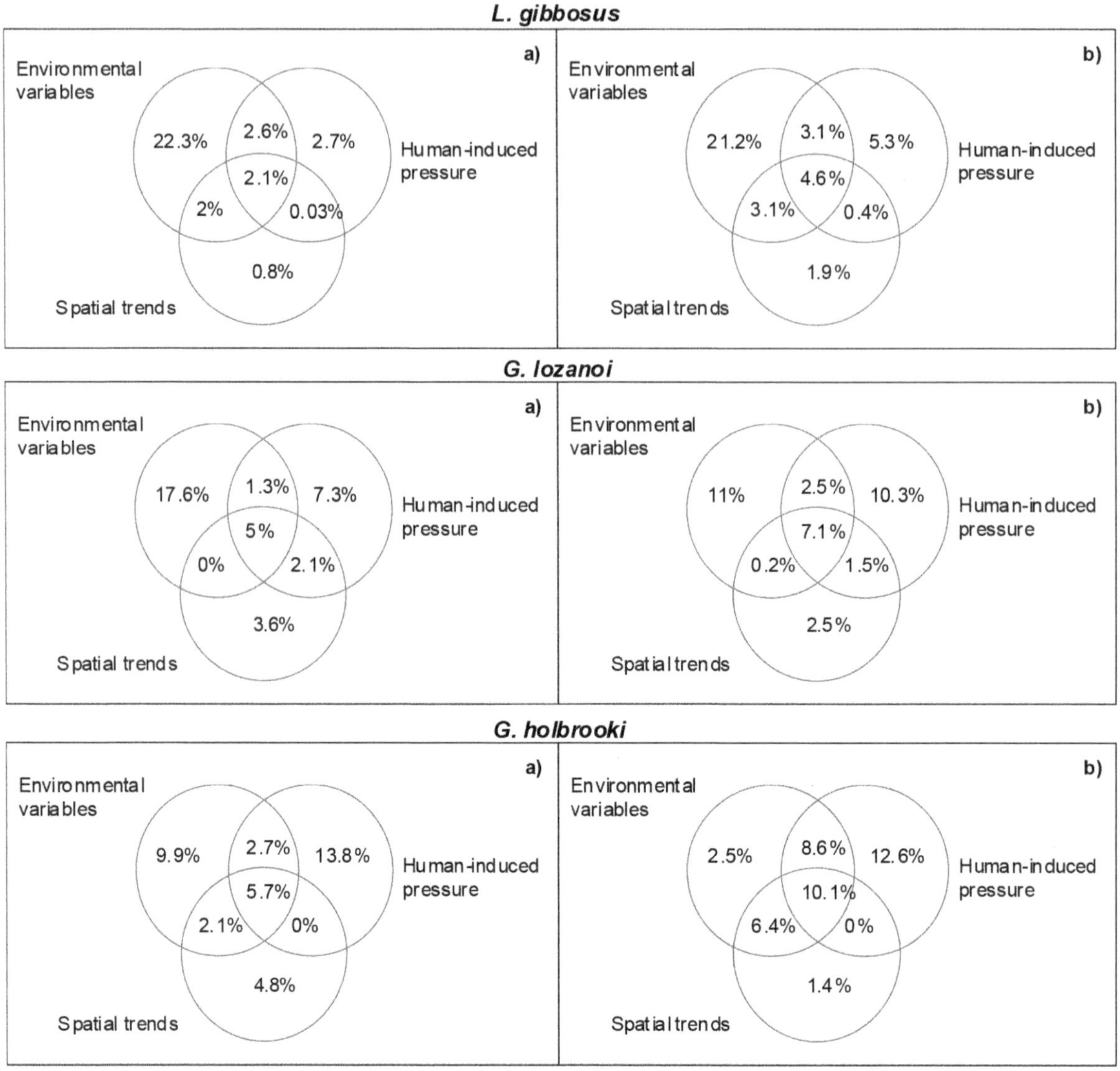

Figure 3. Veen diagrams showing the partition of the variance in the occurrence (a) and mean density (b) of the most abundant NNS explained by environmental variables, human-induced pressure and spatial trends.

TSS and COD also assumed a significant positive influence. Although nutrient/organic loads were retained in both L. gibbosus human pressure GLMs, these variables did not present a relevant decrease of deviance. Total dissolved nitrogen (N) and urban area showed an important relation with G. lozanoi occurrence and density. Sediment load revealed a particular influence on G. holbrooki occurrence, presenting the highest deviance reduction in this model.

For all the models produced, values of VIF were very low (Table 5), indicating the lack of significant multicollinearity between predictors, ensuring that no covariation exists among the variables included in the models, and therefore ensuring the reliability of the conclusions arising from the results.

Discussion

Non-native fish present in Portuguese rivers have different histories. Although rapid spread is possible in a short period, the timing and location of NNS introductions is relevant to the expansion of species distribution and abundance, which explains the relatively narrow distribution and abundance of *A. melas*, *A. alburnus*, and *S. lucioperca*, all recently introduced [9–10,51] at the date of this study, compared with other NNS introduced earlier. The most frequent and abundant species were *L. gibbosus*, *G. holbrooki and G. lozanoi*. In southern rivers the most common and abundant NNS were the former ones while *G. lozanoi* dominated the northern rivers. As such, the discussion will mainly focus on these NNS, which presented the higher invasive potential and are totally established in Portuguese continental waters for more than forty years.

A striking fact regarding the distribution of NNS in this study was their absence from mountain and southern chalk river types. Mountain rivers and small highland streams, with high slope and energy and low productivity (such as M, N4, S4 and N3), seem to be much less vulnerable to invasion by the set of introduced species. The upstream reaches of most Mediterranean basins experience strong seasonal patterns in their environmental conditions. In winter and spring, these streams present high flows whereas in summer they are frequently reduced to small isolated pools. This hydrological regime may prevent the invasion of NNS

Table 5. Summary of the environmental and human pressure GLMs for *L. gibbosus*, *G. lozanoi* and *G. holbrooki* (response variables occurrence and density).

Species	Response variables	Models features	Models AIC	Explanatory variables	Type	Deviance reduction	Coef. sign	VIF	*P*
L. gibbosus	Occurrence	Binomial distribution		Drainage area	Env	76.31	+	2.14	***
		"logit" link function	274.7	% Riffles	Env	21.95	-	1.5	*
				TSS	Pres	21.93	+	1.09	*
			356.5	Nutrient/organic loads	Pres	3.06	+	2.09	*
	Density	Poisson distribution		Drainage area	Env	69.53	+	2.36	***
		"log" link function		% Riffles	Env	28.44	-	1.63	**
			430.9	Mean annual runoff	Env	7.66	-	1.78	*
				Toxic/acid levels	Pres	18.99	-	1.44	*
				COD	Pres	13.08	+	1.31	**
				TSS	Pres	5.66	+	1.12	*
			496.1	Nutrient/organic loads	Pres	2.68	+	2.19	*
G. lozanoi	Occurrence	Binomial distribution		Drainage area	Env	38.40	+	2.21	**
		"logit" link function	216.3	Slope	Env	11.22	+	1.61	*
				Dissolved N	Pres	17.62	+	1.42	**
				Toxic/acid levels	Pres	13.53	-	1.92	*
				COD	Pres	5.39	-	2.27	*
			236.1	TSS	Pres	4.31	+	1.29	**
	Density	Poisson distribution		Drainage area	Env	26.26	+	2.73	**
		"log" link function	286.4	Slope	Env	13.73	+	1.69	*
				COD	Pres	15.06	-	2.17	**
				Urban area	Pres	11.41	+	2.62	*
				TSS	Pres	4.94	+	1.22	*
				Toxic/acid levels	Pres	6.16	-	1.69	*
			293.38	Land use	Pres	6.13	-	2.43	*
G. holbrooki	Occurrence	Binomial distribution		Mean annual temperature	Env	22.88	+	2.12	*
		"logit" link function		Drainage area	Env	14.87	+	2.38	**
			186.1	% Riffles	Env	7.86	-	1.44	*
				Sediment load	Pres	22.06	+	2.31	***
			186.7	Toxic/acid levels	Pres	13.21	-	1.98	***
	Density	Poisson distribution		Mean annual temperature	Env	26.06	+	2.25	*
		"log" link function	247.3	Drainage area	Env	9.48	+	2.29	*
				Toxic/acid levels	Pres	42.29	-	1.77	**
				Sediment load	Pres	9.70	+	1.36	*
			240.8	Artif. lentic water bodies	Pres	6.40	+	2.75	*

Explanatory variables type: environmental (Env) and human-induced pressure (Pres). Significance levels (P) at
*$P<0.05$.
**$P<0.01$.
***$P<0.001$.

which are poorly adapted to high discharge events, including flash flows (e.g. [52–53]). Conversely, NNS occurrence and abundance were particularly meaningful in river types with medium and large drainage areas, both in the northern "cool–warm water" and in the south. The high proportion of non-native fishes in the middle and lower reaches of Mediterranean-climate rivers was also reported in other studies (e.g. [54–56]). In general, the pattern of NNS distribution and abundance among river types followed the pattern of human-induced disturbance. Although human-induced disturbance occurred among all river types, some areas exhibited higher degradation. Overall, the sites located in littoral regions, both in the centre-north and in the southern region, which are associated with higher human density, presented higher degraded conditions and higher occurrence and abundance of non-native fishes. The least disturbed river types were located in higher altitude regions and with small drainage areas, where NNS were absent or present in very low densities. Most of those streams are located in isolated areas with difficult human access, far from the main human pathways. Remote areas with low human disturbance receive much lower propagule pressure than areas hosting immense human settlement [57].

Sites with higher human pressure exhibited significantly higher occurrence, abundance, and percentage of NNS. This pattern was observed in all river types and for all species, excepting those with very low abundance of NNS and *H. facetum*. Several species (*A. alburnus, A. melas, S. lucioperca*) occurred exclusively in impaired sites. The response of the most abundant NNS to the human pressure gradient was shaped by a marked increase in their abundance along the quality classes. Though the same general pattern was observed for the three species, *G. lozanoi* and *G. holbrooki* were particularly responsive to degradation, reaching the highest abundances in the poor quality class.

Non-native freshwater fish have commonly been documented to succeed in degraded aquatic habitats in many areas of the world [28,58–61]. Disturbed systems and communities may attract biological invasions more than pristine ecosystems due to the redistribution of space and energy resources and may promote new vacant niches for the most adaptable and tolerant invaders [11,62]. One of the most important human pressures in Mediterranean-climate rivers is related to nutrient and organic loads, which under the highly favourable climatic conditions increase the aquatic productivity and food availability. The resource availability theory argues that invasibility is directly influenced by nutrient enrichment and resource availability [11,63], which may facilitate biological contamination by reducing resource limitation and therefore competition. The high proportion of NNS in medium and large lowland rivers where productivity is naturally higher corroborates this hypothesis. Moreover, these river types (Littoral, N1>100 km^2 and S1>100 km^2) are also the most affected by hydrological disturbances, flow regulation, connectivity loss, and dam construction. Dams typically create lentic conditions that favour non-native fishes, thus presenting higher abundance in regulated streams than in unregulated ones [64–65]. These river types are particularly vulnerable to non-native fish invasions, as some species, namely *L. gibbosus* and *G. holbrooki*, prefer enriched lentic habitats (e.g. [53,66]).

This study reveals that natural landscape/environmental drivers and human-induced disturbance are both important factors determining the distribution and invasiveness of NNS, and their relative importance vary from species to species.

The occurrence of *G. lozanoi* was mostly explained by environmental variables, while its abundance was equally determined by landscape/habitat and human-disturbance factors. In the Portuguese continental waters *G. lozanoi* populations were particularly abundant in N1>100 km^2 and littoral river types which is consistent with previous work associating this species to lowland rivers with slow flow [67]. Being absent from high altitude rivers and warm waters, this species occurrence was positively related to slope and drainage area (larger areas providing water flow all year long). For this reason, *G. lozanoi* does not occur in the warmer southern and central rivers, which present low or no flow during summer. The species abundance was positively correlated with human-induced hydrological disturbance, habitat modification, water quality degradation (TSS, COD, acidification) and agricultural areas. Other studies also evidenced this species response to anthropogenic hydrological disturbance [64,68].

G. holbrooki is mainly present in southern lowland river types (S1>100 km^2, S1<100 km^2 and S3) with the highest water temperature and conductivity. This species occurrence was positively correlated with temperature and drainage area. Although G. *holbrooki* is able to withstand wide temperature ranges, it prefers warm water temperatures [69–70]. Conversely, the lower temperatures of the northern rivers could limit the species proliferation, due to the effect of latitude on life-history traits, namely the reproductive potential [71]. At the habitat scale, *G. holbrooki* was negatively associated with riffles, that is, it displays a preference for standing or slow flowing waters rich in organic detritus and muddy sediments. High river discharges tend to displace individuals and eliminate populations [52,58,72], as this species has no swimming abilities and behavioural response to fast flowing waters [73]. The invasiveness of *G. holbrooki* seems to be mostly determined by human-induced disturbance factors. Its abundance was strongly related to indicators of disturbance describing local in-stream habitat modifications and organic and sediment loads. The response to degradation is partially related to the species tolerance to a wide range of environmental conditions, including pH (from 3.9 to 8.8, [74]) and dissolved oxygen (0.28 mg/L, [69–70]. The ability to tolerate low dissolved oxygen enables it to survive in anoxic eutrophic waters with high organic and nutrient loads, and it is often the only fish species present in these water bodies (Ilhéu pers. obs.). Moreover, this species is also tolerant to a wide range of pollutants, including organic wastes, phenols, pesticides, and heavy metals, due to the species' phenotypic plasticity [75–76].

L. gibbosus is one of the most widespread NNS in Portugal and a very successful invader. This species was particularly frequent and abundant in warm water rivers of southern and central Portugal – littoral and southern river types, with the exception of southern mountain rivers – while in the north it occurred mainly in larger rivers (N1>100 km^2 and N2). *L. gibbosus* responded positively to a certain degree of anthropogenic disturbance, namely hydrological and habitat modification, and nutrient and organic loads, probably because these conditions lead to an increase in suitable habitats and also offer feeding advantages [77]. The intermediate tolerance of *L. gibbosus* to degradation, also reported by other authors [78], supports the idea that human-induced disturbance is not a requisite for successful invasion by all introduced species [79–80], and that a particular site may contain introduced organisms simply because they were introduced there and natural environmental conditions were favourable to their establishment. Indeed, this species abundance and distribution was mostly explained by pure environmental variables; its density was positively related to drainage area and negatively related with runoff and percentage of riffles, which is consistent with its higher occurrences and abundance in the warm water lowland rivers. Sunfishes exhibit low ability to withstand high current velocities and are often reduced in number by flood events [73,81].

Differences in swimming ability or behavioural response to high current velocities between native and non-native species may be the mechanism responsible for the observed differential removals [65–66]. Furthermore, *L. gibbosus* is known to be invasive in some southern and central European locations but not in England or Norway [82–84], being the climatic conditions the limiting factor in the species invasiveness [85]. This illustrates the high vulnerability of the warm water rivers of southern Europe, where environmental conditions tend to favour the proliferation of this type of species, being high flows the major constraint.

The invasive potential of NNS seems to be determined by different environmental factors that may interact direct and indirectly. However, other factors may also explain the success of many NNS invasions, namely the spatial structure and the biota features, including species life-histories and interspecific interactions [86–88]. Thus further studies integrating all these factors should be performed in order to enhance the ability to predict the vulnerability of ecosystems to invasive species.

Conclusions

This study reveals that natural landscape/environmental drivers are major factors determining the distribution of non-indigenous species in Portuguese rivers, while their abundance seems to be largely promoted by human-induced disturbance. The importance and role of different invasion factors are context dependent, because of the interaction between species traits and environment, regarding habitat, climate match, and human disturbance, and thus the river type approach seems to be useful to predict NNS invasiveness. This study also stresses the high vulnerability of the warm water lowland river types to invasion by NNS where conditions are amplified by human-induced degradation. Although some warm water species may be present in most basins (e.g. *L. gibbosus*), they become successful invaders mainly in large lowland rivers, particularly in the south. In contrast, high energy and low productivity rivers, typically at higher altitudes, seem to resist better to the invasion of NNS even when subjected to anthropogenic disturbance. These are relevant issues for assessment of the ecological status of aquatic systems, river basin management and river rehabilitation programmes.

Acknowledgments

We are grateful to several persons and their teams, who have taken part in data collection and entry: field colleagues from Universidade de Évora, M. T. Ferreira from Instituto Superior de Agronomia/Universidade de Lisboa, P. Raposo de Almeida from Instituto de Oceanografia/ Universidade de Évora, N. Formigo from Universidade do Porto and R. Cortes from Universidade de Trás-os-Montes e Alto Douro. We are also grateful to Dárcio Sousa for GIS help and A. Márcia Barbosa for statistical advice in GLM procedure. Valuable comments and critical review of an earlier draft of this manuscript by Thierry Oberdorff are deeply acknowledged, as well as the insightful comments of three anonymous referees, which have improved the manuscript.

Author Contributions

Conceived and designed the experiments: MI PM JMB. Performed the experiments: MI PM JMB. Analyzed the data: PM MI. Contributed reagents/materials/analysis tools: MI JMB. Wrote the paper: MI PM JMB.

References

1. Clavero M, Garcia-Berthou E (2006) Homogenization dynamics and introduction routes of invasive freshwater fish in the Iberian Peninsula. Ecol Appl 16: 2313–2324.
2. Leprieur F, Beauchard O, Blanchet S, Oberdorff T, Brosse S (2008) Fish invasions in the world's river systems: when natural processes are blurred by human activities. PloS Biol 6: 404–410.
3. Marr SM, Olden JD, Leprieur F, Arismendi I, Ćaleta M, et al. (2013) A global assessment of freshwater fish introductions in Mediterranean-climate regions. Hydrobiologia 719: 317–329.
4. Benejam L, Carol J, Benito J, Garcia-Berthou E (2007) On the spread of the European catfish (*Silurus glanis*) in the Iberian Peninsula: first record in the Llobregat river basin. Limnetica 26: 169–171.
5. Caiola N, de Sostoa A (2002) First record of the Asiatic cyprinid *Pseudorasbora parva* in the Iberian Peninsula. J Fish Biol 61: 1058–1060.
6. Comesaña J, Ayres C (2009) New data on the distribution of pumpkinseed *Lepomis gibbosus* and largemouth bass *Micropterus salmoides*, and of non-endemic Iberian gudgeon *Gobio lozanoi* in the Galicia region (NW Spain). Aquat Invasions 4: 425–427.
7. Miranda R, Leunda PM, Oscoz J, Vilches A, Tobes I, et al. (2010) Additional records of non-native freshwater fishes for the Ebro River basin (Northeast Spain). Aquat Invasions 5: 291–296.
8. Pérez-Bote JP, Romero RR (2009) First record of *Sander lucioperca* (Perciformes, Percidae) in the Alqueva reservoir, Guadiana basin (SW Iberian Peninsula). Limnetica 28: 225–228.
9. Ribeiro F, Chaves ML, Marques TA, Moreira da Costa L (2006) First records of *Ameiurus melas* (Siluriformes, Ictaluridae) in the Alqueva Reservoir, Portugal. Cybium 30: 283–284.
10. Ribeiro F, Gante HF, Sousa G, Filipe AF, Alves MJ, et al. (2009) New records, distribution and dispersal pathways of *Sander lucioperca* (Linnaeus, 1758) in Iberian freshwaters. Cybium 33: 255–256.
11. Davis M, Grime J, Thompson K (2000) Fluctuating resources in plant communities: a general theory of invasibility. J Ecol 88: 528–534.
12. Ruesink JL (2005) Global analysis of factors affecting the outcome of freshwater fish introductions. Conserv Biol 19: 1883–1893.
13. Bomford M, Barry SC, Lawrence E (2009) Predicting establishment success of introduced freshwater fishes: a role for climate matching. Biol Invasions 12: 2559–2571.
14. Duncan RP, Bomford M, Forsyth DM, Conibear L (2001) High predictability in introduction outcomes and the geographical range size of introduced Australian birds: a role for climate. J Anim Ecol 70: 621–32.
15. Williamson M (1996) Biological Invasions. London: Chapman and Hall. 244 p.
16. Fausch KD, Taniguchi Y, Nakano S, Grossman G, Townsend C (2001) Flood disturbance regimes influence rainbow trout invasion success among five holarctic regions. Ecol Appl 11: 1438–1455.
17. Davies KF, Chesson P, Hamson S, Inouye BD, Melbourne BA, et al. (2005) Spatial heterogeneity explains the scale dependence of the native-exotic diversity relationship. Ecology 86: 1602–1610.
18. Blanchet S, Leprieur F, Beauchard O, Staes J, Oberdorff T, et al. (2009) Broad-scale determinants of non-native fish species richness are context-dependent. P Roy Soc Lond B Bio 276: 2385–2394.
19. Gasith A, Resh VH (1999) Streams in Mediterranean-climate regions: abiotic influences and biotic responses to predictable seasonal events. Annu Rev Ecol Syst 31: 51–58.
20. European Commission (2000) Directive 2000/60/EC of the European Parliament and of the Council of 23 October 2000. Establishing a framework for Community action in the field of water policy. Official Journal of the European Communities L327.
21. Leunda PM (2010) Impacts of non-native fishes on Iberian freshwater ichthyofauna: current knowledge and gaps. Aquat Invasions 5: 239–262.
22. Olden JD, Poff NL (2003) Towards a mechanistic understanding and prediction of biotic homogenization. Am Nat 162: 442–460.
23. Rahel FJ (2000) Homogenization of fish faunas across the United States. Science 288: 854–856.
24. Leprieur F, Olden J, Lek S, Brosse S (2009) Contrasting patterns and mechanisms of spatial turnover for native and exotic freshwater fish in Europe. J Biogeogr 36: 1899–1912.
25. Gozlan RE (2010) The cost of non-native aquatic species introductions in Spain: fact or fiction? Aquat Invasions 5: 231–238.
26. Harris JH, Silveira R (1999) Large-scale assessments of river health using an Index of Biotic Integrity with low-diversity fish communities. Freshwater Biol 41: 235–252.
27. Joy MK, Death RG (2004) Application of the index of biotic integrity methodology to New Zealand freshwater fish communities. Environ Manage 34: 415–428.
28. Kennard MJ, Arthington AH, Pusey BJ, Harch BD (2005) Are alien fish a reliable indicator of river health? Freshwater Biol 50: 174–193.
29. Angermeier PL, Smogor RA, Stauffer JR (2000) Regional frameworks and candidate metrics for assessing biotic integrity in Mid-Atlantic highland streams. Trans Am Fish Soc 129: 962–981.
30. Hughes RM, Larsen DP (1988) Ecoregions: an approach to surface water protection. J Water Pollut Control Fed 60: 486–493.
31. Schmutz S, Melcher A, Frangez C, Haidvogl G, Beier U, et al. (2007) Spatially based methods to assess the ecological status of riverine fish assemblages in European ecoregions. Fisheries Manag Ecol 14: 441–452.

32. INAG IP (2008) River Typology in Continental Portugal under the implementation of Water Framework Directive – Abiotic characterization. Lisboa: Instituto da Água, IP (in Portuguese).
33. Miranda R, Oscoz J, Leunda PM, García-Fresca C, Escala MC (2005) Effects of weir construction on fish population structure in the River Erro (North of Spain). Ann Limnol-Int J Lim 41: 7–13.
34. Cabral MJ, Almeida J, Almeida PR, Dellinger TR, Ferrand de Almeida N, et al. (2005) Livro Vermelho dos Vertebrados de Portugal. Lisboa: Instituto da Conservação da Natureza. 659 p.
35. Smith KG, Darwall WRT, editors (2006) The status and distribution of freshwater fish endemic to the Mediterranean basin. Gland, Switzerland, and Cambridge: IUCN. 34 p.
36. FAME (2004) Development, Evaluation and Implementation of a Standardised Fish-based Assessment Method for the Ecological Status of European Rivers-A Contribution to the Water Framework Directive. Final Report (Co-ordinator: Schmutz S). Institute for Hydrobiology and Aquatic Ecosystem Management, University of Natural Resources and Applied Life Sciences, Vienna. Available: http://fame.boku.ac.at. Accessed: 2007 Jul 23.
37. CIS-WFD (2003) Guidance on Establishing Reference Conditions and Ecological Status Class Boundaries for Inland Surface Waters. Final Version. Brussels: EU Common Implementation Strategy for the Water Framework Directive.
38. Clesceri LS, Greenberg AE, Eaton AD (1998) Standard methods for the examination of water and wastewater, 20th Edition. Washington: American Public Health Association, American Water Works Association, Water Environmental Federation.
39. Giller PS, Malmqvist B (1998) The biology of streams and rivers. New York: Oxford University Press. 296 p.
40. INAG IP (2008) Manual for biological assessment of water quality in rivers according to Water Framework Directive. Sampling and analysis protocol for fish fauna. Lisboa: Instituto da Água, I.P. (in Portuguese).
41. CEN (2003) Water Quality–Sampling of Fish with Electricity. European standard–EN 14011:2003. Brussels: European Committee for Standardization.
42. Quade D (1967) Rank analysis of covariance. J Am Stat Assoc 62: 1187–1200.
43. Conover WJ, Iman RL (1982) Analysis of covariance using the rank transformation. Biometrics 38: 715–724.
44. Daniel WW (1987) Biostatistics: A foundation for analysis in the health sciences. New York: Wiley. 734 p.
45. Borcard D, Legendre P, Drapeau P (1992) Partialling out the spatial component of ecological variation. Ecology 73: 1045–1055.
46. Hinch SG, Sommers KM, Collins NC (1994) Spatial autocorrelation and assessment of habitat-abundance relationships in littoral zone fish. Can J Fish Aquat Sci 51: 701–712.
47. Legendre P (1993) Spatial autocorrelation: Trouble or new paradigm. Ecology 74: 1659–1673.
48. Legendre P, Fortin MJ (1989) Spatial pattern and ecological analysis. Vegetation 80: 107–138.
49. Zuur AF, Ieno EN, Smith GM (2007) Analyzing Ecological Data. New York: Springer. 672 p.
50. Andersen MJ, Cribble NA (1998) Partitioning the variation among spatial, temporal and environmental components in a multivariate dataset. Aust J Ecol 23: 158–167.
51. Ribeiro F, Collares-Pereira MJ, Moyle PB (2009) Non-native fish in the fresh waters of Portugal, Azores and Madeira Islands: a growing threat to aquatic biodiversity. Fisheries Manag Ecol 16: 255–264.
52. Meffe GK (1984) Effects of abiotic disturbance on coexistence of predator-prey fish species. Ecology 65: 1525–1534.
53. Bernardo JM, Ilhéu M, Matono P, Costa AM (2003) Interannual variation of fish assemblage structure in a Mediterranean river: implications of streamflow on the dominance of native exotic species. River Res Appl 19: 1–12.
54. Ilhéu M (2004) Patterns of habitat use by freshwater fish in Mediterranean streams. PhD Thesis, University of Évora.
55. Vila-Gispert A, Alcaraz C, García-Berthou E (2005) Life-history traits of invasive fish in small Mediterranean streams. Biol Invasions 7: 107–116.
56. Hermoso V, Clavero M, Blanco-Garrido F, Prenda J (2011) Invasive species and habitat degradation in Iberian streams: an analysis of their role in freshwater fish diversity loss. Ecol. Appl 21: 175–188.
57. Drake JA, Mooney HA, DiCastri F, Grove RH, Kruger FJ, et al. (1989) Biological Invasions: a Global Perspective. New York: Wiley. 525 p.
58. Arthington AH, Hamlet S, Bluhdorn DR (1990) The role of habitat disturbance in the establishment of introduced warm-water fishes in Australia. In: Pollard DA, editor.Introduced and translocated fishes and their ecological effect.Can-Canberra: Australian Government Publishing Service. pp. 61–66.
59. Gido KB, Brown JH (1999) Invasion of North American drainages by alien fish species. Freshwater Biol 42: 387–399.
60. Meador MR, Brown LR, Short T (2003) Relations between introduced fish and environmental Conditions at large geographic scales. Ecol Indic 3: 81–92.
61. Ross R, Lellis WA, Bennett RM, Johnson CS (2001) Landscape determinants of non indigenous fish invasions. Biol Invasions 3: 347–361.
62. Belote RT, Jones RH, Hood SM, Wender B (2008) Diversity-invasibility across an experimental disturbance gradient in Appalachian forests. Ecology 89: 183–192.
63. González AL, Kominoski JS, Danger M, Ishida S, Iwai N et al. (2010) Can ecological stoichiometry help explain patterns of biological invasions? Oikos 119: 779–790.
64. Alexandre CM, Almeida PR (2010) The impact of small physical obstacles on the structure of freshwater fish assemblages. River Res Appl 26: 977–994.
65. Power ME, Dietrich WE, Finlay JC (1996) Dams and downstream aquatic biodiversity: Potential food web consequences of hydrologic and geomorphic change. Environ Manage 20: 887–895.
66. Godinho FN, Ferreira MT (2000) Composition of endemic fish assemblages in relation to exotic species and river regulation in a temperate stream. Biol Invasions 2: 231–244.
67. Doadrio I, Madeira MJ (2004) A new species of the genus Gobio Cuvier, 1816 (Actynopterigii, Cyprinidae) from the Iberian Peninsula and Southwestern France. Graellsia 60: 107–116.
68. Lobón-Cervia J, Montanes C, de Sostoa A (1991) Influence of environment upon the life history of gudgeon, *Gobio gobio* (L.): a recent and successful colonizer of the Iberian Peninsula. J Fish Biol 39: 285–300.
69. Lloyd LN (1984) Exotic Fish: Useful Additions or "Animal Weeds"? J Austr New Guinea Fishes Assoc 1: 31–42.
70. Pyke GH (2005) A review of the biology of *Gambusia affinis* and *G. holbrooki*. Rev Fish Biol Fisher 15: 339–365.
71. Benejam L, Alcaraz C, Sasal P, Simon-Levert G, Garcıa-Berthou E (2009) Life history and parasites of the invasive mosquitofish (*Gambusia holbrooki*) along a latitudinal gradient. Biol Invasions 1: 2265–2277.
72. Galat DL, Robertson B (1992) Response of endangered *Poeciliopsis occidentalis sonoriensis* in the Rio Yaqui drainage, Arizona, to introduced *Gambusia affinis*. Environ Biol Fish 33: 249–264.
73. Minckley WL, Meffe GK (1987) Differential selection by flooding in stream fish communities of the arid American Southwest. In: Matthews WJ, Heins DC, editors.Community and evolutionary ecology of North American stream fishes.Norman: University of Oklahoma Press. pp. 96–104.
74. Brown-Peterson N, Peterson MS (1990) Comparative life history of female mosquitofish, *Gambusia affinis*, in tidal freshwater and oligohaline habitats. Environ Biol Fish 27: 33–41.
75. Andreasen JK (1985) Insecticide resistance in mosquitofish of the lower Rio Grande Valley of Texas - an ecological hazard? Arch Environ Con Tox 14: 573–577.
76. Saiki MK, Martin BA, May TW (2004) Reproductive status of western mosquitofish inhabiting selenium-contaminated waters in the Grassland Water District, Merced County, California. Arch Environ Con Tox 47: 363–369.
77. Almeida D, Almodóvar A, Nicola GG, Elvira B (2009) Feeding tactics and body condition of two introduced populations of pumpkinseed *Lepomis gibbosus*: taking advantages of human disturbances? Ecol Freshw Fish 18: 15–23.
78. Halliwell DB, Langdon RW, Daniels RA, Kurtenbach JP, Jacobson RA (1999) Classification of freshwater fish species of the northeastern United States for use in the development of indices of biological integrity, with regional applications. In: Simon TP, editor.Assessing the Sustainability and Biological Integrity of Water Resource Quality Using Fish Communities.Boca Raton: CRC Press. pp. 301–333.
79. Niemela J, Spence JR (1991) Distribution and abundance of an exotic groundbeetle (Carabidae): a test of community impact. Oikos 62: 351–335.
80. Townsend CR (1996) Invasion biology and ecological impacts of brown trout *Salmo trutta* in New Zealand. Biol Conserv 78: 13–22.
81. Schultz AA, Maughan OE, Bonar SA, Matter WJ (2003) Effects of flooding on abundance of native and nonnative fishes downstream from a small impoundment. N Am J Fish Manage 23: 503–511.
82. Copp GH, Fox MG, Przybylski M, Godinho FN, Vila-Gispert A (2004) Life-time growth patterns of pumpkinseed *Lepomis gibbosus* introduced to Europe, relative to native North American populations. Folia Zool 53: 237–254.
83. García-Berthou E, Alcaraz C, Pou-Rovira Q, Zamora L, Coenders G, et al. (2005) Introduction pathways and establishment rates of invasive aquatic species in Europe. Can J Fish Aquat Sci 62: 453–463.
84. Sterud E, Jorgensen A (2006) Pumpkinseed *Lepomis gibbosus* (Linnaeus, 1758) (Centrarchidae) and associated parasites introduced to Norway. Aquat Invasions 1: 278–280.
85. Villeneuve F, Copp GH, Fox MG, Stakėnas S (2005) Interpopulation variation in growth and life-history traits of the introduced sunfish, pumpkinseed *Lepomis gibbosus*, in southern England. J Appl Ichthyol 21: 275–281.
86. Allen CR, Nemec KT, Wardwell DA, Hoffman JD, Brust M, et al. (2013) Predictors of regional establishment success and spread of introduced non-indigenous vertebrates. Glob Ecol Biogeogr 22: 889–899.
87. Filipe AF, Magalhães F, Collares-Pereira MJ (2010) Native and introduced fish species richness in Mediterranean streams: the role of multiple landscape influences. Divers Distrib 16: 773–785.
88. Olden JD, Poff LN, Bestgen KR (2006) Life-history strategies predict fish invasions and extirpations in the Colorado River Basin. Ecol Monograph 76: 25–40.

14

Biodiversity Loss following the Introduction of Exotic Competitors: Does Intraguild Predation Explain the Decline of Native Lady Beetles?

Chelsea A. Smith*, Mary M. Gardiner

Department of Entomology, The Ohio State University, Ohio Agricultural Research and Development Center, Wooster, Ohio, United States of America

Abstract

Exotic species are widely accepted as a leading cause of biodiversity decline. Lady beetles (Coccinellidae) provide an important model to study how competitor introductions impact native communities since several native coccinellids have experienced declines that coincide with the establishment and spread of exotic coccinellids. This study tested the central hypothesis that intraguild predation by exotic species has caused these declines. Using sentinel egg experiments, we quantified the extent of predation on previously-common (*Hippodamia convergens*) and common (*Coleomegilla maculata*) native coccinellid eggs versus exotic coccinellid (*Harmonia axyridis*) eggs in three habitats: semi-natural grassland, alfalfa, and soybean. Following the experiments quantifying egg predation, we used video surveillance to determine the composition of the predator community attacking the eggs. The extent of predation varied across habitats, and egg species. Native coccinellids often sustained greater egg predation than *H. axyridis*. We found no evidence that exotic coccinellids consumed coccinellid eggs in the field. Harvestmen and slugs were responsible for the greatest proportion of attacks. This research challenges the widely-accepted hypothesis that intraguild predation by exotic competitors explains the loss of native coccinellids. Although exotic coccinellids may not be a direct competitor, reduced egg predation could indirectly confer a competitive advantage to these species. A lower proportion of *H. axyridis* eggs removed by predators may have aided its expansion and population increase and could indirectly affect native species via exploitative or apparent competition. These results do not support the intraguild predation hypothesis for native coccinellid decline, but do bring to light the existence of complex interactions between coccinellids and the guild of generalist predators in coccinellid foraging habitats.

Editor: Heloise Gibb, La Trobe University, Australia

Funding: This project was supported by funding from the OSU-OARDC SEEDS Grant Program (http://www.oardc.ohio-state.edu/seeds/) to Mary Gardiner. The funders had no role in study design, data collection and analysis, decision to publish, or preparation of the manuscript.

Competing Interests: The authors have declared that no competing interests exist.

* E-mail: Smith.7231@osu.edu

Introduction

Declines of many native species have occurred simultaneously with the introduction and spread of non-native organisms, a pattern that implicates competition with exotics as a contributing factor [1,2,3,4]. Often, this is based primarily on correlative evidence and lacks mechanistic understanding [5]. The influence of exotic generalist predators on native population decline is particularly difficult to elucidate, as both direct and indirect competitive interactions occur [6,7,8,9]. For example, the invasion of two wasp species into New Zealand's southern beech forests has directly and indirectly negatively affected native fauna. Following their establishment, populations of invertebrates preyed on by the wasps have declined. Additionally, the kaka, a native parrot that forages on honeydew produced by a native scale has abandoned the invaded area as up to 95% of available honeydew resources are claimed by the wasps [7].

Coccinellids (Coleoptera: Coccinellidae) are an excellent model to study these direct and indirect interactions between native and exotic competitors; they compete for resources and also interact with each other though intraguild predation (IGP) [9,10,11,12,13,14]. Coinciding with the establishment of exotic coccinellids, several multi-year surveys have documented population declines of native coccinellid species including *Coccinella novemnotata*, *Coccinella transversoguttata*, *Adalia bipunctata*, *Hippodamia tredecimpunctata*, and *Hippodamia convergens* [15,16,17,18,19,20,21]. Among the most recently detected declines is that of the convergent lady beetle, *H. convergens*, within the Midwest United States [22,23,24,25]. These declines have led to several hypotheses which try to explain the loss of native coccinellids following exotic introduction, including interference competition via IGP [26,27], exploitative competition for shared prey [20,28], and apparent competition via a shared parasitoid [29]. The majority of research examining native coccinellid decline has focused on interference competition.

Laboratory studies have documented that exotic coccinellids feed on eggs and larvae of coccinellids and other predatory insects [27,30,31,32,33,34]. Field studies documenting the decline of native coccinellids following the introduction of exotic species are limited and the conclusions reached are often conflicted regarding the degree of IGP that occurs between species. For example, Hoogendoorn and Heimpel [35] found that in field cages, the presence of exotic *Harmonia axyridis* larvae did not reduce the survival or weight gain of native *Coleomegilla maculata* larvae. Conversely, Gardiner et al. [24] reported that *C. maculata* incurs significant egg predation in soybean fields; however, the contribution of exotic coccinellids to this mortality was not measured.

Thus, a determination of which animals consume coccinellid eggs and quantification of the incidence of egg predation for different coccinellid species in varied habitats is necessary to understand of the contribution of egg predation to native coccinellid decline. To address this, we compared the extent of egg predation experienced by three coccinellid species: a native experiencing population decline (*H. convergens*), a common native (*C. maculata*), and a common exotic (*H. axyridis*) species. The inclusion of a common native species provided an opportunity to compare egg loss of both a common native and exotic against the declining native species. This study examined levels of egg predation occurring across coccinellid foraging habitats and determined the guild of predators responsible. If egg predation has contributed to native coccinellid decline we expected consumption of declining native coccinellid eggs to exceed that of common native or exotic species. We tested the predictions that 1) exotic coccinellids are the dominant coccinellid egg predator, 2) declining native species incur greater egg predation than common species, and 3) less predation of native eggs will occur in semi-natural grasslands relative to croplands. Previous studies have found native coccinellids to be abundant in semi-natural grassland habitats as well as landscapes comprising of higher proportions of semi-natural grassland habitat implying that they may be reproductively successful there [22,36].

Materials and Methods

Ethics statement

Specific permits were not required for field sites and this study did not involve protected or endangered species. Permission to gather data from fields was obtained from each grower and private landowner.

Study sites

Data were collected within nine counties in 2009 and 2010 throughout Ohio, USA: Delaware (2010 only), Fayette, Huron, Knox, Marion (2009 only), Perry, Putnam, Shelby, and Wayne (Figure S1). Within each county one semi-natural grassland, alfalfa, and soybean site was selected; each was separated by a minimum of four km. Alfalfa and soybean fields were managed by grower-collaborators. Grasslands were owned privately and contained cool and warm season grasses, native forbs, and agricultural weeds. Four plots were established per site where data were collected; all plots were a minimum of 30 m from any field edge (Figure S2).

Rearing of coccinellid species

Coleomegilla maculata and *H. axyridis* adults were collected from overwintering aggregations in Ohio. *Hippodamia convergens* adults were ordered from a commercial supplier, which collects from overwintering aggregations in California (Rincon-Vitova, Ventura, CA, USA). Adult females were placed into individual plastic vials (8.9 cm deep; 4.5 cm diameter) lined with white paper to act as an egg-laying substrate as in Gardiner et al. [24]. All beetles were provided with water, honey, and aphids. Eggs deposited onto the paper substrate were counted and stored in a –80°C freezer until experiments were conducted.

Measuring predation of frozen egg masses

Previous research by Gardiner et al. [24] found no preference among predators for live versus previously-frozen coccinellid eggs. To further determine that previously frozen eggs could be used for field experiments, the consumption of previously-frozen and live egg masses was compared for additional predator taxa (Stylommatophora (slugs), Opiliones, Acrididae, and Tettigoniidae). Predators collected from the field (Wooster, OH, USA) were starved for 12 hours, and placed into individual containers (8 cm deep; 11 cm diameter). Each predator was provided a previously-frozen and fresh egg mass of similar size and species (*H. axyridis* or *H. convergens*). The fresh egg masses were no older than three days. All bioassays were conducted at room temperature, and took place between May 22, 2011 and August 14, 2011. The slugs were moistened with water every 60 minutes to prevent desiccation. The number of eggs remaining within each container was determined every 30 minutes for four hours, and thereafter at 24 and 48 hours.

Egg predation experimental procedure

To quantify egg predation in the three focal habitats, egg loss in Open (predator accessible) and Exclusion (cage used to exclude predators) treatments were compared (Figure S2A). The Exclusion treatment consisted of an egg mass enclosed in a 22 mm diameter mesh cage. This measured background egg loss due to desiccation or dislodgement to ensure that there was not an overestimate of predation events. The Open treatment consisted of an un-caged egg mass. To begin each experiment, egg masses deposited onto 1.5 cm^2 paper squares were counted and attached to vegetation 0.2 – 0.3 m from the soil surface. Two egg masses from each species were present in each plot; one was assigned to the Open treatment and the other to the Exclusion treatment. The number of eggs within a mass ranged from 2 to 54 with a median of 8. The high variability could not be avoided due to the large number of egg masses required for the experiments. All treatments remained in the field for 48 hours, after which the egg masses were collected and the remaining eggs counted. The proportion of eggs remaining was calculated for each egg mass. The presence of predator frass or a hole in the cage mesh indicated that predators were able to gain access to 40 Exclusion cages (across 2009 and 2010). Data from infiltrated cages were excluded from analysis. For each experiment, the proportion of eggs remaining in each treatment was averaged by site for each coccinellid species.

In 2009, one egg predation experiment was conducted during the week of July 20. Predation on *C. maculata* eggs was measured within all 24 field sites (eight of each focal habitat) and predation on *H. axyridis* eggs was measured within nine sites (one of each focal habitat within Shelby, Wayne, and Perry Co., OH). In 2010, experiments were conducted during the weeks of June 8 and July 26. The June 8 experiments were carried out within the grassland and alfalfa habitats. Predation in soybean was not examined since plants had recently or not yet emerged from the ground within our sites. Egg masses of *H. axyridis* and *H. convergens* were deployed in all grassland and alfalfa sites and *C. maculata* in four replicates of each habitat (grassland and alfalfa sites in Knox, Perry, Putnam, and Shelby Co., OH). During the July 26 experiments, predation was measured in all three focal habitats. Eggs of *H. axyridis* and *H. convergens* were deployed in all 24 sites, and *C. maculata* egg masses in all but one site (alfalfa, Huron Co., OH).

Video surveillance of egg predation events

We surveyed the arthropods responsible for egg predation using surveillance systems modified from a design by M. Grieshop [37,38]. Each system consisted of a 4-channel digital video recorder (DVR: model QH25DVR, QSee, Anaheim, CA, USA) and four weather resistant surveillance cameras with infrared light for night vision (model QD28414C4 QSee, Anaheim, CA, USA). Systems were powered with a deep-cycle boat battery (SLI24MDC Xtreme Deep Cycle Marine and Boat, Hartland, WI, USA). Cameras were fitted with aluminum adaptors (Wayne Machine Shop, Wooster, OH, USA) allowing for the attachment of a 10x magnification lens (Figure S3).

Within each site, two DVR systems were placed in the field (n = 8 cameras). Two egg masses of *H. axyridis* and *H. convergens* were observed (n = 4 observations per egg mass species per site). Egg masses were attached to vegetation which was secured to a corrugated plastic stand (5×5 cm) that was anchored to the ground with 12-gauge wire to prevent movement of egg mass (Figure S3). Each camera was fastened to metal t-posts at a height of 0.2 – 0.3 m under a rain guard of white corrugated plastic (30×17 cm) to record activity at each egg mass. The battery powered the system for approximately 24 h and the video was downloaded from the DVR onto a portable hard drive. A total of 128 egg masses were observed with the cameras (66 *H. convergens* and 62 *H. axyridis*); 63 in grassland, 37 in alfalfa, and 28 in soybean from June 18th – August 13th, 2010. From this video, we determined which animals damaged or consumed the coccinellid egg masses. Only the first attack on each egg mass is reported herein.

Coccinellid activity density and relative abundance during egg predation experiments

The activity density and relative abundance of coccinellids present within each site was determined using yellow sticky card traps and sweep sampling during each of the three egg predation experiments. These surveys were conducted to determine if native and/or exotic coccinellids were in the field at the time of the experiments. Unbaited yellow sticky card traps (23×28 cm unfolded) (Pherocon AM, Trécé, Inc. Adair, OK, USA) were attached to step-in fence posts and placed at a height of 0.5 m at each experimental plot (Figure S2C). One seven day catch was collected per plot. Yellow sticky card traps sample the "activity density" of a population since they collect arthropods moving through the site. Four 20-sweep samples were also collected from the fields (one per plot) using a 15" diameter net (Figure S2B). All species counts were averaged across plots for each site.

Aphid Abundance

In 2010 all aphids (Aphidoidea) were counted from the sweep samples and averaged across plots for each site.

Data analysis

A mixed effects repeated measures analysis of variance model (ANOVA) was used to determine if predators had a preference for previously frozen or fresh egg masses. Separate analyses were conducted for each predator (Stylommatophora, Opiliones, Acrididae, and Tettigoniidae). The response variable in these models was the proportion of eggs remaining, which was arcsine($\sqrt{x}$) transformed prior to analysis. Fixed factors were Treatment (fresh or frozen eggs) and Time (0.5, 1, 1.5, 2, 2.5, 3, 3.5, 4, 24, or 48 h); the interaction term Treatment*Time was also included. Data were analyzed using the PROC MIXED procedure in SAS [39].

To determine if egg predation varied among coccinellid species or across habitats a mixed effects ANOVA (PROC MIXED) [39] was conducted for each of the three experiments (June 20, 2009, June 8, 2010, and June 26, 2010) with fixed effects: Treatment (Exclusion and Open), Habitat, and Egg Species. All possible interaction terms among the fixed effects were also included in the models. To meet the assumptions of ANOVA, the response variable, mean proportion of eggs remaining, was arcsine($\sqrt{x}$) transformed prior to analysis. Differences in least squares means were used to compare all variable combinations.

To determine if relative abundance of native and exotic coccinellids varied across habitats during egg experiments a generalized linear model with a negative binomial or Poisson distribution was used depending on which distribution best fit the data (PROC GENMOD) [39]. A Poisson distribution was the best fit for the mean exotic coccinellids on yellow sticky card traps from the July 20, 2009 experiment, and mean native coccinellids in sweep samples from the June 8, 2010 experiment. The negative binomial distribution best fit the data in all other cases. The response variables for these analyses were mean exotic or native coccinellids per plot, and Habitat was the predictor variable.

To determine if the relative abundance of aphids collected via sweep samples varied across habitats a generalized linear model with a negative binomial distribution was used (PROC GENMOD) [39]. The June 8, 2010 and July 26, 2010 experiments were each analyzed separately. The response variable was mean aphids per plot, and the predictor variable was Habitat.

Results

Measuring predation of previously frozen egg masses

We found no difference in consumption of live versus previously-frozen eggs for any predator examined (Stylommatophora (n = 16): $F_{1,138} = 0.06$, $P = 0.812$; Opiliones (n = 22): $F_{1,200} = 0.64$, $P = 0.427$; Acrididae (n = 14): $F_{1,120} = 0.70$, $P = 0.403$; Tettigoniidae (n = 11): $F_{1,89} = 0.85$, $P = 0.359$). For Stylommatophora, Acrididae, and Tettigoniidae, egg predation increased over time (Stylommatophora: $F_{9,138} = 13.48$, $P < 0.001$; Opiliones: $F_{9,200} = 0.42$, $P = 0.9224$; Acrididae: $F_{9,120} = 3.84$, $P < 0.001$; Tettigoniidae: $F_{9,89} = 2.01$, $P = 0.044$). However, there was no significant Treatment*Time interaction for any of the predators (Stylommatophora: $F_{9,138} = 0.35$, $P = 0.957$; Opiliones: $F_{9,200} = 0.06$, $P = 1.000$; Acrididae: $F_{9,120} = 0.33$, $P = 0.965$; Tettigoniidae: $F_{9,89} = 0.17$, $P = 0.997$).

Egg predation experiments

A significant interaction between Treatment (open and exclusion) and Egg Species ($F_{2,53} = 4.97$, $P = 0.011$) was detected for the July 20, 2009 experiments. *Coleomegilla maculata* experienced significant egg predation within all habitats, indicated by fewer eggs remaining in the Open compared to the Exclusion treatment (Figure 1a). *Harmonia axyridis* eggs were significantly reduced relative to the caged control in grassland and soybean but not in alfalfa fields (Figure 1a). The proportion of *C. maculata* and *H. axyridis* eggs remaining did not vary between habitats (Figure 1d). These species experienced similar patterns of egg predation across foraging habitats, with significantly greater egg predation occurring in grasslands than in alfalfa fields. *Coleomegilla maculata* egg masses experienced significantly greater predation in grasslands, than in soybean ($P = 0.002$) and alfalfa ($P < 0.001$). There was no difference in predation of *C. maculata* egg masses within soybean and alfalfa ($P = 0.207$) (Figure 1g). *Harmonia axyridis* experienced greater egg predation in grasslands than alfalfa ($P = 0.019$) with no difference between grassland and soybean fields ($P = 0.149$). *Harmonia axyridis* egg masses in soybean and alfalfa incurred low levels of predation (82.5% and 93.3% of eggs were remaining in soybean and alfalfa respectively), with no difference in the number of eggs removed by predators ($P = 0.275$) (Figure 1g).

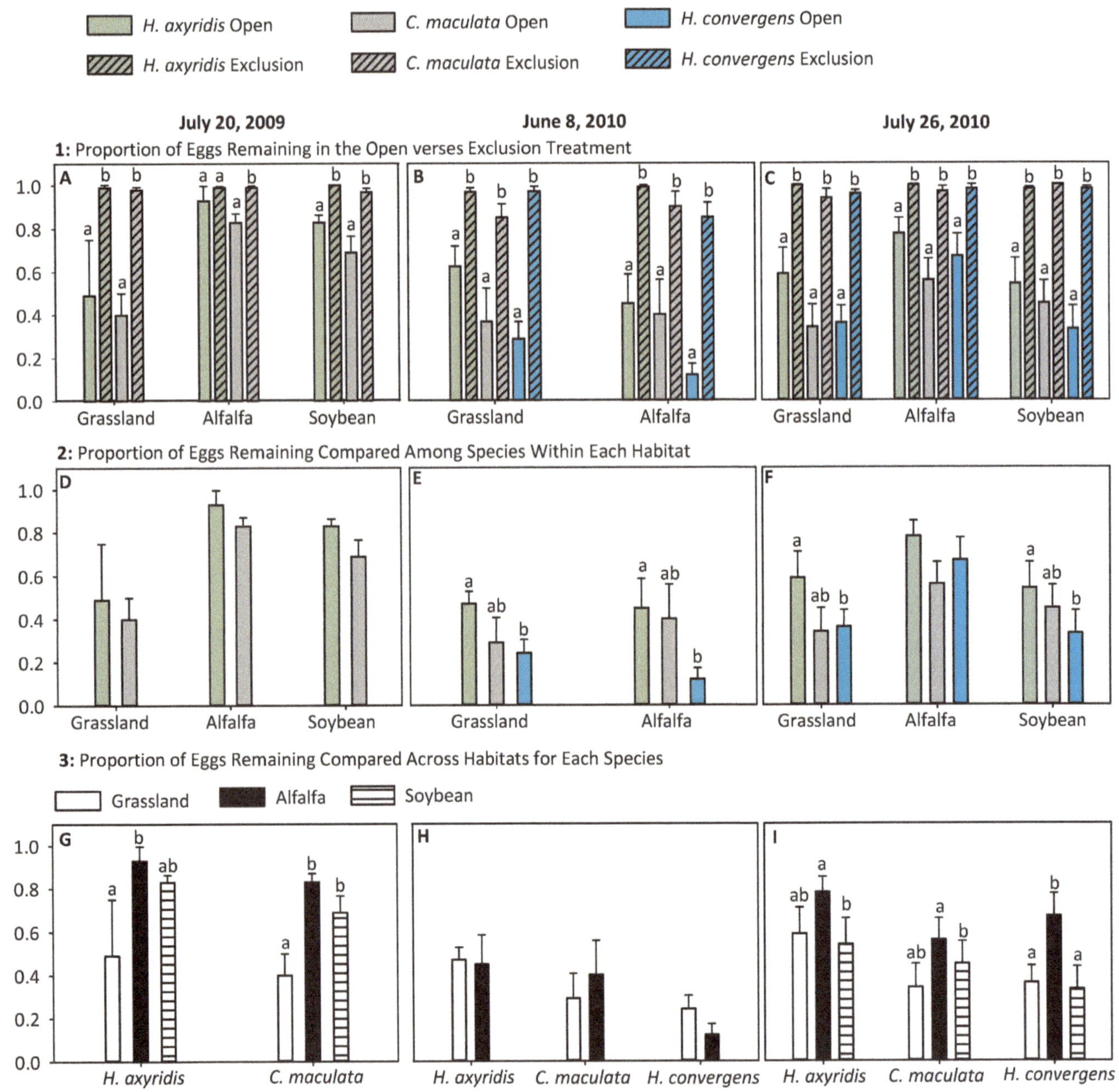

Figure 1. Quantification of coccinellid egg predation. The proportion of eggs remaining (Mean ± SEM) after 48 h in the field for egg predation experiments conducted the weeks of July 20, 2009, June 8, 2010 and July 26, 2010. Here we show the same data presented in three ways for clarity. Each grouping of panels focuses on different comparisons. Panels A-C show differences in the proportion of eggs remaining in Open versus Exclusion treatments. Panels D-F illustrate differences in egg predation across species within each habitat. Panels G-I show differences in egg predation within each species across habitats. Experiments were analyzed separately ($P<0.05$).

During the week of June 8, 2010, *H. convergens*, *C. maculata*, and *H. axyridis* incurred significant egg predation in both grassland and alfalfa habitats ($F_{1,68}=11.6$, $P<0.001$) (Figure 1b). The extent of egg predation that occurred within grasslands versus alfalfa did not differ for any of the three species ($F_{2,68}=1.76$, $P=0.190$) (Figure 1h). However, there were differences in egg predation experienced among species ($F_{1,68}=5.86$, $P=0.005$). Greater predation occurred on *H. convergens* than *H. axyridis* eggs in both grassland ($P=0.007$) and alfalfa ($P=0.016$) (Figure 1e). There were no significant differences in the extent of predation experienced by *C. maculata* and the other two species in either of the habitats (Figure 1e).

During the week of July 26, 2010, all species sustained significant egg predation (Figure 1c). The extent of predation varied among habitats. Predation on eggs of *H. convergens* was greater in soybean ($P=0.002$) and grassland ($P=0.009$) than in alfalfa. There was no difference in *H. convergens* egg predation among grassland and soybean habitats ($P=0.625$) (Figure 1i). Predation on *H. axyridis* eggs was greater in soybean relative to alfalfa ($P=0.048$) (Figure 1i). Predation of *C. maculata* eggs in grassland was significantly higher than in alfalfa ($P=0.049$). There was no difference in *C. maculata* egg predation among soybean and grassland ($P = 0.409$) or soybean and alfalfa ($P = 0.210$) (Figure 1i). Within foraging habitats, differences in the extent of egg predation among species were detected. In soybean fields, *H. convergens* egg masses sustained significantly greater egg predation than *H. axyridis* egg masses ($P = 0.040$) (Figure 1f). However, there was no significant difference in predation experienced by *C. maculata* and the other two focal species in soybean (Figure 1f). In grassland, significantly fewer *C. maculata* eggs remained relative to *H. axyridis* ($P = 0.050$). There was no significant difference between the amount of predation experienced by *H. convergens* and the other two focal species in grassland. In alfalfa there were no differences in predation of coccinellid eggs among any species (Figure 1f). An ANOVA table summarizing the results of the three egg predation experiments is included as a supplementary table (Table S1).

Video observations of egg predation

Of 128 egg masses recorded, 41, 21, and 16 predation events were observed in grassland, alfalfa, and soybean fields respectively. We documented only one instance of a coccinellid preying on an egg mass, and this was an adult *C. maculata*, attacking *H. axyridis* eggs within a grassland habitat. Beyond coccinellids, a diverse community of organisms attacked coccinellid egg masses (Figure 2). In grassland Stylommatophora were a dominant egg predator responsible for 27.3% and 26.3% of the attacks on *H. convergens* and *H. axyridis* eggs respectively. In alfalfa Opiliones were the dominant predator, responsible for 81.1% and 100% of attacks on *H. convergens*, and *H. axyridis* egg masses respectively. In soybean Opiliones were again the dominant predator, responsible for 88.9% and 42.9% of attacks on *H. convergens* and *H. axyridis* egg masses respectively.

Coccinellid activity density and relative abundance

Exotic coccinellids were detected within all habitats during each egg predation experiment and *H. axyridis* was the most common species. During the week of July 20, 2009, the amount of exotic coccinellids collected from yellow sticky card traps varied across habitats ($\chi^2 = 10.68$, d.f. $= 2$, $P = 0.005$). Exotic species had a greater activity density in soybean than grasslands ($\chi^2 = 8.69$, d.f. $= 1$, $P = 0.003$) or alfalfa ($\chi^2 = 4.76$, d.f. $= 1$, $P = 0.029$) (Figure 3a). There was no difference in exotic coccinellid activity density among alfalfa and grasslands ($\chi^2 = 0.80$, d.f. $= 1$, $P = 0.371$) (Figure 3a). During the week of June 8, 2010, exotic coccinellid relative abundance sampled by yellow sticky card traps was similar in alfalfa and grasslands (Figure 3c). During the week of July 26, 2010, there was a greater activity density of exotic coccinellids in alfalfa than grasslands ($\chi^2 = 5.20$, d.f. $= 1$, $P = 0.022$) or soybean ($\chi^2 = 7.15$, d.f. $= 1$, $P = 0.008$) (Figure 3e). There was no difference in the activity density of exotic coccinellids among grasslands and soybean ($\chi^2 = 0.18$, d.f. $= 1$, $P = 0.671$) (Figure 3e).

During the weeks of July 20, 2009, and June 8, 2010, the relative abundance of exotic coccinellids sampled via sweep samples did not differ among habitats (Figure 3b, d). During the week of July 26, 2010 a greater relative abundance of exotic coccinellids was detected in alfalfa than in soybean ($\chi^2 = 18.58$, d.f. $= 1$, $P < 0.001$), and grasslands ($\chi^2 = 18.56$, d.f. $= 1$, $P < 0.001$) (Figure 3f).

Eight native coccinellid taxa were detected throughout the 2009–2010 sampling period: *Psyllobora vigintimaculata*, *Hyperaspis bigeminata*, *Scymnus* sp., *Hippodamia glacialis*, *Brachiacantha ursina*, *Cycloneda munda*, *C. maculata*, and *Hippodamia parenthesis* (Figure 4). No *H. convergens* were detected during sampling in 2009 or 2010, and *C. maculata* was among the most common of the native coccinellids collected. The activity density of native coccinellids

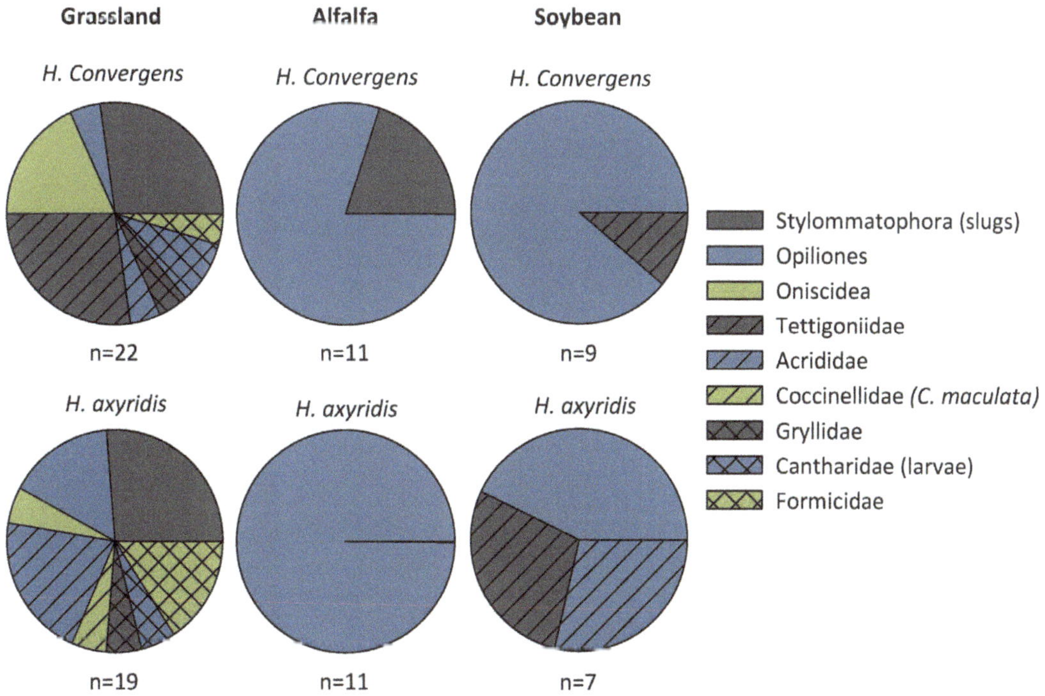

Figure 2. Guild of predators observed attacking coccinellid egg masses. The proportion of egg mass attacks by each predator taxa observed via video surveillance. Of 128 egg masses observed, predators attacked 41 *H. convergens* and 37 *H. axyridis* egg masses. Only the first attacked that occurred on each egg mass observed is reported here. The number of egg masses attacked in each category is reported as "n" under each pie chart.

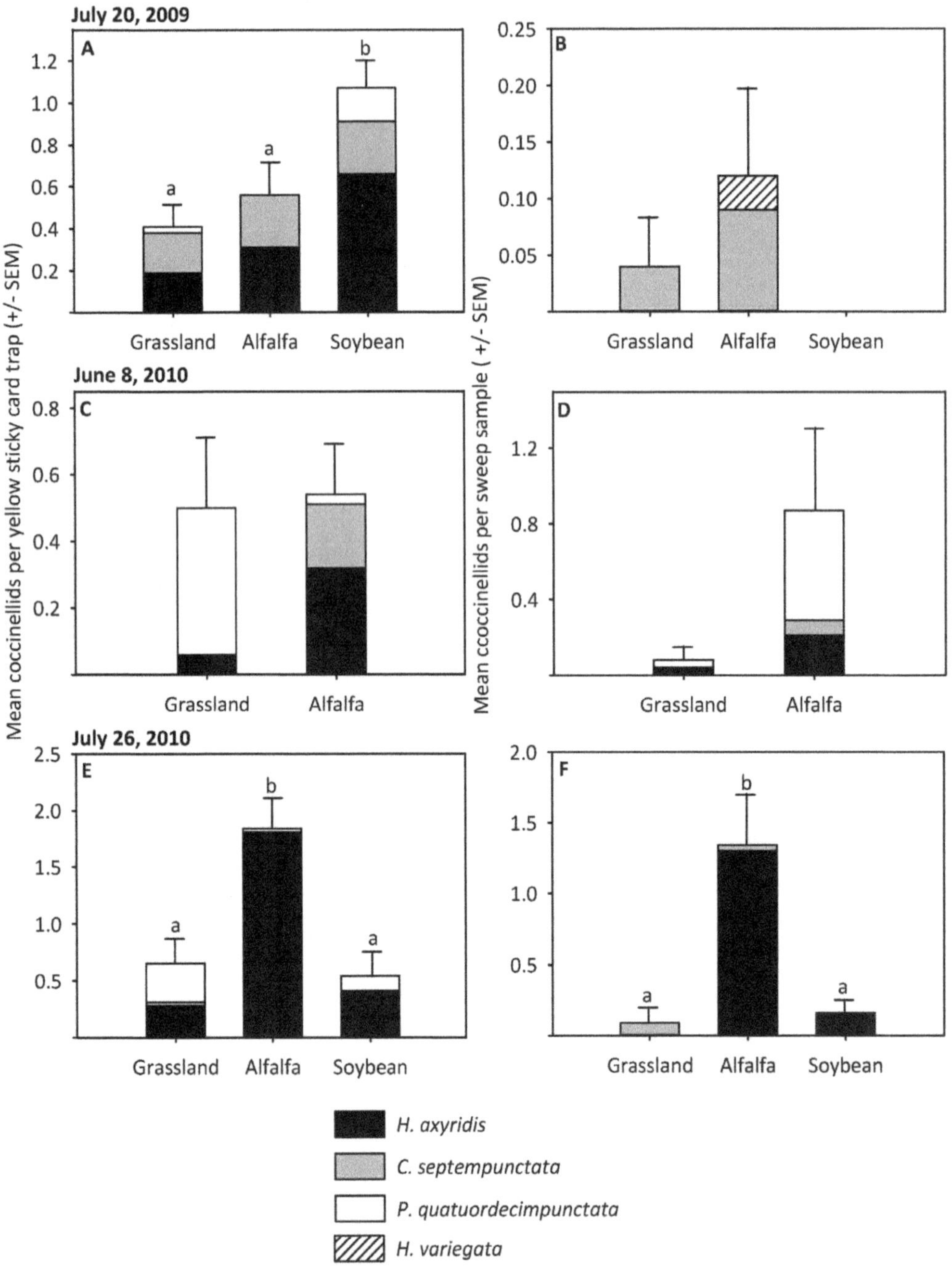

Figure 3. The exotic coccinellid community within field sites. The activity density and relative abundance of exotic coccinellids measured during each egg predation experiment within grassland, alfalfa and soybean habitats ($P<0.05$). A) July 20, 2009 yellow sticky card trap, B) July 20, 2009 sweep sample, C) June 8, 2010 yellow sticky card trap, D) June 8, 2010 sweep sample, E) July 26, 2010 yellow sticky card trap, F) July 26, 2010 sweep sample.

sampled via yellow sticky card traps did not vary across habitats throughout the study (Figure 4a, c, e). The relative abundance of native coccinellids collected via sweep samples also did not vary across habitats during the weeks of July 20, 2009, and June 8, 2010 (Figure 3b, d). During the week of July 26, 2010, there were fewer native coccinellids in sweep samples from grasslands compared to alfalfa ($\chi^2 = 3.90$, d.f. $= 1$, $P = 0.048$), and no native coccinellids were found in soybean (Figure 4f).

Relative abundance of aphids

During the June 8, 2010 experiment, a mean of 74.10±34.10 aphids per sweep sample were found in alfalfa whereas no aphids were recovered from the grasslands. During the July 26, 2010 experiment, alfalfa had a greater relative abundance of aphids than both grassland ($\chi^2 = 55.71$, d.f. $= 1$, $P<0.001$) and soybean ($\chi^2 = 58.53$, d.f. $= 1$, $P<0.001$), with a mean of 164.70±43.90 aphids per sweep sample from alfalfa. Sweep samples from grasslands and soybean contained 0.72±0.37 and 0.19±0.08 aphids per sweep sample respectively.

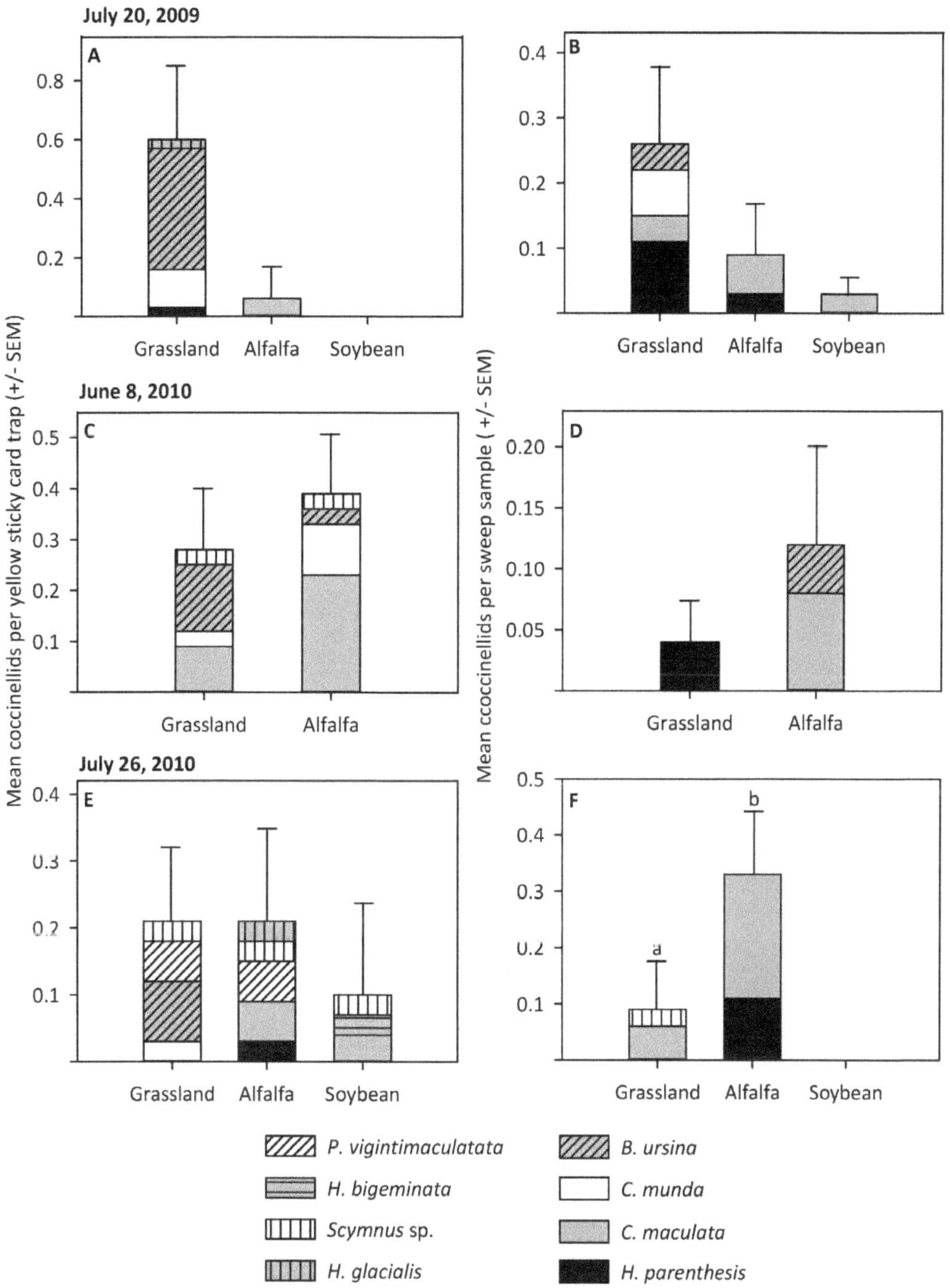

Figure 4. The native coccinellid community within field sites. The activity density and relative abundance of native coccinellids measured during each egg predation experiment within grassland, alfalfa and soybean habitats ($P<0.05$). A) July 20, 2009 yellow sticky card trap, B) July 20, 2009 sweep sample, C) June 8, 2010 yellow sticky card trap, D) June 8, 2010 sweep sample, E) July 26, 2010 yellow sticky card trap, F) July 26, 2010 sweep sample.

Discussion

Worldwide, native coccinellids have declined rapidly since the introduction, establishment, and spread of exotic coccinellids [16,17,19,20,21,40]. Research has focused primarily on direct competitive interactions among native and exotic species [27,30,32,41,42]. The propensity of exotic coccinellids, predominately *H. axyridis*, to act an intraguild predator of native coccinellid eggs and larvae has been confirmed in the laboratory. However, the extent to which declining coccinellids experience egg predation in the field and the guild of predators that contribute to this decline was not reported prior to this study.

Role of exotic coccinellids in egg predation

Our findings did not support the prediction that exotic coccinellids are a dominant predator of declining or common native coccinellid eggs. There were detectable levels of exotic coccinellids within the focal habitats, which were observed entering the screenshots of the video experiment at times, however they were not observed consuming the egg masses. Video surveillance revealed that a diversity of organisms consume native and exotic coccinellid egg masses including harvestmen (Opiliones), ants (Formicidae), slugs, (Stylommatophora), wood lice (Oniscidae), crickets (Gryllidae), soldier beetle larvae (Cantharidae) and long and short horned grasshoppers (Tettigoniidae and Acrididae). Previous observational studies have illustrated that some of these taxa act as egg or larval predators [43,44] but only Formicidae were commonly reported attacking coccinellid eggs elsewhere [45,46,47]. From our observations, Formicidae often appeared to damage eggs but not consume them, a behavior that may be triggered to protect honeydew resources [48,49,50]. The majority of egg predation was due to Opiliones and Stylommatophora; organisms that feed on a diversity of prey but had not previously been known to attack coccinellid eggs [51,52,53,54].

Extent of native and exotic egg predation

We found that this predator guild significantly reduced coccinellid eggs within alfalfa, soybean, and grassland habitats. Across all experiments there was only one incidence where predation on *H. axyridis* eggs did not differ from the Exclusion treatment, and this was in the alfalfa habitat. We also found variation in attack frequency among species and across habitats. If egg predation has contributed to native coccinellid decline we expected consumption of declining native coccinellid eggs to exceed that of common native or exotic species. We found partial support for this prediction. No differences in egg predation among *H. convergens* and *C. maculata* were found. However, both native species' lost a greater proportion of eggs to predation than *H. axyridis* within some, but not all, habitats and time periods examined. The amount of predation on *H. convergens* eggs was high, with up to 88% of eggs removed within 48 hours. Additionally, up to a 33.3% difference in the proportion of *H. convergens* versus *H. axyridis* eggs removed by predators was found.

This leads to the question, why was a diverse generalist predator guild more likely to consume native versus exotic coccinellid eggs? Coccinellids are known to have species specific alkaloids present on the surface and within their eggs which may signal toxicity and deter predators [55]. Coccinellids are often reluctant to consume heterospecific eggs which can decelerate development or cause death [55,56,57]. Although a majority of studies investigate the effect of these alkaloids on heterospecific coccinellids, it is possible that the differences in defensive chemicals could explain why the predator guild we observed tended to consume the *H. convergens* eggs more readily than *H. axyridis* eggs. Kajita et al. [56] examined variation in quantities of alkaloids in coccinellid eggs and found that concentrations can vary significantly, both within the same species and among species.

Although exotic coccinellids are not a dominant predator of native coccinellid egg masses, reduced egg predation may confer an advantage to *H. axyridis* over competitors. Less predation on *H. axyridis* eggs could explain a higher abundance allowing them to consume more available shared resources such as aphids.

Variation in egg predation across habitats

We also found that habitat affected egg predation risk. Across species and experiments, egg predation was either equivalent or greater within grasslands relative to the croplands studied thereby refuting our prediction that native coccinellid egg predation is reduced within grasslands. This prediction was based on the findings of several studies indicating that native coccinellids are abundant in grassland habitats relative to crop fields [58,59,60]. Our study illustrated that alfalfa fields offered the highest reproductive success for both native and exotic coccinellids. Eggs of all species within alfalfa fields were attacked least often in late July in both 2009 and 2010. This may have been due to greater aphid availability, as aphid counts were significantly greater in alfalfa fields in late July 2010, with 164.7 aphids per sweep sample versus fewer than one aphid per sample in the grassland and soybean fields. Extraguild prey abundance can affect intraguild predation among generalist predators [45,61]; a pattern of decreased predation with higher extraguild prey abundance is often reported [42,62,63].

Conclusions and implications

Our findings did not support the hypothesis that egg predation by exotic coccinellids has contributed to native coccinellid decline. This does not verify that exotic coccinellids cannot act as intraguild predators. Despite a lack of evidence for IGP between native and exotic coccinellids presented here, native larval coccinellid alkaloids have been detected in 121 of 590 *H. axyridis* larvae collected from lime trees in Brussels, Belgium. Native alkaloids within *H. axyridis* specimens collected indicate that native coccinellids had been consumed [63]. However, it is unknown if the level of predation was sufficient to affect the targeted native coccinellid population.

Despite the lack of direct competition found in this study, we did find that *H. convergens* sustained higher egg loss relative to *H. axyridis*. We do not have sufficient evidence to conclude that this has contributed to the decline of this species. There are no known measurements of coccinellid egg predation prior to the establishment of exotic coccinellids within the US. Thus, we do not know whether native egg predation rates have changed during the timeframe of decline. Still, it is clear that throughout these experiments, *H. axyridis* often had a reproductive advantage over native species within croplands and grasslands. Reduced egg predation of this exotic coccinellid by generalist predators may have aided its rapid population increase within its invaded range. This advantage may indirectly affect native coccinellids via mechanisms such as enhanced exploitative or apparent competition [20,29].

Supporting Information

Figure S1 Site Map. Map depicting the locations of field sites within Ohio, USA during the 2009-10 study.

Figure S2 Description of experimental set up in field. Four plots were established per site (depicted with circles 1-4). All plots were a minimum of 30 meters from any field edge. Data was collected at these plots including: Egg predation experiment with open and exclusion treatments (A), sweep sampling for coccinellid abundance (B), and yellow sticky card trap sampling for coccinellid activity density (C).

Figure S3 Description of experimental set up for camera experiments. Systems were placed randomly at two of the four plots (described in Figure S2). Systems consisted of a DVR powered by a deep cycle boat battery (B). Four cameras were connected to the DVR by a 28.3m power and video cord. Each camera was focused on a platform where a coccinellid egg mass was placed (A). All cameras were a minimum of 30 meters from any field edge.

Table S1 Egg predation experiments ANOVA table. ANOVA table for the effect of Treatment (Open or Exclusion), Species (*C. maculata*, *H. convergens*, or *H. axyridis*), and Habitat (alfalfa, soybean, or semi-natural grassland) on the proportion of coccinellid eggs remaining after 48 hours of exposure.

Acknowledgments

We thank J.D. Harwood and two anonymous reviewers for their helpful comments on an earlier version of this manuscript. The authors thank S. Prajzner, B. Hunt, H. Edgington, J. Power, and A. Alumai for assistance collecting field data. We thank M.E. Grieshop for providing detailed methods to construct the surveillance camera systems and S.D. Scott for building them.

Author Contributions

Conceived and designed the experiments: CS MG. Performed the experiments: CS MG. Analyzed the data: CS MG. Contributed reagents/materials/analysis tools: MG. Wrote the paper: CS MG.

References

1. Vitousek PM, Dantonio CM, Loope LL, Rejmanek M, Westbrooks R (1997) Introduced species: A significant component of human-caused global change. New Zealand Journal of Ecology 21: 1–16.
2. Wilcove DS, Rothstein D, Dubow J, Phillips A, Losos E (1998) Quantifying threats to imperiled species in the United States. Bioscience 48: 607–615.
3. Clavero M, Garcia-Berthou E (2005) Invasive species are a leading cause of animal extinctions. Trends in Ecology & Evolution 20: 110–110.
4. Pimentel D, Zuniga R, Morrison D (2005) Update on the environmental and economic costs associated with alien-invasive species in the United States. Ecological Economics 52: 273–288.
5. Gurevitch J, Padilla DK (2004) Are invasive species a major cause of extinctions? Trends in Ecology & Evolution 19: 470–474.
6. Bertness MD (1984) Habitat and community modification by an introduced herbivorous snail. Ecology 65: 370–381.
7. Mack RN, Simberloff D, Lonsdale WM, Evans H, Clout M, et al. (2000) Biotic invasions: Causes, epidemiology, global consequences, and control. Ecological Applications 10: 689–710.
8. Snyder WE, Evans EW (2006) Ecological effects of invasive arthropod generalist predators. Annual Review of Ecology Evolution and Systematics 37: 95–122.
9. Crowder D, Snyder W (2010) Eating their way to the top? Mechanisms underlying the success of invasive insect generalist predators. Biological Invasions 12: 2857–2876.
10. Hagen KS (1962) Biology and ecology of predaceous Coccinellidae. Annual Review of Entomology 7: 289–326.
11. Hagen KS (1968) Impact of pathogens, parasites, and predators on aphids. Annual Review of Entomology 13: 325–384.
12. Gordon RD (1985) The Coccinellidae (Coleoptera) of America north of Mexico. Journal of the New York Entomological Society 93: 1–912.
13. Herren HR, Neuenschwander P (1991) Biological control of cassava pests in Africa. Annual Review of Entomology 36.
14. Obrycki JJ, Kring TJ (1998) Predaceous coccinellidae in biological control. Annual Review of Entomology 43. 295–321.
15. Wheeler AG, Hoebeke ER (1995) *Coccinella novemnotata* in northeastern North America - historical occurrence and current status (Coleoptera, Coccinellidae). Proceedings of the Entomological Society of Washington 97: 701–716.
16. Elliott N, Kieckhefer R, Kauffman W (1996) Effects of an invading coccinellid on native coccinellids in an agricultural landscape. Oecologia 105: 537–544.
17. Colunga-Garcia M, Gage SH (1998) Arrival, establishment, and habitat use of the multicolored Asian lady beetle (Coleoptera: Coccinellidae) in a Michigan landscape. Environmental Entomology 27: 1574–1580.
18. Turnock W, Wise I, Matheson F (2003) Abundance of some native coccinellids (Coleoptera: Coccinellidae) before and after the appearance of *Coccinella septempunctata*. The Canadian Entomologist 135: 391–404.
19. Alyokhin A, Sewell G (2004) Changes in a lady beetle community following the establishment of three alien species. Biological Invasions 6: 463–471.
20. Evans EW (2004) Habitat displacement of North American ladybirds by an introduced species. Ecology 85: 637–647.
21. Roy HE, Adriaens T, Isaac NJB, Kenis M, Onkelinx T, et al. (2012) Invasive alien predator causes rapid declines of native European ladybirds. Diversity and Distributions: DOI: 10.1111/j.1472-4642.2012.00883.x.
22. Gardiner MM, Landis DA, Gratton C, Schmidt N, O'Neal M, et al. (2009) Landscape composition influences patterns of native and exotic lady beetle abundance. Diversity and Distributions 15: 554–564.
23. Gardiner M, Tuell J, Isaacs R, Gibbs J, Ascher J, et al. (2010) Implications of three biofuel crops for beneficial arthropods in agricultural landscapes. BioEnergy Research 3: 6–19.
24. Gardiner MM, O'Neal ME, Landis DA (2011) Intraguild predation and native lady beetle decline. PLoS One 6: e23576.
25. Gardiner MM, Allee L, Brown P, Losey JE, Roy H, et al. (2012) Lessons from lady beetles: accuracy of monitoring data from US and UK citizen science programs. Frontiers in Ecology and the Environment.
26. Koch RL (2003) The multicolored lady beetle, *Harmonia axyridis*: A review of its biology, uses in biological control, and non-target impacts. Journal of insect Science 3: 1–16.
27. Cottrell TE (2004) Suitability of exotic and native lady beetle eggs (Coleoptera : Coccinellidae) for development of lady beetle larvae. Biological Control 31: 362–371.
28. Hardin G (1960) The competitive exclusion principle. Science 131: 1292–1297.
29. Hoogendoorn M, Heimpel GE (2002) Indirect interactions between an introduced and a native ladybird beetle species mediated by a shared parasitoid. Biological Control 25: 224–230.
30. Cottrell TE (2005) Predation and cannibalism of lady beetle eggs by adult lady beetles. Biological Control 34: 159–164.
31. Cottrell TE (2007) Predation by adult and larval lady beetles (Coleoptera : Coccinellidae) on initial contact with lady beetle eggs. Environmental Entomology 36: 390–401.
32. Snyder WE, Clevenger GM, Eigenbrode SD (2004) Intraguild predation and successful invasion by introduced ladybird beetles. Oecologia 140: 559–565.
33. Sato S, Yasuda H, Evans EW (2005) Dropping behaviour of larvae of aphidophagous ladybirds and its effects on incidence of intraguild predation: interactions between the intraguild prey, *Adalia bipunctata* (L.) and *Coccinella septempunctata* (L.), and the intraguild predator, *Harmonia axyridis* Pallas. Ecological Entomology 30: 220–224.
34. Gardiner MM, Landis DA (2007) Impact of intraguild predation by adult *Harmonia axyridis* (Coleoptera : Coccinellidae) on *Aphis glycines* (Hemiptera : Aphididae) biological control in cage studies. Biological Control 40: 386–395.
35. Hoogendoorn M, Heimpel GE (2004) Competitive interactions between an exotic and a native ladybeetle: a field cage study. Entomologia Experimentalis Et Applicata 111: 19–28.
36. Rand TA, Louda SA (2006) Spillover of agriculturally subsidized predators as a potential threat to native insect herbivores in fragmented landscapes. Conservation Biology 20: 1720–1729.
37. Teixeira L, Grieshop M, Gut L (2010) Effect of pheromone dispenser density on timing and duration of approaches by peachtree borer. Journal of Chemical Ecology 36: 1148–1154.
38. Grieshop MJ, Werling B, Buehrer K, Perrone J, Isaacs R, et al. (2012) Big brother is watching: studying insect predation in the age of digital surveillance. American Entomologist 58: 172–182.
39. SAS (2008) SAS (R) 9.2 Cary, NC: SAS Institute Inc.
40. Harmon JP, Stephens E, Losey J (2007) The decline of native coccinellids (Coleoptera : Coccinellidae) in the United States and Canada. Journal of Insect Conservation 11: 85–94.
41. Cottrell TE, Yeargan KV (1998) Intraguild predation between an introduced lady beetle, *Harmonia axyridis* (Coleoptera : Coccinellidae), and a native lady beetle, *Coleomegilla maculata* (Coleoptera : Coccinellidae). Journal of the Kansas Entomological Society 71: 159–163.
42. Yasuda H, Evans EW, Kajita Y, Urakawa K, Takizawa T (2004) Asymmetric larval interactions between introduced and indigenous ladybirds in North America. Oecologia 141: 722–731.
43. Brust GE, Stinner BR, McCartney DA (1986) Predator activity and predation in corn agroecosystems. Environmental Entomology 15: 1017–1021.
44. Lundgren JG, Shaw JT, Zaborski ER, Eastman CE (2006) The influence of organic transition systems on beneficial ground-dwelling arthropods and predation of insects and weed seeds. Renewable Agriculture and Food Systems 21: 227–237.
45. Lucas E, Coderre D, Brodeur J (1998) Intraguild predation among aphid predators: characterization and influence of extraguild prey density. Ecology 79: 1084–1092.
46. Phoofolo MW, Obrycki JJ (1998) Potential for intraguild predation and competition among predatory Coccinellidae and Chrysopidae. Entomologia Experimentalis Et Applicata 89: 47–55.
47. Oliver TH, Jones I, Cook JM, Leather SR (2008) Avoidance responses of an aphidophagous ladybird, *Adalia bipunctata*, to aphid-tending ants. Ecological Entomology 33: 523–528.
48. McLain DK (1980) Relationships among ants, aphids and coccinellids on wild lettuce. Journal of the Georgia Entomological Society 15: 417–418.
49. Vinson SB, Scarborough TA (1989) Impact of the imported fire ant on laboratory populations of cotton aphid (*Aphis gossypii*) predators. The Florida Entomologist 72: 107–111.
50. Sloggett JJ, Majerus MEN (2003) Adaptations of *Coccinella magnifica*, a myrmecophilous coccinellid to aggression by wood ants (*Formica rufa* group). II. Larval behaviour, and ladybird oviposition location. European Journal of Entomology 100: 337–344.
51. Newton BL, Yeargan KV (2001) Predation of *Helicoverpa zea* (Lepidoptera: Noctuidae) Eggs and First Instars by *Phalangium opilio* (Opiliones: Phalangiidae). Journal of the Kansas Entomological Society 74: 199–204.
52. Pallant D (1972) The food of the grey field slug, *Agriolimax reticulatus* (Müller), on grassland. Journal of Animal Ecology 41: 761–769.

53. Jennings TJ, Barkham JP (1975) Food of slugs in mixed deciduous woodland. Oikos 26: 211–221.
54. Fox L, Landis BJ (1973) Notes on the predaceous habits of the gray field slug *Deroceras laeve*. Environmental Entomology 2: 306–307.
55. Hemptinne JL, Lognay G, Gauthier C, Dixon AFG (2000) Role of surface chemical signals in egg cannibalism and intraguild predation in ladybirds (Coleoptera : Coccinellidae). Chemoecology 10: 123–128.
56. Kajita Y, Obrycki JJ, Sloggett JJ, Haynes KF (2010) Intraspecific alkaloid variation in ladybird eggs and its effects on con- and hetero-specific intraguild predators. Oecologia 163: 313–322.
57. Rieder JP, Newbold TAS, Sato S, Yasuda H, Evans EW (2008) Intra-guild predation and variation in egg defence between sympatric and allopatric populations of two species of ladybird beetles. Ecological Entomology 33: 53–58.
58. Leather SR, Cooke RCA, Fellowes MDE, Rombe R (1999) Distribution and abundance of ladybirds (Coleoptera : Coccinellidae) in non-crop habitats. European Journal of Entomology 96: 23–27.
59. Evans EW (2000) Morphology of invasion: body size patterns associated with establishment of *Coccinella septempunctata* (Coleoptera : Coccinellidae) in western North America. European Journal of Entomology 97: 469–474.
60. Werling BP, Meehan TD, Robertson BA, Gratton C, Landis DA (2011) Biocontrol potential varies with changes in biofuel-crop plant communities and landscape perenniality. Global Change Biology Bioenergy 3: 347–359.
61. Ingels B, De Clercq P (2011) Effect of size, extraguild prey and habitat complexity on intraguild interactions: a case study with the invasive ladybird *Harmonia axyridis* and the hoverfly *Episyrphus balteatus*. BioControl 56: 871–882.
62. Nóia M, Borges I, Soares AO (2008) Intraguild predation between the aphidophagous ladybird beetles *Harmonia axyridis* and *Coccinella undecimpunctata* (Coleoptera: Coccinellidae): The role of intra and extraguild prey densities. Biological Control 46: 140–146.
63. Hautier L, Martin GS, Callier P, de Biseau JC, Gregoire JC (2011) Alkaloids provide evidence of intraguild predation on native coccinellids by *Harmonia axyridis* in the field. Biological Invasions 13: 1805–1814.

15

Spatial Structure and Distribution of Small Pelagic Fish in the Northwestern Mediterranean Sea

Claire Saraux[1]*, Jean-Marc Fromentin[1], Jean-Louis Bigot[1], Jean-Hervé Bourdeix[1], Marie Morfin[1], David Roos[1], Elisabeth Van Beveren[1], Nicolas Bez[2]

1 IFREMER (Institut Français de Recherche pour l'Exploitation de la MER), Research Unit EME (UMR 212), Sète, France, **2** IRD (Institut de Recherche pour le Développement), Research Unit EME (UMR 212), Sète, France

Abstract

Understanding the ecological and anthropogenic drivers of population dynamics requires detailed studies on habitat selection and spatial distribution. Although small pelagic fish aggregate in large shoals and usually exhibit important spatial structure, their dynamics in time and space remain unpredictable and challenging. In the Gulf of Lions (north-western Mediterranean), sardine and anchovy biomasses have declined over the past 5 years causing an important fishery crisis while sprat abundance rose. Applying geostatistical tools on scientific acoustic surveys conducted in the Gulf of Lions, we investigated anchovy, sardine and sprat spatial distributions and structures over 10 years. Our results show that sardines and sprats were more coastal than anchovies. The spatial structure of the three species was fairly stable over time according to variogram outputs, while year-to-year variations in kriged maps highlighted substantial changes in their location. Support for the McCall's basin hypothesis (covariation of both population density and presence area with biomass) was found only in sprats, the most variable of the three species. An innovative method to investigate species collocation at different scales revealed that globally the three species strongly overlap. Although species often co-occurred in terms of presence/absence, their biomass density differed at local scale, suggesting potential interspecific avoidance or different sensitivity to local environmental characteristics. Persistent favourable areas were finally detected, but their environmental characteristics remain to be determined.

Editor: Konstantinos I. Stergiou, Aristotle University of Thessaloniki, Greece

Funding: Surveys were cofunded by IFREMER and the European Union through the Data Collection Framework (DCF). This study is part of the program EcoPelGol (Study of the Pelagic ecosystem in the Gulf of Lions), funded by France Filière Pêche (FFP). The funders had no role in study design, data collection and analysis, decision to publish, or preparation of the manuscript.

Competing Interests: The present study is part of a more general program investigating changes in small pelagic populations in the Gulf of Lions, named EcoPelGol. This program is partly funded by the interprofessional association France Filière Pêche. The authors are free to fully develop and publish their research without consulting France Filière Pêche, as per previous signed agreement with the association. France Filière Pêche has not yet seen these results, which will be communicated to them once published only. This does not alter the authors' adherence to PLOS ONE policies on sharing data and materials.

* Email: claire.saraux@ifremer.fr

Introduction

Because animal spatial distribution is often strongly associated with population dynamics, spatial indices may provide valuable tools for assessing the status of these populations, and in particular the status of endangered or exploited species [1–4]. At the population scale, spatial distribution can be seen, in the absence of substantial anthropogenic impacts, as the emergent property of habitat selection. In addition to social motivation (presence *vs.* absence of conspecifics; [5]), spatio-temporal aggregation patterns in animal populations may be explained by individuals sharing similar needs and dealing with similar biotic and abiotic pressures, such as prey abundance *vs.* predation risk [6], propitious *vs.* detrimental environmental conditions [7]. Nonetheless, as population density increases, so does intra-specific competition [8], and individuals are expected to spread towards less suitable habitats when a certain threshold is reached ('Basin hypothesis'; [9]). Both population density and its occupation area should then vary with its abundance [10–12]. If their density-dependent and density independent drivers are difficult to disentangle (but see [13]), spatial distributions in themselves offer valuable information for both ecological understanding and management.

Indeed, besides an obvious fundamental interest in terms of population dynamics and marine ecosystem functioning [14], information on fish biomass location is also of crucial importance for stock management. For instance, the effective implementation of Marine Protected Areas (MPAs) requires detailed knowledge on species spatio-temporal dynamics, *i.e.* on temporal variability in fish spatial distributions [15]. In addition, knowledge on interspecific interactions, such as the co-occurrence or repulsion of species based on their spatial distributions, may greatly help scientists to understand competition or predator-prey processes but also policy makers to implement ecosystemic rather than single species management measures, as presently done.

In this study, we investigated the spatio-temporal distribution of small pelagic fish in the Gulf of Lions using a unique dataset of 10 years of acoustic surveys. Because small pelagic fish are key species of the pelagic ecosystem due to their central place in the food web, transferring energy from the lowest trophic levels (plankton) towards top-predators [16], and because their life-history traits (short lifespan, large fecundity) make them strongly dependent on the abiotic environment [17], changes in their distribution and abundance have been a major source of concern for scientists and managers worldwide ([18]; Global Ocean Ecosystem Dynamics program - GLOBEC;).

Over the past years, important changes in the two main target species (the European sardine, *Sardina pilchardus* and the European anchovy, *Engraulis encrasicolus*) biomass [19], along with a shift in the size-distribution of these species towards smaller individuals [20] have been observed in the Northwestern Mediterranean Sea (Gulf of Lions), resulting in important economic losses for fisheries. Both stocks are now considered to have a low biomass and a low fishing mortality by General Fisheries Commission for the Mediterranean [19]. In parallel, a third small pelagic species, the sprat (*Sprattus sprattus*), which is not commercially exploited in the Western Mediterranean has appeared in the system with a steadily increasing biomass since 2007 [19]. This unexpected situation offered us the opportunity to describe and compare simultaneous spatio-temporal distributions of three species sharing the same trophic level, with similar feeding behaviours [21–22], but with completely different trends in biomass, through the analysis of 10 years of scientific acoustic surveys. First, we studied the aggregation patterns and spatial structures of these species separately. Second, we investigated species spatial dynamics by considering spatio-temporal changes in biomass, and defined optimal recurrent areas based on a combination of biomass levels and their variability. Finally, we focused on interspecific relationships by studying the overlap between species at different spatio-temporal scales.

Methods

Ethic statement

The study was conducted in the Gulf of Lions (Longitude in [3.05°; 5.20°] and Latitude in [42.44°; 43.44°]), a public sea area. No sampling was operated from private land and field studies did not involve endangered or protected species. All data used in this study came from PELMED acoustic surveys, which comply with the MEDIAS (Mediterranean Acoustic Survey) protocol. The sampling has been performed under repeated international standardized surveys where the research vessel had full permission to sample from all relevant national public authorities (governments). Acoustic data were collected at a distance, which does not require any particular ethic approval. Further data used in this study came from scientific trawls conducted during these same PELMED surveys. Again, no approval by an ethic committee was required as the targeted species are exploited species and trawling methods done according to international standard trawl surveying.

Data collection and survey design

Every July since 2003, the French Research Institute IFREMER has been carrying out acoustic surveys of the pelagic resources present in the Gulf of Lions, Mediterranean Sea. The summer period of the survey corresponds to contrasted biological periods for our 3 species. Indeed, this is the peak of reproduction for anchovies, while sprats and sardines reproduce in winter [23]. Due to the biological cycle of these species, it is important not to extrapolate our results outside the summer period. Sampling was performed along 9 parallel transects, regularly spaced by 12 nautical miles (nm) (see Figure S1 in File S1). Acoustic data were recorded every 1 nm using multi-frequency echosounders (Simrad EK500 and ER60), while travelling at a constant speed of 8 nm.h^{-1}. All 4 frequencies were visualized during sampling to help deciding when to trawl for species identification. However, only energies from the 38 kHz (typical frequency used for fish) channel were used to estimate fish density. Acoustic data analyses, such as bottom correction, were later performed using *Movies+* [24] and *FishView* IFREMER softwares. Species discrimination and echo-partitioning were performed by the combination of echo trace classification and trawl outputs [25]. Species biomass and abundance were finally estimated from species energy using specific target strength (TS = 20 log(L) - 71.2, where L is the length of the fish for all 3 species, see [26] for more details on acoustic surveys and analyses). Main survey features are summarised in Table 1.

Analyses

We defined and used different spatial indicators and geostatistical methods described in details below. An overview of these indicators, including their calculation formula, representative scale and biological meaning are summarised in Table 2. A key concept in statistics in general and in spatial statistics in particular is the support of the information, i.e. the geographical area over which measures are recorded [27–28]. In this study, the support size was the size of the Elementary Sampling Distance Unit, that is 1 nm. All statistics derived from these data are thus associated to the 1 nm sampling support and, for all of them, their values would have been different if computed at another support. Even though some metrics are labelled like spatial statistics (e.g. space selectivity index, local index of collocation), they are not sensitive to the location of the points in space (i.e. any exchange between two data does not change the result). They are sensitive to the spatial patterns that exist at scales smaller than the support size and that are integrated in the observations. For some of them (e.g. Empirical Orthogonal Functions), it remains possible to represent them as geographical distributions making the confusion even worth.

Aggregation patterns and spatial structure. Presence Area (PA) was calculated as the percentage of sampled points at which the species was found, independently of its abundance.

A space selectivity index was defined on the basis of annual concentration curves [29–30]. Like Lorenz curves (as used in [31]), concentration curves represent the maximum proportion of biomass as a function of the proportion of samples [30]. For instance, a proportion of biomass of 0.6 for a proportion of sampled area of 0.2 would indicate that 60% of population total biomass was found in only 20% of the total area sampled. For a homogeneously distributed population, each proportion of biomass should be found in the same proportion of total area (i.e. 10% of biomass in 10% of area, etc.), so that the annual concentration curve of a homogeneously distributed population equals the first diagonal. It follows that the more concentrated a population, the further the curve falls from the first diagonal, *i.e.* a higher proportion of biomass is situated in a given proportion of the total area. Consequently, the space selectivity index is defined as twice the area between the concentration curve and the first diagonal [29–30], so that the higher the space selectivity index, the more concentrated the spatial distribution of the fish.

Spatial structure was defined as the spatial autocorrelation between values at different locations. In other words, spatial structure considers whether the correlation between biomass in two locations depends on the distance separating them. It was investigated using variograms [27] that calculate the value of the spatial autocorrelation at different lags (distance intervals). As is often the case, the frequency distributions of species biomass were highly skewed with a large proportion of zeros or small values, and few extremely large values contributing importantly to total biomass (coefficients of variation CV were 2.6, 3.8 and 4.2 for anchovies, sprats and sardines, respectively). Besides, the variance

Table 1. Main PELMED survey features from 2003 to 2012.

Year	Sampling dates	Nb of nm	Nb of EDSU	Nb of trawls	Anchovy biomass	Sardine biomass	Sprat biomass
2003	07/07/03-06/08/03	1274	284	27	27 860	126 120	685
2004	05/07/04-04/08/04	1174	285	29	25 953	215 560	786
2005	08/07/05-07/08/05	1462	294	33	15 962	264 024	1 955
2006	09/07/06-08/08/06	1473	288	39	25 658	102 276	772
2007	11/07/07-10/08/07	1500	290	43	13 654	88 297	15
2008	15/06/08-30/07/08	2151	273	60	23 395	91 546	5 002
2009	24/06/09-29/07/09	2173	284	43	30 424	52 977	7 845
2010	24/06/10-29/07/10	2000	276	39	23 514	51 819	15 760
2011	27/06/11-31/07/11	1704	282	42	25 906	44 926	26 638
2012	27/06/12-31/07/12	1172	279	37	39 061	80 537	70 263

Biomasses are indicated in tons, nm stands for nautical mile.

in biomass increased proportionally with mean biomass in our data. This led us to log-transform the studied variables (the optimal Box-Cox transformation was **Y = ln (B+c)**, with ***c*** a positive constant added to insure positive values). This constant was set to the smallest biomass observed in our data, *i.e.* 0.05 tons. While distributions of the log-biomass were still skewed, the proportionality effect disappeared. The log-transformed data was used to calculate empirical variograms for all three species for each year of survey. To account for the anisotropy of the sampling (only along parallel transects) and for a potential anisotropic structure due to a bathymetry gradient, bidirectional variograms (along transects and perpendicularly to them) were calculated and modelled by automatic fitting using a least square method [32]. The temporal variability of the spatial structure was assessed on annually standardised data by comparing annual variograms to the mean variogram (average across all years of survey, pairs being retained only if the two sampling locations belong to the same year) and its 95%-confidence interval (adaptation from [12] for bidirectional variograms). The confidence interval was obtained by simulating 1000 random fields according to the mean modelled variogram using the turning-bands method [33] and estimating the 1000 variograms associated with the sampling points extracted from these random fields.

Species spatial distributions. To capture the spatial patterns of the 3 populations as simply as possible and to investigate year-to-year variations, we calculated the centres of gravity of their biomass along with their associated inertia [34–35]. The centre of gravity represents the mean location of the population, while inertia describes the dispersion of the population around its centre of gravity. Because fish populations are often aggregated in a given area and therefore spatial distributions heterogeneous, we also investigated the presence of spatial patches of high biomass densities, by adapting the recursive algorithm developed by [35]. This was adapted to take into account the anisotropy of our sampling by setting two different threshold distances (one parallel to the transect and another one perpendicular). The sensitivity to the threshold distances was tested using different values. Very small distances did not enable us to identify patches with more than 10% of biomass, while increasing the distance reduced the number of patches to one which comprised the whole presence area. The final distances (i.e. 6 nm in the direction of the transect and 24 perpendicular to it) were set in the interval where the resulting number of patches was stable and were kept constant both across species and years. The number of patches, centres of gravity and inertias of all patches were then used to study year-to-year and interspecific variations.

Kriging maps were produced for each species on a 1 nm*1 nm-grid over the study area, using the modelled mean or annual variograms depending on the results of interannual variability tests. In order to account for anisotropic sampling, we used an anisotropic search of neighbours when conducting kriging, *i.e.* the distance at which neighbours were looked for was smaller in the transect direction than perpendicular to it to make sure that each point comprised neighbours in two different transects.

Two different methods were used to quantify the temporal stability/variability of spatial distributions at 2 different scales (Table 2). First, to assess the stability/variability of the entire distribution, we calculated the Empirical Orthogonal Functions (EOF) on the raw acoustic data for each species. EOF analysis is a decomposition of a spatio-temporal dataset in terms of orthogonal basis functions. It is thus very similar to a Principal Component Analysis, except that it is applied to spatio-temporal data, using time as a descriptor and space (i.e. sampling locations) as objects [36]. EOF was performed on standardized data to compare

Table 2. Spatial indicators used in the study.

Indicator name	Formula	Spatial scale	Time scale	Biological meaning
Presence area				
Presence area	$PA = \frac{\sum 1_i}{n} * 100$	Global	Annual	Area of presence(in %) of the species over the studied area
Aggregation and spatial structure				
Space selectivity index		Global	Annual	Spatial compactness
Annual variograms			Annual	Spatial autocorrelation
Mean variogram			Decadal	Mean Spatial autocorrelation
Spatial distribution				
Centre of gravity	$CG = \frac{\int x\, z(x)\, dx}{\int z(x)\, dx}$	Global	Annual & decadal	Mean location of the species
Inertia	$I = \frac{\int (x-CG)^2\, z(x)\, dx}{\int z(x)\, dx}$	Global	Annual& decadal	Spatial dispersion around CG
Patches		Intermediate	Annual	Concentration of biomass
EOF		Global	decadal	Spatio-temporal variability
CV	$CV_x = \frac{\sigma_{z_a\,(x)}}{Mean(z_{a\,(x)})}$	Local	decadal	Temporal variability
Interspecific relationships/co-occurence				
Overlap	$O_{i,j} = \frac{A_{E_i \cap E_j}}{A_{E_i \cup E_j}}$	Global	Annual decadal	Overlap between the spatial distribution envelops of two species
Overlap of patches	$O_{i,j} = \frac{\sum_{p_i} \sum_{p_j} A_{Ep_i \cap Ep_j}}{A_{\bigcup_{p_i, p_j} E_{p_i} E_{p_j}}}$	Intermediate	Annual decadal	Proportion of patches where both species present
Co-occurrence index	$Co_{i,j} = \frac{\sum 1_{i,j}}{\sum 1_i + \sum 1_j}$	Local	Annual decadal	Level of local co-occurrence between 2 species based on presence/absence
Local index of collocation	$LIC_{ij} = \frac{\int z_i(x)\, z_j(x)\, dx}{\sqrt{\int z_i^2(x)\, dx}\sqrt{\int z_j^2(x)\, dx}}$	Local	Annual decadal	Level of local cooccurence between species based on densities
Dominance index	$Dom_{i,j} = \frac{z_i(x) - z_j(x)}{z_i(x) + z_j(x)}$	Local	Annual decadal	Relative biomass in each point
Mapping				
Kriging	Anisotropic neighbourhood	Global	Annual	Distribution maps
Area classification				
	Variability map * Average map	Global	Decadal	Identification of recurrent, occasional & unfavourable areas

n: Number of sampling points, z(x): biomass density in x, z_a (x): annual biomass density in x.

annual spatial distributions independently of annual total levels of biomass. The first axis (eigen vector) of the EOF is the linear combination of years which maximizes the percentage of the interannual variance of spatial distributions. When all yearly contributions to this first axis share the same sign, the greater the percentage of variance explained by the first EOF, the more persistent the spatial distributions [37;12]. In order to have a local estimation of temporal variability/stability and see whether some areas were more stable than others, we also calculated the Coefficient of Variation (CV) of the 10 annual values in each sampling point. As biomass varied substantially between years, we used relative biomass (i.e absolute biomass divided by the total annual biomass) instead of absolute biomass, so that the observed CV corresponded to a change in geographical location rather than a change in overall biomass.

Finally, we defined recurrent, occasional and unfavourable areas by combining average and variability maps [38]. To produce average and variability maps, we calculated the mean and the standard deviation of the 10 annual kriged maps node by node (1 nm*1 nm). Then, each 1 nm*1 nm pixel was assigned one of the 3 categories (recurrent, occasional and unfavourable) depending on its value in the average and variability maps. Pixels with both low mean and variability (i.e. inferior to the median) were considered unfavourable; those with high mean and low variability recurrent, while pixels with high variability were defined as occasional.

Interspecific indices: collocation index and overlap between species. We quantified the collocation at three different spatial scales in a coherent manner: from a global scale comparing the distribution over the whole study area, to a local scale comparing species at each sampling point, passing by an intermediate scale comparing previously defined patches.

Global overlap was defined based on centres of gravity and associated inertia ellipses (see above), as follows:

$$O_{i,j} = \frac{A_{E_i} \cap E_j}{A_{E_i} \cup E_j}$$

where E_i and E_j are the inertia ellipses of species i and j respectively and A is the area. The index varies from 0 when the two ellipses are totally separated to 1 when the two ellipses are identical.

At the intermediate scale, we used the same index but calculated from the inertia ellipses associated with the patches of the monospecific distributions:

$$O_{i,j} = \frac{\sum_{P_i} \sum_{P_j} A_{Ep_i} \cap Ep_j}{A \bigcup_{P_i, P_j} Ep_i \, Ep_j}$$

where E_{pi} and E_{pj} are the ellipses associated with the patches of species i and j respectively.

To characterize the co-occurence of two species at the local scale, we used two different indices: (*i*) an index of co-occurrence defined as a proportion of sampling points where both species co-occur (taking into account only the sampling points where at least one of the two species is present), and (*ii*) a local index of collocation (LIC; [39]) defined as:

$$LIC_{ij} = \frac{\int z_i(x)\, z_j(x)\, dx}{\sqrt{\int z_i^2(x)\, dx}\sqrt{\int z_j^2(x)\, dx}}$$

where z_i (x) and z_j (x) are the biomass density of the species i and j in the location x. Both indices vary between 0 (no location where the 2 species are found simultaneously) and 1 (the two species are situated in exactly the same locations). Yet, they differ in the fact that the former is based only on presence/absence while the second is based on biomass density.

To evaluate the deviation of these indices from random expectation, we used a randomization test by non-parametric bootstrap of 1000 values. For species of strong local overlap (i.e. anchovies and sardines), we also calculated a dominance index based on the relative difference in biomass of the two species in each point (Table 2). This index varies between -1, where only the species j is present and 1 where only the presence i is present, 0 corresponding to a situation where species i and j are equally abundant [40]. This index enabled us to better represent which species dominate where.

Softwares. Statistics were conducted using R v. 2.15.0 [41]. Geostatistical analyses were performed with the package RGeoS [42]. Spatial data are given in WGS84 coordinate system. Analyses were conducted on both raw and kriged data. As both methods yielded very similar results, we only presented results on raw data for clarity purposes. Kriging was produced only for mapping purposes.

Results

Aggregation patterns and spatial structure

Species were absent from several sampled locations, regardless of whether anchovies (absent from 9.3% of sampled locations), sardines (12.9%) or sprats (51.0%) were considered. However, these absences were not spatially consistent across years. When pooling all data, anchovies and sardines could be observed at least once in all sampled locations; sprats in contrast were never found after the 200 m isobath (Fig. S1 in File S1). During each of the 10 study-years, sardines and anchovies occupied most of the Gulf of Lions, and presence area only varied slightly for these two species (PA: 85.0 to 94.7% and 67.2 to 99.6% for anchovies and sardines, respectively). Variability was far greater for sprats, which were almost completely absent in 2007 (PA = 0.3%), but covered most of the Gulf of Lions in 2012 (PA = 91.8%). Presence area was positively correlated with total log-biomass index for sprats (LM: $R^2_{adj} = 0.87$, $p<0.001$), but not for anchovies ($p = 0.68$) or sardines ($p = 0.10$) (Figure 1), *i.e.* the area occupied by the population expanded with total logbiomass in sprats but not in the other two species. Log-biomass and presence area were far less variable for sardines and anchovies than for sprats (CVs being at least 4 to 7 fold smaller for PA and biomass, respectively). The mean density in presence area also increased with log-biomass for anchovies ($R^2_{adj} = 0.71$, $P = 0.001$) and sprats ($R^2_{adj} = 0.69$, $P = 0.002$). The same trend was observed in sardines, but the probability was lower ($R^2_{adj} = 0.27$, $P = 0.07$).

The degree of aggregation appeared to be different in anchovies compared to sprats or sardines (Figure 1; Figure S2 in File S1). Anchovies were, on average, less concentrated than the two other species. This was confirmed by the space selectivity index, which was lower in anchovies (0.722) than in sardines and sprats (0.850 and 0.867 respectively). Over the 10 study-years, yearly space selectivity indices decreased significantly with increasing log-biomass in anchovies (LM: $R^2_{adj} = 0.43$, $P = 0.02$,) and in sprats ($R^2_{adj} = 0.94$, $P<0.001$), but not in sardines ($P = 0.41$, Figure 1).

Mean variograms of log biomass exhibited spatial structure for the three species in both directions (i.e, increasing variance when the distance between sampling points increased before levelling off; Figure 2), meaning that the spatial distribution was not random and that two close points had a higher probability to have similar values than distant points. While sardine and anchovy variograms shared similar main characteristics, sprat mean variogram differed slightly. In particular, it differed in the range at which variance stabilized (Figure 2). Variance stabilized rapidly in sprats (range = 4 and 12 nm respectively for the first and second spherical component), while the first structure appeared around 15 nm for sardines and anchovies and the second was not apparent at the scale of our study, *i.e.* variance did not stabilize completely within the 50 nm limit that we fixed. Further, the mean structure seemed similar between the two perpendicular directions in sprats, while it differed slightly in anchovies and sardines, though the mean variogram of each direction was mostly included in the confidence interval around the mean variogram of the other direction.

Annual variograms were computed for each of the three species (Figure 2). All exhibited clear spatial structure and were included in the 95% confidence interval around the mean variogram for anchovies and sardines (see Figure 2). This revealed the absence of a year effect on the spatial structure of these 2 species. In sprats however, some variograms (especially along transect) were not included in the confidence interval, suggesting interannual variability in spatial structure.

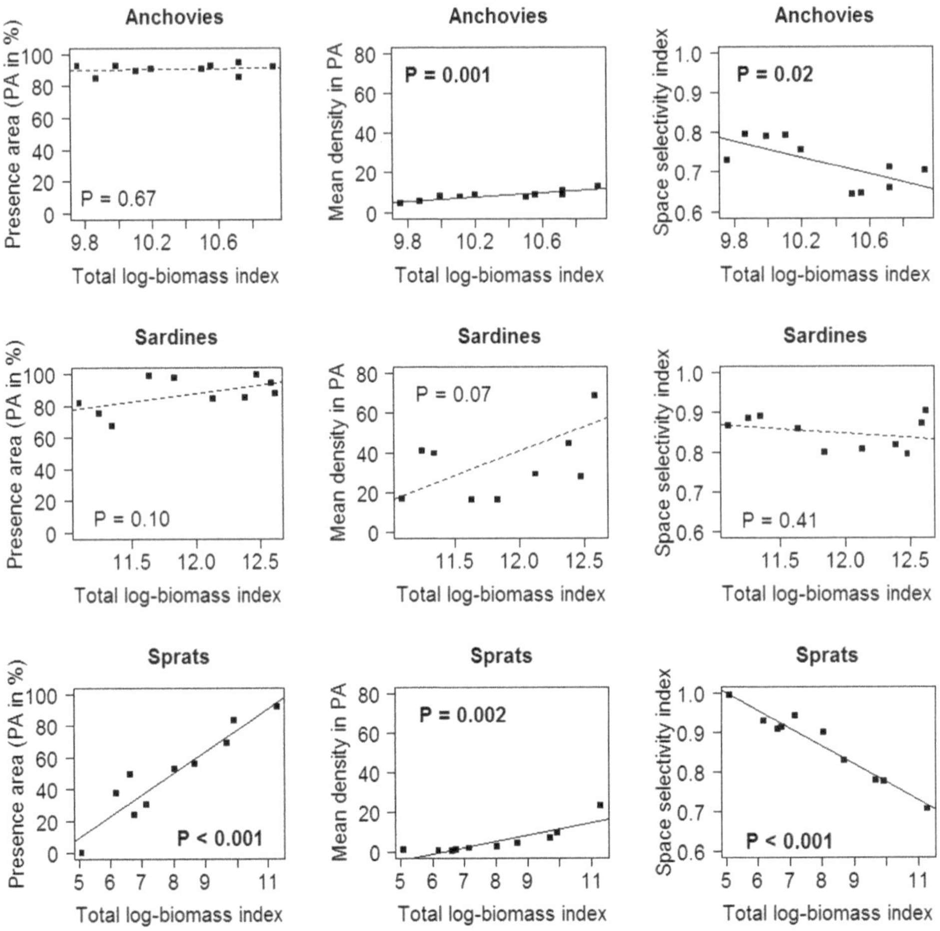

Figure 1. Presence areas, biomass densities and space selectivity indices relatively to total log-biomass indices in anchovies, sardines and sprats. Lines represent the linear regressions. Significant linear regressions are represented by plain lines and Pvalues are indicated in bold. For non-significant relationships, the trend is shown by a dotted line.

Species spatial distribution

Log-biomass distributions. The annual spatial distributions exhibited similarities between species in some years (*e.g.* in 2011 a large area in the middle of the shelf was unoccupied both for anchovies and sardines), but not in others (Figures S3, S4, S5 in File S1). Some consistent differences between species also appeared. For instance, anchovies occupied the centre of the continental shelf, while sardines were more coastal. For the three species, kriged annual distributions also revealed the important interannual variability of these populations both in terms of biomass levels and their spatial repartitions.

Centres of gravity of spatial distributions did not vary much according to species or year and were situated close to the geometrical centre of gravity of the sampled area (8.2±0.8 nm). Interestingly, the associated inertia was high for anchovies and sardines (inertia = 1152±70 nm^2), meaning that the mean distance between a presence point and the centre of gravity of its population was 34 nm. This distance was remarkably close to that between a sampling point and the geometrical centre of gravity of the total area (37 nm). As a consequence, the ellipse associated with this inertia covered most of the sampled area for these two species (Figure S6 in File S1), *i.e.* the populations were dispersed in the entire Gulf of Lions. In sprats, while inertia was quite low in the first 6 years (447±101 nm^2; Figure S6 in File S1) due to their low presence, it reached values almost as high as the ones in sardines and anchovies, in the last 4 years (991±75 nm^2).

The number of patches present per year and per species varied between 1 and 4 (Figure 3). As inferred by the square root of the inertia associated with the patch, the mean distance between a presence point in a patch and its centre of gravity ranged from 6 nm to 17 nm (excluding 2007 in sprats for which the species had only been observed in 1 location). The centres of gravity of anchovy patches were slightly more offshore than those of sardines (14.9±0.8 vs. 11.1±1.0 nm, Wilcoxon test: W = 662, P = 0.001), confirming the tendency of sardines to be more coastal than anchovies.

Instability of the spatial distribution. EOF of the 10 annual maps of each species confirmed the visual inspection highlighting high temporal variations in the spatial distributions. The first component of the EOF explained no more than 20% of the variance (14%, 20% and 17% for anchovies, sardines and sprats, respectively). Further, no trend could be detected from the contributions of each year to the first 2 axes of the EOF. The sign of the correlations between annual maps and the two first axes varied depending on the year for the three species (Figure S7 in File S1).

Coefficients of variation (calculated in each sampling point) were higher in sprats than in sardines and anchovies. Most coefficients of variation were higher than 1 (100%, 92.3% and

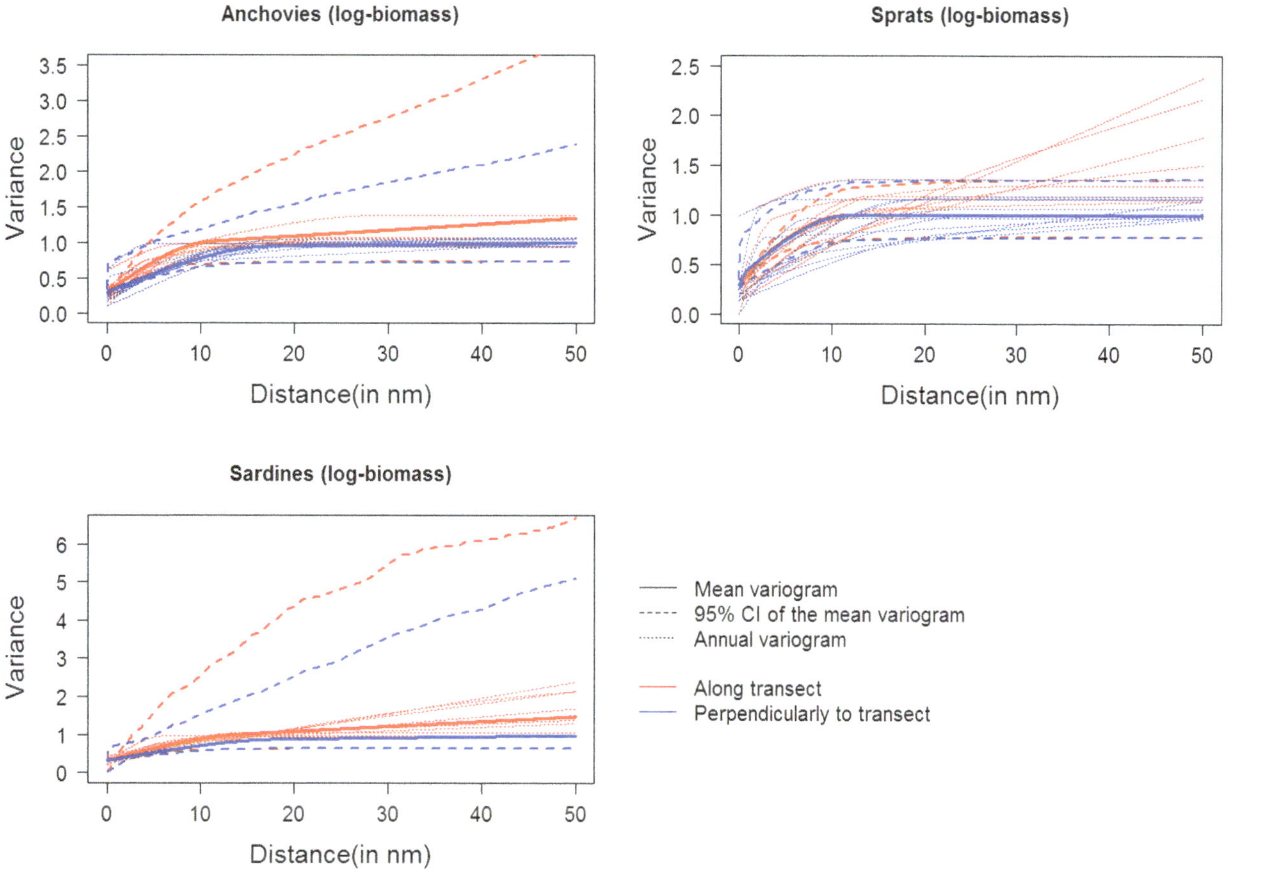

Figure 2. Annual (black) and mean (red) modelled variograms of anchovies, sardines and sprats. The red dotted lines correspond to the 95% confidence interval of the mean variogram deduced from 500 simulations.

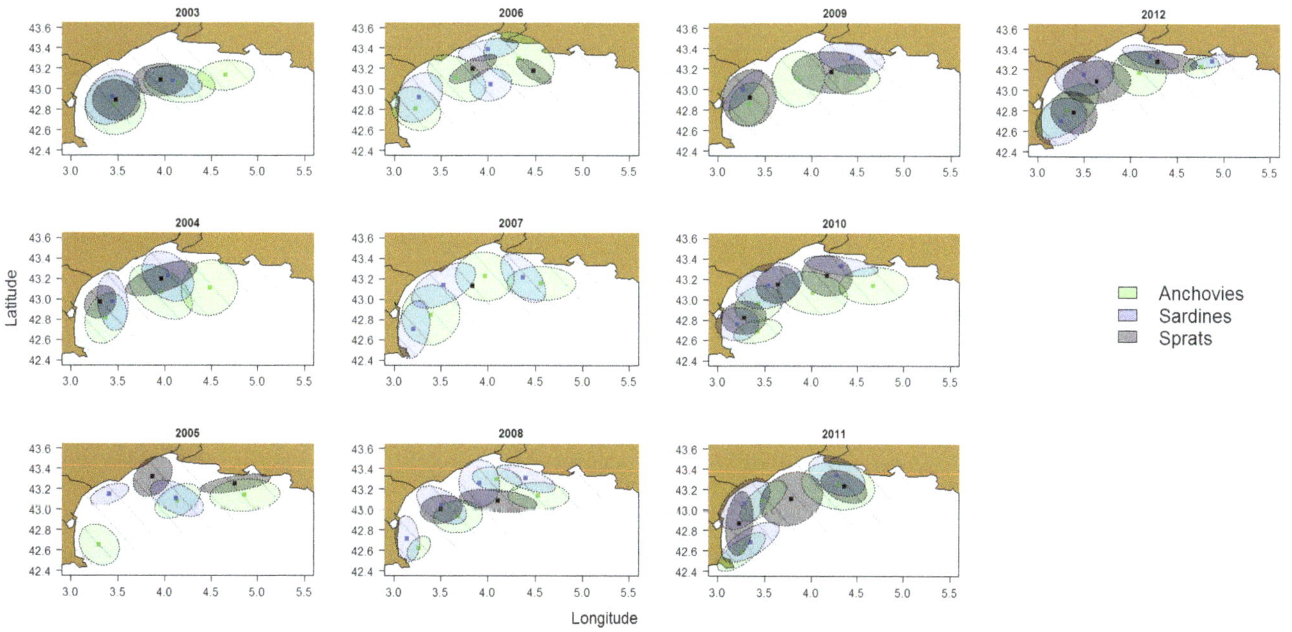

Figure 3. Annual maps of centres of gravity and inertia of patches for anchovies (in green), sardines (in blue) and sprats (in black).

81.6% of CV values were higher than 1 in sprats, sardines and anchovies, respectively), meaning that the distributions were overdispersed and again suggesting high temporal variability of the spatial distributions for all three species.

Recurrent, occasional and unfavourable areas. Average maps confirmed the tendency for sardines to be quite coastal, while anchovies occupied most of the shelf (Figure 4). Sprats seemed in an intermediate position where high biomass areas were situated in the centre of the Gulf of Lions, neither too coastal nor too offshore, but, as noted above, sprat biomass was also more variable. The combination of average and variability maps enabled us to detect a recurrent (persistent) area for anchovies in the centre of the Gulf, slightly West of the Rhone estuary, and, for sardines, in western coastal areas. Recurrent areas were minimal for sprats as the period encompassed some years with a quasi-absence of this species. Finally, deep waters represented unfavourable areas both for sardines and sprats.

Interspecific indices: collocation index and overlap between species

Regardless of the scale at which we investigated it, collocation indices were higher between sardines and anchovies than between any of these 2 species and sprats (Table S1 in File S1), meaning that sardines and anchovies co-occurred more often than they did with sprats. Additionally, collocation indices were higher, on average, at the global scale than at the intermediate scale (*e.g.* between anchovies and sardines 0.53 ± 0.04 *vs.* 0.30 ± 0.04 for the global and intermediate scales, respectively), suggesting that species globally lived in the same areas, but were not always found together in each given sampling location. Collocation indices, whether global, local or intermediate, varied substantially from year-to-year (Table S1 in File S1). Collocation indices at the three different scales were not correlated with each other, except for the global and intermediate indices between sprats and anchovies. For instance in 2004, the index of collocation between sprats and sardines was fairly high at the global scale (2nd highest observed), while the associated LIC was low and not significantly different from random expectation. The two local collocation indices were not correlated for anchovy and sardine association or anchovy and sprat association ($\rho = -0.61$, $P = 0.07$; $\rho = 0.44$, $P = 0.20$ respectively), suggesting that the co-occurrence of species (i.e. species found in the same location), did not necessarily concur with a co-occurrence of their hotspots or peaks of biomass. They were however slightly correlated when looking at the association between sardines and sprats ($\rho = 0.72$, $P = 0.02$).

Finally, although the co-occurrence between sardines and anchovies was high (Table S1 in File S1), in most sampled points one of the two species clearly dominated the other (Figure 5). The

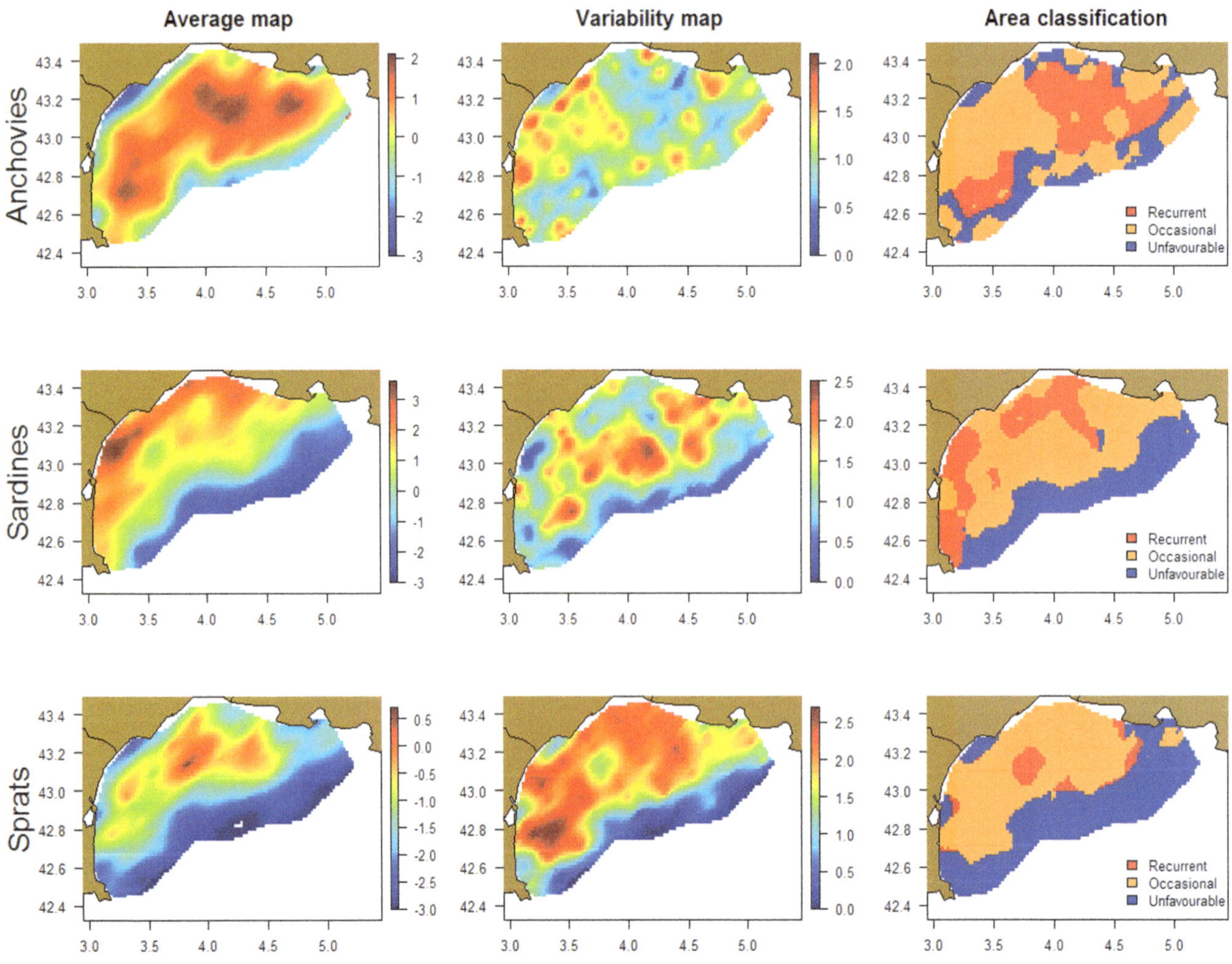

Figure 4. Average and variability maps and area classification for anchovies, sardines and sprats.

dominance index increased with depth, *i.e.* anchovies dominated sardines in deeper offshore waters, while sardines were dominant in shallow waters close to the coast (see also statistics in Figure S8 in File S1). Besides, the dominance index also increased with longitude, anchovies dominating towards the East (see statistics in Figure S8 in File S1).

Discussion

The strong increase in sprat biomass (from almost 0 to 70,000 tons from 2007 to 2012) that has paralleled a strong decline in anchovy and sardine biomass over the past few years [19], highlighted the urgent need to better understand the cross and mutual dynamics of those three species. This study documents the spatio-temporal patterns observed in these three species in the Gulf of Lions. However, it should be noted that surveys were conducted only in summer, corresponding to peak reproduction of anchovies but resting period for sardines and sprats. The results presented here thus correspond to different biological phases for the three species and should not be extrapolated to other non-monitored seasons.

Our main objective was to investigate the temporal variability of biomass and spatial distribution in the three species. Biomass estimates obtained by direct acoustic stock assessment substantially varied during the 10 years of survey. However, sprat was the only species in which a clear trend in biomass could be identified over the study period. For anchovies and sardines, biomass variability corresponded, respectively, either to fluctuations around a low value or a succession of an increasing and decreasing trends. The increase in sprat biomass resulted in both an increase of fish local densities and an expansion of fish towards new areas, in agreement with the McCall basin hypothesis [9]. On the contrary, the Gulf of Lions being a favourable area for small pelagics [43], sardine and anchovy presence area was large and stable over the entire study period and fluctuations in biomass only resulted in fluctuations of local densities. Changes in local densities could be a buffer for fluctuating biomasses of already well implanted species. However, if a species continuously increase or decrease its biomass, this might well not be enough anymore. For instance, sprats which were almost absent from the area at the start of the study had to expand their presence area to cope with their increasing biomass. Inversely, a declining population would see local areas disappear from its presence area as these species are gregarious and have to stay in shoals. Our results may thus support the hypothesis that interannual variability may not be enough to observe contraction/expansion, and that a directional trend would be needed to investigate the link between abundance and distribution [44–45]. Still, using space selectivity indices to investigate aggregation patterns in more details, we found that sprats and anchovies became less selective with increasing biomass, suggesting that they selected less suitable areas rather than increased their local density. This is in contrast with sardines for which the space selectivity index did not vary according to total biomass and densities 5 fold higher could be reached. This could suggest different social constraints/benefits, sardines being able to aggregate in much denser shoals than the other 2 species.

Similarly to what was found in demersal species [12], we highlighted the stability of the spatial structure (i.e. spatial autocorrelation) of small pelagic fish in the Gulf of Lions over time. Only for sprats did some interannual variability appear, but this is probably due to a reduced confidence interval around the mean variogram due to low and distance-independent variance between sampling points in the first half of the study, when sprats were quasi-absent. Spatial structure is defined as the spatial

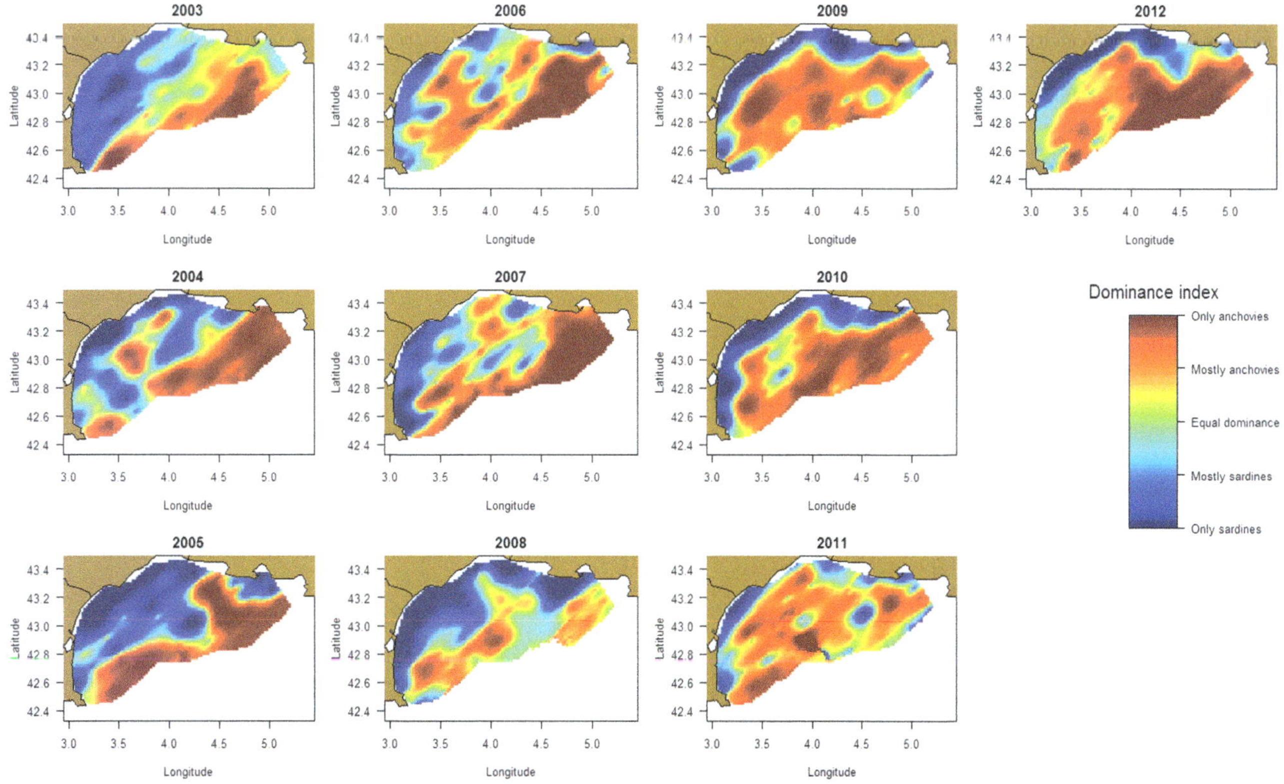

Figure 5. Annual maps of relative biomass between anchovies and sardines.

autocorrelation between values at different locations, i.e. in our case whether biomasses taken at 2 close points were more correlated than biomasses of points situated far apart. From an ecological point of view, spatial structure could thus indicate the occurrence of concentration in certain areas due to similar needs (e.g. habitat preferences) or aggregation patterns due to species social constraints (formation of shoals for instance here). Such stability of the spatial structure could result from important aggregative behaviours of small pelagic species, giving them intrinsic aggregation properties, independently from their abiotic environment.

Though necessary, the study of the spatial structure offers no insight on the locations where these aggregations took place or where the population was absent. In a second step, we thus investigated temporal changes in the geographic locations of biomass. In contrast to the stable spatial structure, the geographical distributions of small pelagics were highly variable from year-to-year. Although such variability is clearly visible when comparing the maps, it is more difficult to quantify it. Therefore, we resorted to spatial indicators (CVs and empirical orthogonal functions) for summarizing annual information and exploring possible spatial shifts in population distribution [35;46]. These indicators confirmed the visual inspection of the annual maps (high interannual variability) and thus the expected mobility of small pelagic fish, which contrasted to the stable and area-specific spatial distributions of demersal species of the same area [12]. Such result was to be expected, as demersal species are restricted to the bottom and very dependent on substrates or bathymetry. By contrast, small pelagic fish live in the water column, are much more mobile and should be affected mostly by dynamic environmental factors, which change from year to year. Previous studies on small pelagic fish habitats conducted in the Mediterranean [43] have identified the Gulf of Lions as a potential favourable habitat for sardines and anchovies. In this smaller-scale study, we tried to identify whether there could be within Gulf variations in habitat quality for small pelagic fish. From the presence/absence data pooled over the entire study period, we saw that the three species could use the entire Gulf of Lions (except past the 200 m isobath for sprat). Nonetheless, it was clear that spatial distributions were not uniform over the Gulf and that they exhibited more than one patch of high biomass. The location of patches varied across years and between species, suggesting that they reacted to some external variable factors. In the future, linking spatio-temporal distributions to environmental variables at a minute scale while accounting for aggregation structures should help in understanding the environmental drivers affecting the population dynamics of those three pelagic species. Previous studies on environmental drivers of small pelagic fish habitats have already been conducted at a large scale in the Mediterranean, showing the importance of temperature and chlorophyll concentration as forcing factors of the spatial distribution (*e.g.* [43]). However, processes that drive the small pelagic populations may be different at smaller scales (i.e. within the Gulf of Lions). In particular, the relative importance of density-dependent *versus* density-independent (environmental) processes on spatial distributions is likely to be dependent on the scale of the study [47]. A study accounting for both density-dependent and independent variables is thus of necessity to understand the identified areas [13].

Despite the important variability in spatial distributions, we defined unfavourable areas close to the coast for anchovies and on the 150–200 m-deep stratum for sprats and sardines. In contrast, we could assess recurrent areas (high average biomass and low variability) for sardines and anchovies in the Gulf of Lions. For sardines, recurrent areas were situated near the coast and in the Western part of the Gulf of Lions. For anchovies, two recurrent areas were identified further from the coast (isobaths of ~70 to 100 m). One was situated on the west part while the other one could be associated with the Rhone river and its plume on the east. As the surveys occurred during peak reproduction for anchovies [23], it is likely these areas corresponded to the spawning grounds of anchovies in this region. Interestingly, these 2 areas are very similar to the spawning areas detected from egg surveys in the 60 s [48]. This gives us important insights on their spawning grounds, exhibiting their stability despite several changes in external variables between these 2 periods (e.g. fishing effort and captures, environment). Also, such information is important for the elaboration of spatially explicit management plan (e.g. MPA, etc.). Indeed, it shows that despite the high mobility of small pelagic fish and high interannual variability in peak biomass locations, some areas consistently offer favourable spawning grounds for this species, making the task easier if one wanted to protect them.

Finally, we investigated interspecific relationships between 3 species sharing similar trophic level (i.e. zooplankton and especially copepods constitute the bulk of their diet at that period, though phytoplankton is also consumed in particular by sardines [21–22]). Different results were obtained according to the scale at which it was studied. Indeed, the index of overlap that we proposed decreased from the global (the population envelop) to the intermediate scale, indicating that if the three species co-habited in the Gulf of Lions, their hotspots were not always situated in the same area. This suggests that global environmental factors may have driven the three species to inhabit the Gulf of Lions, but that local environmental factors and/or inter-specific competition may have resulted in segregation within the Gulf. This was confirmed by the difference obtained between the two local collocation indices used. The collocation index translating species co-occurrence was surprisingly high (especially for anchovies and sardines), revealing that at a 2-nm sampling scale, species almost always co-occurred. The LIC index (based on density rather than presence/absence) had lower values, suggesting that for most sampled locations a given species predominantly occurred in each location (as also indicated by the dominance index). In particular, sardines clearly dominated the first depth strata (0–50 m) and the western part of the Gulf of Lions, while anchovies were more abundant in deeper waters and towards the central and eastern part of the Gulf. Collocation analyses give us interesting information on potential competition of these three planktivorous species, as their co-occurrence is a potential source of competition if/when food becomes limiting.

In summary, this study exhibited the relatively stable aggregation structure of small pelagic species, probably due to their inherent social structure (shoals, etc.). In contrast, it confirmed that the three species were highly mobile and could inhabit every stratum of the Gulf of Lions, though some preferred and unfavourable habitats could be highlighted for each species. The environmental drivers of both these habitats and the variability in locations of peak biomass remain to be investigated to better understand population dynamics. Finally, differences between species aggregative behaviour and habitat preferences were also highlighted, exhibiting potential competition avoidance.

Supporting Information

File S1 Supplementary material. Figure S1. Presence of anchovies, sardines and sprats in the Gulf of Lions. **Figure S2.** Annual aggregation curves for each species. **Figure S3.** Annual maps of log-biomass for anchovies. **Figure S4**. Annual maps of log-biomass for sardines. **Figure S5**. Annual maps of log-biomass for sprats. **Figure S6**. Centres of gravity and inertia. **Figure S7**. Empirical Orthogonal Function analysis. **Figure S8**. Dominance index depending on depth or longitude strata. **Table S1**. Yearly and global collocation indices at three different time scales.

Acknowledgments

We are very grateful to the crew of the Ifremer RV *L'Europe* and everyone who participated to the PELMED surveys between 2003 and 2012. PELMED surveys are cofunded by IFREMER and the European Union through the Data Collection Framework (DCF). This study is part of the program EcoPelGol (Study of the Pelagic ecosystem in the Gulf of Lions), funded by France Filière Pêche (FFP). We also wish to thank Vincent A. Viblanc for his helpful suggestions and his English-proofing of the manuscript. Finally, we thank 2 anonymous reviewers for their helpful comments.

Author Contributions

Conceived and designed the experiments: CS JLB DR. Performed the experiments: JLB JHB DR EVB CS. Analyzed the data: CS NB JMF. Contributed reagents/materials/analysis tools: CS MM. Wrote the paper: CS JMF NB. Made useful suggestions on the manuscript: MM EVB.

References

1. Woillez M, Petitgas P, Rivoirard J, Fernandes P, Hoftstede Rter, et al. (2006) Relationships between population spatial occupation and population dynamics. ICES CM 2006/O.05
2. Agostini VN, Hendrix AN, Hollowed AB, Wilson CD, Pierce SD, et al. (2008) Climate-ocean variability and Pacific hake: a geostatistical modeling approach. Journal of marine systems 71: 237–248.
3. Bacheler NM, Bailey KM, Ciannelli L, Bartolino V, Chan KS (2009) Density-dependent, landscape and climate effects on spawning distribution of walleye Pollock Theragra chalcogramma. Marine Ecology Progress Series 391: 1–12.
4. Cardinale M, Hagberg J, Svedang H, Bartolino V, Gedamke T, et al. (2010) Fishing through time: population dynamics of plaice (Pleuronectes platessa) in the Kattegat-Skagerrak over a century. Population Ecology 52: 251–261.
5. Boulinier T, Danchin E (1997) The use of conspecific reproductive success for breeding patch selection in terrestrial migratory species. Evolutionary Ecology 11: 505–517.
6. Hixon MA, Anderson TW, Buch KL, Johnson DW, McLeod JB, et al. (2012) Density dependence and population regulation in marine fish: a large-scale, long-term field manipulation. Ecological Monographs 82: 467–489.
7. Petitgas P, Alheit J, Peck MA, Raab K, Irigoien X, et al. (2012) Anchovy population expansion in the North Sea. Marine Ecology Progress Series 444: 1–13.
8. Lewis S, Sheratt TN, Hamer KC, Wanless S (2001) Evidence of intra-specific competition for food in pelagic seabird. Nature 412: 816–819.
9. MacCall AD (1990) Dynamic Geography of Marine Fish Populations. Seattle: University of Washington Pr. 153 p.
10. Paloheimo JE, Dickie LM (1964) Abundance and fishing success. Rapport et Proces Verbaux des Reunions du Conseil International pour l'Exploration de la Mer 155: 152–163.
11. Bertrand A, Segura M, Gutiérrez M, Vásquez L (2004) From small-scale habitat loopholes to decadal cycles: a habitat-based hypothesis explaining fluctuation in pelagic fish populations off Peru. Fish and Fisheries 5: 296–316.
12. Morfin M, Fromentin JM, Jadaud A, Bez N (2012) Spatio-temporal Patterns of Key Exploited Marine Species in the Northwestern Mediterranean Sea. PlosOne 7 (5): e37907.
13. Ciannelli L, Bartolino V, Chan KS (2012) Non-additive and non-stationary properties in the spatial distribution of a large marine fish population. Proceedings of the Royal Society B 279: 3635–3642.
14. Ciannelli L, Fisher JAD, Skern-Mauritzen M, Hunsicker ME, Hidalgo M, et al. (2013) Theory, consequences and evidence of eroding population spatial structure in harvested marine fishes: a review. Marine Ecology Progress Series 480: 227–243.
15. Apostolaki P, Milner-Gulland EJ, McAllister MK, Kirkwood GP (2002) Modelling the effects of establishing a marine reserve for mobile fish species. Canadian Journal of Fisheries and Aquatic Sciences 59: 405–415.
16. Cury P, Bakun A, Crawford RJM, Jarre-Teichmann A, Quinones R, et al. (2000) Small pelagics in upwelling systems: patterns of interaction and structural changes in "wasp-waist" ecosystems. ICES J. Mar. Sci. 57: 603–618.
17. Bakun A (1996) Patterns in the ocean: ocean processes and marine population dynamics. University of California Sea Grant, San Diego, California, USA in cooperation with Centro de Investigaciones Biologicas de Noroeste, La Paz, Baja California Sur, Mexico. 323 pp.
18. Checkley DM, Ayon P, Baumgartner TR, Bernal M, Coetzee JC, et al. (2009) Habitats. In: Climate Change and Small Pelagic Fish. A. Checkley, J. Alheit, Y. Oozeki & C. Roy (eds) New York: Cambridge University Press
19. General Fisheries Commission for the Mediterranean (2012) Report of the working group on stock assessment of small pelagic species. Split, Coratia.
20. Van Beveren E, Bonhommeau S, Fromentin JM, Bigot JL, Bourdeix JH, et al. (2014) Rapid changes in growth, condition, size and age of small pelagic fish in the Mediterranean. Marine Biology 161: 1809–1822.
21. Costalago D, Navarro J, Álvarez-Calleja I, Palomera I (2012) Ontogenetic and seasonal changes in the feeding habits and trophic levels of two small pelagic fish species. Marine Ecology Progress Series 460: 169–181
22. Nikolioudakis N, Isari S, Somarakis S (2014) Trophodynamics of anchovy in a non-upwelling system: direct comparison with sardine. Marine Ecology Progress Series 500: 215–229.
23. Palomera I, Olivar MP, Salat J, Sabates A, Coll M, et al. (2007) Small pelagic fish in the NW Mediterranean Sea: An ecological review. Progress in Oceanography 74: 377–396.
24. Weill A, Scalabrin C, Diner N (1993) MOVIES-B: An acoustic detection description software. Application to shoal species classification. Aquatic Living Resources, 6: 255–267
25. Simmonds J, MacLennan D (2005) Fisheries Acoustics, Theory and Practice, 2nd edn. Oxford: Blackwell Publishing, 437 pp.
26. Doray M, Masse J, Petitgas P (2010) Pelagic fish stock assessment by acoustic methods at Ifremer, Internal report, pp.18, *epic.awi.de/33739/1/ifremer-acoustic-methods.pdf*.
27. Matheron G (1963) Principles of geostatistics. Economic Geology 58: 1246–1266.
28. Dungan JL, Perry JN, Dale MRT, Lengendre P, Citron-Pousty S, et al. (2002) A balanced view of scale in spatial statistical analysis. Ecography 25: 626–640.
29. Gini C (1921) Measurement of inequality of income. Economic Journal 31: 22–43.
30. Petitgas P (1998) Biomass-dependent dynamics of fish spatial distributions characterized by geostatistical aggregation curves. ICES Journal of Marine Science 55: 443–453.
31. Myers RA, Cadigan NG (1995) Was an increase in natural mortality responsible for the collapse of northern cod? Canadian Journal of Fish and Aquatic Science 52: 1274–1285.
32. Rivoirard J, Simmonds J, Foote KG, Fernandes P, Bez N (2000) Geostatistics for estimating fish abundance. Blackwell Publishing, Oxford.
33. Chilès J-P, Delfiner P (1999) Geostatistics: Modeling Spatial Uncertainty. 1st ed. New York: Wiley-Interscience. 720 p.
34. Bez N, Rivoirard J (2001) Transitive geostatistics to characterise spatial aggregations with diffuse limits: an application on mackerel ichtyoplankton. Fisheries Research 50: 41–58.
35. Woillez M, Poulard J-C, Rivoirard J, Petitgas P, Bez N (2007) Indices for capturing spatial patterns and their evolution in time, with application to European hake (Merluccius merluccius) in the Bay of Biscay. ICES Journal of Marine Science 64: 537–550.
36. Lorenz EN (1956) Empirical orthogonal functions and statistical weather prediction. Department of Meteorology, MIT, Scientific Report n°1.
37. Korres G, Pinardi N, Lascaratos A (2000) The Ocean Response to Low-Frequency Interannual Atmospheric Variability in the Mediterranean Sea. Part II: Empirical Orthogonal Functions Analysis. Journal of Climate 13: 732–745.
38. Bellier E, Planque B, Petitgas P (2007) Historical fluctuations in spawning location of anchovy (Engraulis encrasicolus) and sardine (Sardina pilchardus) in the Bay of Biscay during 1967–73 and 2000–2004. Fisheries Oceanography 16: 1–15.
39. Bez N, Rivoirard J (2000) Indices of collocation between populations. In: Checkley DM, Hunter JR, Motos L, von der Lingen CD (editors).Workshop on the Use of Continuous Underway Fish Egg Sampler (CUFES) for mapping spawning habitat of pelagic fish. GLOBEC Rep., pp. 48–52.
40. Barange M, Coetzee JC, Twatwa NM (2005) Strategies of space occupation in anchovy and sardine in the southern Benguela: the role of stock size and intra-species competition. ICES Journal of Marine Science 62: 645e654.
41. R Development Core Team (2012) R: A language and environment for statistical computing. R Foundation for Statistical Computing, Vienna, Austria.
42. Renard D, Bez N, Desassis N, Beucher H, Ors F (2013) RGeoS: Geostatistical Package. R package version 9.1.6. (Ecoles des Mines website, available at http://cg.ensmp.fr/rgeos, accessed 2014 October 10th).
43. Giannoulaki M, Iglesias M, Tugores MP, Bonanno A, Patti B, et al. (2013) Characterizing the potential habitat of European anchovy Engraulis encrasicolus in the Mediterranean Sea, at different life-stages. Fisheries Oceanography 22: 69–89.

44. Gaston KJ, Blackburn TM, Gregory RD (1999) Intraspecific abundance–occupancy relationships: case studies of six bird species in Britain. Div. Distr. 5: 197–212. (doi:10.1046/j.1472-4642.1999.00054.x)
45. Fisher JAD, Frank KT (2004) Abundance–distribution relationships and conservation of exploited marine fishes. Mar. Ecol. Prog. Ser. 279: 201–213. (doi:10.3354/meps279201)
46. Atkinson DB, Rose GA, Murphy EF, Bishop CA (1997) Distribution changes and abundance of northern cod (*Gadus morhua*), 1981–1993. Canadian Journal of Fisheries and Aquatic Sciences 54: 132–138.
47. Shepherd T, Litvak M (2004) Density-dependent habitat selection and the ideal free distribution in marine fish spatial dynamics: considerations and cautions. Fish and Fisheries 5: 141–152.
48. Aldebert Y, Tournier H (1971) La reproduction de la sardine et de l'anchois dans le Golfe du Lion. Revue de Travail de l'Institut des Pêches maritimes 35: 57–75.

16

Landscape Management of Fire and Grazing Regimes Alters the Fine-Scale Habitat Utilisation by Feral Cats

Hugh W. McGregor[1,2]*, Sarah Legge[3], Menna E. Jones[1], Christopher N. Johnson[1]

1 School of Biological Sciences, University of Tasmania, Hobart, Tasmania, Australia, **2** Australian Wildlife Conservancy, Derby, Western Australia, Australia, **3** Australian Wildlife Conservancy, Mornington Wildlife Sanctuary, Derby, Western Australia, Australia

Abstract

Intensification of fires and grazing by large herbivores has caused population declines in small vertebrates in many ecosystems worldwide. Impacts are rarely direct, and usually appear driven via indirect pathways, such as changes to predator-prey dynamics. Fire events and grazing may improve habitat and/or hunting success for the predators of small mammals, however, such impacts have not been documented. To test for such an interaction, we investigated fine-scale habitat selection by feral cats in relation to fire, grazing and small-mammal abundance. Our study was conducted in north-western Australia, where small mammal populations are sensitive to changes in fire and grazing management. We deployed GPS collars on 32 cats in landscapes with contrasting fire and grazing treatments. Fine-scale habitat selection was determined using discrete choice modelling of cat movements. We found that cats selected areas with open grass cover, including heavily-grazed areas. They strongly selected for areas recently burnt by intense fires, but only in habitats that typically support high abundance of small mammals. Intense fires and grazing by introduced herbivores created conditions that are favoured by cats, probably because their hunting success is improved. This mechanism could explain why, in northern Australia, impacts of feral cats on small mammals might have increased. Our results suggest the impact of feral cats could be reduced in most ecosystems by maximising grass cover, minimising the incidence of intense fires, and reducing grazing by large herbivores.

Editor: Paul Adam, University of New South Wales, Australia

Funding: The research was funded by the Australian Research Council, the Western Australian Department of Parks and Wildlife (Kimberley Science and Conservation Strategy), and supporters of the Australian Wildlife Conservancy. The funders had no role in study design, data collection and analysis, decision to publish, or preparation of the manuscript.

Competing Interests: The authors have declared that no competing interests exist.

* Email: hugh.mcgregor@australianwildlife.org

Introduction

Predator-prey relationships are strongly influenced by the structure and quality of habitat, principally its vegetation [1,2,3]. Variability in vegetation structure may be used by predators to increase hunting success. For example, lions use dense vegetation to hide their approach from prey [4]. Also, it may be used by prey to help them evade predators, such as elk using woodlands as a refuge from wolves [5]. Changes in habitat structure may therefore shift the relationships between predators and prey [6,7]. Such changes can determine the extent to which some prey are threatened with extinction by heavy predation [2,8].

One of most pervasive impacts on vegetation structure arises from changes to fire and grazing regimes. While drivers of such changes vary immensely, the impacts on fauna communities display some general trends. For example, small mammal populations are especially sensitive, and the vast majority of studies detecting declines in populations in response to either intense fire events or intense grazing [9,10,11,12,13], unless they occur in ecosystems with few predators [14,15]. The underlying mechanisms of these declines remain elusive, but are likely to be indirect rather than through direct effects such as being burnt by the fires or trampled by cattle [16]. Instead, such disturbances may improve habitat for predators in ways that increases their impacts on prey [9,17,18], although no field data are available to confirm this.

Many small mammal species are declining in the savannas of northern Australia, and several are threatened with extinction [18,19]. Declines have been greatest in areas subject to intense fires [20,21] and recent experimental evidence also supports an association of grazing by introduced herbivores (cattle, horses, donkeys, buffalo) with the magnitude of small-mammal decline [22]. Both fire and grazing regimes in northern Australia have intensified substantially over recent decades in ways that could contribute to the contemporary native mammal decline. These changes to fire and grazing have generally made grass communities less complex and more open [23].

Predation by feral cats *Felis catus* may also be contributing to the declines. This is suggested by three lines of evidence. First, the declining species fall within the preferred prey-size range of cats [18,24]. Second, mammal populations in complex rocky habitats have been less affected than those in more productive woodlands and savanna, suggesting a predation effect [25,26]. Finally, populations of declining mammal species are more stable in the absence of cats, on island or in large enclosures [27,28]. However, there is a temporal mismatch between the arrival of cats in northern Australia up to 170 years ago [29], and mammal declines observed in the last 20 years [19].

The apparent mismatch in timing of the early arrival of cats and recent mammal declines could be explained by the hypothesis that cat predation has its largest impact when it interacts with fire and grazing regimes established in the more recent past. Reduction of structural complexity of vegetation and increased openness due to fire and grazing might increase the exposure of small mammals to predators, making prey easier to detect and capture [15]. Small mammals are cats' preferred prey [24,30]. If cats preferentially use the open and relatively simple habitats created by fire and grazing, the result could be a higher predation impacts on small mammals. This has been suggested as a possible mechanism for these declines [19,31], however, until now there has been no evidence.

If cats do favour the conditions created by fire and grazing, this should be revealed by the patterns of movement of individual cats in heterogeneous landscapes with variable effects of fire and grazing. We tested this hypothesis using intensive GPS tracking of a large sample of individual cats; both within and outside of a large 40 300 hectare area that has been destocked of all introduced herbivores (cattle, horses, and donkeys) [22], and spanning contrasting fire patterns (mild control fires or intensive wildfires). A dynamic habitat map was created so that fire and vegetation attributes at any location or point in time could be determined. This was used to generate a parsimonious model of fine-scale habitat selection by cats. We predicted that feral cats would select for open grass cover to improve hunting success. If so, we hypothesise that cats would increase their use of habitats that have been recently burnt or intensely grazed, and that this relationship would be stronger in areas of high small-mammal abundance.

Methods

Study area

Our study area encompassed three large properties in the central Kimberley of north-western Australia (**17°01′S, 126°01′E**, see Fig. 1). One property is managed for commercial cattle production (Glenroy, 1455 km^2) and two are ex-pastoral leases managed for conservation by the Australian Wildlife Conservatory (Mornington and Marion Downs Sanctuaries, 3225 km^2 and 2676 km^2 respectively). Habitats are mostly savanna woodlands with a perennial grass layer, dissected by riparian vegetation along the edges of creeks. The region has a tropical monsoon climate with three broad seasons: the wet (December – March), early dry (April – July) and late dry (August – November). Fire is managed on all three properties to promote biodiversity values. This fire management aims to reduce the incidence of extensive, high-intensity, uncontrolled fires in the late dry season using strategic prescribed burning in the early dry season when fires tend to be small and of low intensity because of weather and condition of the grass layer. In addition, when uncontrolled late dry season fires occur, they are suppressed where possible. All large introduced herbivores (cattle, horses and donkeys) have been removed from a 40,300 ha fenced section of Mornington since 2005 [22], - hereafter referred to as the 'destocked' zone. The only other large mammalian predator in the study area is the dingo *Canis dingo*. Whilst they are controlled elsewhere in Australia, they are not persecuted in the study area, and occurred at a density of ~0.2 per km^2 [32].

Cat capture and tracking

Feral cats were captured between September 2010 and June 2013, using either large wire cage traps, leg-hold traps (soft-jaw, size #1.5) or by spotlighting and netting with the assistance of dogs trained to locate and bail cats up trees. If a cat was either bailed up a tree or required examination of possible injury, it was sedated with Zolotil at a rate of 0.5 cc/kg via intramuscular injection. Cats were fitted with GPS collars (Telemetry Solutions Quantum 4000 enhanced). Cats weighing between 2 and 3.3 kg were fitted with a 70 g collar, and those weighing more than 3.3 kg were fitted with a 100 g collar (<3% of body-weight). Sedated cats were released after full muscle control was regained (4–6 hours later), and non-sedated cats were released as soon as possible (2–5 minutes later). When it was necessary to replace GPS collars, the cats were recaptured using the dogs.

GPS collars were deployed on equal numbers of cats in the stocked and destocked zone, and between burnt and unburnt areas. Within burnt areas, the cats were split evenly between areas with low and high intensity fires (Table 1). The GPS units were programmed to record fixes every 15 minutes for two-day bouts, starting and finishing at 12 pm. These bouts were separated by intervals of one, two or fourteen days. All bouts were timed to commence at least 24 hrs after the cat was handled. Units were programmed to search for a satellite for 60 seconds, and to remain on for at least 5 seconds to refine the fix if there was memory from the last fix, or 15 seconds if not.

Habitat variables

Across the study area, we developed habitat maps relating to fire and grazing, along with any other variable likely to influence cat habitat selection or movement. Where necessary, the maps were updated to make them temporally dynamic, so that attributes at any given time and location could be determined. Most descriptors of habitat related to the ground layer, rather than the tree layer. Nine distinct grass communities common in the region (see Table 2) were mapped by examining colour disjunctions on aerial photos while altering light levels in Photoshop Elements v. 8, tracing the boundaries of discrete polygons onto the aerial photos, then geo-rectifying these boundaries in ArcGIS v.10. For example, spinifex *Triodia* spp. grasslands are uniquely green in dry-season aerial photos, while communities dominated by bluegrass *Dichanthium fecundum* are white. The digital map was ground-truthed at 768 plots (described later); the attribution of grass community was correct at 96% of sites.

Fire scars were initially mapped using monthly Landsat 7 remote-sensing imagery available from the US Geological Survey (2011–2013), and fire boundaries were then refined using aerial photography taken from a helicopter flying approximately 300 m above ground. For each burnt area, we assigned the date of burn and intensity (intense = 100% tree scorch and no ground cover remaining unburnt, or mild = all other fires). Relative to the date of each GPS fix, fire was considered in multiple binary variables at 30, 60, 90, 180, 360 or 600 days since fire. The Australian Wildlife Conservancy's stock-proof fence [22] separated the stocked and destocked areas (see Fig. 1).

A dynamic map was created that estimated grass cover at any given location and time since fire, based on a series of models of the response of the grass layer to fire (given grass community, stocking status). They were created from field data, where vegetation attributes were measured at 768 plots (each 10 m^2) across the study area and duration, spread equally across grass communities (see Table 1) and combinations of mild/intense fire, time since fire, and grazed and destocked areas (total of 96 plots per community). At each plot, we estimated the extent of grass

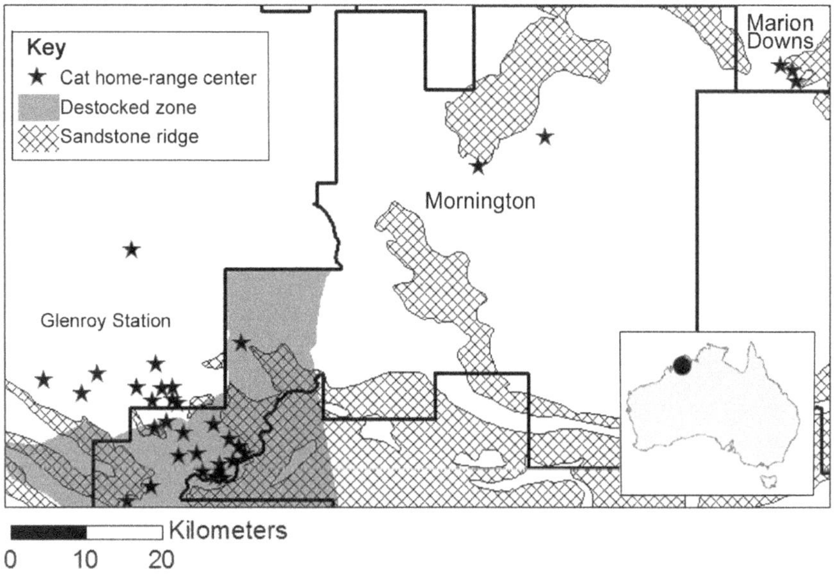

Figure 1. Map of study area in north-west Australia (see inset), including home-range centroids of feral cats used in this study. The dark grey represents the destocked zone.

cover at different heights by adapting a line-intercept method. We inserted a 100 cm pole (diameter of 1.5 cm) vertically through the grass to the ground at 50 points in a systematic grid over the plot. The number of grass intercepts was recorded in height intervals of 0–10 cm, 11–30 cm, and 31–100 cm. This was used to derive grass cover, cover of dense tussocks, and relative biomass at each plot (see Material S1 for more detail on methods). These variables were modelled against grass community, time since last fire, intensity of fire, and stocking status. Models were then used to derive values for all GPS fixes and random points used in discrete choice models (see Material S1).

The influence of grazing on grass biomass was measured as the difference in the average total number of grass intercepts per plot between grass communities in the stocked and destocked sites. We used only unburnt sites for this comparison. These averages were converted into a grazing impact rank. However, these variables would be confounded by correlation as the habitats favourable to cattle would also be favoured by feral cats (e.g. riparian areas), irrespective of impacts of cattle. Therefore, this score or rank was applied to all fixes, grazed or not, and the actual impacts of cattle grazing considered as the difference in this variable's strength between the destocked zone and outside.

Relative small-mammal abundances in different habitats were estimated from the Australian Wildlife Conservancy's annual fauna monitoring data, which is carried out across all three properties [32]. The sample at each site is based on 60 trap-nights using small mammal box traps (Elliott traps) and 12 nights of possum-sized wire traps, with 55 to 64 sites sampled each year - totalling 176 site-years between 2011–3. Small mammals were defined as those weighing between 30 g and 2000 g [18,24], and were predominantly *Rattus tunneyi* and *Pseudomys nanus* (these species comprised 87% of captures). For each year, the average rates of capture of individual small mammals were calculated for each grass community in the grazed and destocked areas (see Table 2). However, as sites were typically larger than the mapped patterns of these grass communities, results from some communities were combined (see Table 2). In the event cats altered their response to a threshold of small mammal abundance, we also included three binary variables of abundance (whether there was more than 1, 2 or 5 small mammal captures per plot) for each habitat.

We measured other spatial and temporal features likely to be important to cats to provide context for their movements. As cats may prefer hunting on edges between open and dense vegetation, we delineated a 25 m buffer either side of any linear boundary where grass cover was <25% on one side and >50% on the other (these were most often fire edges). Water features were mapped in the field, and each assigned a descriptor for the seasons that they contained water (wet season only, wet and early dry, or year-round). This was used to derive distance to known standing water at any given time throughout the study. Elevation and slope were derived from a digital elevation model of 15 m resolution. We

Table 1. The number of cats fitted with GPS collars in each of the different grazing and fire treatments. Destocked means that all introduced herbivores are excluded.

Management	No fire	Mild fire	Intense fire	Total
Stocked	8 (5♂ 3♀)	4 (3♂ 1♀)	4 (3♂ 1♀)	**16**
Destocked	8 (6♂ 2♀)	4 (4♂)	4 (4♂)	**16**
Total	**16**	**8**	**8**	**32**

Table 2. Vegetation communities mapped in the study region.

			Small mammal abundance					
		Grazing rank	Stocked			Destocked		
Grass community	Dominant grasses		2011	2012	2013	2011	2012	2013
Riparian forest[A]	*Chrysopogon fallax*	1	0.2	0.3	0.8	1.2	3.4	9.4
	Mnesithea rottboellioides							
	Various introduced grasses							
Alluvial grasslands[A]	*Crysopogon fallax*	2	0.2	0.3	0.8	1.2	3.4	9.4
	Dichanthium fecundum							
Bluegrass plains	*Aristida* spp.	3	1	1.3	4.6	0.3	2.6	2.9
	Dichanthium fecundum							
Canegrass[A]	*Mnesithea rottboellioides*	4	0.2	0.3	0.8	1.2	3.4	9.4
Mixed woodlands	*Aristida* spp.	5	0	1.3	0.6	0.2	3.8	6
	Dichanthium fecundum							
	Heteropogon contortus							
Sandseep	*Triodia* spp.	6	1.7	1.2	1.2	1.8	7.4	16.8
Hillside woodlands[B]	*Sehima nervosum*	7	0	0.6	1	0.4	3.2	4
Spinifex woodlands[B]	*Triodia* spp.	8	0	0.6	1	0.4	3.2	4
Bare ground	No or little	9	0	0	0	0.2	0.2	0.5

[A] and [B] denote grass communities where the small mammal abundance was grouped, as sites were typically larger than the mapped distributions of these communities.
Vegetation classes are ranked by their preference for grazing by domestic stock, with 9 the most impacted and 1 the least (Grazing rank). Small mammal abundance is number of small mammals captured per 100 trap nights in the stocked and destocked zone, from 2011 to 2013.

created a variable representing the spatial home-range context for each cat, by making a kernel density estimate (smooth cross-validation) for all moving fixes of each cat, and delineating kernel isopleths at 50%, 90%, 95% and 99% contours. Temporal variables included sun time [33], season (wet, early dry or late dry), number of months since the end of the wet season, and minimum nightly temperature (HOBO temperature data logger, Micro-DAQ). Finally, a cat's choice of which habitat to select in successive GPS fixes may be dependent on the type of habitat it was last in. This spatial correlation might arise when an animal selects for certain landscape features and tends to remain within them for some time, rather than making a *de novo* selection at every 15 min interval of their walk. Therefore, we fitted a binary 'carry-over' variable, which described whether the vegetation, grazing and fire values were equal to the previous fix.

Organisation of data

All GPS fixes that were likely to be erroneous, biased or to represent a stationary cat were removed from the analysis. Erroneous fixes were those representing implausible 'spikes' in movement, presumably caused by GPS error [34]. We deemed it unlikely that a cat would suddenly change course and speed, then return to the same area it was in 15 minutes ago, so spikes were defined where fixes met all the following criteria: distances from the last fix were >50 m, difference in distance from the preceding and succeeding fix was <10%, and turning angle >170° [34,35]. The HDOP values were not used to filter fixes, as a test of six GPS collars found no relationship between HDOP and distance from the GPS fix to the known location. Fixes that were potentially biased by human disruption of the behaviour of the cat were removed, being those within three hours of field VHF tracking (this was occasionally carried out in order to download data remotely from the cat GPS collar). Fast-moving fixes were also removed, as it was likely cats were moving away from something, rather than choosing habitat. For this purpose a filtering speed of greater than 2 km per hour was used, as this was where the histogram of speeds between fixes reached an asymptote [35], representing a shift in behaviour mode. A test of GPS error within the open savanna found that 95% of fixes had <5 m error (from 634 fixes on six different collars), so fixes less than 10 m from the preceding fix were classed as stationary. We considered only moving fixes, as cats may have different habitat requirements for resting versus hunting.

Data analysis

Habitat selection by cats was analysed using discrete choice modelling [36]. The range of 'available' habitats was calculated for each fix, and then we compared the option selected by the cat to the available habitats. To find the available habitats, we first constructed probability distributions of a cat's step-length and bearing over 15-minute intervals, then used these probability distributions to select five random points to sample the cat's options [36]. Each GPS fix and associated random points were attributed with the habitat variables of interest using the dynamic vegetation map.

We determined resource selection by cats by creating models with all combinations of variables, including different biologically relevant interaction terms, and comparing them within an information theory framework [37]. This produced a total of 916 models. For each interaction term a model was included with all combinations, or with only significant combination terms retained. No pairs of variables with Pearson's correlation greater than 0.5 were included in the same model. Models for habitat selection were created using standard case-control logistic regression models, and were implemented in R [38] using the 'clogit' command from the 'survival' library. Each individual cat was considered as a mixed effect in the models, using Gaussian frailty [39]. The most parsimonious models of cat habitat selection were chosen as those with an AIC score within two points of the highest-ranked model [37], and only these are presented in the results. The cat's selection is measured as an odds ratio representing the magnitude of change in the odds of selection for each unit of the predictor variable. Differences in the odds ratio are relative only to the other habitat choices immediately available to a cat.

Ethics statement

All data collection fulfilled all legal requirements in Australia, and has been approved by University of Tasmania Animal Ethics Committee (A0011661) and Western Australian Department of Parks and Wildlife Animal Ethics Committee (2010/35), with a Regulation 17 licence to research animals (SF009379). All research was conducted with permission on three pastoral leases; Mornington Wildlife Sanctuary, Marion Downs (both managed by the Australian Wildlife Conservancy, ph; +61 8 9191 7014), and Glenroy Station (ph: +61 8 9191 4703). All three leases are located around 17°01′S, 126°01′E. Field studies did not involve protected or endangered species.

Results

In total, 60 cats were captured between September 2010 and June 2013. Three cats were caught in wire cage traps (265 trap nights), 19 in leg-hold traps (940 trap nights) and 38 by spotlighting/netting with trained cat-dogs (221 hours). GPS collars were placed on 37 cats, and at least one month of GPS data was obtained from 32 cats whose locality was spread equally across stocking and fire management treatments (Table 1). There was a strong male bias in the sample of cats: males comprised 78% of all captured cats (47/60) and 78% of cats from which GPS data were obtained (25/32). Of the GPS-collared cats, four had disappeared and their fates were unknown at the end of the study, nine had died naturally, and the rest were euthanased.

From the 32 cats that provided useable GPS data, we obtained a total of 133 047 GPS fixes. Cats were moving 56% of time. Of these moving fixes, 62% were at night (between sunset and sunrise). Removal of erroneous or biased fixes and those representing high-speed movement left 38 472 choices for habitat between successive 15-minute moving fixes.

From the 916 models generated to describe habitat selection by cats, three were ranked within the candidate model set (the model with the lowest AIC score and two other models within two AIC scores of the top model). The top model carried 41% of the weight, compared to 24% and 20% for the second and third ranked models. Of these, the second and third ranked models were almost identical to the top model, but contained interaction terms that did not decrease the models AIC value. As these terms did not improve the maximum likelihood for these models, only the top model was considered further (Table 3).

The top model included negative selection for grass cover (equivalently, positive selection for open areas) and positive selection for edges (see Table 3). Both variables had a significant interaction with small-mammal density in a binary format (>2 captures per 100 trap nights), showing that cats selected particularly strongly for open areas in habitats with higher density of small mammals.

Fire was represented in the top model with fire scars up to 360 days old, fire scars <90 days old, whether the fire was intense, small-mammal abundance, and interaction terms between these

Table 3. Statistics of the top ranked model of cat habitat selection based on GPS data at 15 minute intervals from 32 individuals.

Variable	Odds ratio	robust SE	Z	Pr(>|z|)	
Grass cover with scarce small mammals[a]	−1.26	0.08	−2.88	0.004	**
Grass cover with abundant small mammals[a]	−1.85	0.1	−6.44	0.0001	***
Bare/grass edge with scarce small mammals[a]	1.2	0.07	2.81	0.005	**
Bare/grass edge with abundant small mammals[a]	1.41	0.06	6.26	<0.0001	***
Fire scar <360 days	−1.32	0.07	−4.09	<0.0001	***
Intense fire scar <360 days old	1.54	0.13	3.29	0.001	**
Intense fire scar <90 days old	−2.11	0.13	−5.67	<0.0001	***
Small mammal abundance (square-root)	3.52	0.21	6.03	<0.0001	***
Small mammal abundance, intense fire scar <360 days	−4.93	0.66	−2.4	0.0163	*
Small mammal abundance, intense fire scar <90 days	10.71	0.79	2.99	0.0028	**
Grass communities ranked on grazing impacts, in:					
- stocked areas during day	5.96	0.14	12.56	<0.0001	***
- destocked areas during day	3.52	0.25	5.07	<0.0001	***
- stocked areas over night	2.64	0.17	5.65	<0.0001	***
- destocked areas over night	2.44	0.16	5.67	<0.0001	***
Water proximity (km) by months into dry season	1.56	0.06	−7.77	<0.0001	***
- as above, by minimum nightly temperature (10°C)	1.13	0.03	3.94	<0.0001	***
Elevation (100m)	−1.93	0.28	−2.32	0.0202	*
Home range isopleth[b], adult female	3.53	0.05	27.17	<0.0001	***
Home range isopleth[b], adult male	2.28	0.1	−8.41	<0.0001	***
Home range isopleth[b], sub-adult	1.06	0.49	−0.12	0.9077	
Same habitat as last fix, if within 95% isopleth	2.06	0.02	41.4	<0.0001	***

[a]Scarce and abundant small mammals are defined as less than or greater than two individuals captured per 100 trap nights at Australian Wildlife Conservancy monitoring sites.
[b]Home range isopleth derived at 50, 90, 95 and 99% contours from kernel density estimator.
The odds ratio is the change in selection likelihood per unit of the variable.

variables. Once the odds ratios of these variables were combined, cats showed strong positive selection for areas within 90 days after an intense fire and where density of small-mammal prey was high (Fig. 2). However, in all other circumstances selection for areas burnt by intense or mild fires was negative.

Vegetation types with greater susceptibility to grazing impacts (e.g. riparian areas, Table 2) were strongly selected for in both the grazed and destocked zones. However, this relationship was significantly stronger in the stocked zone than in the destocked zone, especially during the day (see Fig. 3).

Several other variables influenced selection by cats. Cats chose to move towards water. The odds ratio of moving towards water became progressively stronger throughout the dry season (by 0.56 every month into the dry season) and on days with higher minimum temperature (by 0.13 for each 10°C increase). Cats selected against changes in elevation, with odds declining by 0.93 every 100 m. Cats tended to move into higher-use areas within their home range (determined by kernel density estimates), with an interaction with age and sex. Adult females demonstrated the strongest fidelity to home-range isopleths, while for sub-adults of either sex this was not significant (P = 0.9). Cats were twice as likely to select for the same habitats as the previous fix (P<0.001), assuming the fix was inside a cats' home range (95% isopleth).

Discussion

Our study provides a detailed analysis of the preferences that underlie movement decisions by feral cats in a tropical savanna environment in northern Australia. We show that modifications of habitat produced by grazing and by certain types of fire have strong effects on cat movement behaviour, with the general result that fire and grazing can create habitat conditions which are

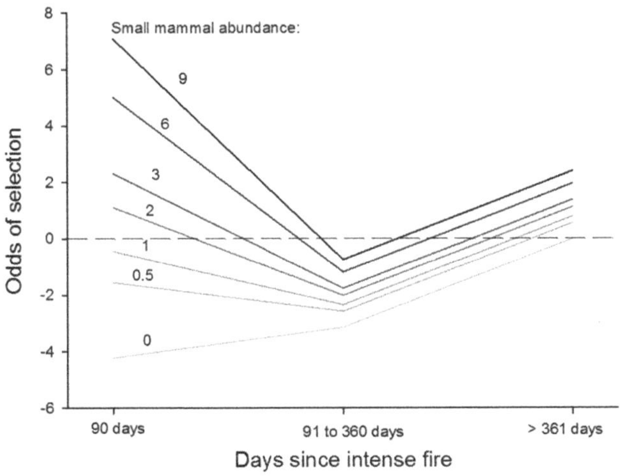

Figure 2. Change in the odds of selection ratios for different days since intense fires, at different average small mammal abundances based on capture rates per 100 trap nights (ranging from 0 to 9, lighter to darker respectively). All other variables in the model are assumed to be constant (see Table 3).

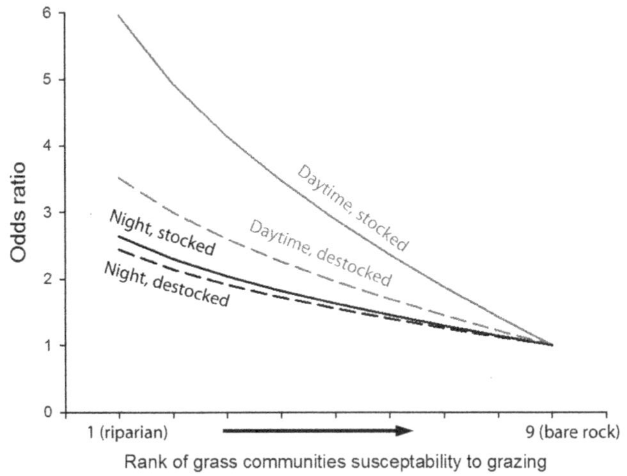

Figure 3. Odds ratios for selection of cats at night (black) and day (grey) in stocked (solid) and destocked (dashed) areas against grass communities ranked by grazing susceptibility. All other variables in model assumed to be constant (see Table 3).

strongly favoured by cats. When faced with choices about where to move, cats consistently selected sites with a more open grass layer, which had recently been subject to intense fires, and which were heavily grazed. Further, the influence of fire intensity on selection of habitat by cats was strongly affected by whether the habitats in question supported high densities of small mammals: sites in mammal-rich habitats that had recently been burnt at high intensity were especially strongly favoured.

This interaction between cat movements, fire and grazing regimes may help explain the recent declines in small mammals in northern Australia. There has been doubt that predation by cats might be driving these declines because of the mismatch in timing between the introduction of cats and small mammal declines, together with the fact that we have no evidence that small-mammal declines have coincided with an increased populations of cats [19]. As our results demonstrate, impacts of cats could have become more severe with the changes in fire and grazing regimes that began around the 1960s [40,41], even if cat density remained constant.

Selection by cats for areas with an open grass layer is a consistent finding from other studies on habitat selection by small felids [42,43], except in situations where moving into open areas leaves them exposed to larger predators [44] or where prey density is low [45]. Our cats' selection for open areas is almost certainly an expression of hunting preferences, as their selection became stronger in areas with higher small-mammal abundances. Although grass cover in itself had a large effect size (odds ratio maximum of 1.8), the effects of grazing and fire patterns on cat movements were much stronger.

Selection for fire scars was strongly dependent on fire intensity and time since fire. In general, cats avoided fire scars up to 360 days after fire. The exception was for recent scars resulting from intense fires in areas where small mammal densities were high, as this created the greatest increase in the odds ratio of any variable in this study. Fire opens up the grass layer thereby leaving prey more exposed and accessible to predators [15]. Intense fires would create pulses where prey would be easy for cats to catch - provided local abundance of prey was high. Cats did not select for recent mild fire scars, even in habitats with high small-mammal abundance. Mild fires typically leave pockets of unburnt vegetation [46], which provide protection for prey [10]. Also, mild fires are often stopped by riparian and alluvial areas [40], and such areas act as refugia for small mammals post-fire [10]. Our results suggest that cats are not able to capitalise on vulnerable small mammals after mild fires, but can after intense fires. This can explain why declines of some threatened native mammals have been so dramatic under regimes of consecutive high intensity fires [20,21].

Grazing by introduced herbivores also affected habitat choice by cats. Certain vegetation types (with more palatable grasses) were more preferred by introduced herbivores than others - typically riparian areas and bluegrass plains [47]. Cats selected these habitats even when they had been destocked, but grazing intensified this preference. Cats may have a stronger selection for these areas in the grazed zone as lower prey densities [22] require them to hunt in areas of relatively higher mammal density in order to obtain enough food. Daytime movements of cats were especially affected by grazing, suggesting that grazing creates favourable conditions for hunting of diurnal prey, such as quail [48].

Adult cat movements at small-scale 15 minute segments were dictated by their overall home range. Females had a stronger home-range fidelity than males, probably due to the importance of staying near dens that are used for rearing young [49]. Sub-adults displayed no such home range fidelity as they were probably in the process of creating and defining their home-ranges.

The top-order predator, the dingo, would also have influenced the movements of cats. Over the area and duration of the study, the dingo populations was abundant (0.2 individuals per km^2) and stable [32], and dingoes would have been a constant threat to cats [50]. A concurrent study in the same area that compared GPS movements of cats and dingoes found that although high use areas of dingoes and cats overlapped, cats were constantly avoiding the locations of individual dingoes [51]. This suggests avoidance of dingoes would have affected the timing of behavioural decisions of cats [52], but not necessarily by excluding them from certain areas [53]. Considering dingoes potential role in trophic regulation [25], further research into the relationship between dingoes, cats, and fire and grazing is warranted.

Conclusions

We demonstrate increased predator activity after intense fires and with grazing by large herbivores, which is likely to increase predation rates on small mammals. Small mammals are the preferred prey of cats, and form a substantial part of the diet of cats in the study region (55% of prey volume, from 33 stomach contents; unpublished data). Furthermore, the preference of cats for open and intensely burned areas depended on small-mammal abundance, and was reduced in habitats where abundance of small mammals was low. These patterns of habitat selection by cats correspond with declines of small mammal populations with intensifying fire and/or grazing regimes in mainland northern Australia, outside complex rocky outcrops [20,21,22,54,55,56,57]. Our work supports the hypothesis that the declines in small mammals across northern Australia are driven by cat predation facilitated by simplification of ground layer structure. However, this evidence is not direct, and other possible mechanisms including trophic alterations [25] and disease [19] may still have a role. Further research needs to measure whether mortality of prey is greater in more open areas. Regardless, the magnitude of the impacts of cats globally [58] suggests that our findings provide a general mechanism for prey decline in ecosystems with grass-dominated understoreys.

The results presented here suggest that manipulation of habitat through careful management of fire and grazing could be used to reduce impacts of cats over large landscapes. Reducing the frequency of intense fires and removing introduced herbivores is likely to be beneficial for small mammals - especially if this management is focussed on naturally mammal-rich habitats, and if it increases ground cover. Vegetation structure is pivotal in creating ideal landscapes for predators to hunt, and/or refuges for prey to hide.

Acknowledgments

Staff and volunteers from the Australian Wildlife Conservancy assisted with field work, in particular Katherine Tuft, Monica Griffith, Danni Lisle, Isabel Connell, and Lily Leahy. Anja Skroblin provided comments.

Author Contributions

Conceived and designed the experiments: HWM SL MJ CJ. Performed the experiments: HWM SL. Analyzed the data: HWM. Contributed reagents/materials/analysis tools: HWM SL CJ. Wrote the paper: HWM SL MJ CJ.

References

1. Lima SL, Dill LM (1990) Behavioural decisions made under the risk of predation: a review and prospectus. Canadian Journal of Zoology 68: 619–640.
2. Didham RK, Tylianakis JM, Gemmell NJ, Rand TA, Ewers RM (2007) Interactive effects of habitat modification and species invasion on native species decline. TRENDS in Ecology & Evolution 22: 490–496.
3. Griffiths D (1975) Prey availability and the food of predators. Ecology 56: 1209–1214.
4. Hopcraft JCC, Sinclair ARE, Packer C (2005) Planning for success: Serengeti lions seek prey accessibility rather than abundance. Journal of Animal Ecology 74: 559–566.
5. Hebblewhite M, Merrill EH, McDonald TL (2005) Spatial decomposition of predation risk using resource selection functions: an example in a wolf-elk predator-prey system. Oikos 111: 101–111.
6. Koenen MT, Leslie D, Gregory MS (1996) Habitat changes and success of artificial nests on an alkaline flat. Wilson Bulletin 108: 292–301.
7. Dickson BG, Beier P (2002) Home-range and habitat selection by adult Cougars in Southern California. Journal of Wildlife Management 66: 1235–1245.
8. Gilliam JF, Fraser DF (1987) Habitat selection under predation hazard: test of a model with foraging minnows. Ecology 68: 1856–1862.
9. Yarnell RW, Scott DM, Chimimba CT, Metcalfe DJ (2007) Untangling the roles of fire, grazing and rainfall on small mammal communities in grassland ecosystems. Oecologia 154: 387–402.
10. Banks SC, Dujardin M, McBurney L, Blair D, Barker M, et al. (2011) Starting points for small mammal population recovery after wildfire: recolonisation or residual populations? Oikos 120: 26–37.
11. Tabenia S, Ojedaa RA (2003) Assessing mammal responses to perturbations in temperate aridlands of Argentina. Journal of Arid Environments 55: 715–726.
12. Jones A (2000) Effects of cattle grazing on North American arid ecosystems: a quantitative review. Western North American Naturalist 60: 155–164.
13. Horn KJ, McMillan BR, St Clair SB (2012) Expansive fire in Mojave Desert shrubland reduces abundance and species diversity of small mammals. Journal of Arid Environments 77: 54–58.
14. Southgate R, Palmer C, Adams MA, Masters P, Triggs B, et al. (1996) Population and habitat characteristics of the Golden Bandicoot (*Isoodon auratus*) on Marchinbar Island, Northern Territory. Wildlife Research 23: 647–664.
15. Conner LM, Castleberry SB, Derrick AM (2011) Effects of mesopredators and prescribed fire on hispid cotton rat survival and cause-specific mortality. Journal of Wildlife Management 75: 938–944.
16. Driscoll DA, Lindenmayer DB, Bennett AF, Bode M, Bradstock RA, et al. (2010) Fire management for biodiversity conservation: key research questions and our capacity to answer them. Biological Conservation 143: 1928–1939.
17. Sutherland EF, Dickman CR (1999) Mechanisms of recovery after fire by rodents in the Australian environment: a review. Wildlife Research 26: 405–419.
18. Fisher DO, Johnson CN, Lawes MJ, Fritz SA, McCallum HI, et al. (2013) The current decline of tropical marsupials in Australia: is history repeating? Global Ecology and Biogeography 23: 181–190.
19. Woinarski JCZ, Legge S, Fitzsimons JA, Traill BJ, Burbidge AA, et al. (2011) The disappearing mammal fauna of northern Australia: context, cause, and response. Conservation Letters 4: 1–10.
20. Firth RSC, Brook BW, Woinarski JCZ, Fordham DA (2010) Decline and likely extinction of a northern Australian native rodent, the Brush-tailed Rabbit-rat *Conilurus penicillatus*. Biological Conservation 143: 1193–1201.
21. Pardon G, Brook BW, Griffiths AD, Braithwaite RW (2003) Determinants of survival for the northern brown bandicoot under a landscape-scale fire experiment. Journal of Animal Ecology 72: 106–115.
22. Legge S, Kennedy MS, Lloyd R, Murphy SA, Fisher A (2011) Rapid recovery of mammal fauna in the central Kimberley, northern Australia, following the removal of introduced herbivores. Austral Ecology 36: 791–799.
23. Liedloff AC, Coughenour MB, Ludwig JA, Dyer R (2001) Modelling the trade-off between fire and grazing in a tropical savanna landscape, northern Australia. Environment International 27: 173–180.
24. Dickman CR (1996) Overview of the impacts of feral cats on Australian native fauna. Sydney: University of Sydney.
25. Radford IJ, Dickman CR, Start AN, Palmer C, Carnes K, et al. (2014) Mammals of Australia's tropical savannas: A conceptual model of assemblage structure and regulatory factors in the Kimberley region. PLos ONE 9: e92341.
26. Oakwood M (2000) Reproduction and demography of the northern quoll, *Dasyurus hallucatus*, in the lowland savanna of northern Australia. Australian Journal of Zoology 48: 519–539.
27. Woinarski JCZ, Ward S, Mahney T, Bradley J, Brennan K, et al. (2011) The mammal fauna of the Sir Edward Pellew island group, Northern Territory, Australia: refuge and death-trap. Wildlife Research 38: 307–322.
28. Frank ASK, Johnson CN, Potts J, Alaric F, Lawes MJ, et al. (2014) Experimental evidence that feral cats cause local extirpation of small mammals in Australia's tropical savanna. Journal of Applied Ecology - online early.
29. Abbott I (2002) Origin and spread of the cat, *Felis catus*, on mainland Australia, with a discussion of the magnitude of its early impact on native fauna. Wildlife Research 29: 51–74.
30. Kutt AS (2012) Feral cat (Felis catus) prey size and selectivity in north-eastern Australia: implications for mammal conservation. Journal of Zoology 287: 292–300.
31. Ziembicki MR, Woinarski JCZ, Mackey B (2013) Evaluating the status of species using Indigenous knowledge: Novel evidence for major native mammal declines in northern Australia. Biological Conservation 157: 78–92.
32. Legge S, Webb T, Smith J, Tuft K, James A (2014) The ecological health of Mornington and Marion Downs Wildlife Sanctuaries, 2013 Report. Perth, WA: Australian Wildlife Conservancy.
33. Torrea I, Díazb M, Martínez-Padillac J, Bonalb RD, Viñuelae J, et al. (2007) Cattle grazing, raptor abundance and small mammal communities in Mediterranean grasslands. Basic and Applied Ecology 8: 565–575.
34. Bjørneraas K, van Moorter B, Rolandsen CM, Herfindal I (2010) Screening global positioning system location data for errors using animal movement characteristics. Journal of Wildlife Management 74: 1361–1366
35. Recio MR, Mathieu R, Virgos E, Seddon PJ (2014) Quantifying fine-scale resource selection by introduced feral cats to complement management decision-making in ecologically sensitive areas. Biological Invasions 1: 1–13.
36. Cooper AB, Millspaugh JJ (1999) The application of discrete choice models to wildlife resource selection studies. Ecology 80: 566–575.
37. Burnham KP, Anderson DR (1998) Model selection and multimodel inference: A practical information-theoretic approach: Springer Science & Business Media, Inc.
38. R Development Core Team (2008) R: a language and environment for statistical computing. Vienna: R Foundation for Statistical Computing.
39. Therneau TM, Grambsch PM, Pankratz S (2003) Penalized survival models and frailty. Journal of Computational and Graphical Statistics 12: 156–175.
40. Russell-Smith J, Yates CJ, Edwards A, Allan GE, Cook GD, et al. (2003) Contemporary fire regimes of northern Australia, 1997–2001: change since Aboriginal occupancy, challenges for sustainable management. International Journal of Wildland Fire 12: 283–297.
41. Woinarski JCZ, Mackey B, Nix HA, Traill BJ (2007) The nature of northern Australia: Natural values, ecological processes and future prospects. Canberra, ACT: ANU Press.
42. Moseby KE, Stott J, Crisp H (2009) Movement patterns of feral predators in an arid environment – implications for control through poison baiting. Wildlife Research 36: 422–435.
43. Rajaratnam R, Sunquist M, Rajaratnam L, Ambu L (2007) Diet and habitat selection of the leopard cat (*Prionailurus bengalensis borneoensis*) in an agricultural landscape in Sabah, Malaysian Borneo. Journal of Tropical Ecology 23: 209–217.
44. Genovesi P, Besa M, Toso S (1995) Ecology of a feral cat *Felis catus* population in an agricultural area of northern Italy. Wildlife Biology 1: 233–237.
45. Hall LS, Kasparian MA, Van Vuren D, Kelt DA (2009) Spatial organization and habitat use of feral cats (*Felis catus L.*) in Mediterranean California. Mammalia 64: 19–28.
46. Russell-Smith J, Ryan PG, Cheal DC (2002) Fire regimes and the conservation of sandstone heath in monsoonal northern Australia: frequency, interval, patchiness. Biological Conservation 104: 91–106.

47. Tomkins N, O'Reagain P (2007) Global positioning systems indicate landscape preferences of cattle in the subtropical savannas. The Rangeland Journal 29: 217–222.
48. Brown RL (1982) Effects of livestock grazing on Mearns Quail in southeastern Arizona. Journal of Range Management 35: 727–732.
49. Krofel M, Skrbinšek T, Kos I (2013) Use of GPS location clusters analysis to study predation, feeding, and maternal behavior of the Eurasian lynx. Ecological Research 28: 103–116.
50. Moseby KE, Neilly H, Read JL, Crisp HA (2012) Interactions between a top order predator and exotic mesopredators in the Australian rangelands. International Journal of Ecology 2012.
51. Brook LA (2013) Predator guild interactions in northern Australia: behaviour and ecology of an apex predator, the dingo *Canis lupus dingo*, and an introduced mesopredator, the feral cat *Felis catus*. Townsville, QLD: James Cook University.
52. Brook LA, Johnson CN, Ritchie EG (2012) Effects of predator control on behaviour of an apex predator and indirect consequences for mesopredator suppression. Journal of Applied Ecology 49: 1278–1286.
53. Wang Y, Fisher DO (2012) Dingoes affect activity of feral cats, but do not exclude them from the habitat of an endangered macropod. Wildlife Research 39: 611–620.
54. Woinarski JCZ, Armstrong M, Brennan K, Fisher A, Griffiths AD, et al. (2010) Monitoring indicates rapid and severe decline of native small mammals in Kakadu National Park, northern Australia. Wildlife Research 37: 116–126.
55. Kutt AS, Woinarski JCZ (2007) The effects of grazing and fire on vegetation and the vertebrate assemblage in a tropical savanna woodland in north-eastern Australia. Journal of Tropical Ecology 23: 95–106.
56. Andersen AN, Cook GD, Corbett LK, Douglas MM, Eager RW, et al. (2005) Fire frequency and biodiversity conservation in Australian tropical savannas: implications from the Kapalga fire experiment. Austral Ecology 30: 155–167.
57. Legge S, Murphy S, Heathcote J, Flaxman E, Augusteyn J, et al. (2008) The short-term effects of an extensive and high-intensity fire on vertebrates in the tropical savannas of the central Kimberley, northern Australia. Wildlife Research 35: 33–43.
58. Loss SR, Will T, Marra PP (2012) The impact of free-ranging domestic cats on wildlife of the United States. Nature Communications 4: 1396.

17

Are Flying-Foxes Coming to Town? Urbanisation of the Spectacled Flying-Fox (*Pteropus conspicillatus*) in Australia

Jessica Tait[1,3], Humberto L. Perotto-Baldivieso[1], Adam McKeown[2], David A. Westcott[3]*

1 School of Energy, Environment and Agrifood, Cranfield University, Cranfield, Bedfordshire, United Kingdom, 2 CSIRO Sustainable Land and Water, Smithfield, QLD, Australia, 3 CSIRO Sustainable Land and Water, Atherton, QLD, Australia

Abstract

Urbanisation of wildlife populations is a process with significant conservation and management implications. While urban areas can provide habitat for wildlife, some urbanised species eventually come into conflict with humans. Understanding the process and drivers of wildlife urbanisation is fundamental to developing effective management responses to this phenomenon. In Australia, flying-foxes (Pteropodidae) are a common feature of urban environments, sometimes roosting in groups of tens of thousands of individuals. Flying-foxes appear to be becoming increasingly urbanised and are coming into increased contact and conflict with humans. Flying-fox management is now a highly contentious issue. In this study we used monitoring data collected over a 15 year period (1998–2012) to examine the spatial and temporal patterns of association of spectacled flying-fox (*Pteropus conspicillatus*) roost sites (camps) with urban areas. We asked whether spectacled flying-foxes are becoming more urbanised and test the hypothesis that such changes are associated with anthropogenic changes to landscape structure. Our results indicate that spectacled flying-foxes were more likely to roost near humans than might be expected by chance, that over the period of the study the proportion of the flying-foxes in urban-associated camps increased, as did the number of urban camps. Increased urbanisation of spectacled flying-foxes was not related to changes in landscape structure or to the encroachment of urban areas on camps. Overall, camps tended to be found in areas that were more fragmented, closer to human habitation and with more urban land cover than the surrounding landscape. This suggests that urbanisation is a behavioural response rather than driven by habitat loss.

Editor: Elissa Z. Cameron, University of Tasmania, Australia

Funding: This research was supported by a grant from Bat Conservation International to JT and funding to DAW from the Commonwealth of Australia's Department of the Environment, the National Environmental Research Program and the Rural Industries Research Development Corporation (through funding from the Commonwealth of Australia, the State of New South Wales and the State of Queensland under the National Hendra Virus Research Program). The funders had no role in study design, data collection and analysis, decision to publish, or preparation of the manuscript.

Competing Interests: The authors have declared that no competing interests exist.

* Email: David.Westcott@csiro.au

Introduction

By 2030 five billion humans are expected to live in urban areas while the global urban footprint is predicted to expand by 163% from c. 727,000 km^2 to c. 1,527,000 km^2 [1]. While already significant ecosystems in their own right, such urban expansion will place increasing pressure on other land uses, threatening native species and ecosystems, and becoming potent sources of invasive species and pathogens [2]. For example, much urban expansion is currently occurring in sensitive areas for biodiversity, e.g. in coastal lowlands or close to protected areas [1]. This means that consideration of urban systems is increasingly important in conservation planning and management. Urbanisation generally has a negative net effect on biodiversity, with many species becoming rare or locally extinct, in particular specialists, slow reproducers, and disturbance-sensitive species [2,3]. However, some species, often generalists, readily adapt to the urban landscape, some even reaching higher abundances in cities than in natural vegetation [4,5]. Although the presence of wildlife in urban areas can enhance human quality of life, some urban animal populations can prove problematic due to their impacts on amenity, damage or their role as vectors of disease [6,7]. Understanding how particular species respond to urbanisation and identifying the processes leading to these responses is fundamental if we are to successfully manage the interaction between urbanisation, biodiversity and human welfare.

Urbanisation influences species distribution, abundance and movement. The urban mosaic can prove an attractive habitat for a wide range of taxa due to abundant food and shelter [2,3,8,9]. Urban areas can provide a refuge from hunting or predation pressure [10], from environmental disturbances such as drought or fire, as well as a more stable resource supply whilst natural vegetation is recovering [11]. Wildlife adapt to urban areas in a variety of ways, e.g. by adjusting their foraging, anti-predator behaviour, breeding behaviour and taking advantage of the climate associated with urban areas [12,13,14,15]. With increased urbanisation of wildlife populations comes increased contact with humans and attendant increases in opportunity for conflict, including amenity impacts such as noise, smell and vegetation damage [16], hazards such as vehicle collision or attacks on pets [17,18], and risk of disease transmission [6,7]. These 'human-wildlife conflicts' can lead to intense disagreement over the management of habitat and wildlife in and around urban areas [19].

Flying-foxes are large (up to c. 1 kg), colonially roosting bats which readily adapt to urban ecosystems. Roost sites (hereafter called camps) are found in and around many Australian towns and cities [20,21,22,23]. Hypotheses for why flying-foxes might use urban areas include loss of native habitat and urban expansion [20,24], changes in resource distribution due to plantings [25,26,27] and urban effects on local climate [21]. Evidence from other species suggests that urban areas may also provide refuge from predation [27], disturbance events such as droughts [28], bushfires [29] and post-cyclone effects [22,30,31]. It may also be possible that urban areas may be attractive because they offer a movement advantage, e.g. increase the ease of manoeuvring in flight due to the open nature of the habitat or ease of navigation due to landmarks and lighting, e.g. [32].

Roosting by flying-foxes in urban and peri-urban areas can result in contact and conflict with humans. The greatest concern is impact on amenity. It is not uncommon for flying-fox camps to contain 50,000 individuals [22,27], but even much smaller camps can be potent point sources of noise, odour and faeces, particularly when they occur within metres of residences. While smaller camps are often tolerated, larger camps in particular become a focus of community disquiet.

Alongside impact on amenity are concerns about the risks of disease transmission [33]. Bats host a high diversity of viruses and some of these are of agricultural and human health significance [33,34]. It would appear that in cases such as the paramyxoviruses (which include diseases such as measles, distemper, mumps, parainfluenza, Newcastle disease), these diseases have been present in bats for very long periods and have switched hosts to other mammals, including humans [34]. In Australia, there are two diseases of particular concern that are known to be carried by flying-foxes: Hendra virus and Australian bat lyssavirus. It has been suggested that the dynamics of diseases such as Hendra in flying-fox populations and the pattern of spillover events from flying-foxes to horses, may be influenced by urbanisation through the effects it has on habitat loss and through that, on resource distribution, connectivity between groups of flying-foxes, and increased interactions between flying-foxes and horses [35].

The negative impacts of flying-foxes, and the ever more strident calls for their 'control' that these impacts create, are at odds with the significance of their ecological role and conservation status. Flying-foxes are significant pollinators and seed dispersers in most vegetation types in their range [24,36]. Furthermore, two species, the spectacled flying-fox (*Pteropus conspicillatus*) and the grey-headed flying-fox (*P. poliocephalus*), are listed as vulnerable under Australia's Environment Protection and Biodiversity Conservation Act (1999). The resultant divergent and strongly held opinions on flying-foxes creates persistent tension between those who wish to see the animals conserved and those demanding they be controlled. Flying-fox management is a contentious and politicised issue in Australia.

To date, despite urbanised flying-foxes being a major management issue in northern and eastern Australia, there are few studies of urbanisation of flying-foxes. Identifying whether urbanisation of flying-fox populations is in fact occurring and the nature of its drivers is a fundamental step in developing effective management solutions. The aim of this study is to determine whether spectacled flying-foxes are urbanising and whether landscape features or change are associated with this. We use data from 15 years of monitoring of the spectacled flying-fox population to examine the spatial and temporal patterns of association of spectacled flying-fox camps with urban areas in the main part of their Australian range, the Wet Tropics of north eastern Australia. Specifically we (1) ask whether spectacled flying-foxes are becoming more urbanised, (2) we test the hypotheses that any shift to urban areas is associated with anthropogenic changes to landscape structure or to an increase in the size or number of urban camps, and (3) whether the landscape characteristics of camp sites differ from those of the surrounding landscape and how this has changed over the study period.

Materials and Methods

Ethics Statement

This research was conducted under Animal Ethics Approvals from the CSIRO Ecosystem Sciences Animal Ethics Committee and complied with the Australian Code of Practice for the Care and Use of Animals for Scientific Purposes (2004; 2013). The work required no interaction with or handling of flying-foxes. The research was conducted under Scientific Purposes Permit #WTK03462308 and a 173P Authorisation from the Queensland Parks and Wildlife Service. The research was conducted on 50 separate land tenures. Details of the location of each camp have been lodged as a part of the National Flying-Fox Monitoring Program (NFFMP; http://www.environment.gov.au/node/16393) and camp locations, tenure, access information and camp sizes can be obtained from either from the NFFMP or from the authors.

Study Species

The spectacled flying fox is a phytophagous species, feeding primarily on floral resources and fruits in a wide range of vegetation communities, including closed forest, gallery forest, eucalypt open forest and woodland, coastal Melaleuca swamps, mangroves, vegetation in urban settings, and commercial fruit crops [37]. This highly mobile species forages at night and can disperse seeds and pollen over large distances [24]. By day the animals roost in camps, with an unknown proportion roosting solitarily or in small groups throughout the year [22]. In Australia, the spectacled flying-fox is found in the Wet Tropics of Queensland World Heritage Area between Townsville and Cooktown with small outlier populations north in the Iron and McIllwraith Ranges on Cape York [22,24] and to the south at Finch Hatton, near Mackay [38].

Study Site

The study included all known current and past spectacled flying-fox camps in the Wet Tropics region of Queensland, an area of approximately 9,000 km^2, on the north-east coast of Queensland (Figure 1). The region has a diverse, fragmented terrain of coastal plains and extensive uplands, typically 600–900 masl [39]. The vegetation is a complex mosaic of closed canopy rainforest and open eucalypt woodlands, with tropical savannas and grasslands on its drier margins and with clearing for agriculture on the wet fertile coastal floodplains, mid-montane tablelands and on the drier western slopes [39]. The region has a human population of approximately 252,000 [40]. Urbanisation is focused on several regional centres, Cairns, Mareeba, Atherton, Innisfail and Ingham, but there are also numerous small communities throughout the region.

Camp data

Data on the location and sizes of all flying-fox camps in the region were obtained during regular monitoring programs begun in 1998 and continuing today. Over the course of the study the number of camps surveyed increased from 30 to 50 as new camp

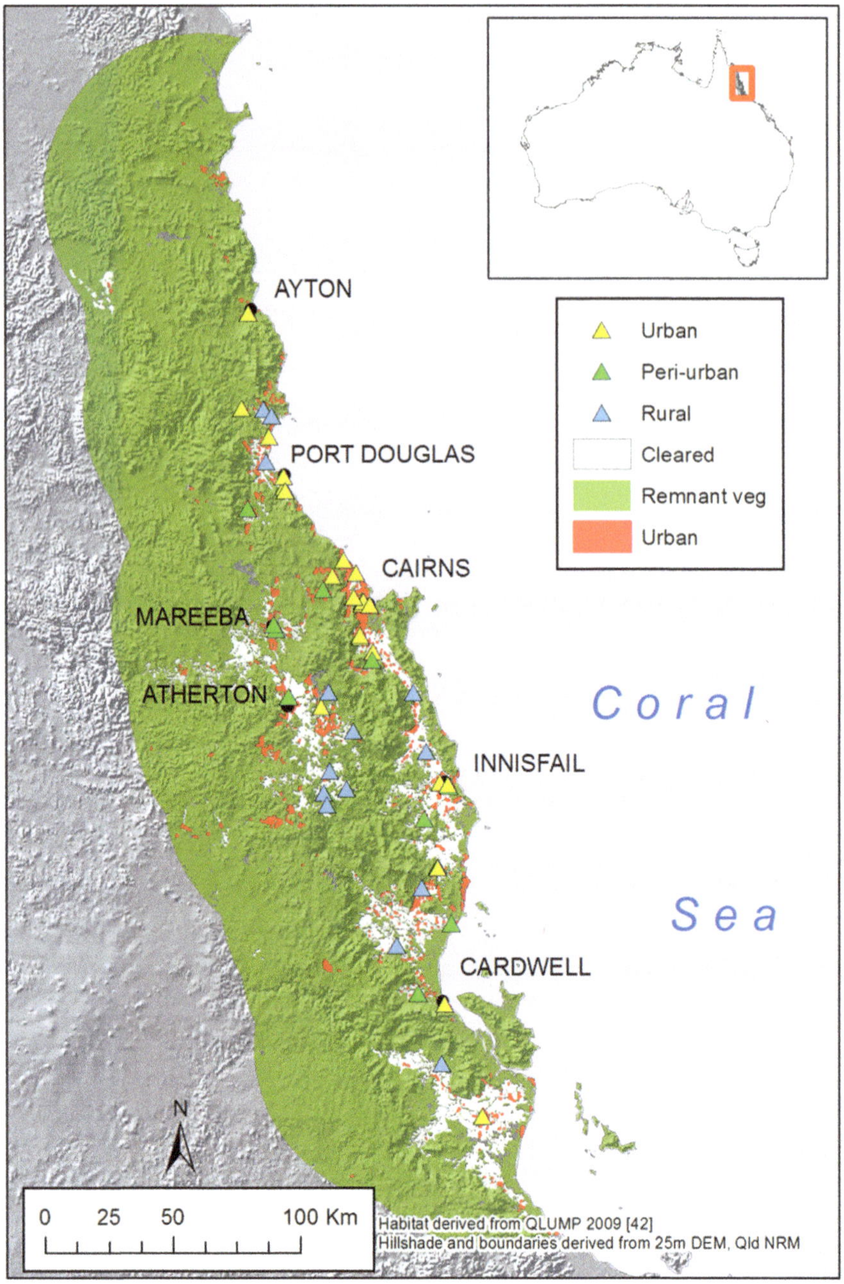

Figure 1. Location of the study area in the Wet Tropics Region of Northern Queensland, Australia. Spectacled flying-fox camps (triangles), towns (black dots) and urban areas (red shading) are also shown. Habitat mapping is derived from QLUMP 2009 [42] and the hillslopes and shading from the Qld. Dept. Natural Resources and Mines 25 m DEM (http://www.dnrm.qld.gov.au/mapping-data/data/topographic).

locations were identified. This increase resulted largely from the inclusion of historical camps. Once identified camps were not dropped from the surveys. These surveys were conducted under three programs. In 1998 and 1999 surveys were conducted in March and November while from 2000 to 2003 surveys were conducted in November only. These surveys were conducted by the Queensland Parks and Wildlife Service and involved positioning counters around the perimeter of camps to count the animals as they flew out of the camp at dusk, i.e. fly-out counts (see [41] for a more detailed description). Since May 2004 monthly, daytime, walk-through surveys of every camp in the study region have been conducted. In small camps (<1000 individuals), surveyors attempted to count all flying-foxes in a camp. In larger camps the density of individuals was assessed by counting the number of roosting individuals in randomly-selected roost trees, the average of these was then extrapolated to give a camp size estimate by counting the number of roost trees. Each regional survey was typically completed within three consecutive days to minimise the chance of inter-camp movements and any resultant recounting of individuals [22]. The use of these different survey methods in the early and later phases of the data collection has the potential to introduce biases into the data. To avoid such issues our analyses do not rely on direct comparisons of abundance estimates derived from different methods. This is achieved either by restricting analyses to data derived from a single method or by calculating proportions based on samples collected using a single method, i.e. within years, and only comparing the proportions across years.

How urbanised is the spectacled flying-fox population in the Wet Tropics?

Count data from 1998–2012 was analysed to assess the proportion of the population associated with urban areas. Camps were assigned to three categories (urban, peri-urban and non-urban) depending on their distance from urban areas – defined as human habitation. Urban camps were defined as those surrounded by urban land use according to Queensland Government's land use classification (primary attributes), QLUMP 2009 [42] and on-ground assessment. Peri-urban camps were defined as those adjacent to urban land cover. Non-urban camps were those more than 250 m from any urban land use - a distance at which there is usually little concern about the presence of flying-foxes. When referring to urban and peri-urban camps together we use the term urban-associated.

To examine patterns of urbanisation across years we used data from all surveys, i.e. from 1998–2012. To examine seasonal patterns of camp use we restricted the analyses to the monthly data obtained between 2004 and 2012. To detect any seasonal trends in urbanisation, the monthly percentage of the population in urban-associated camps, consistency of occupation (proportion of surveys in which a camp was occupied), mean camp size and total population surveyed were analysed using monthly survey data from 2004 onwards. The proportion of the population using identified camps is known to vary through the year with a lower proportion in camps mid-year and a higher proportion during the warmer months when mating, birthing and raising of young occurs. We use the term population to refer to that part of the population using camps at any point in time.

Are spectacled flying-foxes being driven into urban areas by landscape change?

There were two QLUMP land use sampling periods during the time of the study, 1999 and 2009. These two sampling periods fell conveniently near the beginning and end of our study and so were used to document changes in land cover and use during the study period. QLUMP primary landuse attribute "intensive uses" was used as the first filter to determine urbanisation and the tertiary attribute was used to differentiate between non-urban uses, urban and rural residential uses. Percentage land cover assigned to particular land uses was extracted from QLUMP and overlaid with Regional Ecosystem data describing vegetation types [42] to produce the land cover categories: True urban, Rural Residential, Cleared, Rainforest, Sclerophyll and Other.

To determine whether areas surrounding camp sites had changed over time, landscape metrics which convey key information about landscape spatial structure were calculated in Fragstats 4.0 [43] for circular buffer zones with a diameter of 3.3 km around camps, this being half the average nearest neighbour distance between camps. These metrics were chosen to provide a description of the structure of the landscape in which camps occurred and were: forest mean patch area (MPA) a measure of the average size of forest patches in the buffer, forest patch density (PD) or the number of patches in the buffer, edge density (ED) or the length of patch edge as a function of area and a measure of the amount of interior patch habitat relative to edge habitat, and percentage urban cover (% urban) [44]. These measures for 1999 and 2009 were compared using Mann-Whitney U tests. In order to test whether camps were found in more fragmented areas than would be expected by chance, buffers around camps and $n = 33$ random points in the landscape were compared with Mann-Whitney U tests.

Results

Surveys of spectacled flying-fox campsites across the Wet Tropics of Queensland from 1998–2012 indicate that the majority of the population were roosting in camps associated with urban areas (Figure 2). In each monthly survey between 5 and 16 camps were occupied (mean = 10, S.D. = 2.3). While a total of 30 camps were surveyed in the initial years and this number increased to 50 in the final years of the monitoring there was no significant increase over time in the number of camps found to be occupied each year ($r_p = 0.38$, $p = 0.16$, $n = 15$) (Table S1 in File S1). There was an increase over time in the number of urban camps occupied each year ($r_p = 0.64$, $p < 0.01$, $n = 15$) but no significant trend in the numbers of peri-urban or non-urban camps ($p > 0.05$ for both). The net result was an increase in urban-associated camps ($r_p = 0.5$, $p < 0.03$, $n = 15$). The mean percentage of the population found in urban and peri-urban camps across the 111 surveys was 82% (±26 S.D.), though this varied across all surveys from 59% in the early years to 99% in the later years.

There was a distinct seasonal pattern in the proportion of the population that was encountered in urban-associated camps with this proportion consistently reaching high levels in May and June before declining slowly there after (Figure 3). The proportion of the population found in urban-associated camps was greatest during the period of the year when the population count was at its lowest, May–November (Figure 4), and was significantly greater than during the rest of the year (Mann-Whitney U test, $Z = -4.95$, $p < 0.0001$). During this May–November period all camp types were less consistently occupied (Mann-Whitney U test, Non-urban: $Z = -3.56$, $p < 0.05$; Urban-associated: $Z = -3.58$, $p < 0.05$), and had smaller mean camp sizes (Mann-Whitney U test, Non-urban: $Z = 6.06$, $p < 0.05$; Urban-associated: $Z = 4.64$, $p < 0.05$), though the decrease in non-urban camp size is far more pronounced than that of the urban camps (Figure 5). Urban camps were occupied in 32% of month's surveyed (n = 16 camps), peri-urban camps 22% (n = 22) and non-urban camps 10% (n = 20).

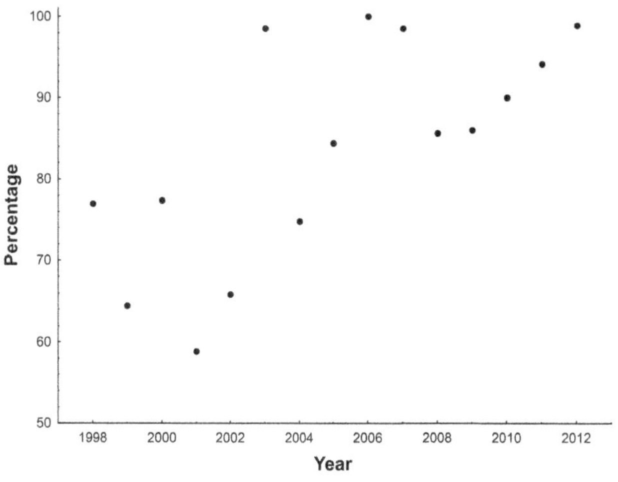

Figure 2. The percentage of the spectacled flying-fox population of the Wet Tropics found in urban associated camps during November surveys in each year of the study.

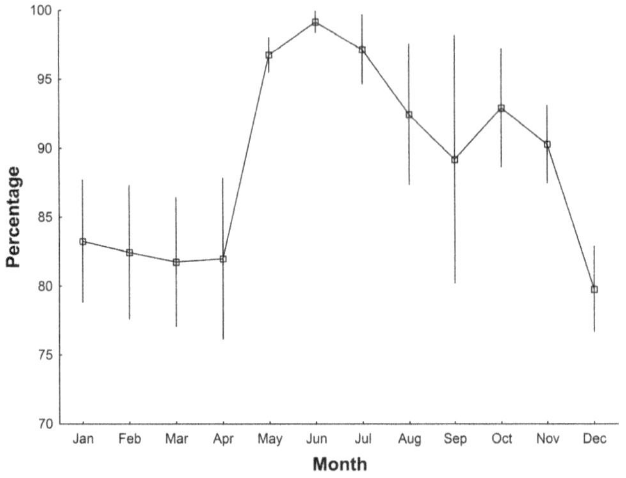

Figure 3. Changes through the year in the percentage (±S.E.) of the population occupying urban-associated camps. Monthly means calculated with survey data from 2004–2012.

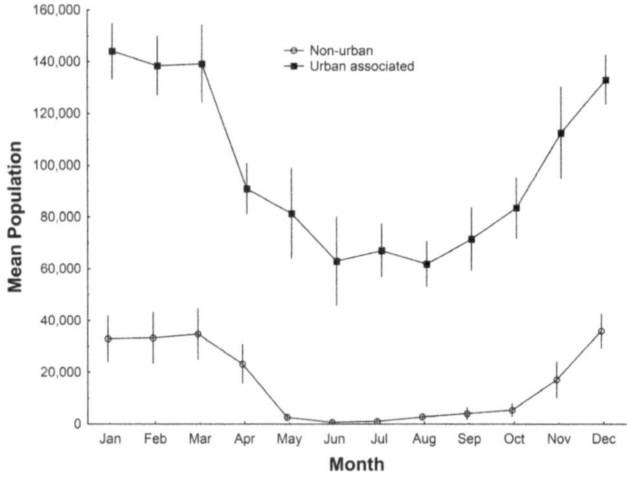

Figure 4. Changes through the year in the mean monthly population occupying urban associated and non-urban camps across the Wet Tropics calculated from 2004–2012 data. Monthly means calculated with survey data from 2004–2012.

There was an increase in the proportion of the population recorded using urban and peri-urban camps over the study period (Figure 2). This was the case irrespective of whether we considered the proportions recorded during November surveys only, the month that was surveyed in all years and in which the greatest proportion of the population are encountered in camps, ($r_p = 0.69$, $p<0.005$, $n = 15$), those for March and November surveys only, i.e. the months surveyed in all years except 2000–2003 ($r_p = 0.54$, $p<0.01$, $n = 26$), or for all months for which we have data ($r_p = 0.28$, $p<0.005$, $n = 111$). This trend appears to be due to a decrease in the number of non-urban camps used in any month (November only – $r = -0.81$, $p<0.001$, $n = 15$; March and November – $r = -0.78$, $p<0.001$, $n = 26$; all months – $r = -0.57$, <0.001, $n = 111$) and an increase in the number of urban camps used per month over time (November only - $r = 0.7$, $p< 0.002$, $n = 15$, [Figure 6], November and March - $r = 0.75$, $p< 0.001$, $n = 26$, all months – $r = 0.56$, $p<0.001$, $n = 111$).

A comparison of the landscape context of random points and the camps indicated that camps are significantly closer to urban areas than would be expected by chance (Wilcoxon signed rank test, $Z = 33.26$, $p<0.0001$). This effect did not arise during the study due to changes in where flying-foxes roosted. Individual camps did not move closer to urban areas over the period of the study. Nor did changes over time in the sub-set of camps occupied result in camps on average becoming closer to urban areas. Nor was the association of camps with urban areas due to urban expansion during the study. Urban land cover increased over the period of 1999 to 2009 from 1.3% to 1.4%, however this was not a statistically significant effect at the scale of the region (Table S2 in File S1). While there was a greater percentage of urban land cover around camps than around randomly chosen points in the landscape (Mann-Whitney U test, $Z = 5.41$, $p<0.0001$; Figure 7), comparison of land cover in 1999 and 2009 indicates no change in the extent of urban land use in the areas surrounding recorded camp locations (Mann-Whitney U test, $Z = 0.31$, $p = 0.76$). Similarly, areas surrounding occupied campsites did not become more fragmented over time than the randomly chosen locations, with no significant changes in MPA, PD or ED over time ($p>0.05$ in all cases). Despite this, camps do occur in locations that are more fragmented than random points in the landscape, with a smaller MPA, a significantly greater ED and PD (Mann-Whitney U tests, $Z = 4.11$, 3.54 and 4.55, respectively, $p<0.001$ for all; Figure 7).

Discussion

We found that spectacled flying-foxes commonly roost near humans, as is the case for other *Pteropus* species in Australia, e.g. *P. poliocephalus, P. alecto, P. scapulatus* [20,21,23], and elsewhere, e.g. *P. giganteus* [45], and *P. dasymallus* [46]. Over the period of our study the majority of the spectacled flying-foxes found in camps were found in urban-associated camps, these camps were more consistently occupied than non-urban camps, and the proportion of the counted population encountered in urban camps increased. Furthermore, while the number of non-urban camps declined through the period of the study, the number of urban camps increased.

Of the hypotheses proposed for the association between flying-foxes and urban areas, our data only allow direct assessment of hypotheses related to the effect of changes in landscape structure.

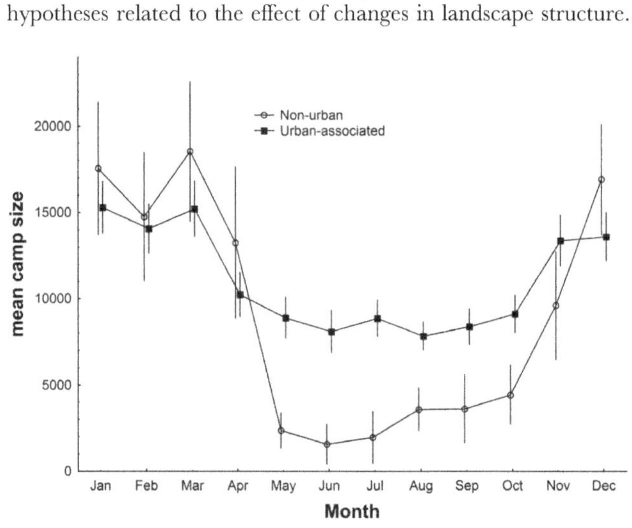

Figure 5. Changes through the year in the mean monthly camp size for urban-associated and non-urban camps. Calculated from 2004–2012 data.

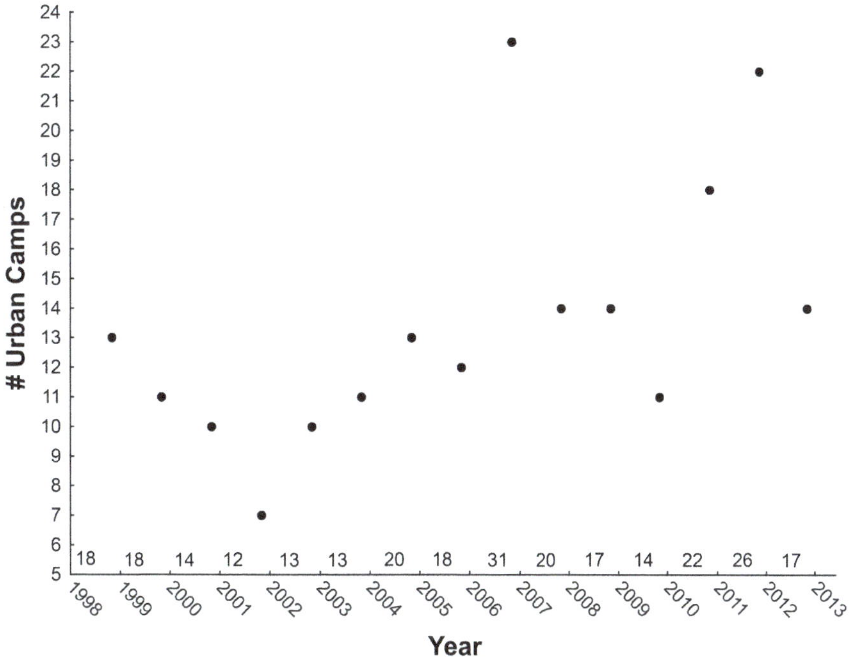

Figure 6. The number of urban associated camps recorded in surveys conducted in November of each year of the study. The number of occupied camps recorded in each year is indicated on the x axis.

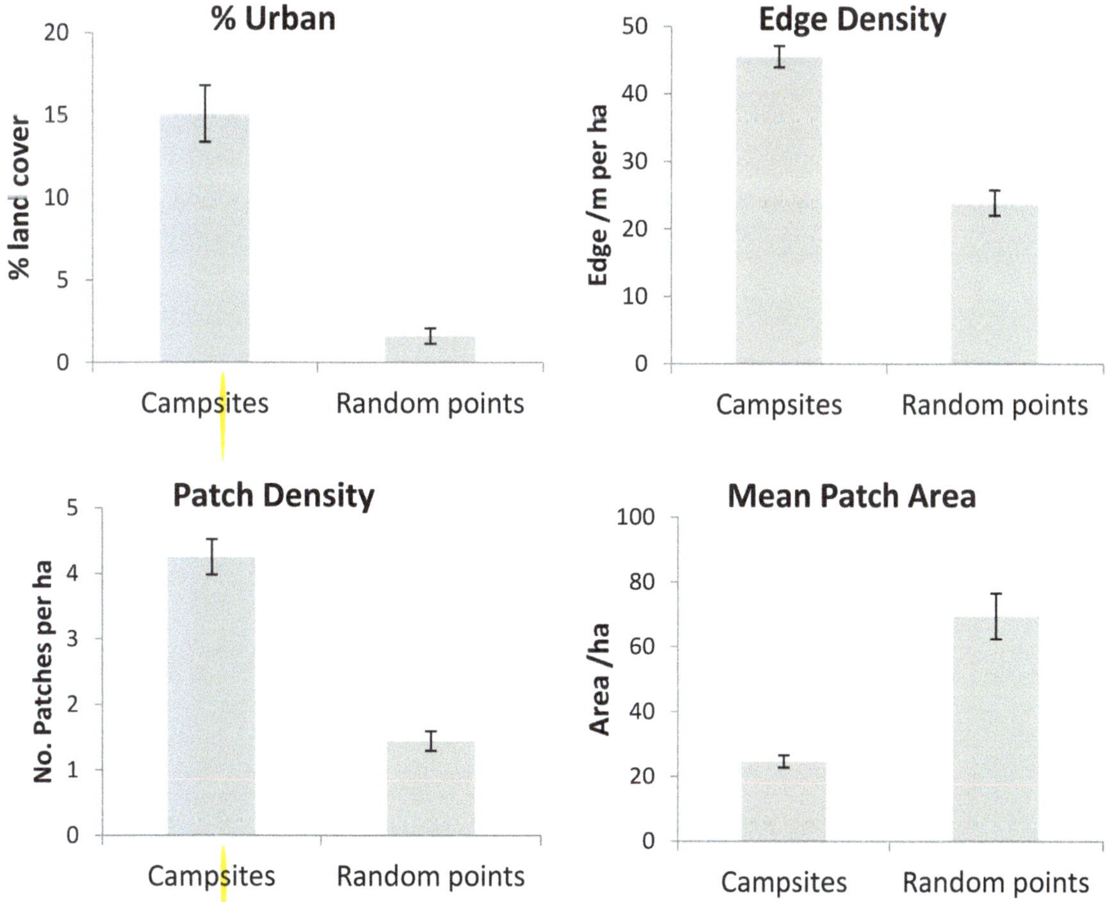

Figure 7. Landscape metrics (mean, ±SE) for buffer zones with a 3.3 km diameter surrounding all recorded campsites and random points in the landscape showing: percentage urban land cover, edge density, patch density and mean patch area.

Given that we found i) no significant changes in land cover at the scale of the landscape as a whole or, ii) in the immediate vicinity of camps, and, iii) no change in the proximity of individual camps to urban areas over the duration of the study, our findings for spectacled flying-foxes do not support the hypotheses that urbanisation is occurring due to urban expansion towards camps, habitat loss, roost site loss or habitat fragmentation [20,47,48,49]. Whether this also holds for other *Pteropus* species remains to be seen.

Our data does allow us to indirectly consider some alternative hypotheses for flying-fox urbanisation. Peaks in the urban-associated percentage of the population in 1998, 2000, 2003, 2006–7 and 2011–12 and are apparent in the November, November and March and all months' samples. The peaks suggest that there are periodic shifts in the distribution of the population towards urban areas. Although the drivers of these temporary shifts are unknown, explanations could include the attraction of fruiting or flowering events near urban areas [24], or disturbance events such as droughts [28], fires [29], cyclones [22,30,31], and human disturbance or culling at non-urban roosts or orchards [24]. While there were two major cyclonic disturbances during the study period, in 2006 and 2010 and preceding the peaks in 2007 and 2011 respectively, no disturbance events can be associated with the other peak years. Significantly, these cyclones also precede peaks in the number of occupied camps (Table S1 in File S1). Furthermore, changes in permitting for mitigation of flying-fox damage in orchards mean that there has been a decrease in disturbance and culling at non-urban associated camps over the period of the study [37]. Rather than being driven by increasing or episodic disturbance, the long-term trend of increasing urbanisation since 1998 suggests a longer-term population shift towards urban areas is occurring.

Another explanation for these results could be bias in the sampling of urban and non-urban camps; urban camps are easier to locate, access and survey and therefore potentially more likely to be monitored. Such bias is unlikely to explain our results for the following reasons. First, camps are never dropped from our monitoring, even when they haven't been occupied for long periods, so there is no shift of monitoring to an urban focus that might result from easier sampling. Second because non-urban camps are more likely to be overlooked than urban camps, we would expect an initial bias towards urban camps that would decrease as the population became better known through broad-scale searches, reports from the public [22] and telemetry studies [36,50]. Instead we have seen the shift towards urban camps increase. Consequently, we feel confident that our results are not due to a sampling bias favouring urban camps.

Our data shows a seasonal pattern of change in the size of the urban-associated proportion of the population encountered in camps, with a greater proportion found in these camps from May–Nov than Dec-Apr (Fig. 4). This period is also when the population count is at its lowest, and camps of all types are less reliably occupied and have a smaller mean size. It is possible that without strong social reasons for aggregating (at this time mating has finished and females are pregnant), the animals disperse, possibly in association with reduced or more widely dispersed foraging resources. The decrease in non-urban camp size at this time of the year was far more pronounced than that of urban-associated camps, suggesting urban-associated camps may be core population centres (Fig. 5).

The urbanisation of spectacled flying-foxes documented here has significant implications for how the management of this, and other flying-fox species, is approached. Our results support the suggestion that flying-foxes are becoming increasingly urbanised and suggest that the conflict their presence in urban areas engenders is not going to go away. The lack of evidence for loss of habitat or roosting sites as a driver of this shift further suggests that spectacled flying-foxes are not being forced into urban areas, raising the possibility that their move is a behavioural response to the advantages offered by such locations. If this is the case then it is difficult to argue that moving problem urban camps on through the use of disturbance is likely to have any significant negative impacts on the population. Despite this, there is little evidence that past attempts to move urban camps have been successful or cost effective [23] and newspaper reports from Far North Queensland over the last century suggest that the common use of lethal methods was ineffective in deterring spectacled flying-foxes from urban areas (DAW, unpubl. data). We believe that this points to a need to explore new management options, particularly the identification of options that facilitate the co-existence of humans and flying-foxes. Management of the human side of the conflict is likely to prove more cost effective and successful. Identifying the actual drivers of urbanisation of flying-foxes will be significant for understanding and managing this process.

Supporting Information

File S1 Table S1. Patterns of occupancy of camps of different types over the duration of the study. **Table S2.** Proportion of the landscape in each landcover category across the period of the study, 1999 and 2009. There are no significant differences between the proportions in each category in 1999 and 2009 ($p = 0.05$).

Acknowledgments

Louise Shilton contributed to data collection. This paper was improved by comments from J. Parsons, C. Pavey and three anonymous reviewers.

Author Contributions

Conceived and designed the experiments: JT AM HLPB DAW. Performed the experiments: AM DAW. Analyzed the data: JT AM DAW. Contributed reagents/materials/analysis tools: JT AM HLPB DAW. Wrote the paper: JT AM HLPB DAW.

References

1. Seto KC, Fragkias M, Güneralp B, Reilly MK (2011) A Meta-Analysis of Global Urban Land Expansion. PLoS One 6: e23777.
2. Magle SB, Hunt VM, Vernon M, Crooks KR (2012) Urban wildlife research: Past, present, and future. Biol. Conserv. 155: 23–32.
3. Pickett STA, Cadenasso ML, Grove JM, Boone CG, Groffman PM, et al. (2011) Urban ecological systems: Scientific foundations and a decade of progress. Journal of Environmental Management 92: 331–362.
4. Moller AP, Diaz M, Flensted-Jensen E, Grim T, Ibanez-Alamo JD, et al. (2012) High urban population density of birds reflects their timing of urbanization. Oecologia 170: 867–875.
5. Davis A, Taylor CE, Major RE (2012) Seasonal abundance and habitat use of Australian parrots in an urbanised landscape. Landsc. Urban Plan. 106: 191–198.
6. Brearley G, Rhodes J, Bradley A, Baxter G, Seabrook L, et al. (2013) Wildlife disease prevalence in human-modified landscapes. Biol. Rev. Cambridge Philosophic. Soc. 88: 427–442.
7. Bradley CA, Altizer S (2007) Urbanization and the ecology of wildlife diseases. Trends Ecol. Evol. 22: 95–102.
8. Johnston R (2001) Synanthropic birds of North America. In: Marzluff J, Bowman R, Donnelly R, editors. Avian ecology and conservation in an urbanizing world. Massachusetts: Kluwer Academic Publishers. 49–69.

9. Puth LM, Burns CE (2009) New York's nature: a review of the status and trends in species richness across the metropolitan region. Divers. Distrib. 15: 12–21.
10. DeStefano S, DeGraaf RM (2003) Exploring the ecology of suburban wildlife. Frontiers in Ecology and the Environment 1: 95–101.
11. Davis A, Taylor CE, Major RE (2011) Do fire and rainfall drive spatial and temporal population shifts in parrots? A case study using urban parrot populations. Landsc. Urban Plan. 100: 295–301.
12. Slabbekoorn H, Peet M (2003) Birds sing at a higher pitch in urban noise: Great tits hit the high notes to ensure that their mating calls are heard above the city's din. Nature Clim. Change 424: 267–267.
13. McKinney RA, Raposa KB (2013) Factors influencing expanded use of urban marine habitats by foraging wading birds. Urban Ecosystems 16: 411–426.
14. Jokimäki J, Kaisanlahti-Jokimäki M-L, Suhonen J, Clergeau P, Pautasso M, et al. (2011) Merging wildlife community ecology with animal behavioral ecology for a better urban landscape planning. Landsc. Urban Plan. 100: 383–385.
15. Zuckerberg B, Bonter DN, Hochachka WM, Koenig WD, DeGaetano AT, et al. (2011) Climatic constraints on wintering bird distributions are modified by urbanization and weather. Journal of Animal Ecology 80: 403–413.
16. Urbanek RE, Allen KR, Nielsen CK (2011) Urban and suburban deer management by state wildlife-conservation agencies. Wildlife Society Bulletin 35: 310–315.
17. Forman R, Sperling D, Bissonette J, Clevenger A, Cutshall C (2003) Road ecology: science and solutions. Washington DC: Island Press. 504 p.
18. Curtis P, Hadidian J (2010) Responding to human–carnivore conflicts in urban areas. In: Gerth S, Riley S, Cypher B, editors. Urban carnivores: ecology, conflict, and conservation. Baltimore: The Johns Hopkins University Press. 201–212.
19. Redpath SM, Young J, Evely A, Adams WM, Sutherland WJ, et al. (2013) Understanding and managing conservation conflicts. Trends Ecol. Evol. 28: 100–109.
20. Markus N, Hall L (2004) Foraging behaviour of the black flying-fox (Pteropus alecto) in the urban landscape of Brisbane, Queensland. Wildl. Res. 31: 345–355.
21. Parris KM, Hazell DL (2005) Biotic effects of climate change in urban environments: The case of the grey-headed flying-fox (Pteropus poliocephalus) in Melbourne, Australia. Biol. Conserv. 124: 267–276.
22. Shilton LA, Latch PJ, McKeown A, Pert P, Westcott DA (2008) Landscape-scale redistribution of a highly mobile threatened species, Pteropus conspicillatus (Chiroptera, Pteropodidae), in response to Tropical Cyclone Larry. Austral Ecology 33: 549–561.
23. Roberts BJ, Eby P, Catterall CP, Kanowski J, Bennett G (2011) The outcomes and costs of relocating flying-fox camps: insights from the case of Maclean, Australia. In: Law B, Eby P, Lunney D, Lumsden L, editors. The Biology and Conservation of Australasian Bats. Mosman, NSW, Australia: Royal Zoological Society of NSW. 277–287.
24. Hall L, Richards G (2000) Flying foxes: Fruit and Blosson Bats of Australia. Sydney, NSW: University of NSW.
25. Eby P (1991) Seasonal movements of gray-headed flying-foxes, *Pteropus poliocephalus* (Chiroptera, Pteropodidae), from 2 maternity camps in northern New South Wales. Wildl. Res. 18: 547–559.
26. Eby P, Richards G, Collins L, Parry-Jones K (1999) The distribution, abundance and vulnerability to population reduction of a nomadic nectarivore, the grey-headed flying-fox *Pteropus poliocephalus* in New South Wales, during a resource concentration. Australian Zoologist 31: 240–253.
27. Parry-Jones KA, Augee ML (2001) Factors affecting the occupation of a colony site in Sydney, New South Wales by the Grey-headed Flying-fox Pteropus poliocephalus (Pteropodidae). Austral Ecology 26: 47–55.
28. Tidemann CR, Nelson JE (2004) Long-distance movements of the grey-headed flying fox (Pteropus poliocephalus). Journal of Zoology 263: 141–146.
29. Jenkins RKB, Andriafidison D, Razafimanahaka HJ, Rabearivelo A, Razafindrakoto N, et al. (2007) Not rare, but threatened: the endemic Madagascar flying fox Pteropus rufus in a fragmented landscape. Oryx 41: 263–271.
30. Craig P, Trail P, Morrell TE (1994) The decline of fruit bats in American Samoa due to hurricanes and overhunting. Biol. Conserv. 69: 261–266.
31. Esselstyn JA, Amar A, Janeke D (2006) Impact of Post-typhoon Hunting on Mariana Fruit Bats (Pteropus mariannus) 1. Pacific science 60: 531–539.
32. Meade J, Biro D, Guilford T (2005) Homing pigeons develop local route stereotypy. Proceedings of the Royal Society B: Biological Sciences 272: 17.
33. Halpin K, Hyatt AD, Plowright RK, Epstein JH, Daszak P, et al. (2007) Emerging viruses: Coming in on a wrinkled wing and a prayer. Clinical Infectious Diseases 44: 711–717.
34. Drexler JF, Corman VM, Muller MA, Maganga GD, Vallo P, et al. (2012) Bats host major mammalian paramyxoviruses. Nat Commun 3: 796.
35. Plowright RK, Foley P, Field HE, Dobson AP, Foley JE, et al. (2011) Urban habituation, ecological connectivity and epidemic dampening: the emergence of Hendra virus from flying foxes (Pteropus spp.). Proc. R. Soc. B-Biol. Sci. 278: 3703–3712.
36. Westcott DA, Dennis AJ, McKeown A, Bradford M, Margules C (2001) The Spectacled Flying fox, Pteropus conspicillatus, in the context of the world heritage values of the Wet Tropics World Heritage Area. Atherton, Queensland: CSIRO. 75 p.
37. Queensland Department of Environment and Resource Management (2010) National recovery plan for the spectacled flying fox Pteropus conspicillatus. Report to the Department of Sustainability, Environment, Water, Population and Communities, Canberra.
38. Parsons JG, Blair D, Luly J, Robson SKA (2009) Bat Strikes in the Australian Aviation Industry. J. Wildl. Manage. 73: 526–529.
39. Stork N, Turton S (2009) Living in a dynamic tropical forest Landscape. Oxford: Blackwell Publishing Ltd.
40. Queensland Treasury and Trade (2012) Population and Dwelling Profile, Far North Queensland. Office of Economic and Statistical Research.
41. Garnett S, Whybird O, Spencer HJ (1999) The Conservation status of the Spectacled Flying-fox Pteropus conspicillatus in Australia. Australian Zoologist 31: 38–54.
42. Queensland Government Information Service (2012).
43. McGarigal K, Cushman S, Ene E (2012) FRAGSTATS v4: Spatial Pattern Analysis Program for Categorical and Continuous Maps.
44. Perotto-Baldivieso HL, Melendez-Ackerman E, Garcia MA, Leimgruber P, Cooper SM, et al. (2009) Spatial distribution, connectivity, and the influence of scale: habitat availability for the endangered Mona Island rock iguana. Biodiversity and Conservation 18: 905–917.
45. Mahmood-ul-Hassan M, Gulraiz TL, Rana SA, Javid A (2010) The diet of Indian flying-foxes (Pteropus giganteus) in urban habitats of Pakistan. Acta Chiropt. 12: 341–347.
46. Nakamoto A, Kinjo K, Izawa M (2007) Food habits of Orii's flying-fox, Pteropus dasymallus inopinatus, in relation to food availability in an urban area of Okinawa-jima Island, the Ryukyu Archipelago, Japan. Acta Chiropt. 9: 237–249.
47. Williams NSG, McDonnell MJ, Phelan GK, Keim LD, Van der Ree R (2006) Range expansion due to urbanization: Increased food resources attract Grey-headed Flying-foxes (Pteropus poliocephalus) to Melbourne. Austral Ecology 31: 190–198.
48. Lunney D, Moon C (1997) Flying-foxes and their camps in the rainforest remnants of north-east NSW. In: Dargavel J, editor. Australia's everchanging forests III. Canberra: Centre for Resource and Environmental Studies, Australian National University. 247–277.
49. van der Ree R, McDonnell MJ, Temby I, Nelson J, Whittingham E (2006) The establishment and dynamics of a recently established urban camp of flying foxes (Pteropus poliocephalus) outside their geographic range. Journal of Zoology 268: 177–185.
50. McKeown A, Westcott DA (2012) Assessing the accuracy of small satellite transmitters on free-living flying-foxes. Austral Ecology 37: 295–301.

18

Running the Gauntlet: Regional Movement Patterns of *Manta alfredi* through a Complex of Parks and Fisheries

Elitza S. Germanov*, Andrea D. Marshall

Marine Megafauna Foundation, Truckee, California, United States of America

Abstract

Manta rays (Genus *Manta*) are economically important for fisheries and tourism in Indonesia. These species have been listed by the International Union for the Conservation of Nature Red List as Vulnerable to extinction; therefore, human exploitation of manta rays must be regulated. A better understanding of the habitat use and movement patterns of manta rays in Indonesia is needed in order to employ effective conservation measures. To gain better insight into the movements of *Manta alfredi* we used 'Manta Matcher', an online database with an integrated automated matching algorithm, to compare photographs from 2,604 encounters of *M. alfredi* collected by recreational divers and dive operators throughout Indonesia over a nine-year period. This photographic comparison revealed that manta rays migrated between regional sanctuaries such as Nusa Penida, the Gili Islands, and the Komodo National Park (up to 450 km straight-line distance). The areas between these sanctuaries are heavily fished and trafficked by ships, and when manta rays travel through these regions they risk being fished and injured by ship strikes. These long-range manta ray movements suggest connectivity between *M. alfredi* populations in neighboring islands and raise concerns about the future management of regional populations. It is recommended that a national conservation strategy be developed to protect the remaining populations in the country.

Editor: Michael L. Fine, Virginia Commonwealth Univ, United States of America

Funding: The authors received no specific funding for this work. This work was enabled by large scale volunteer efforts.

Competing Interests: The authors have declared that no competing interests exist.

* Email: germanov@dal.ca

Introduction

Manta rays (genus *Manta*) are large pelagic filter-feeding, elasmobranchs that are economically important for both tourism [1] and fisheries throughout many parts of their range [2,3]. Manta rays are classified as Vulnerable to extinction according to the International Union for Conservation of Nature (IUCN) Red List [4,5] due to their slow growth rates, low fecundity [6] and globally rising pressure from fisheries [3,7]. Manta rays aggregate at many locations within the Indonesian archipelago [3,7,8]. Indonesia has the fourth highest number of known tourism sites for viewing manta rays worldwide and rates as one of the top countries in the world for manta ray watching tourism [1]. Manta ray tourism in Indonesia is valued at over $15 million USD per year [1]. Indonesia also has some of the most aggressive targeted fisheries for manta rays [2,3,9,10]. Fisheries most negatively impact the sizes of manta ray populations [1,3]. The reduced number of manta rays will likely negatively impact the burgeoning manta ray tourism industry [1].

To protect the manta ray tourism economy in Indonesia, three manta ray sanctuaries were established in Raja Ampat (11,655 km^2), West Manggarai including Komodo National Park (7,000 km^2), and Nusa Penida (200 km^2). In 2013, manta rays were listed on Appendix II of the Convention on International Trade in Endangered Species (CITES) as a preemptive measure against potentially unsustainable fisheries. As a member country, party to this convention, in early 2014 Indonesia further declared manta rays a protected species, prohibiting the catch of *Manta* species throughout the entire exclusive economic zone of the country (Seas and Fisheries decree number 4). Indonesian and CITES regulations will help curb fishing related threats within the region, but sound conservation strategies are also needed to protect decreasing manta ray populations [7]. Information about the distribution, movement patterns, and population sizes of manta rays is needed in order to create effective conservation management strategies.

Understanding the migratory range of manta rays (*M. alfredi*) in Indonesia is particularly important, as major fisheries often occur near to marine sanctuaries. A previous study using acoustic telemetry, showed that *M. alfredi* (referred to then as *M. birostris*) exhibit considerable site fidelity through localized re-sightings within areas of the Komodo National Park [11]. This study also demonstrated that there was considerable movement (of up to approximately 33.5 km apart) between the individual sites that coincided with seasonal monsoon-shifts [11]. While this was a good preliminary effort to learn more about the habitat use of *M. alfredi* within the Komodo National Park, this study was spatially limited by the technology used leaving many unanswered questions. For instance, (34%) of the tagged individuals spent a minimum of 2 months and a maximum of 9 months outside of the range of the acoustic receiver array (which was set up within the park) suggesting that *M. alfredi* may also travel to and spend significant amounts of time in other areas [11]. Recent studies have reported *M. alfredi* movements greater than 380 km (straight-line distance) in places like Mozambique and Australia [5,8,12–14]. Major fishing grounds and known manta ray landing ports, Tanjung Luar, Lombok [3] and Lamakera [3,10], are within 380 km of *M. alfredi* sanctuaries in West Manggarai and Nusa Penida (Figure 1). Thus it is plausible that the migratory range of *M. alfredi* may overlap with heavily fished areas in Indonesia.

The aim of this study was to determine the connectivity and rate of movement of manta rays between the manta ray sanctuaries of Nusa Penida, West Mangarrai province of Flores including the Komodo National Park and Raja Ampat. This study used a large-scale photo-identification (photo-ID) approach to monitor manta ray movements using publically contributed photographs. Photo-ID methodology, where by animals are identified using their unique ventral patterning, has been successfully used in a large number of manta ray population and movement studies to date [6,8,13,15–17]. Photo-ID data usage does have limitations [18], but it is still a cost-effective and minimally invasive tool for gathering large amounts of data, especially when coupled with citizen science efforts [8,12,13,19,20]. This is the first study to use 'Manta Matcher', a global collaborative photo-ID matching database, to compare regional photo-ID databases. 'Manta Matcher' uses an automated biometrics algorithm to accelerate the photo-identification process [21]. Using 'Manta Matcher' enabled the consolidation of existing citizen-scientist collected data in Indonesia into an easily accessible database, thereby uncovering long-distance *M. alfredi* movements and highlighting the need for more effective conservation strategies of this migratory species.

Methods

Study locations

Photographs of the ventral surfaces of individual *M. alfredi* were collected year round by local dive operators and recreational divers at various manta ray aggregation sites throughout Indonesia (Figures 1, 2 and 3) from 2006–February 2014. Data that recently became available on manta ray INNLP0229A in both June and July 2014 were also included in this study. Photographs were collected opportunistically with effort mostly concentrated at sites in Nusa Penida (NP), West Mangarrai & Komodo National Park (WM & K) and Raja Ampat (RA) regions (Table 1). Limited data from Pulau Weh in Aceh province, Pulau Sangalaki in Kalimantan and the Gili Islands was also collected. Monitored sites represent a range of habitat used by *M. alfredi*, including cleaning stations, feeding aggregation areas, and reproductive grounds.

The majority of photographic data comes from the monitoring of 16 separate sites within the NP, WM & K and RA regions. Photographs within the NP region were mostly collected from three nearby sites: Small Manta Bay, Big Manta Bay and Manta Point (Figure 2 inset A.1). The two furthest sites are located at approximately 12.4 km straight-line distance apart. The photographs submitted for the WM & K region were collected from 7 sites out of 15 regularly visited locations (Figure 2 inset A.2). The majority of photographs available for the WM & K region were collected from Karang Makassar and Mawan Island, both located in the central-north region of the Komodo National Park, and Manta Alley to the south of Komodo Island. Mawan Island and Manta Alley are respectively located 4.4 km to the east and approximately 33.5 km (straight line) to the south of Karang Makassar (Figure 2 inset A.2). Most of the monitored sites in the RA region are in close proximity to one another (Figure 3). Manta Sandy, Manta Ridge, Arborek, and Manta Heaven, are all within a 3.5 km straight-line distance. Two further sites from which photographs were obtained, Blue Magic and Dayang, are respectively located approximately 23 km south and 24 km east from Manta Sandy.

Photographic identification

Manta rays can be individually distinguished by their unique ventral markings [15,16], allowing for the application of photo-ID techniques. Photographs of the standardized area [2,8,13,15–17,22] on the ventral surface of manta rays were taken from still photography with or without flash, or from video screen grabs, as described in [8,13,15,16]. Species identification was performed using key morphological features described by Marshall *et al.* (2009) [22]. Whenever possible, manta rays were sexed and their maturity status assessed based on clasper size and/or clasper scarring for males, and pregnancy bulge and/or presence of pectoral fin mating scars for females as outlined in [6]. Other distinguishing features such as color morphology, scars and injuries were also noted.

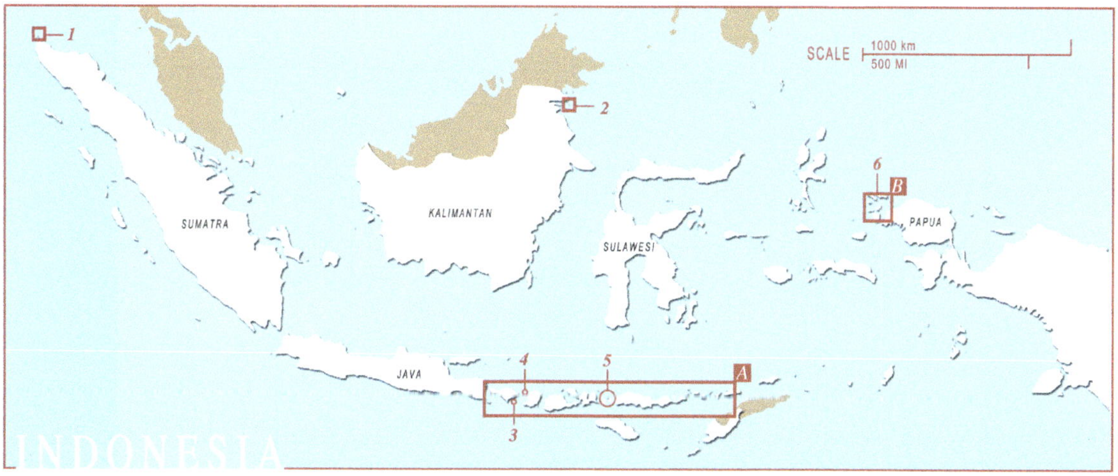

Figure 1. The locations of *M. alfredi* encounters in Indonesia submitted to Manta Matcher. Sites are numbered as follows: 1) Pulau Weh 2) Palau Sangalaki 3) Nusa Penida 4) Gili Islands 5) West Manggarai & Komodo National Park 6) Raja Ampat. (A) Connectivity area and commercial fisheries area (see Figure 2 for enlargement). (B) Region surveyed in Raja Ampat (see Figure 3 for enlargement).

Table 1. The survey effort in Nusa Penida, West Manggarai & Komodo, and Raja Ampat regions including sighting records and number of individuals identified.

Region	Year	Nusa Penida	West Manggarai & Komodo	Raja Ampat	Total
Total Sighting Records	2006–14	2007	426	171	2604
Yearly Sighting Records	2006–10	83	-	-	83
	2011	84	49	-	133
	2012	544	37	-	581
	2013	1092	329	53	1474
	2014	203	11	113	327
M. alfredi identified (N)	2006–14	417	303	100	820

'Manta Matcher'

Photographs of individual manta ray encounters were submitted online by the authors and volunteers to 'Manta Matcher' (www.mantamatcher.org), the Wildbook for manta rays created by WildMe. Individual encounter profiles were completed for each entry (see Table S1 for encounter profiles of key individuals), which included the photographer and/or submitter, date, location and GPS coordinates of sightings. Whenever possible sex, maturity status, color morphology and any other distinguishing characteristics were added. Only clear, good-quality images in which individual rays could be identified using the standardized ventral spot pattern area were submitted to the 'Manta Matcher' biometrics algorithm [21]. When necessary images were first post processed using Adobe Lightroom 3.3 or Photoshop CS5 to enhance visualization of the ventral spot patterns by adjusting brightness and contrast prior to uploading. Images were cropped to isolate the critical identification region (the gill area and abdomen) in a head up orientation [21]. Newly submitted photographs were compared against both the global and regional Indonesian database of submitted images. Overall confidence of results in a scale of 1 (good) and 0 (poor) were automatically generated for every query along with a ranked list of matched hits [21]. Manta rays were subsequently matched by eye from the generated list of ranked hits allowing for regional and global comparisons. Screen grabs of positive matches were saved electronically.

Ethics statement

The authors deemed that the nature of the work (analysis of publically contributed photographs taken opportunistically by recreational divers) did not require any approval or permits regarding human or animal ethics. When photographs came from areas within marine sanctuaries or national parks the burden of

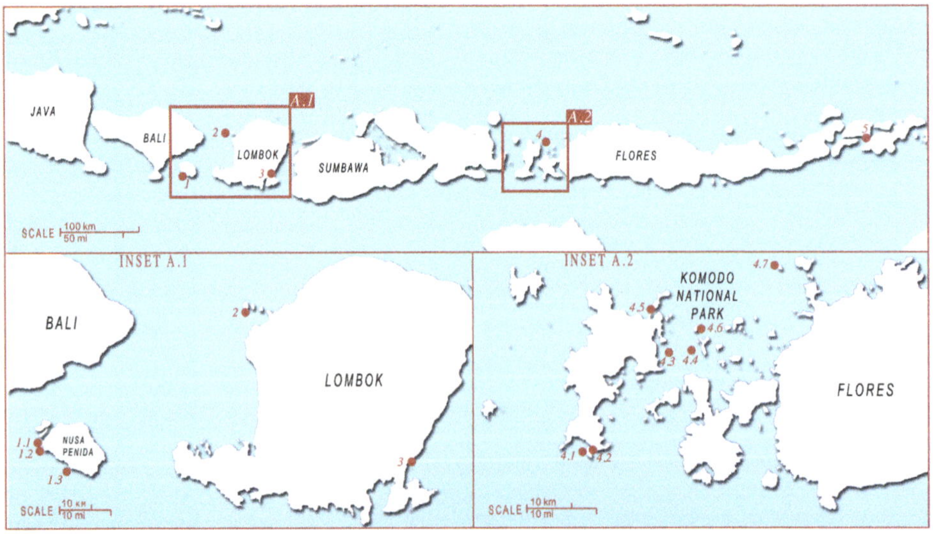

Figure 2. The connectivity area encompassing Nusa Penida, West Manggarai & Komodo regions and the nearby commercial manta ray landing ports. Regions and fishing ports are numbered as follows: 1) Nusa Penida 2) Gili Islands 3) Tunjung Luar fishing port 4) Komodo National Park 5) Lamakera fishing port. (A) Manta ray monitoring sites in Nusa Penida and Lombok and nearby Tujung Luar fishing port: 1.1) Small Manta Bay 1.2) Big Manta Bay 1.3) Manta Point 2) Gili Islands 3) Tanjung Luar fishing port. (B) Manta ray monitoring sites in West Mangarrai and Komodo National Park: 4.1) Manta Alley 4.2) German Flag 4.3) Karang Makassar/Manta Point 4.4) Mawan 4.5) The Cauldron 4.6) Tatawa Besar 4.7) Sabolon Kecil.

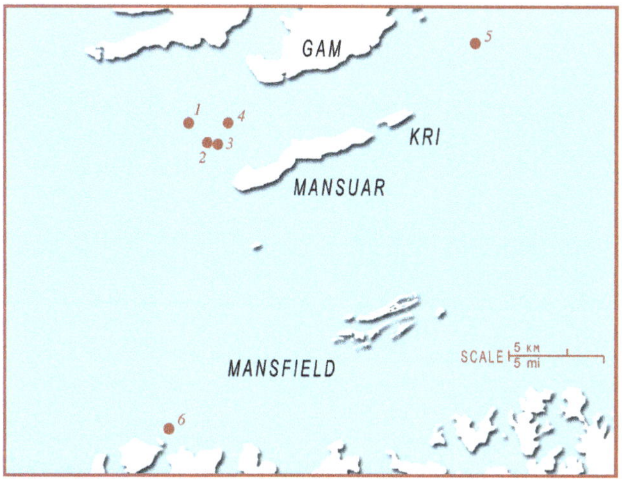

Figure 3. A map of the Raja Ampat region to indicate the location of different monitoring sites. Sites are numbered as follows: 1) Manta Sandy 2) Jetty Arborek 3) Blue Magic 4) Manta Ridge 5) Manta Heaven 6) Dayang.

possession of suitable diving and entrance permits was placed on the individual submitting the photographs.

Results

Identified Individuals

A total of 820 -individual *M. alfredi* were identified from 2,604 sightings within the NP, WM & K and RA regions of Indonesia between 2006 and February 2014 (Figure 1). Re-sighting data on one individual (INNLP0229A) sighted in June and July 2014 were also included. Specific sighting rates are listed in Table 1. In the WM & K region 100 out of 303 identified manta rays (33.1%) were sighted more than once, an average of 2.7 times per individual during the study period. In the NP region 274 out of 419 manta rays (65.4%) identified during the study period were sighted more than once, an average of 6.9 times per individual. During the study period 100 individual manta rays were positively identified in the RA region. In this region 30 out of 100 manta rays (30%) were sighted more than once, an average of 3.1 times per individual.

Long-range Movements and Regional Connectivity

The photo-ID comparisons revealed long-range and inter-regional movements of individual manta rays. In the RA region, one male manta ray (INRA0050A) moved between Manta Sandy and Dayang within 3 days and returned to Aborek within 5 days later, a roundtrip straight-line distance of approximately 48 km within 8 days (Figure 3). Furthermore, two female manta rays (Figure 4A, B and Table 2) sighted in the NP region were later sighted in the waters off the Gili Islands, Lombok (Figure 2 inset A. 1), a minimum straight-line distance of approximately 80 km within 131 days (INNLP0057A) and 343 days (INNLP0074A). One of these individuals (INNLP0074A) also subsequently made a return journey back to NP within 38 days.

Moreover, during the study period, three manta rays (Figures 5, 6 and 7), over seven occasions (Table 2), moved between NP and WM & K, a straight-line distance of approximately 450 km (Figure 2). One of these manta rays, a mature female (INNLP0031A), moved from NP to WM & K and then back again to NP (Figure 5), the fastest of the migrations occurring in 157 days or less (Table 2). This manta ray was then re-sighted in the NP region on several occasions during the remainder of the study period but was not re-sighted back in the WM & K region. A mature male ray (INNLP0229A) made this inter-regional journey from NP to the south region of WM & K and then back to NP with the fastest of the migrations occurring in 44 days or less (Figure 6 and Table 2). The fastest migration on record was a mature female ray (INNLP0059A), which traveled from NP to WM & K twice during the study period (Figure 7), once within a 33-day period (Table 2).

Discussion

Movement Patterns of *M. alfredi* in Indonesia

A collaborative regional photo-ID database comparison indicated that certain individual manta rays might regularly move

Table 2. The sighting records of individual rays migrating between Nusa Penida and Gili Islands or West Manggarai & Komodo regions.

Identification #	Nusa Penida	Gili Islands	West Manggarai & Komodo	Time between Resightings (days)	Approximate Distance (km)
INNLP0031A	2009-Aug-18⇒		⇒2013-Apr-03	1324	450
	2013-Sep-07←		←2013-Apr-03	157	450
INNLP0057A	2012-Sep-22⇒	⇒2013-Jan-31		131	80
INNLP0059A	2008-Dec-25⇒		⇒2013-Mar-24	1550	450
	2013-Jun-03←		←2013-Mar-24	72	450
	2013-Dec-07⇒		⇒2014-Jan-09	33	450
INNLP0074A	2012-Mar-06⇒	⇒2013-Feb-12		343	80
	2013-Mar-22←	*2013-Feb-12		38	80
INNLP0229A	2012-Sep-13⇒		⇒2013-Jun-10	270	435
	2014-Jul-18←		←2014-Jun-04	44	435

Arrows highlight direction of movements.
⇒west to east movements;
←east to west movements.

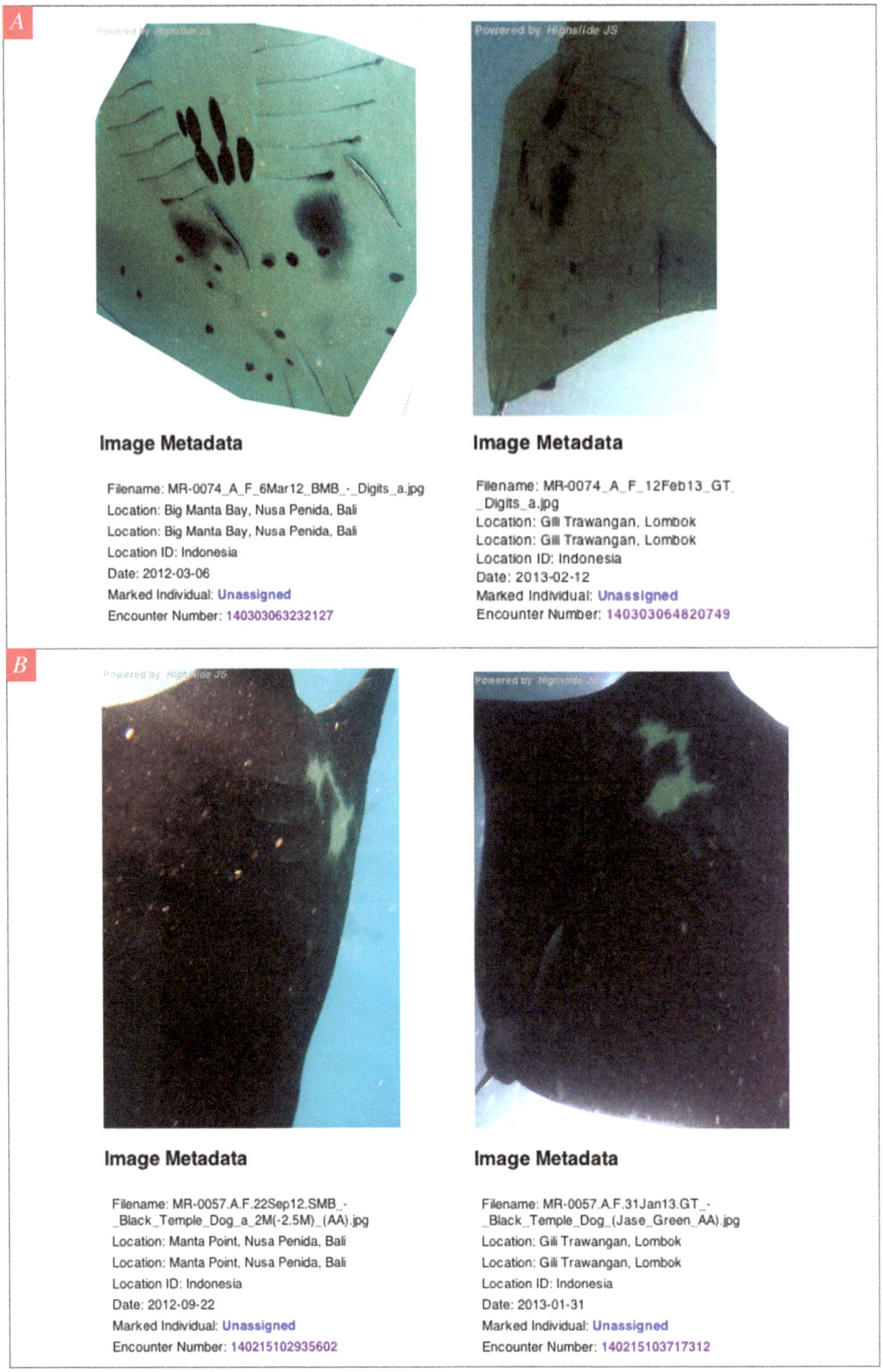

Figure 4. Screen grabs of photographs uploaded to the 'Manta Matcher' online database documenting *M. alfredi* movements from Nusa Penida (NP), Bali to Gili Islands (GI), Lombok. A. Manta ID code INNLP0074A. B. Manta ID code INNLP0057A.

between aggregation sites within the manta ray sanctuaries of WM & K, NP and Gili Islands. These results support studies in other parts of the world [8,11,13,16,17] that have indicated the potential for regional mobility in this species. Moreover, data from this study support the growing concept that *M. alfredi* are capable of rapid, long distance migrations. While to date no international movements have been recorded, the round-trip migrations exhibited by manta rays in southern-central Indonesia suggest that at least certain individuals would be capable of international exchange in some parts of their range. Considering their global

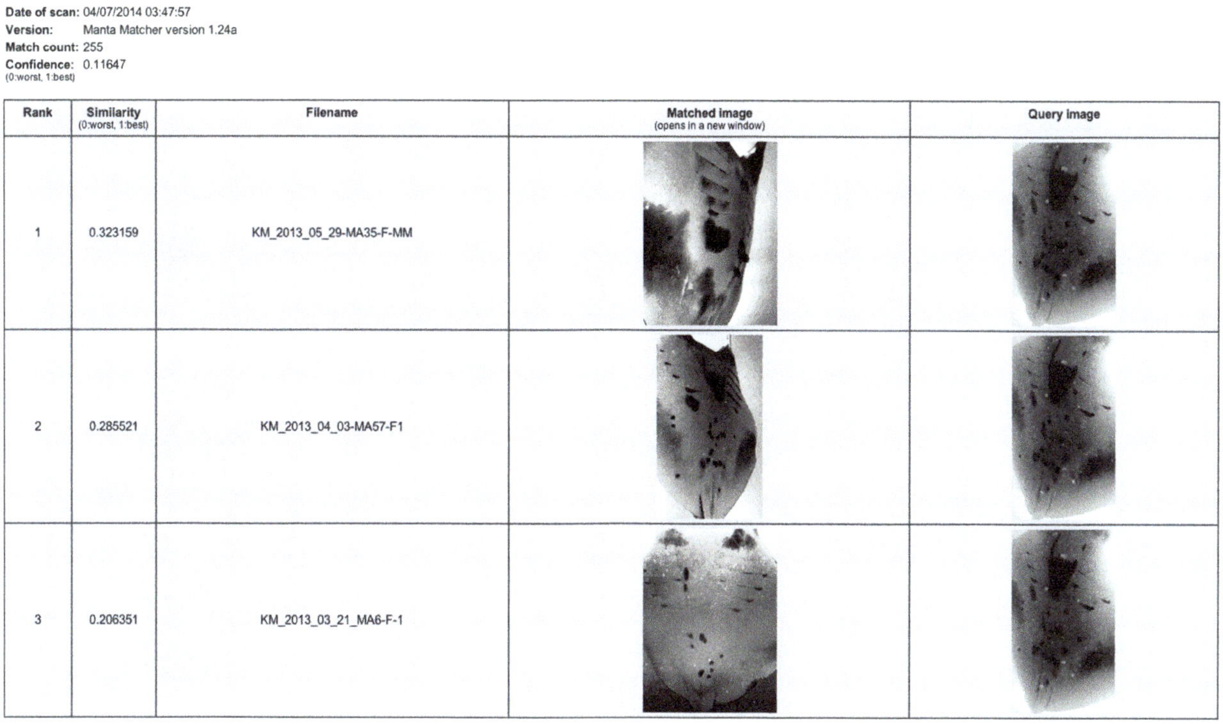

Figure 5. The 'Manta Matcher' matching algorithm screen grabs for INNLP0031A showing the top three matched images. Ranked hit 2 (re-sighting in WM & K) is a true match for the query image of ray INNLP0031A (initial sighting in NP).

conservation status and CITES listing it may be appropriate to include *M. alfredi* on the Convention for Migratory Species Act (CMS) to help support regional management plans between neighboring countries.

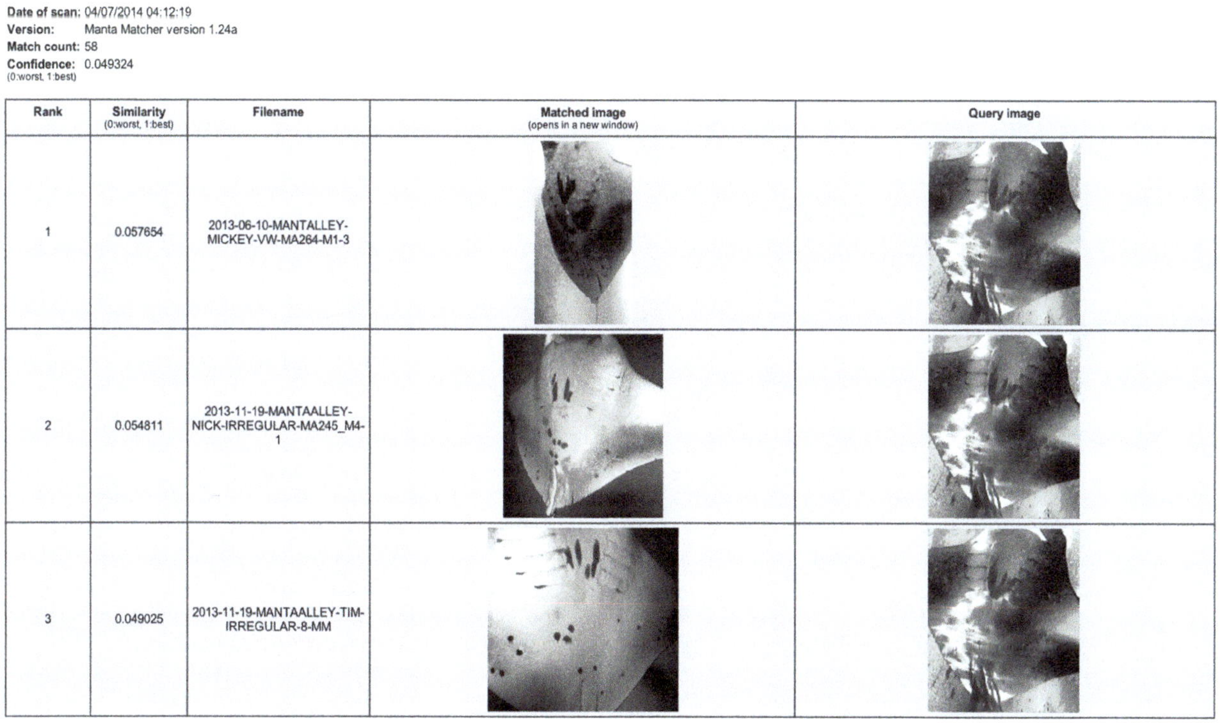

Figure 6. The 'Manta Matcher' matching algorithm screen grabs for INNLP0229A showing the top three matched images. Ranked hit 1 (re-sighting in WM & K) is a true match for the query image of ray INNLP0229A (initial sighting in NP).

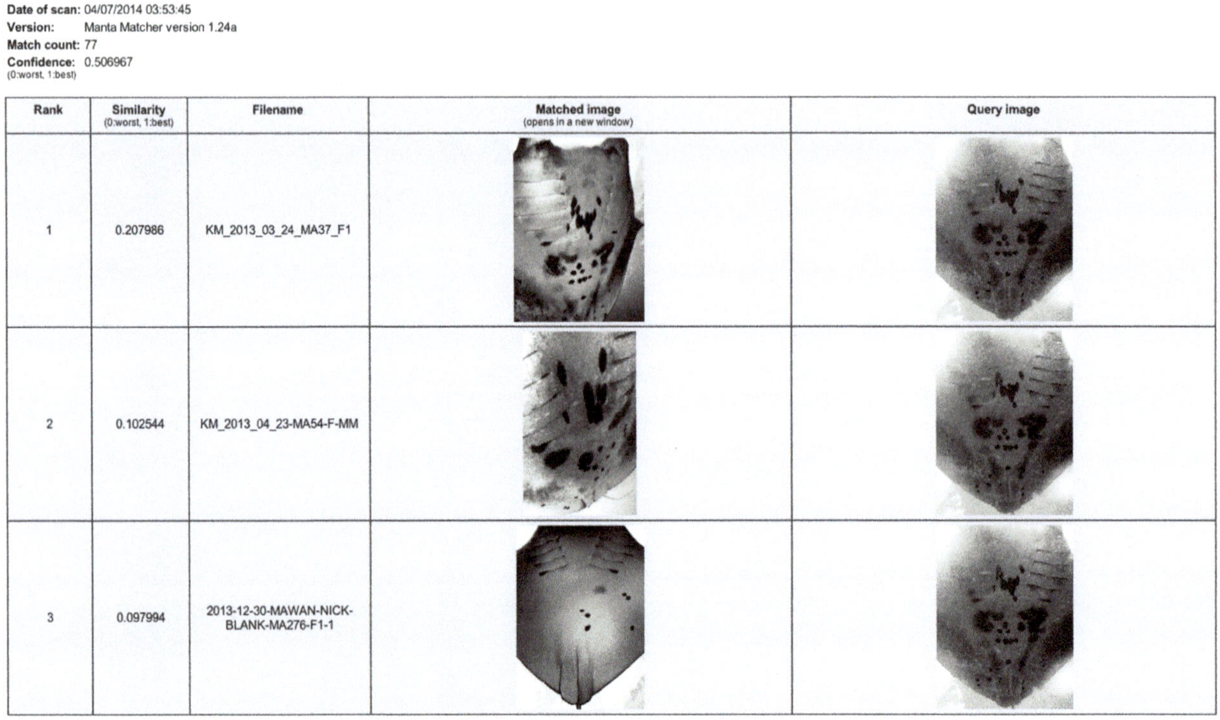

Date of scan: 04/07/2014 03:53:45
Version: Manta Matcher version 1.24a
Match count: 77
Confidence: 0.506967
(0:worst, 1:best)

Rank	Similarity (0:worst, 1:best)	Filename	Matched image (opens in a new window)	Query image
1	0.207986	KM_2013_03_24_MA37_F1		
2	0.102544	KM_2013_04_23-MA54-F-MM		
3	0.097994	2013-12-30-MAWAN-NICK-BLANK-MA276-F1-1		

Figure 7. The 'Manta Matcher' matching algorithm screen grabs for INNLP0037A showing the top three matched images. Ranked hit 1 (re-sightings in WM & K) is a true match for the query image of ray INNLP0037A (initial sighting in NP).

Long-term studies in Australia [8,12,13], Hawaii [17], Maldives [16] and Mozambique [15], have all noted highly localized re-sighting rates for *M. alfredi*, clear diurnal patterns in their inshore and offshore habitat use, and in some cases seasonal or cyclical migratory behavior. The present study showed that some manta rays are making inter-island movements of up to 450 km straight-line distance within the Indonesian Archipelago being one of the furthest reported straight-line distance migrations for this species. In one specific case a movement of this distance was made in as little as 33 days, the fastest straight-line movement for this species on record. *Manta alfredi* have been documented to make long-distance movements, with the longest traveled straight-line distance reported using photo-ID to be 650 km over a six-month period [8]. In most instances these movements involved individuals transiting along continuous continental coastlines rather than across island chains [8,13]. However, a recent study using satellite telemetry off eastern Australia found that one manta ray traveled 2,441 km in 118 days and moved 155 km offshore [14]. While physical barriers, including open expanses of sea in some island nations [17] may restrict *M. alfredi,* we are now left to consider that *M. alfredi* under certain circumstances or in some locations may undertake regular longer distance movements. Understanding the regular migratory routes of *M. alfredi* in Indonesia could have important implications to national management plans for this species.

The reasons underpinning movements between aggregation sites within Indonesia are yet to be fully determined but could be linked with environmental (seasons, tidal influence, temperature, moon phase), biological (food availability, mating), or ecological (critical habitat use, habit disruption) factors. Dewar *et al.* (2008) suggest that the north to south movements within the Komodo National Park coincide with seasonal changes and monsoon shifts and the associated reduction of productivity [11]. Jaine *et al.* (2012) and Anderson *et al.* (2011) also provide evidence that manta ray visitations to Lady Elliot Island, Australia [12] and movements in the Maldives [23], respectively, are seasonally driven and linked to productivity. A comprehensive study of the feeding ecology of this species in Indonesia is therefore warranted and could assist in the understanding of their migratory behaviors in the region.

Highlighting threats

Marine protected areas are seen as the hallmark of conservation and their benefits have been extensively reviewed in [24–26]. Creating comprehensive protection for migratory species can often prove challenging, with many marine species using both inshore aggregation areas and offshore habitats [14,27–28]. Regional management strategies for these species therefore need to incorporate seasonal or regular migratory behavior. This study demonstrated connectivity between several aggregation areas for *M. alfredi* in Indonesia and highlighted the need to work towards more comprehensive regional protection for these species in an area that is currently fraught with anthropogenic threats.

Considering the likely trajectory of individuals moving between aggregation sites, manta rays may be crossing important shipping corridors, increasing their likelihood for boat strike. One of the documented migrations in this study required an individual to pass through the busy shipping corridor located between Mansuar and Mansfield Islands in the RA region. Manta rays transiting from Nusa Penida to the Gili Islands would likely be crossing the Lombok Strait, a heavily used shipping corridor between the islands of Bali and Lombok. Graham *et al.* [27] highlighted similar concerns for manta rays using inshore habitats off the Yucatan, with busy shipping lanes overlapping the areas being utilized by satellite tagged individuals.

More over, in 2012, Indonesia ranked as the 3rd most aggressive fishing nation for manta rays [3]. One of its most productive manta ray fishing ports (Tanjung Luar) lies directly in between NP and WM & K regions, while the other (Lamakera) lies about 380 km east of WM & K [3]. Given their migratory ability, it is reasonable to assume that *M. alfredi* using NP, the Gili Islands, and WM & K regions are at risk from targeted fishing when traveling outside of these discrete protected sanctuaries. Recent legislation has shown that Indonesia is taking the right steps forward to safeguarding their manta rays by prohibiting fishing throughout their entire exclusive economic zone (an area of over 6 million square kilometers). In reality however, it may be a long time before all manta ray fisheries in Indonesia are completely shut down. Additional regional research on the movements and habitat use of manta species will further elucidate trends in their behavior aiding to strengthen management strategies.

Comprehensive management of manta ray populations within Indonesia would be a critical step towards developing a sustainable tourism industry for these animals in the country. Annually manta ray related tourism brings in more revenue to Indonesia (>15 million US$) than manta ray fisheries (approximately US$442,000) [1]. Conceivably manta ray tourism is still a growing industry in Indonesia as manta rays are encountered in many regions throughout the country [1]. Given the economic opportunities and benefits it is recommended that a national management strategy for this species be drafted immediately to protect remaining populations within the country.

'Manta Matcher' as a collaborative tool

Data collection for this study was enabled by the participation of citizen scientists. The sampling range and the amount of available data for this study were both enhanced by combining the work of local dive operators and interested members of the public. Citizen science data is increasingly used in scientific studies [8,12,13,19,20] as it is a cost effective and relatively fast way to collect large amounts of data. The 'Manta Matcher' was collaboratively developed as a tool for researchers to store, organize and analyze datasets. The automated matching algorithm built into the online interface considerably reduces the time it takes to match new data to existing catalogues and increases the accuracy of the photo-ID matching process. As the first public global online database for manta rays, 'Manta Matcher' provides the ideal platform for regional collaborations by consolidating different data collection efforts and allowing for more comprehensive studies to take place. Using 'Manta Matcher' we were able to create regional catalogues and subsequently cross-reference these catalogues with ease as well as search for matches with any previously contributed data. The 'Manta Matcher' database for Indonesia has quickly risen to over 820 individuals, making it one of the largest composites of identified *M. alfredi* in the world. Given the effectiveness of a matching tool like 'Manta Matcher' and its ability to facilitate regional and global research efforts, it is recommended that this global program become the standard for manta ray researchers.

Acknowledgments

We are indebted to H. Mitchell and P. Bassett for their tremendous contribution enabling this project. We are thankful to G. Winstanley for his support regarding 'Manta Matcher'. We are grateful to the numerous citizen scientists, dive industry professionals and dive operators who have enabled this large-scale data collection. In particular, we would like to acknowledge D. Arriaga, J. Arriaga, T. Birkett, J. Green, R. Horner, L. Johnston, K. Klingeler, P. Kuijper, N. Longfellow, J. Marlow, F. Nompas, N. Prest and Barefoot Conservation for their contributions, as well as F. Claassen, M. Cobussen and L. Ellevog for the tremendous assistance in data processing and analysis. We are thankful to L. Couturier, J. Musick and one anonymous reviewer for their comments on the manuscript. We thank P. Bassett and V. Brakalova for their figure design and formatting. ADM would like to thank A. Rooney and J. Artendale for the generous funding which supported the development of 'Manta Matcher'.

Author Contributions

Conceived and designed the experiments: ESG ADM. Analyzed the data: ESG ADM. Performed the experiments: ESG HMM PMB ADM. Contributed reagents/materials/analysis tools: ESG HMM PMB ADM. Contributed to the writing of the manuscript: ESG ADM.

References

1. O'Malley MP, Lee-Brooks K, Medd HB (2013) The global economic impact of manta ray watching tourism. PLoS ONE 8: e65051.
2. Couturier L, Marshall A, Jaine F, Kashiwagi T, Pierce S, et al. (2012) Biology, ecology and conservation of the Mobulidae. J Fish Biol 80: 1075–1119.
3. Heinrichs S, O'Malley M, Medd HB, Hilton P (2011) The Global Threat to Manta and Mobula Rays. Manta Ray of Hope. Available: http://www.mantarayofhope.com/downloads/The-Global-Threat-to-Manta-and-Mobula-Rays.pdf. Accessed 2013 Apr 23.
4. Marshall A, Bennett MB, Kodja G, Hinojosa-Alvarez S, Galvan-Magana F, et al. (2011) *Manta birostris*. IUCN Red List of Threatened Species. Version 2011.2.
5. Marshall A, Kashiwagi T, Bennett MB, Deakos M, Stevens G, et al. (2011) *Manta alfredi*. IUCN Red List of Threatened Species. Version 2012.2.
6. Marshall A, Bennett M (2010) Reproductive ecology of the reef manta ray *Manta alfredi* in southern Mozambique. J Fish Biol 77: 169–190.
7. Ward-Paige CA, Davis B, Worm B (2013) Global population trends and human use patterns of manta and mobula rays. PLoS ONE 8: e74835.
8. Couturier LIE, Dudgeon CL, Pollock KH, Jaine FRA, Bennett MB, et al. (2014) Population dynamics of the reef manta ray *Manta alfredi* in eastern Australia. Coral Reefs 33: 329–342.
9. White WT, Giles J, Dharmadi, Potter IC (2006) Data on the bycatch fishery and reproductive biology of mobulid rays (Myliobatiformes) in Indonesia. Fish Res 82: 65–73.
10. Marshall AD, Conradie J (2014) Manta Fishery Solor. Lamankera 2014 Report of the Marine Megafauna Foundation. Available: http://www.marinemegafauna.org/wp-content/uploads/2014/06/Fishery-Report-Lamankera-May-2014-web.pdf. Accessed 2014 June 29.
11. Dewar H, Mous P, Domeier M, Muljadi A, Pet J, et al. (2008) Movements and site fidelity of the giant manta ray, *Manta birostris*, in the Komodo Marine Park, Indonesia. Mar Biol 155: 121–133.
12. Jaine FRA, Couturier LIE, Weeks SJ, Townsend KA, Bennett MB, et al. (2012) When Giants Turn Up: Sighting Trends, Environmental Influences and Habitat Use of the Manta Ray *Manta alfredi* at a Coral Reef. PLoS One 7: e46170.
13. Couturier LIE, Jaine FRA, Townsend KA, Weeks SJ, Richardson AJ, et al. (2011) Distribution, site affinity and regional movements of the manta ray, *Manta alfredi* (Krefft, 1868), along the east coast of Australia. Mar Freshw Res 62: 628–637.
14. Jaine FRA, Rohner CA, Weeks SJ, Couturier LIE, Bennett MB, et al. (2014) Movements and habitat use of reef manta rays off eastern Australia: Offshore excursions, deep diving and eddy affinity revealed by satellite telemetry. Mar Ecol Prog 510: 73–86.
15. Marshall AD, Dudgeon CL, Bennett MB (2011) Size and structure of a photographically identified population of manta rays *Manta alfredi* in southern Mozambique. Mar Biol 158: 111–1124.
16. Kitchen-Wheeler A, Ari C, Edwards AJ (2011) Population estimates of Alfred mantas (*Manta alfredi*) in central Maldives atolls: North Male, Ari and Baa. Environ Biol Fishes 93: 557–575.
17. Deakos MH, Baker JD, Bejder L (2011) Characteristics of a manta ray Manta *alfredi* population off Maui, Hawaii and implications for management. Mar Ecol Prog Ser 420: 245–260.
18. Marshall AD, Pierce SJ (2012) The use and abuse of photographic identification in sharks and rays. J Fish Biol 80: 1361–1379.
19. Bonney R, Cooper CB, Dickinson J, Kelling S, Phillips T, et al. (2009) Citizen science: A developing tool for expanding science knowledge and scientific literacy. BioScience 59: 977–984.

20. Miller-Rushing A, Primack R, Bonney R (2012) The history of public participation in ecological research. Front Ecol Environ 10: 285–290.
21. Town C, Marshall A, Sethasathien N (2013) Manta Matcher: Automated photographic identification of manta rays using keypoint features. Ecol Evol 3: 1902–1914.
22. Marshall AD, Compagno LJV, Bennett MB (2009) Redescription of the genus *Manta* with resurrection of *Manta alfredi* (Krefft, 1868) (Chondrichthyes; Myliobatoidei; Mobulidae). Zootaxa 2301: 1–28.
23. Anderson RC, Adam MS, Goes JI (2011) From monsoons to mantas: Seasonal distribution of *Manta alfredi* in the Maldives. Fish Oceanogr 20: 104–113.
24. Grüss A, Kaplana DM, Guénette S, Roberts CM, Botsford LW (2011) Consequences of adult and juvenile movement for marine protected areas. Biol Cons 144: 692–702.
25. Game ET, Grantham HS, Hobday AJ, Pressey RL, Lombard AT, et al. (2009) Pelagic protected areas: The missing dimension in ocean conservation. Trends Ecol Evol 24: 360–369.
26. Gaston KJ, Jackson SE, Cantú-Salazar L, Cruz-Piñón G (2008) The ecological performance of protected areas. Annu Rev Ecol Evol Syst 39: 93–113.
27. Graham RT, Witt MJ, Castellanos DW, Remolina F, Maxwell S, et al. (2012) Satellite tracking of manta rays highlights challenges to their conservation. PLoS ONE 7: e36834.
28. Rohner CA, Pierce SJ, Marshall AD, Weeks SJ, Bennett MB, et al. (2013) Trends in sightings and environmental influences on a coastal aggregation of manta rays and whale sharks. Mar Ecol Prog Ser 482: 153–168.

Adjacent Habitat Influence on Stink Bug (Hemiptera: Pentatomidae) Densities and the Associated Damage at Field Corn and Soybean Edges

P. Dilip Venugopal*, Peter L. Coffey, Galen P. Dively, William O. Lamp

Department of Entomology, University of Maryland, College Park, Maryland, United States of America

Abstract

The local dispersal of polyphagous, mobile insects within agricultural systems impacts pest management. In the mid-Atlantic region of the United States, stink bugs, especially the invasive *Halyomorpha halys* (Stål 1855), contribute to economic losses across a range of cropping systems. Here, we characterized the density of stink bugs along the field edges of field corn and soybean at different study sites. Specifically, we examined the influence of adjacent managed and natural habitats on the density of stink bugs in corn and soybean fields at different distances along transects from the field edge. We also quantified damage to corn grain, and to soybean pods and seeds, and measured yield in relation to the observed stink bug densities at different distances from field edge. Highest density of stink bugs was limited to the edge of both corn and soybean fields. Fields adjacent to wooded, crop and building habitats harbored higher densities of stink bugs than those adjacent to open habitats. Damage to corn kernels and to soybean pods and seeds increased with stink bug density in plots and was highest at the field edges. Stink bug density was also negatively associated with yield per plant in soybean. The spatial pattern of stink bugs in both corn and soybeans, with significant edge effects, suggests the use of pest management strategies for crop placement in the landscape, as well as spatially targeted pest suppression within fields.

Editor: Youjun Zhang, Institute of Vegetables and Flowers, Chinese Academy of Agricultural Science, China

Funding: This work is supported by the United Soybean Board, http://unitedsoybean.org/ (GPD PDV); Maryland Soybean Board, http://mdsoy.com/ (GPD); Maryland Grain Producers Utilization Board, http://www.marylandgrain.com/ (GPD); Cosmos Club Foundation, http://www.cosmosclubfoundation.org/ (PDV); Northeast Sustainable Agriculture Research and Education - Graduate Student grant - GNE-12-047, http://www.nesare.org/ (PDV, WOL, GPD); Arthur B. Gahan Fellowship, Dept. of Entomology, Univ. of Maryland, http://entomology.umd.edu/academics/graduate/awardopportunities/gahanfellowshipfund (PDV); and United States Department of Agriculture - National Institute of Food & Agriculture Hatch Project #MD-ENTM-1016, http://www.csrees.usda.gov/ (WOL). Funding for Open Access provided by the UMD Libraries Open Access Publishing Fund. The funders had no role in study design, data collection and analysis, decision to publish, or preparation of the manuscript.

Competing Interests: The authors confirm that funding from the commercial source professional association (United Soybean Board Maryland Soybean Board, Maryland Grain Producers Utilization Board) does not alter the authors' adherence to PLOS ONE policies on sharing data and materials.

* Email: venugopal.dilip@gmail.com

Introduction

Agricultural fields are components within a heterogeneous landscape that strongly connect to and interact with surrounding habitats [1]. The movement of insects between natural and agricultural habitats has important implications for agricultural ecosystem function [2]. The movement of pest insects among seasonal crop resources is often non-random and directional as pest species disperse and colonize crops [3]. This movement may result in insect pests immigrating into an agricultural field in an aggregated manner in specific areas within the field [1]. In such cases, species-specific characteristics such as host range, vagility, chemical ecology, and host developmental status influence the spatial pattern of a pest population within crop fields, often resulting in the aggregation of pests along field edges as they disperse between habitats. The seasonal availability and suitability of source and recipient habitats in relation to the life stages of a mobile, polyphagous insect pest influence the dispersal dynamics of pests from sources to recipient habitats [2,4,5]. Knowledge about insect pest movement among habitats in the landscape, and the subsequent colonization of plants within crop fields, may inform risk of infestation by an insect pest prior to their subsequent population increase as well as provide opportunities for pest management [1].

Stink bugs in the family Pentatomidae are major pests of economically important crops globally [6], and are considered important pests in soybean *Glycine max* (L.) Merr. producing areas of the world [7]. While stink bugs caused economic losses in the southern United States, they were not considered serious pests of crops in the mid-Atlantic region until recently. The most common stink bugs in agricultural fields in the mid-Atlantic are *Chinavia hilaris* (Say 1832) and *Euschistus servus* (Say 1832), but these species have had little economic impact in the region [8,9]. The recent explosion in populations of the invasive brown marmorated stink bug, *Halyomorpha halys* (Stål 1855) however, has led to significant economic and ecological impacts.

Since its discovery near Allentown, Pennsylvania, USA, *H. halys* has been detected in 41 states within United States, and local populations and detections from Europe (Switzerland, France, Canada, Germany, Italy, and Liechtenstein) have also been reported [10]. This polyphagous stink bug has a wide range of host plants including tree fruits, vegetables, field crops, ornamental plants, and native vegetation in its native and invaded ranges. Since 2010, serious crop losses have been reported for apples, peaches, sweet corn, peppers, tomatoes, and row crops such as field corn and soybeans in the mid-Atlantic region [11]. *Halyomorpha halys* is also a nuisance pest in human dwellings. In this context, information on the movement of stink bugs into crops from adjacent managed and natural habitats, and the associated spatially-explicit crop damage, has direct implications for integrated pest management.

The dispersal and movement of various stink bug species between crops and other habitats has been addressed by many studies in the context of dispersal between habitats, adjacent habitat influences on densities at field edges, and their relationship to crop damage [12–23]. However, these studies mainly pertain to stink bug communities in crops in the southern U.S. Currently only anecdotal reports of high *H. halys* abundance in the edges of fields adjacent to woodlots [11] are available. These reports are consistent with the documented use of woody trees as a food source and as overwintering sites by *H. halys* [24]. This factor coupled with use of buildings as overwintering sites suggests that these habitats, which support high local stink bug populations that eventually invade soybean later in the season, are likely to play an important role in stink bug densities.

Many stink bug species cause significant seed quality and yield losses in field corn, *Zea mays* L., and soybean [19,23–29], and stink bugs are also associated with the transmission of plant diseases [32–34]. However, few studies quantify spatially-explicit field crop damage in relation to patterns of stink bug densities [35]. Soybean is one of the preferred hosts for *H. halys* [36], and both field corn and soybean constitute a very high proportion of overall crop area in the mid-Atlantic region and throughout the United States [37]. Research efforts aimed at determining the role of adjacent habitat in influencing stink bug dispersal, population density, and pattern of settlement into crop, contribute to the development of management strategies of *H. halys* in row crops.

In this study, our objectives were a) document the species composition and within-field distribution of stink bugs in field corn and soybean, b) examine the influence of adjacent managed and natural habitats on the density of stink bugs, and c) relate stink bug density to seed quality in field corn and soybean, and pod development and yield in soybean. We expected *H. halys* to be the most abundant stink bug in our study based on previous reports of stink species composition in mid-Atlantic row crops [8]. We predicted higher density of stink bugs along wooded habitats and buildings than open habitats as they provide host plants and over-wintering refuge [24]. We also predicted high density of stink bugs at the field edge, reducing with distance into the field interior as observed by anecdotal reports for *H. halys* [11].

Methods

Ethics statement

The field studies were conducted at facilities provided by the University of Maryland and the USDA Beltsville Agricultural Research Center where we had permission to conduct research. Stink bugs were enumerated *in situ* and were not collected, and no endangered or protected species were involved.

Field selection and stink bug sampling strategy

The study was conducted at the USDA Beltsville Agricultural Research Center at Beltsville, MD (39°01′50N; 76°53′28W) and University of Maryland Research and Education Center facilities at Beltsville (39°00′45N; 76°49′32W), Clarksville (39°15′18N; 76°55′51W) and Keedysville, MD (39°30′37N; 77°43′57W). At these sites, field corn (76.2 cm row spacing) and full season soybean (17.8 cm row spacing) fields with a portion of their perimeter directly adjacent to wooded areas (henceforth wooded habitats), buildings (buildings, houses and barns; henceforth building habitats), crops (alfalfa, sorghum, and vegetable crops; henceforth crop habitats) and open, non-crop areas that were grassy, untilled land (henceforth open habitats) were sampled in 2012 and 2013. Corn fields were chosen as one of the adjacent habitat types *in lieu* of crops for several soybean fields. In each field, the sampling layout included four transects spaced 20 m apart. Each transect was marked for sampling plots at distances 0, 1.5, 3, 4.5, 6, 9, 12, and 15 m from the edge to field interior. Stink bugs were enumerated at each plot by carefully examining 10 consecutive corn plants and later converted to densities based on planting density, or all plants within a semicircular area of 0.5 m radius (1.57 m^2) in soybean.

Visual counts of stink bug adults, nymphs, and egg masses of *H. halys*, *E. servus*, *C. hilaris*, *Murgantia histrionica* (Hahn 1834), and *Thyanta custator* (Fabricius 1803) were recorded and converted to densities. Fields and plots were sampled each week between mid-July and mid-August in field corn, and mid-August and late-September in soybean. Sampling coincided with the kernel development stages of corn (R2–R5; blister – dent; [38]) and the seed development stages of soybean (R4–R7; full pod to physiological maturity; [39]), which are associated with high *H. halys* and other stink bug species densities [8,11,40]. Details on the number of corn and soybean field edges with different adjacent habitats, and the sampling dates during 2012 and 2013 are provided in Table 1. Including the repeated sampling of fields, a total of 4,315 field corn plots in 31 fields, and 2,968 soybean plots in 26 fields across all sites were sampled for stink bugs during 2012 and 2013.

Assessing seed quality in field corn and soybean

To relate stink bug density to ear damage in corn, eight fields with the highest observed count of stink bugs in 2013 and with different adjacent habitats were selected. Of these fields, 3, 3, 1 and 1 were adjacent to wooded, building, crops and open habitats, respectively. In each field, 6–10 consecutive corn ears were collected at each sampling plot prior to harvest maturity and stored in cloth bags for drying. Planting details of the fields used for assessing corn damage are provided in Table 2. For each ear, the following data were recorded once at the end of the season: 1) number of kernels damaged by stink bugs (identified by a characteristic puncture scar typically surrounded by a discolored cloudy marking); 2) number of collapsed kernels due to stink bug damage (this type of damage was carefully examined to distinguish between kernels damaged by stink bugs and dusky sap beetles, *Carcophilus lugubris* Murray); 3) number of kernel rows around the ear; 4) length of one kernel row (mm); and 5) average width of individual kernels (mm). With the individual ear measurements, the total number of kernels was derived by dividing the kernel row length by the width of a kernel times the number of rows. Data were then summed across all ears of a plot and stink bug damage was expressed as the percentage of damaged and collapsed kernels in relation to total number of kernels per sample. A total of 2,326 ears of corn from 252 sampling plots across 8 fields were assessed for stink bug damage.

To relate stink bug density to soybean pod development prior to harvest, samples of 10 consecutive plants at each plot of seven soybean fields in 2012–2013 were examined once at the end of the season *in situ* to count the total numbers of pods with 3 or more

Table 1. Details on field corn and soybean field edges with different adjacent habitats sampled for stink bugs and the sampling occasions at each field in Maryland, USA during 2012–2013.

Crop	Year	Site	Adjacent habitats (number of field edges)	Sampling dates (frequency)
Field Corn	2012	Beltsville	woods (4), buildings (3), crops (1), open (4)	10 July–15 Aug (7–10 days)
		Clarksville	woods (1), buildings (1), crops (3)	10 July–15 Aug (7 days)
	2013	Clarksville	woods (3), buildings (2), crops (3), open (1)	18 July–22 Aug (7 days)
		Keedysville	woods (1), buildings (1), crops (1), open (2)	16 July–20 Aug (7 days)
		Overall	woods (9), buildings (7), crops (8), open (7)	10 July–22 Aug (7–10 days)
Soybean	2012	Beltsville	woods (2), buildings (3), corn (1), open (2)	23 Aug–20 Sept (7–10 days)
		Keedysville	woods (2), buildings (1), corn (2), open (1)	30 Aug–26 Sept (7–10 days)
	2013	Beltsville	woods (1), buildings (1), corn (1), open (1)	13 Aug–06 Sept (7 days)
		Clarksville	buildings (1), corn (1), open (1)	16 Aug–12 Sept (5–7 days)
		Keedysville	woods (2), buildings (1), corn (1), open (1)	15 Aug–18 Sept (5–7 days)
		Overall	woods (7), buildings (7), corn (6), open (6)	13 Aug–26 Sept (5–10 days)

fully developed seeds (full pods), pods with fewer than 3 seeds (half pods), and flat, immature pods (flat pods). For standardization, the proportions for each pod type were calculated for each sample. Pod quality data were collected from 64 plots in 2 fields adjacent to wooded habitat at Keedysville. Planting details for the fields used for assessing soybean seed damage are provided in Table 2. Of the 7 fields sampled for seed quality data, 2 were adjacent to buildings, and 5 fields were adjacent to wooded habitats, and all had the highest counts of stink bugs observed in each year. Just prior to harvest, 20 plants each from 154 plots in the fields were collected, stored in mesh bags, and allowed further drying for optimal thrashing. Seeds were removed from pods for each sample by a stationary motor-driven thrasher (RL Brownfield-Swanson Machine Company; Model B1). Dirt, chaff, or un-thrashed pods were removed, and the remaining seed samples were then weighed to measure yield (measured as grams/20 plants).

To assess seed quality, subsamples of 200–300 seeds were removed from each sample, counted, and weighed to calculate test weight (expressed as weight per 100 seeds). Seed samples in 2012 were sieved to separate smaller, immature seeds (<0.3 cm), whereas these smaller seeds were not removed from subsamples in 2013. In both years, seeds were individually examined and categorized into six groups as follows: 1) stink bug damaged seed, distinguished by a puncture scar and often surrounded by a discolored cloudy area; 2) moldy seed, characterized by having milky white or grayish crusty growth on surface, sometimes with cracks and fissures; 3) shriveled seed that appeared wrinkled and often undersized; 4) purple seed recognized as purple or pink areas on the seed coat due to the fungus *Cercospora kikuchii* Matsumoto & Tomoy 1925 [41]; 5) green seed showing discolored green tissue in cross section, rather than the normal yellow; and 6) normal, undamaged seed. Seeds with characteristics of more than one

Table 2. Details on the field corn and soybean fields used for analyzing grain and seed damage in Maryland, USA during 2012–2013.

Crop	Year	Site	Field ID	Variety	Planting Date	Density/acre
Field Corn	2013	Clarksville	Corn1	P1319HR (113)	2May13	26,000
			Corn2	DKC61-21 (111)	15May13	26,000
			Corn3	P1319HR (113)	2May13	26,000
			Corn4	P1319HR (113)	2May13	26,000
			Corn5	DKC61-21 (111)	16May13	26,000
			Corn6	NK74R3000GT (114)	16May13	26,000
		Keedysville	Corn7	Doebler's 633HXR (110)	23Apr13	26,000
			Corn8	Doebler's 633HXR (110)	23Apr13	26,000
Soybean	2012	Beltsville	Soy1	AG3030 (3.0)	11May12	155,555
			Soy2	AG3030 (3.0)	11May12	155,555
		Keedysville	Soy3	Doebler 3809RR (3.8)	26May12	180,000
			Soy4	Doebler 3809RR (3.8)	4Jun12	180,000
			Soy5	Doebler 3809RR (3.8)	4Jun12	180,000
	2013		Soy6	SCS9360RR (3.5)	22May13	180,000
			Soy7	Doebler 3809RR (3.8)	27May13	180,000

For each of the corn and soybean varieties, the corn relative maturity in days and the soybean maturity groups respectively, have been provided in parenthesis.

damage category were assigned to whichever category seemed dominant, except for stink bug damaged seeds which were all assigned to that category. To standardize across samples, the percentage of seeds in each category was calculated. Soybean seed quality data were collected for a total of 154 plots from 4 fields in 2012 and 2 fields in 2013.

Statistical analyses

Adjacent habitat and distance from edge influences. The influence of adjacent habitat and distance from field edge on the density and distribution of stink bugs was analyzed by Generalized Linear Mixed Models (GLMMs) based on Laplace approximation, with a Poisson-lognormal error distribution and log link function [42]. All analyses were performed with fields as replicates and the sampling plots along transects within the field as subsamples. For corn and soybean data, analyses were performed on –data pooled over all stink bug life stages. For each of these datasets, GLMMs were performed on the data pooled across species, years, and study sites, and on data from each study site pooled across years. Sampling points within each field edge was treated as a random factor to control for repeated measurement [43]; adjacent habitat, distance from edge, and their interaction were the fixed effects, and stink bug density was the response variable. For the overall data models, study site and year were also treated as random effects.

Model building and selection procedures for the mixed effects modeling followed the procedures used by Zuur and others [44]. First, several candidate models, each with different random effects but identical fixed effects, were tested to choose the optimal random effect model using a combination of AIC and BIC values for selection criteria [36]. For all optimum fixed effect models, an initial full model analysis including individual and interactive effects of adjacent habitat (4 levels - woods, buildings, mix crop/corn, and open) and distance from edge was performed. Specifically we designed a model matrix that directly estimated the intercepts and the slopes for each level of adjacent habitat (extension of means parameterization; see pages 61–64 in [45]) in relation to distance from field edge. The significance of the fixed effects was determined by Wald χ^2 tests. If a significant interaction was found these estimates, directly interpreted as the intercept and slope of the regression of stink bug density on distance for each of the adjacent habitat types, were then simultaneously compared through post-hoc comparisons with a Bonferroni correction. Models were evaluated for assumption appropriateness by testing for over-dispersion and correlations among random effect terms, and by visualizing variances in a location-scale plot with superimposed loess fit [42].

Relating stink bug density and seed damage. Influence of stink bug density on damage to corn kernels was assessed using generalized linear models (GLMs) with Poisson or quasi-Poisson error distribution and log link function [46]. Percentage of collapsed and stink bug damaged seeds were used as the response variables and mean stink bug density, the explanatory variable. For significant results, the coefficient of determination was calculated by Nagalkerke's pseudo r^2 statistic [47].

Linear regression was used to assess the influence of stink bug density on soybean pod development. To meet normality assumptions, response variables (% flat or full pods) was square root transformed prior to analysis. Influence of stink bug density on soybean seed quality was assessed by linear mixed models (LMMs) with year as a random effect to account for minor differences in grading seed size protocols. LMMs related stink bug density to the percentage of seeds in each category of seed quality. Influence of stink bug density on soybean yield per plot was assessed by LMMs with field as a random effect to account for differences in soybean variety and other field conditions among sites. Response variables (% seeds in various damage categories, or yield) were log or square root transformed to meet normality requirements and the significance of the fixed effect was determined by Wald t-tests. Diagnostic plots of the models visualizing within-group residuals (standardized residuals vs fitted values, normal Q-Q plots, and histograms of residuals) and estimated random effects (normal Q-Q plots and pairs-scatter plot matrix) were used to assess model appropriateness. The coefficients of determination for the LMMs, based on the likelihood-ratio test, were calculated using Nagalkerke's pseudo r^2 statistic [47]. Patterns in damage to corn kernels, soybean pods, and seeds at different distances from the edge, in relation to stink bug density, were visualized by plotting average values of damage and stink bug density aggregated by distance.

All statistical analyses were performed in R program [48] and associated statistical packages. GLMMs were performed with package 'lme4' [49] and LMMs with package 'nlme' [50]. Multiple comparisons of means for GLMMs were computed with R packages 'contrast' [51] and 'multcomp' [52]. GLMMs and LMMs estimated coefficients were extracted through package 'effects' [53], converted to densities/m^2 and plotted using 'ggplot2' [54]. Coefficient of determination (pseudo r^2) for the GLM was calculated with package 'rms' [55], and with package 'MuMin' [56] for LMMs.

Results

Species composition and density

A total of 9440 individuals (66% nymphs; 34% adults) of four phytophagous stink bug species (*E servus*, *H. halys*, *C. hilaris*, and *M. histrionica*) were recorded in field corn, of which *H. halys* accounted for 97% of the total. Species composition varied among study sites and crop systems. *Halyomorpha halys* comprised 57% of the sampled populations in corn at Beltsville, followed by *E. servus* (35%), whereas *H. halys* accounted for ~97% of all stink bugs at Clarksville and Keedysville. In soybean, a total of 9867 individuals (68% nymphs; 32% adults) of five phytophagous stink bug species (*E. servus, H. halys, C. hilaris, M. histrionic, T. custator*) were recorded, of which *H. halys* accounted for 93% of the total. *Halyomorpha halys* comprised 83–85% of the stink bug numbers in soybean at Beltsville, while greater than 92% were *H. halys* at Clarksville and Keedysville. Results obtained from the statistical analyses hence pertain mainly to patterns of *H. halys* density, since this species constituted ~95% of all observed stink bugs in both field corn and soybean.

Adjacent habitat and distance from field edge influences

The trends for density among adjacent habitats at different distances from the edge observed in adult and nymph data sets were broadly similar and were hence pooled for the analyses, and we report only these results.

Field corn. For the analysis of overall stink bug data from field corn edges, the random effects used for the GLMM included the field, study site and year. Results showed significant interactive influences of adjacent habitat and distance from edge on stink bug density (Wald $\chi^2 = 492.6$, DF = 8, $P < 0.001$). Simultaneous comparison of intercepts and slopes among adjacent habitats accounting for the influence of distance from edge revealed significant differences. Overall, the density of stink bugs at different distance from the edge was significantly higher in fields adjacent to wooded habitats compared to density in fields adjacent to buildings and open habitats (Fig. 1A). Density between wooded and crop habitats did not significantly differ, although mean densities of stink bugs were consistently higher adjacent to wooded habitats (see Fig. S1 for variance around estimated means). Similarly stink bug density did not differ significantly between crop habitats and buildings although higher mean density was observed along buildings.

For site level GLMMs, significant interactive influences of adjacent habitat and distance from edge on stink bug density were observed for all the sites - Clarksville (Wald $\chi^2 = 426.9$, DF = 8, $P<0.001$), Keedysville (Wald $\chi^2 = 367.9$, DF = 8, $P<0.001$) and Beltsville (Wald $\chi^2 = 43.1$, DF = 3, $P<0.001$). Multiple comparisons of means of stink bug density at Clarksville showed similar trends to the results of analyses of data pooled over all study sites, with density in fields adjacent to wooded habitats higher than density adjacent to buildings and open habitats (Fig. 1C). Similar to overall data, at Clarksville, stink bug density did not differ significantly between wooded and crop habitats, although mean densities of stink bugs were consistently higher adjacent to wooded habitats. Similarly stink bug density did not differ significantly between crop habitats and buildings although higher mean density was observed along buildings. At Keedysville however, stink bug density was highest along fields adjacent to crops, and the slopes and intercepts were significantly different among adjacent habitats (Fig. 1B). Additionally, corn fields at Keedysville had higher stink bug density along the outside rows (0 m) adjacent to crops than adjacent to woods. At Beltsville, with the least stink bug density, multiple comparisons did not reveal significant differences in stink bug density among adjacent habitats, after accounting for the distance from edge influence (Fig. 1D; also see Fig. S1 for variance around estimated means for each site). The raw means for each site and year (Fig. S2) are provided as supplementary information.

Soybean. The GLMM analysis of overall stink bug data treated field as the only random effect and showed significant interactive influences of adjacent habitat and distance on stink bug density (Wald $\chi^2 = 717.2$, DF = 8, $P<0.001$). Multiple comparisons of means showed that density of stink bugs was significantly higher at all distances in fields adjacent to wooded habitats compared to density in fields adjacent to buildings and open habitats (Fig. 2A). Stink bug density adjacent to wooded habitats was also consistently higher than density in fields adjacent to corn

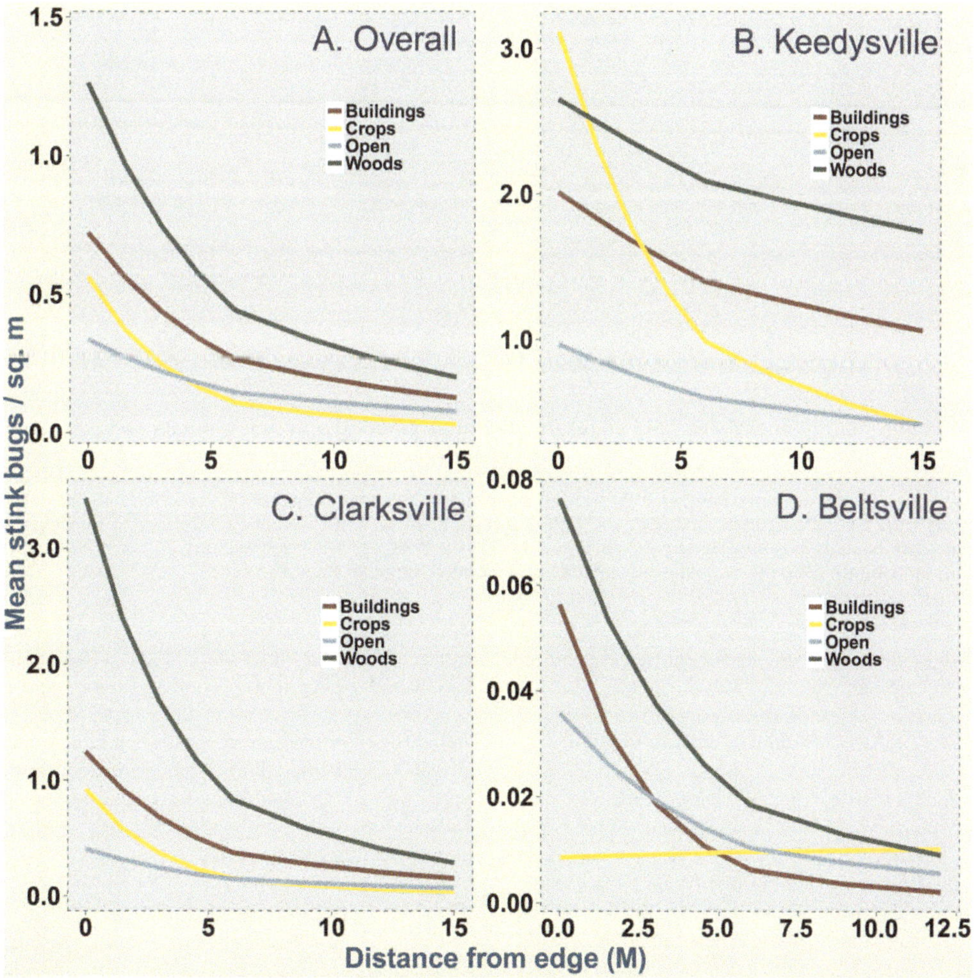

Figure 1. Mean stink bug density in field corn in relation to different adjacent habitats and distance from the field edge. Estimates derived from Poisson-lognormal GLMMs are plotted for overall stink bug data pooled over all study sites (A), Keedysville (B), Clarksville (C) and Beltsville (D). Values presented here have been back transformed from their original link function estimated model coefficients. Multiple comparison of means with a bonferroni correction ($\alpha = 0.05$) showed significant differences in: overall (A) - wooded habitats and buildings and, wooded and open habitats; Keedysville (B) – all adjacent habitats significantly different from each other; and Clarksville (C) - wooded habitats and buildings and, wooded and open habitats. For Beltsville (D), all multiple means comparisons were non-significant.

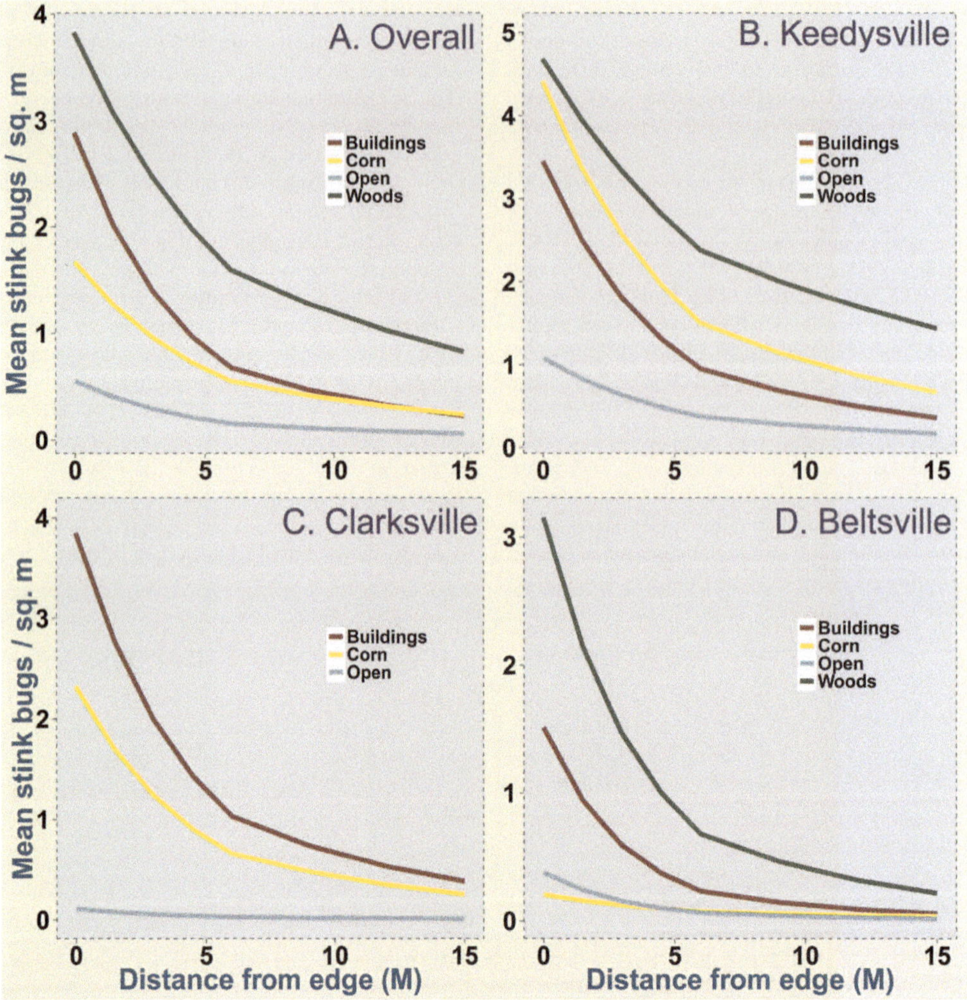

Figure 2. Mean stink bug density in soybean field edges in relation to different adjacent habitats and distance from field edge. Estimates derived from Poisson-lognormal GLMMs are plotted for overall stink bug data pooled over all study sites (A), Keedysville (B), Clarksville (C) and Beltsville (D). Values presented here have been back transformed from its original link function estimated model coefficients. Multiple comparison of means with a bonferroni correction ($\alpha = 0.05$) showed significant differences in: overall (A) - wooded habitats and buildings, wooded and open habitats, buildings and corn habitats, buildings and open habitats; Keedysville (B) – wooded habitats and buildings, wooded and open habitats, buildings and corn habitats, buildings and open habitat; Clarksville (C) - buildings and open habitats, and corn and open habitats; Beltsville (D) – buildings and open habitats, buildings and wooded habitats, and wooded and corn habitats.

field habitats, but differences were not significant (see Fig. S3 for variances around estimated means). Across all the adjacent habitat types the highest density of stink bugs was recorded at the immediate field edge (0 m), and was lowest at 15 m from edge ($< 1/m^2$).

GLMMs performed by study site on overall stink bug data showed significant interactive influence of adjacent habitat and distance from edge on density at all the sites - Keedysville (Wald $\chi^2 = 485.4$, DF = 8, $P<0.001$), Clarksville (Wald $\chi^2 = 216.63$, DF = 5, $P<0.001$) and Beltsville (Wald $\chi^2 = 279.2$, DF = 8, $P< 0.001$). Multiple means comparisons for Keedysville data showed significantly higher densities adjacent to wooded habitats buildings and open habitats (Fig. 2B). The intercept and slopes for buildings and corn, and buildings and open areas were also significantly different. At Clarksville, buildings and corn habitats had significantly higher density of stink bugs than open areas (Fig. 2C). Model estimated stink bug densities for buildings and corn habitats were not significantly different. At Beltsville, wooded habitats and buildings had significantly higher stink bug density than corn as adjacent habitat (Fig. 2D). The slope and intercept were significantly different for wooded habitats and buildings. Unlike the other sites where stink bugs were fewest adjacent to open habitats, at Beltsville, corn as adjacent habitat harbored the lowest density of stink bugs. The variance around the estimated density of stink bugs in soybean fields for each site (Fig. S3), and the raw means for each site and year (Fig. S4) are provided as supplementary information.

Corn and soybean seed damage

For field corn, results from the quasi-Poisson GLM showed a significant positive association between percentage of stink bug damaged kernels and mean stink bug density ($y = 0.57+0.15x$, n = 252, $P<0.001$, pseudo $r^2 = 0.47$). A Poisson GLM showed that the percentage of collapsed kernels was not significantly associated with mean stink bug density ($y = -6.75+0.14x$, n = 252, $P = 0.50$, pseudo $r^2 = 0.17$). For soybean pod development data, regression analysis revealed that the percentage of full pods was negatively influenced by mean stink bug density ($y = 5.9-0.17x$, n = 63, $P<$

0.001, $r^2 = 0.51$), while percentage of flat pods (square root transformed) was positively influenced ($y = 2.18+0.26x$, $n = 63$, $P < 0.001$, $r^2 = 0.63$). Results of LMMs analyzing each seed quality category (Table 3) showed a significant positive association of mean stink bug density and purple stained seeds, percentage of stink bug damaged seeds, percentage of immature, shriveled and moldy seeds, and overall percentage of damaged seeds.

Overall percentage of normal, undamaged soybean seeds and yield had a significant negative relationship with stink bug density (Table 3). The overall seed damage by stink bugs in both corn (Fig. 3A) and soybean (Fig. 3C), and their impact on soybean pod development (Fig. 3D), were highest at field edge and declined gradually towards the field interior. Furthermore, soybean yields (grams/20 plants) were lowest at field edge, gradually increasing from field edge to highest yields at 12 and 15 m from the edge (Fig. 3B).

Discussion

Seasonal availability and suitability of source and recipient habitats influence dispersal dynamics of pest populations [2,4,5]. Aggregation of stink bugs at the field edge of target crops results from their directional movement among a sequence of hosts in an area in response to deteriorating suitability of the host crops or native vegetation and increasing suitability of the target crop plants [57]. Our results provide strong evidence that *H. halys* density exhibits a clear edge effect in both field corn and soybean. Across all adjacent habitats, density of *H. halys* and other stink bugs was highest within the first 3 meters, reaching lowest levels between 9–15 m from the edge, least at 15 m from the edge ($<1/m^2$). In addition to host suitability, species specific behaviors contribute to the resulting aggregations of stink bugs along crop edges. The strong edge effect exhibited by *H. halys* is similar to the within-field infestation pattern reported for other native stink bug species in U.S. crops [12,15,17–19,23,58]. We demonstrated that the aggregations are associated with specific kinds of damage to both corn grain and soybean seeds, thus suggesting that sitenspecific management at the field and farmscape levels may improve crop protection from stink bugs.

The timing of infestations during mid to late July in corn, and then later colonization of soybean fields in August, suggests that the majority of *H. halys* adults were offspring of the generation which developed on host plants earlier in the season (June/July). *Halyomorpha halys* is known to feed on a wide range of cultivated and wild hosts (up to 170 species) [59], of which many tree and shrub species are likely present in wooded habitats. Particularly high density of *H. halys* was observed (personal field observations, Dilip Venugopal) in soybean fields bordering wooded habitats with tree of heaven (*Ailanthus altissima* Swingle), princess tree (*Paulownia tomentosa* Baill.), and black cherry (*Prunus serotina* Ehrhart), all which support high population densities of reproducing *H. halys* [59,60]. Similarly, earlier reports show that *C. hilaris* populations in soybean are greater adjacent to wooded borders with black cherry and elderberry (*Sambucus canadensis* L.) [13]. Additionally, *E. servus* populations are greater in cotton fields with adjacent woods containing many oak species (*Quercus* spp.) and black cherry [18].

This study also addressed the influence of adjacent habitats on stink bug density and quantified differences in density at various distances from the field edges in soybean and corn crops. We found that adjacent habitats, particularly wooded habitats, influenced the densities of *H. halys* and other stink bugs. In both corn and soybeans, fields adjacent to wooded habitats, across all study sites and distances from field edge, consistently harbored

Table 3. Statistical results of LMMs for analyzing the relationship between stink bug density and various soybean seed damage categories and yield.

Response variable	Data Transformation	Intercept	Intercept SE	Estimate	SE	DF	Wald t	Pval	psuedo r^2
% normal seeds	None	75.8	8.04	−2.11	0.25	145	8.28	<0.001	0.30
% stink bug damaged seeds	Square Root	3.41	0.5	0.07	0.01	145	4.58	<0.001	0.12
% purple damaged seeds	log	1.39	0.14	0.09	0.01	145	9.99	<0.001	0.44
% moldy + shriveled – immature seeds	Square Root	2.59	0.47	0.09	0.02	148	5.87	<0.001	0.19
% all damaged seeds	Square Root	4.78	0.63	0.18	0.02	145	9.03	<0.001	0.35
Total Yield (grams/20 plants)	Square Root	17.1	1.04	−0.20	0.04	140	4.67	<0.001	0.13

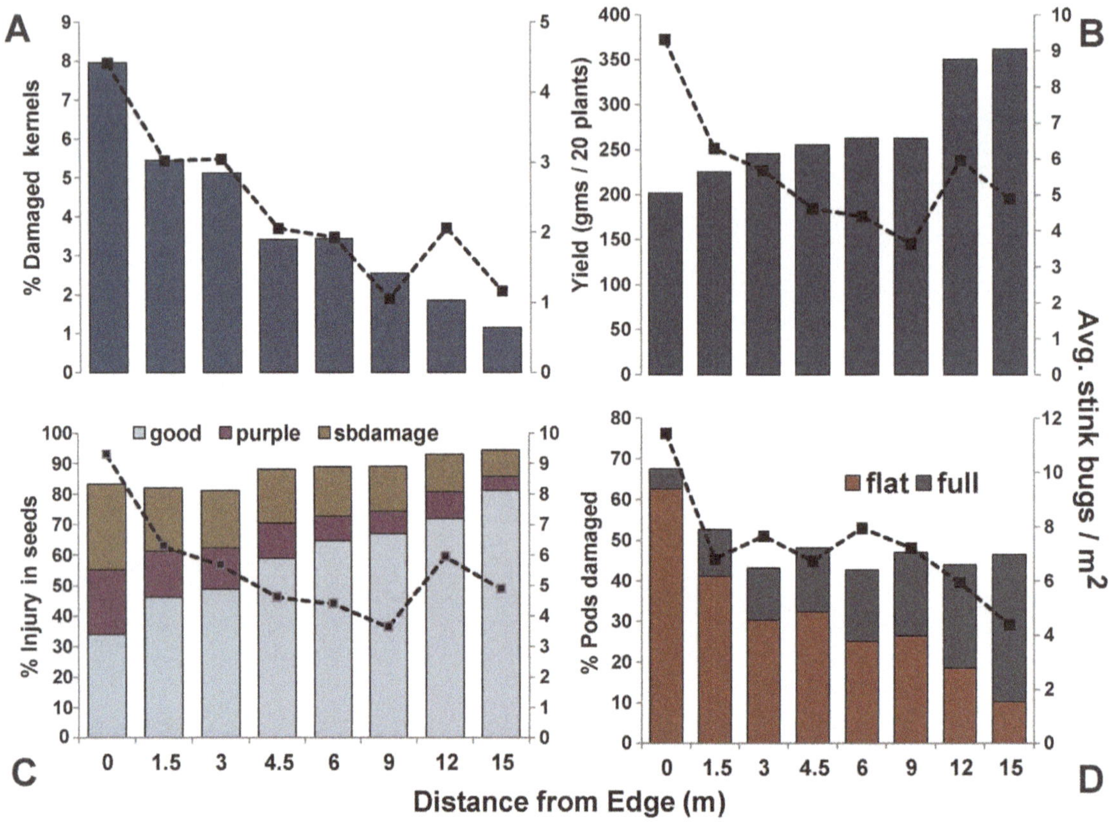

Figure 3. Patterns of kernel damage in field corn (A), soybean yield (B), soybean seed damage by category (C), and soybean pod development (D) in relation to mean stink bug density at different distance from field edge. The proportions of soybean seeds in each seed quality category (stink bug damaged, purple damaged, and normal seeds) and pod types (flat and full) are also provided. Mean stink bug abundance are denoted by the dashed lines which represent the second y axis.

significantly higher densities of stink bugs than in fields adjacent to open habitats. Also, stink bug density in fields adjacent to wooded habitats was consistently higher than in fields adjacent to buildings and corn habitats. These results suggest that wooded habitats play an important role in serving as sources of stink bug populations that colonize crops.

Patterns of infestation along field edges of corn and soybean differed among study sites, partly influenced by overall stink bug density and other adjacent habitats. Stink bug density in both row crops at Keedysville was consistently (3–5 times) greater than the mean density at the other two study sites, and this could be attributed to the higher overall densities of *H. halys* observed in the Western Maryland region over the past 5 years. In addition to wooded habitats, building and other crop habitats served as sources of colonizing adults in row crops. *H. halys* often utilize buildings as overwintering sites and thus these structures would likely influence stink bug populations earlier in the spring when post-diapause adults are moving to host plants, thereby supporting high local populations that eventually invade soybean later in the season. At Keedysville, stink bug density was higher in corn fields adjacent to alfalfa and in soybean fields adjacent to corn than in corn or soybean fields adjacent to building habitats. However, at Clarksville and Beltsville adjacent building habitats showed a higher density of stink bugs than crop habitats.

As with other studies, our results highlight the role of other adjacent cultivated crops in influencing stink bug populations. For example, adjacent fields of alfalfa, field corn, and other cultivated borders have been reported as a sources contributing to higher densities of stink bugs in tomato, cotton, sorghum, and peanut fields [12,16,19,23]. However, differences in the relative influence of adjacent habitats in our study could be related to overall stink bug density at each of the study sites. For example, stink bug density did not differ significantly among adjacent habitats in field corn at Beltsville where *H. halys* densities were lowest. Moreover, the influence of the landscape heterogeneity on stink bug density could extend to larger spatial scales beyond habitats adjacent to a crop. Since insect population dynamics and spatial patterns are affected by regional landscape context and species traits such as dispersal ability [61], distribution, and density of *H. halys* may depend on habitat and other environmental characteristics at spatial scales greater than the local agricultural field [62,63]. Differences in landscape structure among our study sites may explain the higher density of stink bugs at Keedysville, and specifically the role of adjacent crop habitats as sources of stink bugs in field corn and soybean.

We related the various corn and soybean damage measurements to stink bug density within plots. As expected, stink bug damage to corn kernels increased with stink bug density. The percentage of damaged kernels reached levels up to 8% at the field edge to less than 3% between 9–15 m from field edge, and was positively correlated with the stink bug density within plots. The percentage of collapsed kernels was negligible and not significantly influenced by stink bug density. Based on findings by earlier studies [25,64], neither kernel damage, ear weight or grain weight are affected beyond tasseling stage (VT) from feeding damage by *E. servus* or *N. viridula*. Although *H. halys* density can be high along edges of corn fields, our result suggests that *H. halys* kernel quality loss is restricted to about 10 m from the field edge.

Halyomorpha halys populations in soybean had a significant impact on pod development, with the percentage of flat pods significantly increasing with increasing stink bug density. Concomitantly, the proportion of fully developed pods significantly decreased with increasing stink bug density. Changes in the development and maturation of soybean pods due to *H. halys* feeding have been recently documented [35], showing that most severe pod loss occurred at the R4 (full pod) growth stage. Observed effects on pod and seed development with higher stink bug density were similar to damage caused by other stink bug species [26–31]. Our finding that higher stink bug density is related to increased proportions of moldy and purple stained seeds suggest a potential role of *H. halys* in transmitting or facilitating various pathogens; however, this needs to be further investigated experimentally. Our study found a significant, yet weak negative association between soybean yield and stink bug density per plot. In contrast, recent field research using cages addressed the effects of *H. halys* feeding on soybean growth, and did not detect a significant relationship between H. *halys* stink bug densities and yield loss [35]. Soybean field studies to compare yields of insecticide-treated and untreated plots are needed to establish the relationship between soybean yield losses and stink bug density.

Knowledge of how adjacent habitats influence *H. halys* populations and the within-field distribution has several implications in stink bug management. First, our results indicate that scouting corn and soybean fields for decision-making is more efficient if efforts initially concentrated on field edges bordering wooded habitats where there is a greater likelihood of colonization and higher risk of infestation. Secondly, the infestation patterns of stink bug communities dominated by *H. halys* are primarily edge-centric, and population densities beyond 12 m are generally low ($<1/m^2$). Based on our results, if insecticides are to be applied, edge-only applications, particularly along wooded and building habitats, could reduce the cost of control while preventing damage caused by stink bugs in field corn and soybean. Preliminary studies show that treating just 12 m into soybean field prevented further invasion by *H. halys* and other stink bugs (Ames Herbert, personal communication). The edge-only treatment prevented reinvasion and also resulted in an 85–95% reduction in insecticide used compared with whole-field treatments [11]. Results presented here showing highest stink bug density and associated damage limited to the immediate field edge provide validity for the edge-only treatment. While experimental research on effective insecticide application strategies are currently underway, based on our findings, we suggest that integrated pest management programs for the stink bug complex in field crops should include farmscape-level planning - crop location with regards to adjacent habitats, and targeted interventions in the form of edge-only treatments to prevent seed quality and yield losses.

Acknowledgments

The farm managers and staff at each of the research facilities - Kevin Conover (Beltsville), Tim Ellis (Keedysville), David Justice and Mike Dwyer (Clarksville), and David Swain (USDA–BARC) - all extended excellent support and assistance for conducting the study. We thank Terry Patton for his help to identify sampling fields and for data collection. We thank the summer technicians, Jesse Ditillo, Jessie Saunders, Jake Bodart, Taylor Schulden, Kyle Runion, and Emily Zobel for data collection.

Author Contributions

Conceived and designed the experiments: PDV GPD WOL. Performed the experiments: PDV GPD PLC. Analyzed the data: PDV. Contributed reagents/materials/analysis tools: PDV. Contributed to the writing of the manuscript: PDV PLC GPD WOL.

References

1. Nestel D, Carvalho J, Nemny-Lavy E (2004) The spatial dimension in the ecology of insect pests and its relevance to pest management. In: Horowitz PDAR, Ishaaya PDI, editors. Insect pest management. Springer Berlin Heidelberg. 45–63.
2. Ekbom BS, Erwin ME, Robert Y (2000) Interchanges of insects between agricultural and surrounding landscapes. Kluwer Netherlands. 266 p.
3. Stinner RE, Barfield CS, Stimac JL, Dohse L (1983) Dispersal and movement of insect pests. Annu Rev Entomol 28: 319–335. doi:10.1146/annurev.en.28.010183.001535.
4. Kennedy GG, Margolies DC (1985) Mobile arthropod pests: management in diversified agroecosystems. Bull Entomol Soc Am 31: 21–35.
5. Kennedy GG, Storer NP (2000) Life systems of polyphagous arthropod pests in temporally unstable cropping systems. Annu Rev Entomol 45: 467–493. doi:10.1146/annurev.ento.45.1.467.
6. Panizzi AR (1997) Wild hosts of Pentatomids: ecological significance and role in their pest status on crops. Annu Rev Entomol 42: 99–122. doi:10.1146/annurev.ento.42.1.99.
7. Panizzi AR, Slansky F (1985) Review of phytophagous Pentatomids (Hemiptera: Pentatomidae) associated with soybean in the Americas. Fla Entomol 68: 184–214.
8. Nielsen AL, Hamilton GC, Shearer PW (2011) Seasonal phenology and monitoring of the non-native *Halyomorpha halys* (Hemiptera: Pentatomidae) in soybean. Environ Entomol 40: 231–238. doi:10.1603/EN10187.
9. Hooks CRR (2011) Stink bugs and their soybean obsession. Agron News 2: 1–4.
10. CABI (2014) *Halyomorpha halys* [original text by Leskey TC, Hamilton GC, Biddinger DC, Buffington ML, Dicekhoff C, et al.]. In: Invasive Species Compendium. Wallingford, UK: CAB International. www.cabi.org/isc. Available: http://www.cabi.org/isc/?compid=5&dsid=27377&loadmodule=datasheet&page=481&site=144. Accessed 19 March 2014.
11. Leskey TC, Hamilton GC, Nielsen AL, Polk DF, Rodriguez-Saona C, et al. (2012) Pest status of the brown marmorated stink bug, *Halyomorpha halys* in the USA. Outlooks Pest Manag 23: 218–226. doi:10.1564/23oct07.
12. Toscano NC, Stern VM (1976) Dispersal of *Euschistus conspersus* from alfalfa grown for seed to adjacent crops. J Econ Entomol 69: 96–98.
13. Jones WA, Sullivan MJ (1982) Role of host plants in population dynamics of stink bug pests of soybean in South Carolina. Environ Entomol 11: 867–875.
14. Outward R, Sorenson CE, Bradley JR (2008) Effects of vegetated field borders on arthropods in cotton fields in eastern North Carolina. J Insect Sci 8. Available: http://www.ncbi.nlm.nih.gov/pmc/articles/PMC3061576/.
15. Tillman PG, Northfield TD, Mizell RF, Riddle TC (2009) Spatiotemporal patterns and dispersal of stink bugs (Heteroptera: Pentatomidae) in peanut-

cotton farmscapes. Environ Entomol 38: 1038–1052. doi:10.1603/022.038.0411.
16. Toews MD, Shurley WD (2009) Crop juxtaposition affects cotton fiber quality in Georgia farmscapes. J Econ Entomol 102: 1515–1522. doi:10.1603/029.102.0416.
17. Pease CG, Zalom FG (2010) Influence of non-crop plants on stink bug (Hemiptera: Pentatomidae) and natural enemy abundance in tomatoes. J Appl Entomol 134: 626–636. doi:10.1111/j.1439-0418.2009.01452.x.
18. Reay-Jones FPF (2010) Spatial and temporal patterns of stink bugs (Hemiptera: Pentatomidae) in wheat. Environ Entomol 39: 944–955. doi:10.1603/EN09274.
19. Reeves RB, Greene JK, Reay-Jones FPF, Toews MD, Gerard PD (2010) Effects of adjacent habitat on populations of stink bugs (Heteroptera: Pentatomidae) in cotton as part of a variable agricultural landscape in South Carolina. Environ Entomol 39: 1420–1427. doi:10.1603/EN09194.
20. Reisig DD (2011) Insecticidal management and movement of the brown stink bug, Euschistus servus, in corn. J Insect Sci 11: Article 168. doi:10.1673/031.011.16801.
21. Herbert JJ, Toews MD (2011) Seasonal abundance and population structure of brown stink bug (Hemiptera: Pentatomidae) in farmscapes containing corn, cotton, peanut, and soybean. Ann Entomol Soc Am 104: 909–918. doi:10.1603/AN11060.
22. Olson DM, Ruberson JR, Zeilinger AR, Andow DA (2011) Colonization preference of *Euschistus servus* and *Nezara viridula* in transgenic cotton varieties, peanut, and soybean. Entomol Exp Appl 139: 161–169. doi:10.1111/j.1570-7458.2011.01116.x.
23. Tillman PG (2011) Influence of corn on stink bugs (Heteroptera: Pentatomidae) in subsequent crops. Environ Entomol 40: 1159–1176. doi:10.1603/EN10243.
24. Lee D-H, Short BD, Joseph SV, Bergh JC, Leskey TC (2013) Review of the biology, ecology, and management of *Halyomorpha halys* (Hemiptera: Pentatomidae) in China, Japan, and the Republic of Korea. Environ Entomol 42: 627–641. doi:10.1603/EN13006.
25. Ni X, Da K, Buntin GD, Cottrell TE, Tillman PG, et al. (2010) Impact of brown stink bug (Heteroptera: Pentatomidae) feeding on corn grain yield components and quality. J Econ Entomol 103: 2072–2079. doi:10.1603/EC09301.
26. Daugherty DM, Neustadt MH, Gehrke CW, Cavanah LE, Williams LF, et al. (1964) An evaluation of damage to soybeans by brown and green stink bugs. J Econ Entomol 57: 719–722.
27. Todd JW, Turnipseed SG (1974) Effects of southern green stink bug damage on yield and quality of soybeans. J Econ Entomol: 421–426.
28. McPherson RM, Newsom LD, Farthing BF (1979) Evaluation of four stink bug species from three Genera affecting soybean yield and quality in Louisiana. J Econ Entomol 72: 188–194.
29. Brier HB, Rogers DJ (1991) Susceptibility of soybeans to damage by *Nezara viridula* (l.) (Hemiptera: Pentatomidae) and *Riptortus serripes* (f.) (Hemiptera: Alydidae) during three sages of pod development. Aust J Entomol 30: 123–128. doi:10.1111/j.1440-6055.1991.tb00403.x.
30. McPherson RM, Douce GK, Hudson RD (1993) Annual variation in stink bug (Heteroptera: Pentatomidae) seasonal abundance and species composition in Georgia soybean and its impact on yield and quality. J Entomol Sci 28: 61–72.
31. Corrêa-Ferreira BS, De Azevedo J (2002) Soybean seed damage by different species of stink bugs. Agric For Entomol 4: 145–150. doi:10.1046/j.1461-9563.2002.00136.x.
32. Clarke RG, Wilde GE (1971) Association of the green stink bug and the yeast-spot disease organism of soybeans. III. Effect on soybean quality. J Econ Entomol: 222–223.
33. Mitchell PL (2004) Heteroptera as vectors of plant pathogens. Neotrop Entomol 33: 519–545. doi:10.1590/S1519-566X2004000500001.
34. Medrano EG, Esquivel J, Bell A, Greene J, Roberts P, et al. (2009) Potential for *Nezara viridula* (Hemiptera: Pentatomidae) to transmit bacterial and fungal pathogens into cotton bolls. Curr Microbiol 59: 405–412. doi:10.1007/s00284-009-9452-5.
35. Owens DR, Herbert Jr DA, Dively GP, Reisig DD, Kuhar TP (2013) Does feeding by *Halyomorpha halys* (Hemiptera: Pentatomidae) reduce soybean seed quality and yield? J Econ Entomol 106: 1317–1323. doi:10.1603/EC12488.
36. Hoebeke ER, Carter ME (2003) *Halyomorpha halys* (Stål) (Heteroptera: Pentatomidae): a polyphagous plant pest from Asia newly detected in North America. Proc Entomol Soc Wash 105: 225–237.
37. USDA National Agricultural Statistics Service (2014) Crop production. Available: http://www.usda.gov/nass/PUBS/TODAYRPT/crop0614.pdf. Accessed 14 June 2014.
38. Hanway JJ (1963) Growth stages of corn (*Zea mays*, L.). Agron J 55: 487. doi:10.2134/agronj1963.00021962005500050024x.
39. Fehr WR, Caviness CE, Burmood DT, Pennington JS (1971) Stage of development descriptions for soybeans, *Glycine Max* (L.) Merrill. Crop Sci 11: 929. doi:10.2135/cropsci1971.0011183X001100060051x.
40. Schumann FW, Todd JW (1982) Population dynamics of the southern green stink bug (Heteroptera: Pentatomidae) in relation to soybean phenology. J Econ Entomol 75: 748–753.
41. Walters HJ (1980) Soybean leaf blight caused by *Cercospora kikuchii*. Plant Dis 64: 961–962. doi:10.1094/PD-64-961.
42. Bolker BM, Brooks ME, Clark CJ, Geange SW, Poulsen JR, et al. (2009) Generalized linear mixed models: a practical guide for ecology and evolution. Trends Ecol Evol 24: 127–135. doi:10.1016/j.tree.2008.10.008.
43. Pinheiro JC, Bates DM (2000) Mixed effects models in S and S-Plus. New York: Springer. 560 p.
44. Zuur AF, Ieno EN, Walker N, Saveliev AA, Smith GM (2009) Mixed effects models and extensions in ecology with R. New York: Springer. 580 p.
45. Kéry M (2010) Introduction to WinBUGS for ecologists: bayesian approach to regression, ANOVA, mixed models and related analyses. Academic Press. 322 p.
46. Ver Hoef JM, Boveng PL (2007) Quasi-Poisson Vs. negative binomial regression: How should we model overdispersed count data? Ecology 88: 2766–2772. doi:10.1890/07-0043.1.
47. Nagelkerke NJD (1991) A note on a general definition of the coefficient of determination. Biometrika 78: 691–692. doi:10.1093/biomet/78.3.691.
48. R Development Core Team (2011) R: A language and environment for statistical computing. The R foundation for statistical computing, Vienna, Austria. Available: http://www.R-project.org/.
49. Bates D, Maechler M, Bolker B, Walker S (2013) lme4: Linear mixed-effects models using Eigen and S4. Available: http://cran.r-project.org/web/packages/lme4/index.html. Accessed 25 December 2013.
50. Pinheiro J, Bates DM, Debroy S, Sarkar D, R Core Team (2013) nlme: Linear and nonlinear mixed effects models R package version 3.1-117. Available: http://cran.r-project.org/web/packages/nlme/index.html. Accessed 25 August 2014.
51. Kuhn M, Weston S, Wing J, James JF, Thaler T (2013) contrast: A collection of contrast methods. Available: http://cran.r-project.org/web/packages/contrast/index.html. Accessed 25 December 2013.
52. Hothorn T, Bretz F, Westfall P (2008) Simultaneous inference in general parametric models. Biom J 50: 346–363.
53. Fox J, Weisberg S, Friendly M, Hong J, Andersen R, et al. (2013) effects: effect displays for linear, generalized linear, multinomial-logit, proportional-odds logit models and mixed-effects models. Available: http://cran.r-project.org/web/packages/effects/index.html. Accessed 25 December 2013.
54. Wickham H (2009) ggplot2: elegant graphics for data analysis. New York: Springer. Available: http://had.co.nz/ggplot2/book.
55. Harrell FE (2013) rms: Regression modeling strategies. Available: http://cran.r-project.org/web/packages/rms/index.html. Accessed 25 December 2013.
56. Bartoń K (2013) MuMIn: Multi-model inference. Available: http://cran.r-project.org/web/packages/MuMIn/index.html. Accessed 25 December 2013.
57. Todd JW (1989) Ecology and behavior of *Nezara viridula*. Annu Rev Entomol 34: 273–292.
58. Tillman PG (2010) Composition and abundance of stink bugs (Heteroptera: Pentatomidae) in corn. Environ Entomol 39: 1765–1774. doi:10.1603/EN09281.
59. BMSB IPM Working Group & Northeastern IPM Center (2013) Host plants of the brown marmorated stink bug in the U.S. Available: http://www.stopbmsb.org/where-is-bmsb/host-plants/. Accessed 2 January 2014.
60. Nielsen AL, Hamilton GC (2009) Life history of the invasive species *Halyomorpha halys* (Hemiptera: Pentatomidae) in northeastern United States. Ann Entomol Soc Am 102: 608–616. doi:10.1603/008.102.0405.
61. Tscharntke T, Brandl R (2004) Plant-insect interactions in fragmented landscapes. Annu Rev Entomol 49: 405–430. doi:10.1146/annurev.ento.49.061802.123339.
62. Thies C, Steffan-Dewenter I, Tscharntke T (2003) Effects of landscape context on herbivory and parasitism at different spatial scales. Oikos 101: 18–25. doi:10.1034/j.1600-0706.2003.12567.x.
63. Tscharntke T, Rand TA, Bianchi F (2005) The landscape context of trophic interactions: insect spillover across the crop-noncrop interface. Ann Zool Fenn 42: 421–432.
64. Negrón JF, Riley TJ (1987) Southern green stink bug, *Nezara viridula* (Heteroptera: Pentatomidae), feeding in corn. J Econ Entomol 80: 666–669.

High Bee and Wasp Diversity in a Heterogeneous Tropical Farming System Compared to Protected Forest

Christof Schüepp[1,2]*, Sarah Rittiner[1], Martin H. Entling[2]

1 Institute of Ecology and Evolution, Community Ecology, University of Bern, Bern, Switzerland, 2 University of Koblenz-Landau, Ecosystem Analysis, Landau, Germany

Abstract

It is a globally important challenge to meet increasing demands for resources and, at the same time, protect biodiversity and ecosystem services. Farming is usually regarded as a major threat to biodiversity due to its expansion into natural areas. We compared biodiversity of bees and wasps between heterogeneous small-scale farming areas and protected forest in northern coastal Belize, Central America. Malaise traps operated for three months during the transition from wet to dry season. Farming areas consisted of a mosaic of mixed crop types, open habitat, secondary forest, and agroforestry. Mean species richness per site (alpha diversity), as well as spatial and temporal community variation (beta diversity) of bees and wasps were equal or higher in farming areas compared to protected forest. The higher species richness and community variation in farmland was due to additional species that did not occur in the forest, whereas most species trapped in forest were also found in farming areas. The overall regional species richness (gamma diversity) increased by 70% with the inclusion of farming areas. Our results suggest that small-scale farming systems adjacent to protected forest may not only conserve, but even favour, biodiversity of some taxonomic groups. We can, however, not exclude possible declines of bee and wasp diversity in more intensified farmland or in landscapes completely covered by heterogeneous farming systems.

Editor: Mari Moora, University of Tartu, Estonia

Funding: This research project was funded by the International Tropical Conservation Foundation ITCF, Kerzers, Switzerland (www.shipstern.org) and the Natural History Museum of Bern, Switzerland (Naturhistorisches Museum der Burgergemeinde Bern, www.nmbe.ch). The funders had no role in study design, data collection and analysis, decision to publish, or preparation of the manuscript, except for the requirement of ITCF that at least a part of the research project takes place at Shipstern Nature Reserve.

Competing Interests: The authors have declared that no competing interests exist.

* E-mail: christof.schueepp@iee.unibe.ch

Introduction

Importance of Modified Land for Biodiversity Conservation

Although a considerable amount (>13%) of the world's terrestrial surface is nowadays designated as nationally or internationally protected areas [1] the effectiveness of protected areas in biodiversity conservation is limited, especially in the tropics: First, financial support for basic management activities to stop illegal and harmful human activities within parks is often lacking [2]. Second, about 12% of terrestrial vertebrates, mostly endemics, are not covered within the current protected area network [3]. And third, the long-term effectiveness of reserves depends strongly on human population density and activities in the surroundings [4]. Thus, more recently research on biodiversity conservation in the tropics has moved away from strictly focussing on protected areas [5] to considering also the importance of anthropogenic habitats within a landscape mosaic [6–9].This change of focus is crucial when taking into account increased competition for land, due to both increasing demands of resources (caused by human demography and wealth) and, at the same time, increasing loss of agricultural land due to climate change and urbanisation [10].

Effects of Human Land-Modifications on Biodiversity

The 'intermediate disturbance hypothesis' postulates maximum diversity at intermediate regimes of disturbance [11]. Although the validity of the underlying mechanisms is currently discussed [12], this hypothesis has been widely tested in marine [13] and forest ecosystems [14], and can be applied to agricultural systems [15]. Increased anthropogenic disturbance in agricultural landscapes generally leads to declines in species diversity [8,16–19]. However, if land modification is moderate and leads to a more heterogeneous landscape it can also increase biodiversity [8,20–23]. It remains an important question to what extent natural habitats can be modified without decreasing diversity [24,25] or, rephrased, how many species from natural habitats are retained in modified habitats with a certain level of disturbance [7]. In this study, we investigated a small-scale farming system with heterogeneous land-use practices embedded in a tropical forest landscape.

Spatial and Temporal Community Variation (Beta Diversity)

Spatial heterogeneity within one habitat type can lead to high variation of community composition between study sites [26–28]. If spatial heterogeneity is higher in a natural compared to a modified habitat, focussing on mean species richness per site can lead to serious underestimation of the conservation value of the more natural habitat. Therefore, several authors emphasize the value of reporting community variation (beta diversity) over the whole study area [17,25,27,29]. Furthermore, habitat loss and land-use change can lead to the dominance of disturbance-tolerant generalist species [30] and thus increase the similarity of communities (effects on bees [31], on plants [32], on bees and wasps [33]). The same is true for temporal heterogeneity: Strong temporal heterogeneity in biodiversity of a more natural habitat can lead to a serious underestimation of its conservation value, if sampling is temporally restricted to one season [33,34]. In this study, we collected insects over a spatiotemporal climatic gradient and included beta diversity in our analysis (see 'Materials and Methods').

Conservation Research on Arthropods

Despite their overriding diversity and their importance for humankind through providing ecosystem services such as pollination or pest control [35–37], arthropods are widely neglected in conservation policies [38]. However, conservation efforts based on other taxa, such as plant or vertebrates, are often inappropriate for arthropods because of low cross-taxon congruence [6,7,27,31,39]. Effective conservation science therefore needs to incorporate more arthropod research [25]. Within arthropods, responses to habitat modification may depend strongly on life-history traits of species. Predators and parasitoids of higher trophic levels may be more strongly affected by land modification than their hosts or prey [40–44]. Therefore, we investigated three trophic groups of bees and wasps (Insecta: Hymenoptera): Bees (Apidae, Colletidae, Halictidae, Megachilidae) are important pollinators of wild and crop plants [35,36]. Paper wasps (Vespidae) predate on insects and play a role in biological pest control [45]. Spider wasps (Pompilidae) are secondary carnivores (fourth to higher trophic level) by using spiders as larval food [45]. Besides trophic level, degree of sociality and nesting requirements in arthropods may influence the susceptibility to habitat modification [46–50]. If species with different life-history traits are not analysed separately, effects of land modification may be masked by opposite reactions. Here, we analysed separately the three trophic groups mentioned above and controlled for opposing effects of different degrees of sociality and nesting requirements in bees.

Hypotheses

The main goal of this study was to quantify the impact of farming areas on biodiversity. In contrast to previous studies, we selected farming areas embedded in a landscape with a high proportion of protected forest and low level of anthropogenic disturbance. We focussed on three groups of bees and wasps at different trophic levels, of which two (paper and spider wasps) are very rarely used in biodiversity assessments, and for which human impact through land modification is basically unknown. Because farming areas were relatively small and heterogeneous in their land-use practices, we expected higher diversity (richness, abundance and evenness) of bees and wasps in farming areas. We further tested if different degrees of sociality and nesting requirements in bees lead to different responses to land modification. Finally, we hypothesized a higher spatiotemporal community variation (beta diversity) in forest compared to farmland because of naturally occurring heterogeneity between forest sites in the northeast and southwest of our study region (changes in forest structure and plant community; see section 'Study Area' in 'Materials and Methods').

Materials and Methods

Ethics Statement

Insects were collected and exported under the Belizean Forest Department research permit no. CD/60/3/09 (35), issued on November 9, 2009. Additionally, we got oral or written approvals for insect sampling from all private land owners (eight farmers and two owners of private protected areas). Bees and wasps are not legally protected in Belize.

Study Area

The study was conducted in northern coastal Belize, Central America, south and west of the village Sarteneja in Corozal District (N 18°12′-18°20′/W 88°07′-88°16′). The study sites were spread over an area of about 400 km^2 and along a climatic gradient from dry and hot 'Yucatecan low semi-deciduous forest' in the northeast to cooler and more humid 'Yucatecan medium-sized semi-evergreen forest' and '*Cohune*-palm dominated forest' in the southwest [51]. All forest types were relatively dry, low in stature, and seasonal as compared to most of Central America. Besides forest, the main habitat types of the area were savannah, brackish water lagoons, mangroves, some human settlements, and small-scale farming areas. Mean annual rainfall ranged from approx. 1500 mm in the north to 2000 mm in the west and south [52] and mean temperature during the study period differed between 24.3°C in the north and 22.3°C in the west and south (measured using DS1923 Hygrochron iButtons, Maxim, Sunnyvale, USA).

Study Sites

Both the study sites in the forest (seven sites) and in small scale farming areas (eight sites) were equally distributed over the entire study area and along the whole climatic gradient of the study area (Fig. S1). Forest and farming sites were interspersed to avoid problems of spatial autocorrelation. Five out of the seven study sites in forest lay within the protected area of Shipstern Nature Reserve (www.shipstern.org), one lay within a smaller private reserve (www.wildtracksbelize.org), and one in an unused forest close to the private land of the same organisation. Although forest sites were under protection for many years, in the past all of them were subject to some human disturbance (possibly including selective logging) and natural disturbance (hurricanes). Therefore we use the term 'natural' or 'protected forest' instead of 'primary forest'. Forest sites were located between 120 and 140 m from the next forest edge of farmland or clearings for forest roads. Farming areas consisted of a mosaic of mixed crop types, open land, agroforestry, and secondary forest. Within the vicinity of traps in farmland, we monitored more than 30 species of crops. Most crops are grown for subsistence. However, some cash crops (e.g. plantain, banana, bean, and onions) are sold to local markets, and recently Mahogany trees were planted to sell to the international timber market. Farmland size ranged from 3–80 ha and conversion from forest into farmland took place 8–70 years ago. All farming areas were fully surrounded by natural forest. Traps in farming areas were set at least 100 m from the next natural forest edge. The mean distance between two study sites was 9.9 km (min: 440 m, max: 17 km). The minimum distance between two study sites of the same habitat type was 1.3 km.

Trapping Methods

One Malaise trap (B&S Entomological Services, UK) was installed at each study site. Malaise traps are an efficient method for sampling flying insects [53] and also adequate in tropical forest systems [54]. Traps operated from December 2009 to February 2010, covering both the wet season with heavy rain falls in December and the beginning of the dry season in February. All traps were sampled weekly. Every second week a small container (6×5×5 cm) with orchid scents was hung into the Malaise trap to attract male orchid bees (Apidae: Euglossini), which are otherwise rarely collected [55]. Male orchid bees gather chemical compounds from orchids and other plant families and display them during courtship [55]. We used three of the most commonly used attractants: cineole, eugenol (each 0.4 ml/week), and methyl salicylate (0.8 ml/week). Scents constantly dispersed over one week through four small holes in the container (2 mm diameter each). Although, to our knowledge, negative (repellent) effects of attractants on non-orchid bee species has never been reported, we added scents to traps only every second week to avoid possible exclusion of some sensitive bee species. Whereas the presence of attractants had a very strong positive effect (almost hundredfold increase) on abundance of orchid bees (linear mixed effect model with scent as fixed and study sites as random factor: $p<0.001$) it had no effect on the abundance of non-orchid bees (same model: $p=0.17$). Because trapping efficiency of Malaise traps depend on vegetation structure the traps in farming areas were set up in locations characterised by understory vegetation and a closed canopy to mimic the same vegetation structure as in forest.

Determination of Bees and Wasps

Collected bees and wasps were either mounted (bees and spider wasps) or stored in 85% ethanol (paper wasps), and identified by specialists (see Acknowledgements). More than 88% of all individuals were determined to species level (72 species). The remaining samples were assigned to morphospecies (36 morphospecies) because revisions of many neotropic bee and wasp genera are still lacking. Hereafter, morphospecies and species are referred to simply as *species*. Bees were assigned to four groups of sociality (solitary, social, eusocial, and cleptoparasitic) and three group of nesting requirements (cavity-nesting, ground-nesting, and wood-nesting) [45,49,56,57] (Table S1). 'Social' comprised all live forms between strictly solitary and eusocial, namely communal, semi-social, and primitively eusocial. Bees in the genera *Augochlora*, *Ceratina*, *Euglossa*, and *Lasioglossum* were classified as social, although this is not certain for every species. Nesting of cleptoparasitic bees was defined according to the nests of their hosts. All collected specimens, except some vouchers, are archived in the Natural History Museum of Bern, Switzerland (Naturhistorisches Museum der Burgergemeinde Bern).

Statistical Analysis

Differences in species richness and abundance between forest and farming areas were analysed using Generalized Linear Models (GLM) for overdispersed count data (quasi-poisson errors with a log link function), and differences in Simpson's Evenness $E_{1/D}$ [58] were analysed using non-parametric Wilcoxon rank sum test because residuals could not be normalised with any transformation. Because the observed number of species is sensitive to the number of individuals sampled [59] we performed individual-based rarefaction curves, using the open source software EstimateS 8.2.0 [60]. Forest and farmland were compared using sample-based rarefaction curves, where individuals were set as samples [6]. Rarefaction curves and 95% confidence intervals were calculated using Mao Tau estimator. Significance (at $p<0.05$) was inferred if the total observed richness of the habitat with the smaller sample (habitat type with lower number of individuals) fell outside the confidence interval of the larger sample.

To determine temporal community variation (temporal beta diversity) we performed additive partitioning of species diversity [61], using spatial replication (study sites) and temporal variation (weeks) [33]. Two major criticisms exist against additive partitioning: First, beta diversity is not independent of alpha, thus formulating a problem if alpha diversities of compared habitats are very different [62]. Second, the measure of additive beta diversity loses its resolution for datasets with many samples sharing few

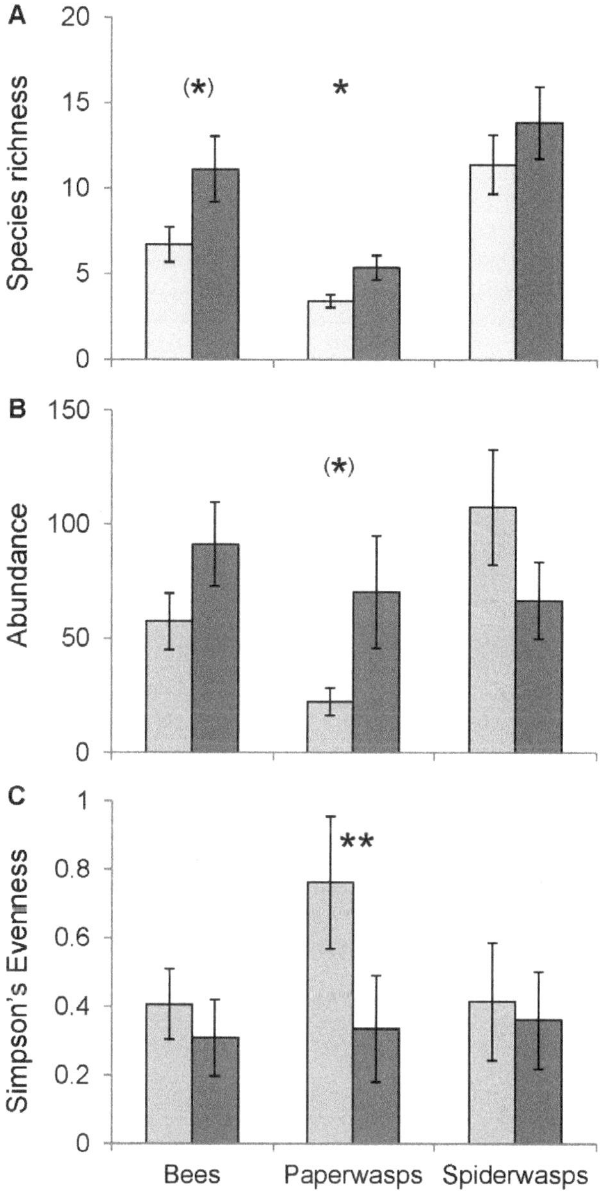

Figure 1. Biodiversity of bees and wasps. Differences in (A) species richness (mean no. of species per study site), (B) abundance (mean no. of collected individuals per site), and (C) Simpson's Evenness per site for bees, paper wasps, and spider wasps between protected forest (light grey) and heterogeneous farmland (dark grey). Error bars show standard error of the mean. Significance levels: ** $p<0.01$, * $p<0.05$, (*) $p<0.1$.

species [63]. We consider these limitations of minor importance for the current study, because alpha diversity of forest and farmland were within the same order of magnitude and because a considerable number of species were shared between samples. Because in additive partitioning spatial beta diversity is not replicated and therefore not testable, we instead compared homogeneity of multivariate dispersions between forest and farmland based on Sørensen (presence/absence) and Bray-Curtis (abundance included) dissimilarities (Method V4 in [63]). Differences in temporal species variation (additive partitioning) and spatial homogeneity of multivariate dispersions between forest and farmland were tested using Generalised Least Squares Models

(GLS), taking into account heterogeneity of residual variances [64]. Diagnostic plots of GLS indicated normal distribution of residuals. Apart from rarefaction curves, all analyses were carried out in the open source software R 2.12.2 [65], using the packages vegan [66] and nlme [67].

Results

Local (Alpha) Diversity

In total, 1133 bees of 43 species, 720 paper wasps of 19 species and 1288 spider wasps of 46 species were collected (Table S1). Species richness of paper wasps was higher in farming areas compared to protected forest ($t_{1,13} = 2.43$; $p = 0.030$) and bees showed a trend in the same direction ($t_{1,13} = 2.01$; $p = 0.066$) (Fig. 1A). Species richness of spider wasps was not significantly different between farming areas and forest ($t_{1,13} = 0.88$; $p = 0.40$). The higher species richness in farming areas was largely due to species trapped in farmland that did not occur in the forest. Conversely, a high percentage of species trapped in forest was also found in farming areas (72% of bees, 85% of paper wasps, and 77% of spider wasps) (Fig. 2). Abundance did not differ significantly between forest and farmland in any of the three groups (bees: $t_{1,13} = 1.50$; $p = 0.16$; paper wasps: $t_{1,13} = 1.95$; $p = 0.073$; spider wasps: $t_{1,13} = -1.39$; $p = 0.19$; Fig. 1B). Simpson's evenness of paper wasps was significantly higher in farming areas than in forest ($W_{7,8} = 53$; $p = 0.002$), but bees and spider wasps did not differ between habitat types (bees: $W_{7,8} = 40$; $p = 0.19$; spider wasps: $W_{7,8} = 34$; $p = 0.54$; Fig. 1C). Rarefaction curves show higher species richness in farmland for bees and spider wasps but not for paper wasps (Fig. S2). Bees showed consistently higher species richness and abundance in farming areas compared to forest in all degrees of sociality and nesting requirements (Fig. 3). Overall species richness (gamma diversity) was 70% higher in farmland (Fig. 4A–C, gamma diversity indicated by total height of bars).

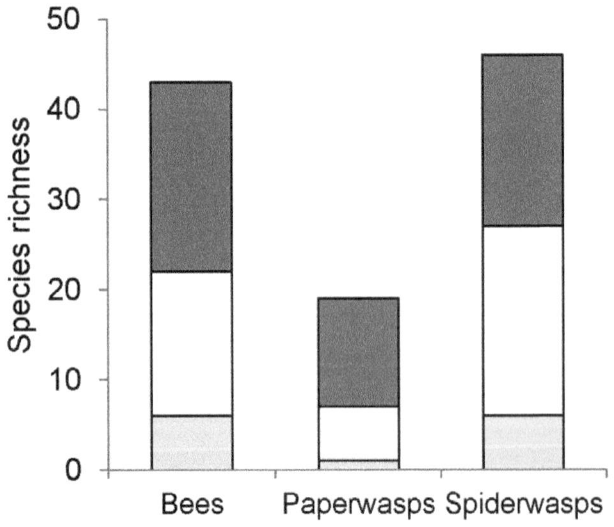

Figure 2. Exclusive and shared species across habitat types. Pooled species richness (over 3 months and 15 study sites) of bees, paper wasps, and spider wasps exclusively trapped in protected forest (light grey), in heterogeneous farmland (dark grey), and trapped in both habitat types (white).

Community Variation (Beta Diversity)

Bees and paper wasps showed a trend towards higher temporal community variation (beta diversity) in farmland compared to forest (bees: $t_{1,13} = 1.96$; $p = 0.072$; paper wasps: $t_{1,13} = 1.85$; $p = 0.087$; Fig. 4A,B). Temporal community variation of spider wasps did not show a significant pattern ($t_{1,13} = 1.56$; $p = 0.139$; Fig. 4C). Spatial community variation (beta diversity) based on presence/absence data (Sørensen) was significantly higher in farmland for wasps (paper wasps: $t_{1,13} = 5.40$; $p < 0.001$; spider wasps: $t_{1,13} = 2.67$; $p = 0.019$; Fig. 4E,F) but this trend was not statistically significant for bees ($t_{1,13} = 1.87$; $p = 0.085$; Fig. 4D). Spatial community variation (beta diversity) based on abundance data (Bray-Curtis) did not differ between forest and farmland in any of the three taxa ($|t_{1,13}| < 1.15$; $p > 0.269$; Fig. 4G–I).

Discussion

Local (Alpha) Diversity

Our results based on linear models and rarefaction curves suggest that small-scale farming systems do not only conserve but even favour biodiversity of some taxonomic groups. Species richness of bees and wasps was higher in farmland areas, first because most species occurring in adjacent forest were also found within farmland, and second because many exclusive species inhabited farmland that were not trapped in forest (Fig. 2). In other words, the proportion of species trapped in forest inhabiting or using farmland was high, indicating the compatibility of this farming system with bee and wasp conservation [7]. The studied farmland type matches perfectly the description of favourable agriculture for pollinators by Kremen [68]: "Positive effects of agriculture on pollinator communities may be more likely to occur in regions in which the presence of agriculture increases habitat heterogeneity, such as farming landscapes that include relatively small field sizes, mixed crop types within and between fields, and patches of noncrop vegetation, such as hedgerows, fallow field, meadows, and seminatural wood or shrublands". In our study region, a natural forest matrix surrounded the small-scale farming areas within dispersal and foraging distance of investigated bees and wasps. Because distance to natural habitat strongly decreases richness and abundance of bees [18,50] it remains unclear to what extent species trapped in forest may maintain viable populations in landscapes covered completely by these farmland habitats [9]. Even if bee and wasp species richness was equal or higher within farmland, and most species trapped in forests were also found in farming areas, forest may nevertheless be a vital component within the life cycle of some or many of the species and may serve as constant source for colonisation of farmland. In accordance with our results, two studies on plants, vertebrates and invertebrates showed that a considerable number of forest species persist in secondary forest, agroforestry, and even pastures if the surrounding landscape comprises high amounts of mature forest [69,70].

In addition to the desirable features of the studied agricultural system, bees and wasps in tropical forest may be less sensitive to human disturbance compared to other taxa [6,27]. This emphasises that our results cannot be generalised to other taxa. At least, in our study effects of land modification were more or less consistent between investigated groups of bees and wasps. The same was true in a study on bees and wasps in Ecuador [71]. Furthermore, all groups of sociality and nesting requirements in bees were positively affected by land modification (Fig. 3), in contrast to previous findings where effects of habitat types or levels of disturbance on bee communities depended on life history traits [46,47,49].

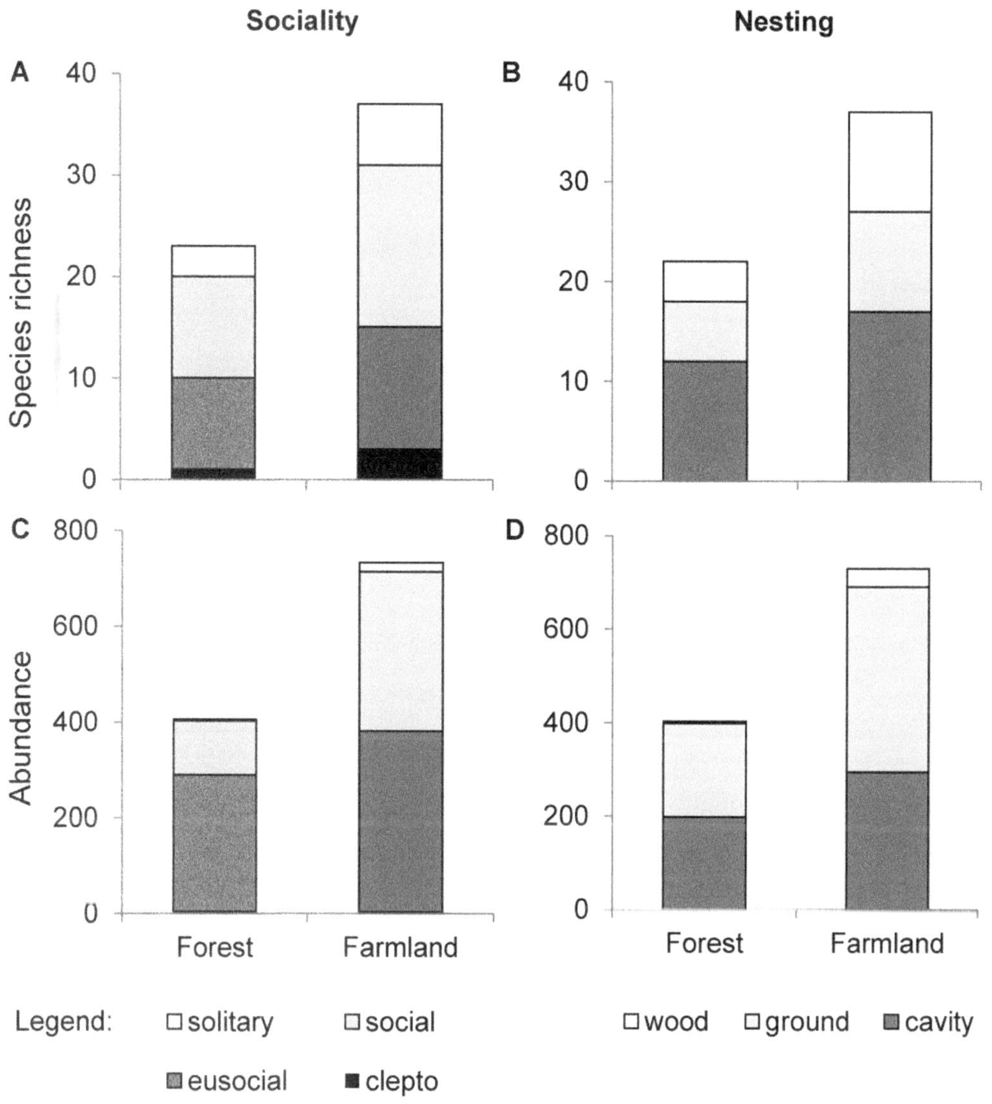

Figure 3. Degree of sociality and nesting requirements in bees. Differences in (A, B) species richness and (C, D) abundance of bees between protected forest and heterogeneous farmland (pooled data over 3 months and 15 study sites). Bees are separated into four degrees of sociality (solitary, social, eusocial, and cleptoparasitic) and three groups of nesting requirements (wood-nesting, ground-nesting, cavity nesting). For assignment of bees see Table S1.

Community Variation (Beta Diversity)

In contrast to our hypothesis, spatial community variation (beta diversity) of bees and wasps was equal or higher in farming areas compared to forest (Fig. 4A–F). High beta diversity within a study region can be a statistical artefact due to low local alpha diversity [72]. However, in our study we can exclude such an inflation of beta diversity because both alpha and beta diversity were increased in farmland. Habitat heterogeneity is a major driver of community variation [26–28]. Thus, differences in farming areas due to variable land-use practices could have led to higher beta diversity in farmland. High management diversity between different agroforestry plots within a study region appeared to be responsible for high beta diversity of bees in Indonesia [73]. Spatial community variation disappeared when analyses were performed including abundances (Fig. 4G–I), showing that effects of beta diversity were driven by the high percentage of species with very low abundances (29% of species are singletons).

Temporal community variation of bees and paper wasps tended to be higher in farmland than in forest, in contrast to a study in Ecuador [33] where lower temporal beta diversity was found in more disturbed habitat types. However, in that study lower beta diversity occurred only in open habitat types, whereas we compared protected forest against a farming system consisting of a considerable amount of woody habitat. In conclusion, the increase in the regional species pool (gamma diversity) by 70% through species only occurring in farming areas (Fig. 4A–C) was composed of both, higher local diversity (alpha) and higher spatial and temporal community variation (beta diversity).

Effects of Human Disturbance on Bees and Wasps

We found that species richness of bees and wasps was favoured in farmland areas compared to protected forest and we conclude that bees and wasps occurring in natural forest were not susceptible to the extent and disturbance level of the studied farming system [7]. Our results are in line with the 'intermediate

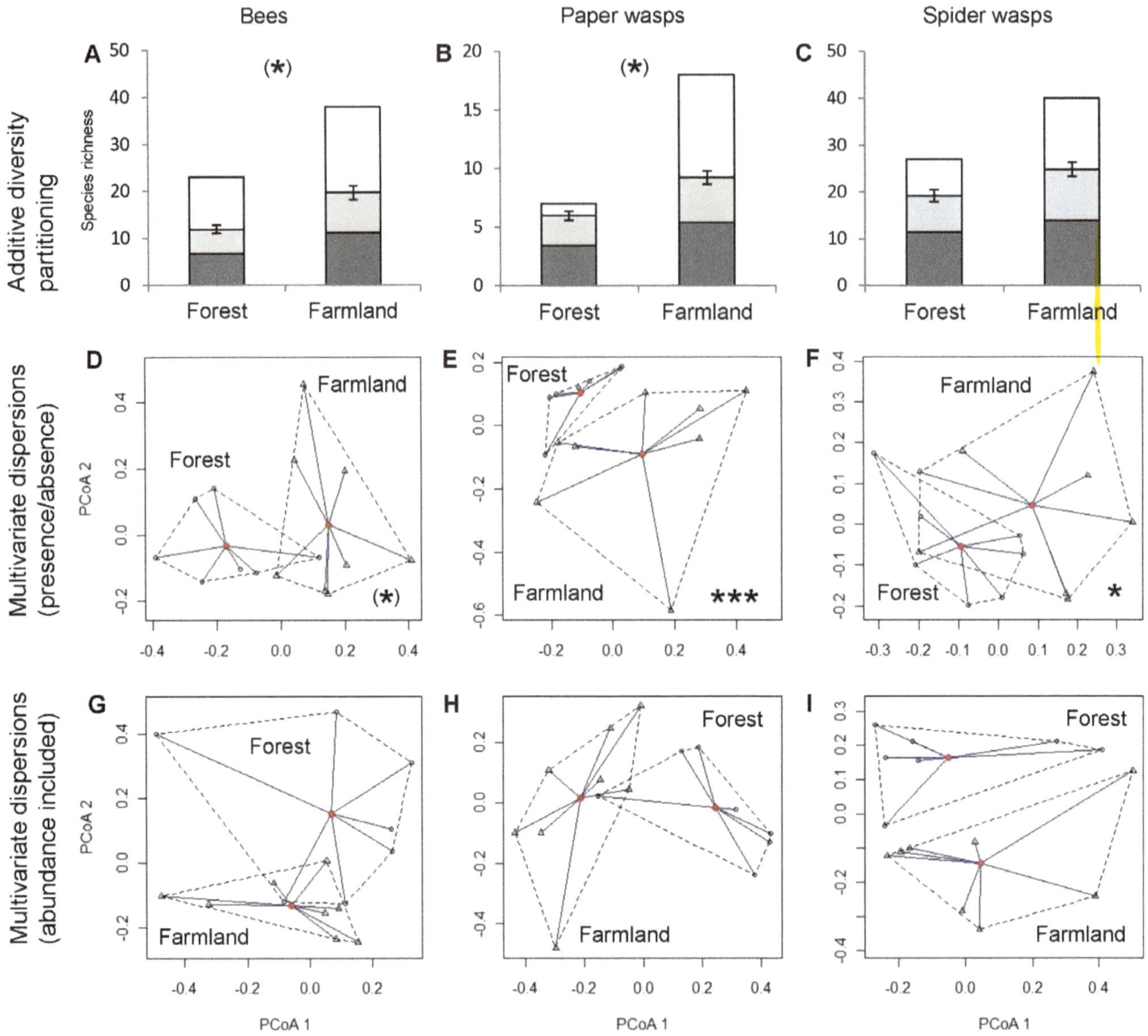

Figure 4. Community variation (beta diversity) of bees and wasps. (A–C) Alpha diversity (dark grey), temporal beta diversity (light grey), and spatial beta diversity (white) compared between protected forest and heterogeneous farmland based on additive diversity partitioning. Temporal beta diversity was replicated and statistically tested. Error bars show standard error of the mean. Total height of bars indicates overall (gamma) diversity within forest or farmland. (D–I) Spatial beta diversity, measured as distance to centroid (homogeneity of multivariate dispersions), was based on (D–F) Sørensen dissimilarities (presence/absence) and (G–I) Bray-Curtis dissimilarities (abundance included). Points (forest) and triangles (farmland) represent species composition of one study site. Larger distances between study sites and the mean composition of a habitat type (unbroken lines) indicate higher difference in species composition. Dashed lines connect all sites of one habitat type. Significance levels: *** $p<0.001$, * $p<0.05$, (*) $p<0.1$.

disturbance hypothesis' [11] predicting higher biodiversity at higher levels of disturbance in generally undisturbed, relatively natural landscapes. Farming areas create additional habitats and benefit biodiversity through increased heterogeneity of resources within foraging distance [8,22,49,74,75]. To the extreme, disturbed areas may even lead to a spillover of rich farmland bee communities into forest remnants [76,77]. However, this may not necessarily be beneficial considering the competitive advantageous of some abundant generalist farmland species. The resilience of forest bees to disturbance may hold as long as some elements of native habitats remain in the disturbed areas [47,78]. However, in very intensively managed landscapes, additional disturbance leads to a reduction of diversity [8,16]. Due to the presence of naturally occurring disturbance in Belizean forests caused by hurricanes, forest communities may be less sensitive to disturbance by logging, as proposed for butterflies by Lewis [79].

From a conservation perspective, it is debatable if species inhabiting modified habitats but not occurring in native forest are of high concern. Disturbed areas may be first invaded by disturbance-tolerant widespread or even non-native generalist species, which are of low conservation value [6,23]. These species may become dominant components in communities of disturbed habitats. In Britain, continuing habitat degradation led to an increased dominance of generalist butterfly species [30]. The lower

Simpson's Evenness of paper wasps in farmland compared to forest reflects the high dominance [80] of the most abundant species in farming areas, namely *Polybia occidentalis nigratella* (Table S1). However, this species is also the most abundant species in forest and it is still an open research question how important community evenness is for ecosystem functioning [17]. Comprehensive comparisons of the biogeography and conservation value of species between farming areas and forest are not possible in our study due to lack of sound information on species distributions, rarity, and specialisation, and because some individuals could only be determined to morphospecies.

Conclusions

Our results show that agricultural systems exist, where local (alpha) diversity and variation in communities (beta diversity) are similar or higher than in protected forest and where species occurring in forests also use these farmlands. We conclude that heterogeneous and small-scale farming areas embedded in a matrix of protected forest are compatible with biodiversity conservation for some taxonomic groups. However, it remains unclear to what extent these systems would retain forest species if protected forest in the surrounding was further reduced or if existing farmland areas are further intensified. Finally, we do not conclude that a transfer of forest into heterogeneous small-scale farming systems is desirable. But, if necessary to meet demands of local populations, it can be tolerated from the perspective of bee and wasp conservation.

Supporting Information

Figure S1 Study area. Google Earth [81] map showing the study area in northern Belize, Central America. Study sites in heterogeneous farmland are marked in red, sites in protected forest in yellow.

Figure S2 Individual-based rarefaction curves. Solid lines are rarefaction curves for (A) bees, (B) paper wasps, and (C) spider wasps in protected forest (grey) and heterogeneous farmland (black) and dotted lines are 95% confidence intervals for the larger sample, i.e. the habitat type with the higher total amount of individuals (see 'Materials and Methods' for details).

Table S1 Species list. Bees, paper wasps, and spider wasps with total abundances in protected forest and heterogeneous farmland. Bees are assigned to different life-history traits (LHT).

Acknowledgments

We are grateful to eight farmers providing their land and to the team of Shipstern Nature Reserve for assistance in the field. We are very thankful for professional determination of bees and wasps: Bees were determined by John Ascher, paper wasps by James Carpenter (both AMNH New York, USA), and spider wasps by Raymond Wahis (Gembloux Agro-Bio Tech, Belgium). Jason Tylianakis, Dirk Sanders, and four anonymous reviewers provided helpful comments on earlier versions of this manuscript.

Author Contributions

Conceived and designed the experiments: CS MHE. Performed the experiments: CS SR. Analyzed the data: CS. Contributed reagents/materials/analysis tools: CS. Wrote the paper: CS.

References

1. Coad L, Burgess ND, Loucks C, Fish L, Scharlemann JPW, et al. (2009) The ecological representativeness of the global protected areas estate in 2009: progress towards the CBD 2010 target. UNEP-WCMC, WWF-US and ECI, University of Oxford.
2. Bruner AG, Gullison RE, Rice RE, da Fonseca GAB (2001) Effectiveness of parks in protecting tropical biodiversity. Science 291: 125–128.
3. Rodrigues ASL, Andelman SJ, Bakarr MI, Boitani L, Brooks TM, et al. (2004) Effectiveness of the global protected area network in representing species diversity. Nature 428: 640–643.
4. Wittemyer G, Elsen P, Bean WT, Burton ACO, Brashares JS (2008) Accelerated human population growth at protected area edges. Science 321: 123–126.
5. Soulé ME, Sanjayan MA (1998) Ecology: Conservation targets: Do they help? Science 279: 2060–2061.
6. Barlow J, Gardner TA, Araujo IS, Avila-Pires TC, Bonaldo AB, et al. (2007) Quantifying the biodiversity value of tropical primary, secondary, and plantation forests. Proceedings of the National Academy of Sciences of the United States of America 104: 18555–18560.
7. Gardner TA, Barlow J, Chazdon R, Ewers RM, Harvey CA, et al. (2009) Prospects for tropical forest biodiversity in a human-modified world. Ecology Letters 12: 561–582.
8. Tscharntke T, Klein AM, Kruess A, Steffan-Dewenter I, Thies C (2005) Landscape perspectives on agricultural intensification and biodiversity - ecosystem service management. Ecology Letters 8: 857–874.
9. Chazdon RL, Peres CA, Dent D, Sheil D, Lugo AE, et al. (2009) The potential for species conservation in tropical secondary forests. Conservation Biology 23: 1406–1417.
10. Godfray HCJ, Beddington JR, Crute IR, Haddad L, Lawrence D, et al. (2010) Food security: The challenge of feeding 9 billion people. Science 327: 812–818.
11. Connell JH (1978) Diversity in tropical rain forests and coral reefs - High diversity of trees and corals is maintained only in a non-equilibrium state. Science 199: 1302–1310.
12. Fox JW (2012) The intermediate disturbance hypothesis should be abandoned. Trends in Ecology & Evolution doi: 10.1016/j.tree.2012.08.014.
13. Svensson JR, Lindegarth M, Siccha M, Lenz M, Molis M, et al. (2007) Maximum species richness at intermediate frequencies of disturbance: Consistency among levels of productivity. Ecology 88: 830–838.
14. Molino JF, Sabatier D (2001) Tree diversity in tropical rain forests: A validation of the intermediate disturbance hypothesis. Science 294: 1702–1704.
15. Bruggisser OT, Schmidt-Entling MH, Bacher S (2010) Effects of vineyard management on biodiversity at three trophic levels. Biological Conservation 143: 1521–1528.
16. Kremen C, Williams NM, Thorp RW (2002) Crop pollination from native bees at risk from agricultural intensification. Proceedings of the National Academy of Sciences of the United States of America 99: 16812–16816.
17. Tscharntke T, Tylianakis JM, Rand TA, Didham RK, Fahrig L, et al. (2012) Landscape moderation of biodiversity patterns and processes - eight hypotheses. Biological Reviews 87: 661–685.
18. Chacoff NP, Aizen MA (2006) Edge effects on flower-visiting insects in grapefruit plantations bordering premontane subtropical forest. Journal of Applied Ecology 43: 18–27.
19. Tilman D (1999) Global environmental impacts of agricultural expansion: The need for sustainable and efficient practices. Proceedings of the National Academy of Sciences of the United States of America 96: 5995–6000.
20. Vulliamy B, Potts SG, Willmer PG (2006) The effects of cattle grazing on plant-pollinator communities in a fragmented Mediterranean landscape. Oikos 114: 529–543.
21. Winfree R, Griswold T, Kremen C (2007) Effect of human disturbance on bee communities in a forested ecosystem. Conservation Biology 21: 213–223.
22. Andren H (1994) Effects of habitat fragmentation on birds and mammals in landscapes with different proportions of suitable habitat - a review. Oikos 71: 355–366.
23. Brown KS (1997) Diversity, disturbance, and sustainable use of Neotropical forests: insects as indicators for conservation monitoring. Journal of Insect Conservation 1: 25–42.
24. Godfray HCJ (2011) Food and biodiversity. Science 333: 1231–1232.
25. Scales BR, Marsden SJ (2008) Biodiversity in small-scale tropical agroforests: a review of species richness and abundance shifts and the factors influencing them. Environmental Conservation 35: 160–172.
26. Tuomisto H, Ruokolainen K, Aguilar M, Sarmiento A (2003) Floristic patterns along a 43-km long transect in an Amazonian rain forest. Journal of Ecology 91: 743–756.
27. Kessler M, Abrahamczyk S, Bos M, Buchori D, Putra DD, et al. (2009) Alpha and beta diversity of plants and animals along a tropical land-use gradient. Ecological Applications 19: 2142–2156.
28. Duivenvoorden JF, Svenning JC, Wright SJ (2002) Ecology - Beta diversity in tropical forests. Science 295: 636–637.

29. Basset Y, Missa O, Alonso A, Miller SE, Curletti G, et al. (2008) Changes in arthropod assemblages along a wide gradient of disturbance in Gabon. Conservation Biology 22: 1552–1563.
30. Warren MS, Hill JK, Thomas JA, Asher J, Fox R, et al. (2001) Rapid responses of British butterflies to opposing forces of climate and habitat change. Nature 414: 65–69.
31. Dormann CF, Schweiger O, Augenstein I, Bailey D, Billeter R, et al. (2007) Effects of landscape structure and land-use intensity on similarity of plant and animal communities. Global Ecology and Biogeography 16: 774–787.
32. Vellend M, Verheyen K, Flinn KM, Jacquemyn H, Kolb A, et al. (2007) Homogenization of forest plant communities and weakening of species-environment relationships via agricultural land use. Journal of Ecology 95: 565–573.
33. Tylianakis JM, Klein AM, Tscharntke T (2005) Spatiotemporal variation in the diversity of hymenoptera across a tropical habitat gradient. Ecology 86: 3296–3302.
34. Summerville KS, Crist TO (2005) Temporal patterns of species accumulation in a survey of Lepidoptera in a beech-maple forest. Biodiversity and Conservation 14: 3393–3406.
35. Klein AM, Vaissiere BE, Cane JH, Steffan-Dewenter I, Cunningham SA, et al. (2007) Importance of pollinators in changing landscapes for world crops. Proceedings of the Royal Society B-Biological Sciences 274: 303–313.
36. Kremen C, Williams NM, Aizen MA, Gemmill-Herren B, LeBuhn G, et al. (2007) Pollination and other ecosystem services produced by mobile organisms: a conceptual framework for the effects of land-use change. Ecology Letters 10: 299–314.
37. Ehrlich PR, Ehrlich AH (1992) The value of biodiversity. Ambio 21: 219–226.
38. Cardoso P, Erwin TL, Borges PAV, New TR (2011) The seven impediments in invertebrate conservation and how to overcome them. Biological Conservation 144: 2647–2655.
39. Axmacher JC, Liu Y, Wang C, Li L, Yu Z (2011) Spatial α-diversity patterns of diverse insect taxa in Northern China: Lessons for biodiversity conservation. Biological Conservation 144: 2362–2368.
40. Tscharntke T, Steffan-Dewenter I, Kruess A, Thies C (2002) Characteristics of insect populations on habitat fragments: A mini review. Ecological Research 17: 229–239.
41. Bailey D, Schmidt-Entling MH, Eberhart P, Herrmann JD, Hofer G, et al. (2010) Effects of habitat amount and isolation on biodiversity in fragmented traditional orchards. Journal of Applied Ecology 47: 1003–1013.
42. Schüepp C, Herrmann JD, Herzog F, Schmidt-Entling MH (2011) Differential effects of habitat isolation and landscape composition on wasps, bees, and their enemies. Oecologia 165: 713–721.
43. Cagnolo L, Valladares G, Salvo A, Cabido M, Zak M (2009) Habitat fragmentation and species loss across three interacting trophic levels: Effects of life-history and food-web traits. Conservation Biology 23: 1167–1175.
44. Davies KF, Margules CR, Lawrence KF (2000) Which traits of species predict population declines in experimental forest fragments? Ecology 81: 1450–1461.
45. Hanson PE, Gauld ID (1995) The hymenoptera of Costa Rica. New York: Oxford University Press.
46. Klein AM, Steffan-Dewenter I, Tscharntke T (2003) Fruit set of highland coffee increases with the diversity of pollinating bees. Proceedings of the Royal Society of London Series B-Biological Sciences 270: 955–961.
47. Brosi BJ, Daily GC, Shih TM, Oviedo F, Duran G (2008) The effects of forest fragmentation on bee communities in tropical countryside. Journal of Applied Ecology 45: 773–783.
48. Potts SG, Biesmeijer JC, Kremen C, Neumann P, Schweiger O, et al. (2010) Global pollinator declines: trends, impacts and drivers. Trends in Ecology & Evolution 25: 345–353.
49. Jha S, Vandermeer JH (2010) Impacts of coffee agroforestry management on tropical bee communities. Biological Conservation 143: 1423–1431.
50. Ricketts TH, Regetz J, Steffan-Dewenter I, Cunningham SA, Kremen C, et al. (2008) Landscape effects on crop pollination services: are there general patterns? Ecology Letters 11: 499–515.
51. Bijleveld CFA (1998) The vegetation of Shipstern Nature Reserve. Kerzers, Switzerland: International Tropical Conservation Foundation.
52. Meerman JC, Clabaugh J (2010) Biodiversity and environmental resource data system of Belize. Available: http://www.biodiversity.bz. Accessed 26 November 2012.
53. Campbell JW, Hanula JL (2007) Efficiency of Malaise traps and colored pan traps for collecting flower visiting insects from three forested ecosystems. Journal of Insect Conservation 11: 399–408.
54. Missa O, Basset Y, Alonso A, Miller SE, Curletti G, et al. (2009) Monitoring arthropods in a tropical landscape: relative effects of sampling methods and habitat types on trap catches. Journal of Insect Conservation 13: 103–118.
55. Roubik DW, Hanson PE, editors (2004) Orchid bees of tropical America. Santa Domingo de Heredia: Instituto Nacional de Biodiversidad (INBio).
56. Calvillo LM, Ramirez VM, Parra-Tabla V, Navarro J (2010) Bee diversity in a fragmented landscape of the Mexican neotropic. Journal of Insect Conservation 14: 323–334.
57. Michener CD (2007) The bees of the world. Baltimore: Johns Hopkins University Press.
58. Smith B, Wilson JB (1996) A consumer's guide to evenness indices. Oikos 76: 70–82.
59. Colwell RK, Chao A, Gotelli NJ, Lin SY, Mao CX, et al. (2012) Models and estimators linking individual-based and sample-based rarefaction, extrapolation and comparison of assemblages. Journal of Plant Ecology-Uk 5: 3–21.
60. Colwell RK (2009) EstimateS: Statistical estimation of species richness and shared species from samples. Version 8.2.0. Available: http://purl.oclc.org/estimates. Accessed 26 November 2012.
61. Lande R (1996) Statistics and partitioning of species diversity, and similarity among multiple communities. Oikos 76: 5–13.
62. Jost L (2007) Partitioning diversity into independent alpha and beta components. Ecology 88: 2427–2439.
63. Anderson MJ, Crist TO, Chase JM, Vellend M, Inouye BD, et al. (2011) Navigating the multiple meanings of beta diversity: a roadmap for the practicing ecologist. Ecology Letters 14: 19–28.
64. Zuur AF (2009) Mixed effects models and extensions in ecology with R. New York, N.Y.: Springer. 574 p.
65. R Development Core Team (2011) R: A language and environment for statistical computing. R Foundation for Statistical Computing, Vienna, Austria. ISBN 3-900051-07-0, Available: http://www.R-project.org/. Accessed 26 November 2012.
66. Oksanen J, Blanchet FG, Kindt R, Legendre P, O'Hara RB, et al. (2011) vegan: Community ecology package. R package version 1.17-8. Available: http://CRAN.R-project.org/package=vegan. Accessed 26 November 2012.
67. Pinheiro J, Bates D, DebRoy S, Sarkar D (2012) nlme: linear and nonlinear mixed effects models. R package version 3.1–104. Available: http://cran.r-project.org/web/packages/nlme//index.html. Accessed 26 November 2012.
68. Kremen C (2008) Crop pollination services from wild bees. In: James RR, editor. Bee pollination in agricultural ecosystems. New York, N.Y.: Oxford University Press.
69. Pardini R, Faria D, Accacio GM, Laps RR, Mariano-Neto E, et al. (2009) The challenge of maintaining Atlantic forest biodiversity: A multi-taxa conservation assessment of specialist and generalist species in an agro-forestry mosaic in southern Bahia. Biological Conservation 142: 1178–1190.
70. Gascon C, Lovejoy TE, Bierregaard RO, Malcolm JR, Stouffer PC, et al. (1999) Matrix habitat and species richness in tropical forest remnants. Biological Conservation 91: 223–229.
71. Tylianakis JM, Klein AM, Lozada T, Tscharntke T (2006) Spatial scale of observation affects alpha, beta and gamma diversity of cavity-nesting bees and wasps across a tropical land-use gradient. Journal of Biogeography 33: 1295–1304.
72. Karp DS, Rominger AJ, Zook J, Ranganathan J, Ehrlich PR, et al. (2012) Intensive agriculture erodes beta-diversity at large scales. Ecology Letters 15: 963–970.
73. Hoehn P, Steffan-Dewenter I, Tscharntke T (2010) Relative contribution of agroforestry, rainforest and openland to local and regional bee diversity. Biodiversity and Conservation 19: 2189–2200.
74. Benton TG, Vickery JA, Wilson JD (2003) Farmland biodiversity: is habitat heterogeneity the key? Trends in Ecology & Evolution 18: 182–188.
75. Eltz T, Bruhl CA, van der Kaars S, Linsenmair KE (2002) Determinants of stingless bee nest density in lowland dipterocarp forests of Sabah, Malaysia. Oecologia 131: 27–34.
76. Hagen M, Kraemer M (2010) Agricultural surroundings support flower-visitor networks in an Afrotropical rain forest. Biological Conservation 143: 1654–1663.
77. Schleuning M, Farwig N, Peters MK, Bergsdorf T, Bleher B, et al. (2011) Forest fragmentation and selective logging have inconsistent effects on multiple animal-mediated ecosystem processes in a tropical forest. PLoS ONE 6: e27785.
78. Samejima H, Marzuki M, Nagamitsu T, Nakasizuka T (2004) The effects of human disturbance on a stingless bee community in a tropical rainforest. Biological Conservation 120: 577–587.
79. Lewis OT (2001) Effect of Experimental Selective Logging on Tropical Butterflies. Conservation Biology 15: 389–400.
80. Magurran AE (2007) Measuring biological diversity. Malden, Mass.: Blackwell Science. 256 p.
81. Google (2011) Google Earth 6.1.0.5001. Google Inc. Available: http://www.google.com/earth/index.html. Accessed 17 Oct 2011.

Permissions

All chapters in this book were first published in PLOS ONE, by The Public Library of Science; hereby published with permission under the Creative Commons Attribution License or equivalent. Every chapter published in this book has been scrutinized by our experts. Their significance has been extensively debated. The topics covered herein carry significant findings which will fuel the growth of the discipline. They may even be implemented as practical applications or may be referred to as a beginning point for another development.

The contributors of this book come from diverse backgrounds, making this book a truly international effort. This book will bring forth new frontiers with its revolutionizing research information and detailed analysis of the nascent developments around the world.

We would like to thank all the contributing authors for lending their expertise to make the book truly unique. They have played a crucial role in the development of this book. Without their invaluable contributions this book wouldn't have been possible. They have made vital efforts to compile up to date information on the varied aspects of this subject to make this book a valuable addition to the collection of many professionals and students.

This book was conceptualized with the vision of imparting up-to-date information and advanced data in this field. To ensure the same, a matchless editorial board was set up. Every individual on the board went through rigorous rounds of assessment to prove their worth. After which they invested a large part of their time researching and compiling the most relevant data for our readers.

The editorial board has been involved in producing this book since its inception. They have spent rigorous hours researching and exploring the diverse topics which have resulted in the successful publishing of this book. They have passed on their knowledge of decades through this book. To expedite this challenging task, the publisher supported the team at every step. A small team of assistant editors was also appointed to further simplify the editing procedure and attain best results for the readers.

Apart from the editorial board, the designing team has also invested a significant amount of their time in understanding the subject and creating the most relevant covers. They scrutinized every image to scout for the most suitable representation of the subject and create an appropriate cover for the book.

The publishing team has been an ardent support to the editorial, designing and production team. Their endless efforts to recruit the best for this project, has resulted in the accomplishment of this book. They are a veteran in the field of academics and their pool of knowledge is as vast as their experience in printing. Their expertise and guidance has proved useful at every step. Their uncompromising quality standards have made this book an exceptional effort. Their encouragement from time to time has been an inspiration for everyone.

The publisher and the editorial board hope that this book will prove to be a valuable piece of knowledge for researchers, students, practitioners and scholars across the globe.

List of Contributors

Lee A. Vierling
Department of Forest, Rangeland, and Fire Sciences, McCall Outdoor Science School, University of Idaho, Moscow, Idaho, United States of America

Kerri T. Vierling
Department of Fish and Wildlife Sciences, University of Idaho, Moscow, Idaho, United States of America

Patrick Adam
Environmental Science, University of Idaho, Moscow, Idaho, United States of America

Andrew T. Hudak
Rocky Mountain Research Station, US Forest Service, Moscow, Idaho, United States of America

Stephanie Pau
Department of Geography, Florida State University, Tallahassee, Florida, United States of America

Christopher J. Still
Forest Ecosystems and Society, Oregon State University, Corvallis, Oregon, United States of America

Diego Alvarez-Berastegui
Balearic Islands Coastal Observing and Forecasting System (SOCIB), Palma de Mallorca, Balearic Islands, Spain

Joaquín Tintoré
Balearic Islands Coastal Observing and Forecasting System (SOCIB), Palma de Mallorca, Balearic Islands, Spain
Institute for Advanced Studies (IMEDEA), Consejo Superior de Investigaciones Científicas (CSIC) and the University of the Balearic Islands (UIB). Esporles, Spain

Lorenzo Ciannelli
College of Earth, Ocean, and Atmospheric Sciences (CEOAS), Oregon State University, Corvallis, Oregon, United States of America

Alberto Aparicio-Gonzalez, Patricia Reglero, Manuel Hidalgo, Jose Luis López-Jurado and Francisco Alemany
Instituto Español de Oceanografía - Centre Oceanogràfic de les Balears (COB-IEO), Palmade Mallorca, Balearic Islands, Spain

Antonio López-Carretero and Cecilia Díaz-Castelazo
Instituto de Ecología A.C., Xalapa, Veracruz, México

Karina Boege
Instituto de Ecología, Universidad Nacional Autónoma de México, México D.F., Meé xico

Víctor Rico-Gray
Instituto de Neuroetología, Universidad Veracruzana, Xalapa, Veracruz, México

Bibek Yumnam, Yadvendradev V. Jhala, Qamar Qureshi, Swati Saini and Y. Srinivas
Wildlife Institute of India, Chandrabani, Dehradun 248001, India

Robert C. Fleischer
Smithsonian Conservation Biology Institute, National Zoological Park, 3001 Connecticut Avenue, Washington, D.C. 20008, United States of America

Jesus E. Maldonado
Smithsonian Conservation Biology Institute, National Zoological Park, 3001 Connecticut Avenue, Washington, D.C. 20008, United States of America
Department of Vertebrate Zoology, National Museum of Natural History, Smithsonian Institution, Washington, D.C. 20013, United States of America

Rajesh Gopal
National Tiger Conservation Authority, Bikaneer House, Shah Jahan Road, New Delhi 110011, India

Mathieu L. Bourbonnais and Trisalyn A. Nelson
Spatial Pattern Analysis and Research Lab, Department of Geography, University of Victoria, Victoria, British Columbia, Canada

Marc R. L. Cattet
Canadian Cooperative Wildlife Health Centre, Western College of Veterinary Medicine, University of Saskatchewan, Saskatoon, Saskatchewan, Canada

Chris T. Darimont
Applied Conservation Science Lab, Department of Geography, University of Victoria, Victoria, British Columbia, Canada

Gordon B. Stenhouse
Foothills Research Institute, Hinton, Alberta, Canada

Karen E. Mabry
Department of Biology, New Mexico State University, Las Cruces, New Mexico, United States of America

Katherine R. Smith
Department of Biology, New Mexico State University, Las Cruces, New Mexico, United States of America
California Department of Fish and Wildlife, Stockton, California, United States of America

Laureen Barthman-Thompson
California Department of Fish and Wildlife, Stockton, California, United States of America

William R. Gould
Department of Economics, Applied Statistics and International Business, New Mexico State University, Las Cruces, New Mexico, United States of America

Nicole A. Hill, Neville Barrett and Justin Hulls
Institute for Marine and Antarctic Studies, University of Tasmania, Hobart, Tasmania, Australia

Emma Lawrence
Digital Productivity Flagship, Commonwealth Scientific and industrial Research Organisation (CSIRO), Brisbane, QLD, Australia

Jeffrey M. Dambacher and Keith R. Hayes
Digital Productivity Flagship, Commonwealth Scientific and industrial Research Organisation (CSIRO), Hobart, Tasmania, Australia

Scott Nichol
Geoscience Australia, Canberra, ACT, Australia

Alan Williams
Oceans and Atmosphere Flagship, Commonwealth Scientific and industrial Research Organisation (CSIRO), Hobart, Tasmania, Australia

Stephen Brown
Manomet Center for Conservation Sciences, Manomet, Massachusetts, United States of America

Hector Galbraith
Manomet Center for Conservation Sciences, Manomet, Massachusetts, United States of America
National Wildlife Federation, Springfield, Massachusetts, United States of America

David W. DesRochers
Department of Natural Sciences, Dalton State College, Dalton, Georgia, United States of America

J. Michael Reed
Department of Biology, Tufts University, Medford, Massachusetts, United States of America

Pierre W. Froneman
Department of Zoology & Entomology, Rhodes University, Grahamstown, South Africa

Ryan Daly
Department of Zoology & Entomology, Rhodes University, Grahamstown, South Africa
Port Elizabeth Museum at Bayworld, Humewood, Port Elizabeth, South Africa

Malcolm J. Smale
Port Elizabeth Museum at Bayworld, Humewood, Port Elizabeth, South Africa
Department of Zoology, Nelson Mandela Metropolitan University, Port Elizabeth, South Africa

Paul D. Cowley
South African Institute for Aquatic Biodiversity (SAIAB), Grahamstown, South Africa

Jakub Horak
Department of Forest Protection and Entomology, Faculty of Forestry and Wood Sciences, Czech University of Life Sciences Prague, Prague, Czech Republic

John D. Hedley
Environmental Computer Science Ltd., Tiverton, Devon, United Kingdom

Kathryn McMahon
School of Natural Sciences and Centre for Marine Ecosystems Research, Edith Cowan University, Joondalup, Western Australia

Peter Fearns
Department of Imaging and Applied Physics, Curtin University of Technology, Perth, Western Australia

João Manuel Bernardo
Departamento de Paisagem Ambiente e Ordenamento, Escola de Ciênciase Tecnologia, Universidade Évora, Évora, Portugal

Maria Ilhéu and Paula Matono
Departamento de Paisagem Ambiente e Ordenamento, Escola de Ciânciase Tecnologia, Universidade de Évora,Évora, Portugal
Instituto de Ciências Agráriase Ambientais Mediterrânicas, Universidade de Évora, Évora, Portugal

Chelsea A. Smith, Mary M. Gardiner
Department of Entomology, The Ohio State University, Ohio Agricultural Research and Development Center, Wooster, Ohio, United States of America

Claire Saraux, Jean-Marc Fromentin, Jean-Louis Bigot, Jean Hervé Bourdeix, Marie Morfin, David Roos and Elisabeth Van Beveren
IFREMER (Institut Franc¸ais de Recherche pourl'de la MER), Research Unit EME (UMR 212), Séte, France

Nicolas Bez
IRD (Institut de Recherche pour le Développement), Research Unit EME (UMR 212), Séte, France

Menna E. Jones, Christopher N. Johnson
School of Biological Sciences, University of Tasmania, Hobart, Tasmania, Australia

Hugh W. McGregor
School of Biological Sciences, University of Tasmania, Hobart, Tasmania, Australia
Australian Wildlife Conservancy, Derby, Western Australia, Australia

Sarah Legge
Australian Wildlife Conservancy, Mornington Wildlife Sanctuary, Derby, Western Australia, Australia

Humberto L. Perotto-Baldivieso
School of Energy, Environment and Agrifood, Cranfield University, Cranfield, Bedfordshire, United Kingdom

Jessica Tait
School of Energy, Environment and Agrifood, Cranfield University, Cranfield, Bedfordshire, United Kingdom
CSIRO Sustainable Land and Water, Atherton, QLD, Australia

Adam McKeown
CSIRO Sustainable Land and Water, Smithfield, QLD, Australia

David A. Westcott
CSIRO Sustainable Land and Water, Atherton, QLD, Australia

Elitza S. Germanov and Andrea D. Marshall
Marine Megafauna Foundation, Truckee, California, United States of America

P. Dilip Venugopal, Peter L. Coffey, Galen P. Dively and William O. Lamp
Department of Entomology, University of Maryland, College Park, Maryland, United States of America

Sarah Rittiner
Institute of Ecology and Evolution, Community Ecology, University of Bern, Bern, Switzerland

Martin H. Entling
University of Koblenz-Landau, Ecosystem Analysis, Landau, Germany

Christof Schüepp
Institute of Ecology and Evolution, Community Ecology, University of Bern, Bern, Switzerland
University of Koblenz-Landau, Ecosystem Analysis, Landau, Germany

Index